D0202516

Finite Element Modeling for Stress Analysis

Finite Element Modeling for Stress Analysis

Robert D. Cook
University of Wisconsin - Madison

JOHN WILEY & SONS, INC.

New York · Chichester · Brisbane · Toronto · Singapore

ACQUISITIONS EDITOR	Charity Robey
MARKETING MANAGER	Susan Elbe
PRODUCTION EDITOR	Ken Santor
TEXT DESIGNER	Lee Goldstein
COVER DESIGNER	Carol Grobe
MANUFACTURING MANAGER	Susan Stetzer
ILLUSTRATION	Sigmund Malinowski

This book was set in Times Roman by General Graphic Services and printed
and bound by Hamilton Printing. The cover was printed by Phoenix Color Corporation.

Recognizing the importance of preserving what has been written, it is a
policy of John Wiley & Sons, Inc. to have books of enduring value published
in the United States printed on acid-free paper, and we exert our best
efforts to that end.

Library of Congress Cataloging in Publication Data:

Cook, Robert Davis.

Finite element modeling for stress analysis / Robert D. Cook.

p. cm.

Includes index.

ISBN 0-471-10774-3

1. Finite element method—Data processing. 2. Structural analysis (Engineering)—Data processing. I. Title.

TA347.F5C665 1994 94-34421

624.1′76—dc20 CIP

Printed in the United States of America

10 9 8 7 6 5 4 3 2 1

Preface

This book is intended for beginning courses in finite elements (FE) that are oriented toward *users* of the method. The courses envisioned emphasize the behavior of FE and include computational work in which problems are solved by means of commercial software and the computed results are critically examined. The instructor may often sit with students at the computer to offer advice and to monitor their skill in modeling and assessment of results. The courses would use computational problems as vehicles to teach proper use of FE, rather than use FE as a way to solve certain problems. The book presents a modest amount of theory, discusses the nature of FE solutions, offers modeling advice, suggests computational problems, and emphasizes the need for checking the computed results. Problem areas treated are common in mechanical engineering and related disciplines. Suggested computational problems include topics often treated in a second course in stress analysis, such as spinning disks and elastic foundations. The computational problems usually have simple geometry, so that FE may be emphasized rather than details of data preparation. Some instructors especially those who teach more advanced students, may wish to devise problems of a more "real world" nature, despite their greater complexity.

Several commercial FE programs are available for use on microcomputers and workstations. This book is not tailored to any particular FE program and therefore does not discuss the formalisms of input data preparation. Suitable software will have most of the following features: capability in static stress analysis, structural dynamics, vibration, and heat transfer; a good library of elements; some node and element generation capability; help screens; plotting and animation of displaced shapes; contour plotting of computed stresses without nodal averaging. The software must be easy to use, at the expense of versatility if necessary, so that time will not be wasted in learning procedures peculiar to a certain code but having little to do with insight into the FE method.

Many powerful analytical tools are readily available in the form of computer software. Engineers do not have time to study the theory of all these tools, and undergraduates usually study theory with little enthusiasm. For undergraduate and graduate students alike, it appears that study of only the *theory* of FE confers no ability in the *use* of FE. Theory cannot be ignored, however; an engineer must understand the nature of the analytical method as well as the physical nature of the phenomenon to be studied because computer implementation makes it all too easy to choose inappropriate options or push an analytical method beyond its limits of applicability. Fortunately, the user of FE software need not understand all its details. Mainly, the FE user should grasp the physical problem, understand how FE's behave, know the limitations of the theory on which they are based, and be able and willing to check results for correctness. The checking phase relies more on physical understanding of the problem than on knowledge of FE.

The presentation in this book presumes a knowledge of elementary matrix algebra and the level of physical understanding that a *good* student should have after completing a first course in mechanics of materials. This is adequate preparation for a one-semester course in the practice of FE, during which students will inevitably be exposed to concepts of stress analysis not treated in an elementary mechanics of materials course. The understanding they gain by working with these problems will be primarily physical but will be helpful if theory is to be studied subsequently. In my opinion students in a beginning course learn theory only if forced to do so, and then with little understanding of it. Only later, when the nature of a problem area has become familiar, can theory be understood and its practical value appreciated. These remarks are not intended to imply that the book is unsuitable for students who have advanced knowledge of stress analysis theory. In my experience, a student at any level may be deficient in physical understanding, and graduate students make many of the modeling mistakes also made by undergraduates.

The beginning course I teach is taken by seniors. We currently discuss most of Chapters 1 through 7 and the first four articles of Chapter 9. Isoparametric elements and Sections 5.5 and 6.6 are omitted. For this course I find that previous exposure to the theories of elasticity, plates, shells, and vibrations is not necessary because the essential physical behavior of such problems is easily grasped: flat plates can stretch or bend; curved plates (shells) can simultaneously stretch and bend; examples of vibration are commonplace (e.g., a bell). If courses in these areas were prerequisites, few would enroll in the FE course. Students would then have education in neither FE nor problems to which FE analyis is applied. Yet after graduation they will use FE whether or not they are prepared to do so.

In addition to serving as the primary text in a first FE course, the book should be useful as an adjunct text in a second FE course that considers theory in more detail, and in other courses such as vibrations where the solution of practical problems is considered important. It is in this context that the latter part of Chapter 9 (Vibration and Dynamics) and Chapter 10 (Nonlinearity in Stress Analysis) seem most appropriate. Practicing engineers as well as students may find that the book contains useful suggestions for modeling and solution strategy.

Several reviewers of the manuscript made many good suggestions. Their contributions are gratefully acknowledged. Thanks are also due to Pat Grinyer, who made it unnecessary for me to update my technical typing skills.

Robert D. Cook

Madison, Wisconsin
July 1994

Contents

Notation

Symbols most often used in stress analysis appear in the following list. Matrices and vectors are denoted by boldface type.

LATIN SYMBOLS

A	Cross-sectional area
B	Element strain displacement matrix; $\boldsymbol{\varepsilon} = \mathbf{Bd}$
C	Constraint matrix, damping matrix
D,d	Nodal d.o.f., structure (global) and element, respectively
$\bar{\mathbf{D}}$	Amplitudes of structure (global) d.o.f. in vibration
d.o.f.	Degrees of freedom
E	Material property matrix, as in $\boldsymbol{\sigma} = \mathbf{E}\boldsymbol{\varepsilon}$
E	Elastic modulus
f	Cyclic frequency of vibration, $f = \omega/2\pi$
G	Shear modulus
I	Unit (or identity) matrix
I	Moment of inertia of cross-sectional area
J	Jacobian matrix of an isoparametric element
K,k	Stiffness matrix, structure (global) and element, respectively
L	Length
M,m	Mass matrix, structure (global) and element, respectively
N	Element shape (or interpolation) function matrix
p	Pressure
q	Distributed load along a line or on a surface
R	Vector of nodal loads applied to a structure
T	A transformation matrix
T	Temperature; also period of vibration ($T = 1/f$)
t	Thickness or time
u	Vector of displacement components, $\mathbf{u} = \{u \quad v \quad w\}$
u,v,w	Components of displacement at an arbitrary material point
V	Volume
z	Vector of scale factors of vibration modes

GREEK SYMBOLS

β_i	Generalized coordinate (amplitude of a displacement mode)
α	Coefficient of thermal expansion
$\boldsymbol{\varepsilon}$	Vector of strains; for example, $\boldsymbol{\varepsilon} = \{\varepsilon_x \quad \varepsilon_y \quad \gamma_{xy}\}$ in the xy plane
η	An error measure, applied to the computed stress field
$\theta_x,\theta_y,\theta_z$	Rotation angles about x, y, and z axes, respectively
ν	Poisson's ratio
ξ	Damping ratio c/c_c in dynamic analyses
ξ, η, ζ	"Natural" coordinates used for isoparametric elements
ρ	Mass density or radius of curvature
$\boldsymbol{\sigma}$	Vector of stresses; for example, $\boldsymbol{\sigma} = \{\sigma_x \quad \sigma_y \quad \tau_{xy}\}$ in the xy plane
σ_e	von Mises or "effective" stress
$\boldsymbol{\phi}$	Modal matrix; its columns are vibration modes $\bar{\mathbf{D}}_i$
ω	Natural frequency of vibration (radians per second)

CHAPTER 1

Introduction

This chapter introduces concepts and procedures that are discussed in detail in subsequent chapters. The finite element (FE) analysis procedure described in Section 1.3 is used in example applications at the ends of Chapters 2, 3, 6, 7, 8, 9, and 10. Chapter 1 closes with a review of elementary matrix algebra, which is used throughout the book.

1.1 THE FINITE ELEMENT METHOD

The FE method was developed more by engineers using physical insight than by mathematicians using abstract methods. It was first applied to problems of stress analysis and has since been applied to other problems of continua. In all applications the analyst seeks to calculate a *field quantity*: in stress analysis it is the displacement field or the stress field; in thermal analysis it is the temperature field or the heat flux; in fluid flow it is the stream function or the velocity potential function; and so on. Results of greatest interest are usually peak values of either the field quantity or its gradients. The FE method is a way of getting a *numerical* solution to a *specific* problem. A FE analysis does not produce a formula as a solution, nor does it solve a class of problems. Also, the solution is approximate unless the problem is so simple that a convenient exact formula is already available.

An unsophisticated description of the FE method is that it involves cutting a structure into several elements (pieces of the structure), describing the behavior of each element in a simple way, then reconnecting elements at "nodes" as if nodes were pins or drops of glue that hold elements together (Fig. 1.1-1). This process results in a set of simultaneous algebraic equations. In stress analysis these equations are equilibrium equations of the nodes. There may be several hundred or several thousand such equations, which means that computer implementation is mandatory.

A more sophisticated description of the FE method regards it as piecewise polynomial interpolation. That is, over an element, a field quantity such as displacement is interpolated from values of the field quantity at nodes. By connecting elements together, the field quantity becomes interpolated over the entire structure in piecewise fashion, by as many polynomial expressions as there are elements. The "best" values of the field quantity at nodes are those that minimize some function such as total energy. The minimization process generates a set of simultaneous algebraic equations for values of the field quantity at nodes. Matrix symbolism for this set of equations is $\mathbf{KD} = \mathbf{R}$, where $\mathbf{D}$ is a vector of unknowns (values of the field quantity at nodes), $\mathbf{R}$ is a vector of known loads, and $\mathbf{K}$ is a matrix of known constants. In stress analysis $\mathbf{K}$ is known as a "stiffness matrix."

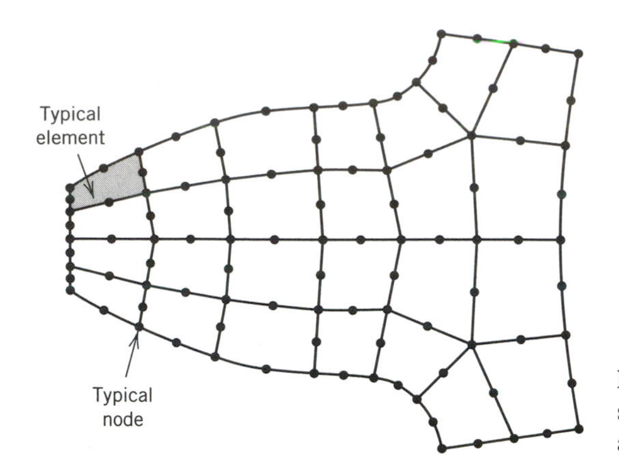

Fig. 1.1-1. A coarse-mesh, two-dimensional model of a gear tooth. All nodes and elements lie in the plane of the paper.

The power of the FE method is its versatility. The structure analyzed may have arbitrary shape, arbitrary supports, and arbitrary loads. Such generality does not exist in classical analytical methods. For example, temperature-induced stresses are usually difficult to analyze with classical methods, even when the structure geometry and the temperature field are both simple. The FE method treats thermal stresses as easily as stresses induced by mechanical load, and the temperature distribution itself can be calculated by FE.

Preprocessing and Postprocessing. The theory of FE includes matrix manipulations, numerical integration, equation solving, and other procedures carried out automatically by commercial software. The user may see only hints of these procedures as the software processes data. The user deals mainly with *preprocessing* (describing loads, supports, materials, and generating the FE mesh) and *postprocessing* (sorting output, listing, and plotting of results). In a large software package the analysis portion is accompanied by the preprocessor and postprocessor portions of the software. There also exist stand-alone pre- and postprocessors that can communicate with other large programs. Specific procedures of "pre" and "post" are different in different programs. Learning to use them is often a matter of trial, assisted by introductory notes, manuals, and on-line documentation that accompanies the software. Also, vendors of large-scale programs offer training courses. Fluency with pre- and postprocessors is helpful to the user but is unrelated to the accuracy of FE results produced. This book emphasizes how to use the FE method properly, not how to use pre- and postprocessors.

FE Method and the Typical User. The typical user of the FE method asks what kinds of elements should be used, and how many of them? Where should the mesh be fine and where may it be coarse? Can the model be simplified? How much physical detail must be represented? Is the important behavior static, dynamic, nonlinear, or what? How accurate will the answers be, and how can they be checked? One need not understand the mathematics of FE to answer these questions. However, a competent user must understand how elements behave in order to choose suitable kinds, sizes, and shapes of elements, and to guard against misinterpretations and unrealistically high expectations. A user must also realize that the FE method is a way of implementing a mathematical theory of physical behavior. Accordingly, assumptions and limitations of theory must not be violated by what we ask the software to do. In some dynamic and nonlinear analyses, algorithms by which theory is implemented must be understood, to avoid choosing an inappropriate algorithm, and to avoid interpreting results produced by algorithmic quirks or limitations as actual physical behavior. Despite all this understanding it is still easy to make mistakes in describing a problem to the computer program. Therefore it is also essential that a competent user have a good physical grasp of the problem so that errors in computed results can be detected and a judgment made as to whether the results are to be trusted or not. An analyst unable to do even a crude pencil-and-paper analysis of the problem probably does not know enough about it to attempt a solution by FE!

A Short History of FE Method. In a 1943 paper, the mathematician Courant described a piecewise polynomial solution for the torsion problem [1.1].* His work was not noticed by engineers and the procedure was impractical at the time due to the lack of digital computers. In the 1950s, work in the aircraft industry introduced the FE method to practicing engineers. A classic paper described FE work that was prompted by a need to analyze delta wings, which are too short for beam theory to be reliable [1.2]. The name "finite element" was coined in 1960 [1.3, 1.4]. By 1963 the mathematical validity of the FE method was recognized and the method was expanded from its structural beginnings to include heat transfer, groundwater flow, magnetic fields, and other areas. Large general-purpose FE software began to appear in the 1970s. By the late 1980s the software was available on microcomputers, complete with color graphics and pre- and postprocessors. By the mid-1990s roughly 40,000 papers and books about the FE method and its applications had been published.

Overview of the Remainder of the Book. Chapter 2 considers elements for bar and beam problems and discusses the mathematical structure of the FE method (the "stiffness method"). Plane problems are treated in Chapter 3. Chapter 4 discusses special methods for element formulation and linear static analysis. After studying Chapters 1 through 4 the reader should have enough background to profit from a thorough discussion of how to use the FE method properly, with attention to planning the model, detecting errors, and verifying results. This material appears in Chapter 5 and is an elaboration of Section 1.3. Chapters 6 and 7 discuss general solids, solids of revolution, plates, and shells. Temperature distribution is considered in Chapter 8, with emphasis on its use in thermal stress analysis. Vibration and other dynamic problems occupy Chapter 9. Chapter 10 is devoted to nonlinear problems and buckling. Example applications of the FE method appear near the ends of most chapters.

1.2 ELEMENTS AND NODES

Finite elements resemble fragments of the structure. Nodes appear on element boundaries and serve as connectors that fasten elements together. In Fig. 1.2-1, elements are triangular or quadrilateral areas and nodes are indicated by dots. Except for element midside nodes along *AED* and nodes at *A*, *B*, and *E*, each node acts as a connector between two or more elements. All elements that share a node have the same displacement components at that node. Lines in Fig. 1.2-1 indicate boundaries between elements. Thus we see elements with corner nodes only and elements with side nodes as well. Such a mixture of element types is neither necessary nor common but serves the present discussion.

Superficially, it appears that a FE structure can be produced by sawing the actual structure apart and then pinning it back together at nodes. Clearly, such an assemblage would be weak and unrepresentative of the actual structure because of strain concentrations at nodes, sliding of elements on one another, and even gaps that would appear be-

*Numbers in brackets indicate references listed at the back of the book.

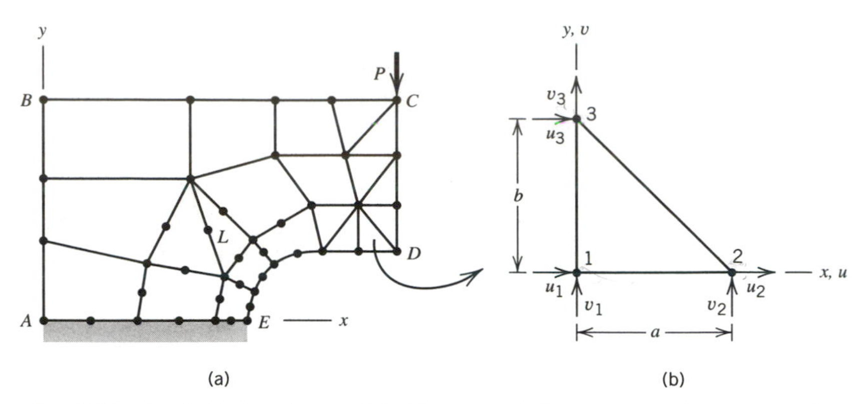

Fig. 1.2-1. (a) A flat bracket modeled by several element types (more types than would actually be used for this problem). (b) One of the elements, a "constant strain triangle". All nodes and elements lie in the plane of the paper.

tween some elements. To avoid these defects and to permit convergence toward exact results as more and more elements are used in the FE model, each element is restricted in its mode of deformation. This leads us to ask what kind of behavior can be expected of each element type. The question is answered repeatedly in subsequent chapters. For now we discuss only the following abbreviated examples of plane elements, which are discussed in more detail in Chapter 3.

Consider the plane triangular element in Fig. 1.2-1b. It does not matter that the origin of coordinates has been moved from its position in Fig. 1.2-1a. The x and y direction components of displacement of an arbitrary point within the element are given the names u and v. In the three-node triangular element each is restricted to be a linear polynomial in x and y:

$$u = \beta_1 + \beta_2 x + \beta_3 y \tag{1.2-1a}$$

$$v = \beta_4 + \beta_5 x + \beta_6 y \tag{1.2-1b}$$

where the β_i are called "generalized coordinates." They can be regarded as displacement amplitudes. As examples, in Eq. 1.2-1a, β_1 is the amplitude of rigid-body displacement, and β_2 and β_3 are amplitudes of linearly varying displacement, all in the x direction. Alternative forms of Eqs. 1.2-1 can be written by expressing the β_i in terms of nodal displacements u_1, v_1, u_2, v_2, u_3, and v_3. To do so for the element in Fig. 1.2-1b we make the following substitutions in Eqs. 1.2-1:

$$\begin{aligned} u &= u_1 \quad \text{and} \quad v = v_1 \quad \text{at} \quad x = 0 \quad \text{and} \quad y = 0 \\ u &= u_2 \quad \text{and} \quad v = v_2 \quad \text{at} \quad x = a \quad \text{and} \quad y = 0 \\ u &= u_3 \quad \text{and} \quad v = v_3 \quad \text{at} \quad x = 0 \quad \text{and} \quad y = b \end{aligned} \tag{1.2-2}$$

Thus, for the element in Fig. 1.2-1b, alternative forms of Eqs. 1.2-1 are found to be

$$u = \left(1 - \frac{x}{a} - \frac{y}{b}\right)u_1 + \frac{x}{a}u_2 + \frac{y}{b}u_3 \tag{1.2-3a}$$

$$v = \left(1 - \frac{x}{a} - \frac{y}{b}\right)v_1 + \frac{x}{a}v_2 + \frac{y}{b}v_3 \tag{1.2-3b}$$

In either form, Eqs. 1.2-1 or 1.2-3, the displacement field $u = u(x, y)$ and $v = v(x, y)$ has six *degrees of freedom*, abbreviated d.o.f. That is, six quantities define the deformed configuration, namely, the six β_i in Eqs. 1.2-1 or the three u_i and three v_i in Eqs. 1.2-3. In Chapter 3 we will explain that strains are displacement gradients. Therefore

$$\begin{aligned} \varepsilon_x &= \frac{\partial u}{\partial x} & &\text{hence } \varepsilon_x = \beta_2 \\ \varepsilon_y &= \frac{\partial v}{\partial y} & &\text{hence } \varepsilon_y = \beta_6 \\ \gamma_{xy} &= \frac{\partial u}{\partial y} + \frac{\partial v}{\partial x} & &\text{hence } \gamma_{xy} = \beta_3 + \beta_5 \end{aligned} \tag{1.2-4}$$

This three-node element is called a "constant strain triangle" because none of the strains varies over the element. This means that the element has a very limited response—it could not represent the linear strain field of pure bending, for example—but at least there will be no strain concentrations at nodes. Also, from Eqs. 1.2-3 we can conclude that element sides will remain straight after deformation. For example, set $x = 0$ to examine side 1–3 in Fig. 1.2-1b; thus u becomes linear in y and depends only on d.o.f. u_1 and u_3. The same will be true along this side in the adjacent element. Because deformed sides remain straight, elements will not gap apart or overlap when load is applied. Similarly, we can show that v along side 1-3 is linear in y and depends only on v_1 and v_3, whether we examine the element on the left or the element on the right of side 1–3. Summing up, it is possible to demonstrate that the triangular element can display constant strain states and will deform in a way that is compatible with its neighbors. The same can be demonstrated for other shapes and types of element. It can be shown that these properties allow exact results to be approached as a mesh is refined; that is, as more and more elements are used to model a structure.

Let us also consider briefly a six-node triangle, such as element L somewhat above E in Fig. 1.2-1a. It has three vertex nodes and three midside nodes. In terms of generalized coordinates β_i, its displacement field is

$$\begin{aligned} u &= \beta_1 + \beta_2 x + \beta_3 y + \beta_4 x^2 + \beta_5 xy + \beta_6 y^2 \\ v &= \beta_7 + \beta_8 x + \beta_9 y + \beta_{10} x^2 + \beta_{11} xy + \beta_{12} y^2 \end{aligned} \tag{1.2-5}$$

Deformed shapes of sides can be straight or parabolic. Some tedious algebra shows that the deformed shape of a side depends on d.o.f. of nodes attached to that side but does not depend on d.o.f. of nodes *not* attached to that side. Accordingly, the element will be compatible with its neighbors because adjacent elements share the same nodes and d.o.f. along a common side. By applying the differentiation used in Eqs. 1.2-4, we see that the six-node element contains constant and linear terms in its strain field. Therefore this element can model constant strain states and also linear strain states that arise in pure bending. Clearly, it is a more competent element than the constant strain triangle. It is also more complicated, which suggests another choice faced by the user of FE: Is it better to use many simple elements or a few complicated elements? We postpone this matter, as the answer is neither short nor simple.

The foregoing discussion is couched in stress analysis terminology. In plane stress analysis the displacement field is a vector field because it has two components, $u = u(x, y)$ and $v = v(x, y)$. In other applications the field may be a scalar field, $\phi = \phi(x, y)$ in two-dimensional problems and $\phi = \phi(x, y, z)$ in three-dimensional problems, where ϕ represents temperature in a heat conduction problem, voltage in an electric field problem, and so on. To restate the foregoing equations in scalar field terms, one may discard equations that contain v and replace u by ϕ in equations that remain.

Equations such as Eqs. 1.2-1 and 1.2-5 constitute the "basis" of a finite element. What remains is to manipulate the basis to generate a "stiffness matrix" that describes element behavior, connect elements together to produce the FE model, apply loads, impose support conditions, solve for nodal d.o.f., and use the d.o.f. to compute strains and finally stresses. Some of these procedures are primarily computational and others require that the analyst make decisions. Subsequent chapters contain a more complete discussion of these matters.

Classification of Stress Analysis Problems. Elements summarized above are used for *plane problems*, in which there is negligible variation of displacement and stress in the z direction, that is, in the direction normal to the analysis plane. If displacements and stresses may vary in a general way with all three coordinates, the object may be called a *3D solid*. The special case of a solid having axial symmetry (like a bell) is usually called a *solid of revolution*. Loads may or may not be axially symmetric. A flat plate that carries in-plane loads is a plane problem, but if the plate is loaded laterally so that it bends it is called a *plate bending problem* or simply a *plate problem*. Floor slabs and highway slabs are examples of plates. Note that thickness must be much less than span if the object is to be analyzed as a plate. If a plate is curved it becomes a *shell*. Water tanks and compressed air tanks are commonly seen shells. Shells can carry both in-plane loads and lateral loads; thus plane deformation and bending deformation usually appear simultaneously in a shell. Elements have been devised for all these problems. Thus there are plane elements, general solid elements, axisymmetric solid elements, plate elements, and shell elements. In addition, there are elements for bars and beams and many specialty elements for elastic foundations, crack tips, pipe bends, and more.

1.3 MODELING THE PROBLEM AND CHECKING RESULTS

Modeling is the simulation of a physical structure or physical process by means of a substitute analytical or numerical construct. It is not simply preparing a mesh of nodes and elements. Modeling requires that the physical action of the problem be understood well enough to choose suitable kinds of elements, and enough of them, to represent the physical action adequately. We want to avoid badly shaped elements and elements too large to represent important variations of the field quantity. At the other extreme we want to avoid the waste of analyst time and computer resources associated with *over*-refinement, that is, using many more elements than needed to adequately represent the field and its gradients. Later, when the computer has done the calculations, we must check the results to see if they are reasonable. Checking is very important because it is easy to make mistakes in describing the problem to the software. The following discussion is a brief survey of these matters. Further discussion appears in subsequent chapters.

Support conditions are very important but are often misrepresented. Consider the problem of Fig. 1.2-1 again. Support along *AE* is portrayed as rigid, meaning that nodes along *AE* are not allowed to move at all. This is probably unrealistic. No support is infinitely stiff. It would be better to enlarge the FE model so that there are finite elements below *AE* to represent the elasticity of the foundation. However, perhaps the intent is to analyze a C-shaped part that has *AE* as an axis of symmetry, and reduce effort by modeling only the upper half. Supports suited to this situation appear in Fig. 1.3-1a. These supports are placed at all nodes along *AE*. Node *A* is fixed and other nodes along *AE* are allowed to move in only the x direction. Thus we prevent rigid-body motion in the xy plane and keep *AE* a straight line as symmetry requires.

The mixture of element types in Fig. 1.2-1 is unusual, but otherwise is the mesh layout good? We cannot say for sure without knowing more about how elements behave. However, by *anticipating the results* we can see that the mesh grading looks reasonable. Stresses near *B* will be low and of little interest. Indeed, theory says that stresses *at B* are zero because it is a point where two free surfaces intersect at an interior angle of less than 180°. Accordingly, a coarse mesh near *B* is acceptable: stresses near *B* may have a large *percentage* error but this does not matter if stresses near *B* are small. The same is true near *D*, so perhaps the mesh near *D* is more detailed than necessary. At *C* the stresses are theoretically infinite because of the concentrated load *P*. In reality, one cannot apply a load that is truly concentrated at a point. Probably load *P* is a convenient way of representing a load that is actually distributed over a small span, and stresses near *C* are not the object of study, so the modeling near *C* is acceptable. Stresses near *E* are probably the stresses of concern. There the stresses and stress gradients are expected to be large, so the model properly displays a finer mesh and/or more competent elements in this area.

The FE method calculates nodal displacements, then (in present software) uses the displacement information to calculate strains and finally stresses. If displacements are incorrect, stresses will probably be incorrect. Accordingly, we should examine the computed displacements first. Without calculation, we anticipate that the displaced shape of our example structure will be as shown in Fig. 1.3-1b. If the computed result is substantially different from this we suspect an error in our model. The software will permit us to display the displaced shape superposed on the original shape, with displacements scaled up so that they are easily visible. Additionally, we can *animate* the displaced shape, so that the model appears to be vibrating slowly between its deformed and undeformed posi-

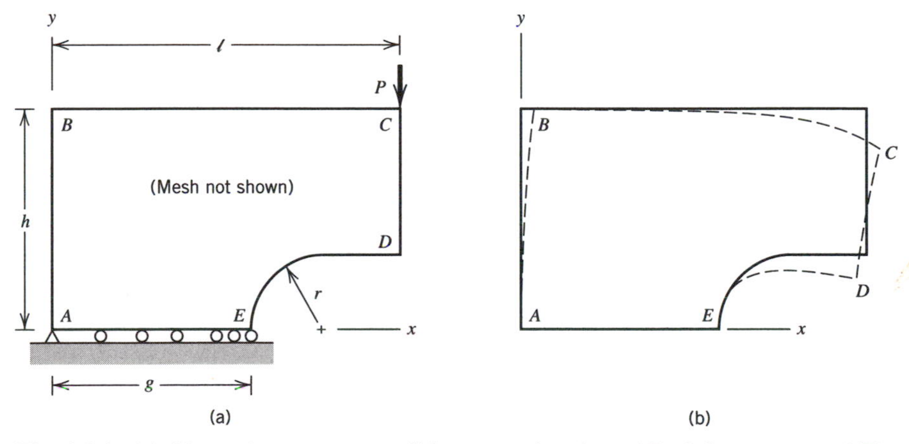

Fig. 1.3-1. (a) Alternative support conditions at nodes along *AE* of the structure of Fig. 1.2-1. (b) Dashed lines show the anticipated deformation, greatly exaggerated.

tions. Thus we can easily see, for example, if nodes along AE move in only the x direction as was intended.

As for stresses, the software will plot them either as contour lines or bands of different colors. A stress contour line connects points that have the same stress. Users may elect an option in the software that calls for averaging of stresses. This means that stresses from individual elements are averaged at nodes before plotting, so that stress contours have no discontinuities between elements. *This is poor practice* because it removes information useful to the analyst. As we will see, unaveraged stresses are usually discontinuous across interelement boundaries. A contour plot that displays significant interelement discontinuities warns that a finer mesh is needed. This point is made in Fig. 1.3-2. In Fig. 1.3-2c, nothing betrays a lack of perfection but small changes of direction where contour lines cross interelement boundaries.

The separate stress contour plots (one for σ_x, one for σ_y, etc.) are examined in turn. Based on experience, physical intuition, and knowledge of theory (including statics, mechanics of materials, and possibly more), it is possible to describe the expected stresses qualitatively. For the problem of Fig. 1.3-1 we expect the following (this is not an exhaustive list):

- σ_y is a large compressive stress near E
- σ_y is tensile near A but smaller in magnitude than σ_y near E
- σ_x is compressive but small in magnitude between A and E
- σ_x and τ_{xy} are very small along AB because of the free surface condition (computed stresses will not be exactly zero because the solution is approximate)

Significant departures from these expectations warn of trouble with the model or shortcomings in physical understanding of the problem. *Discrepancies must be corrected or logically explained before the results can be trusted.*

The analyst should also obtain analytical or experimental results for comparison with FE results. For the problem of Fig. 1.3-1 this task is easy. Cross section AE is loaded by direct force and by bending. The elementary formula for stress in straight beams should provide a fair approximation, the formula for stress in curved beams should provide a good approximation, and tabulated results are available [1.5]. Indeed, FE analysis is probably not needed for this problem. For many problems, approximate solutions can be obtained from tabulated formulas in standard textbooks and handbooks [1.5]. Much of this information is available as software, which makes it far easier to use. However, if this

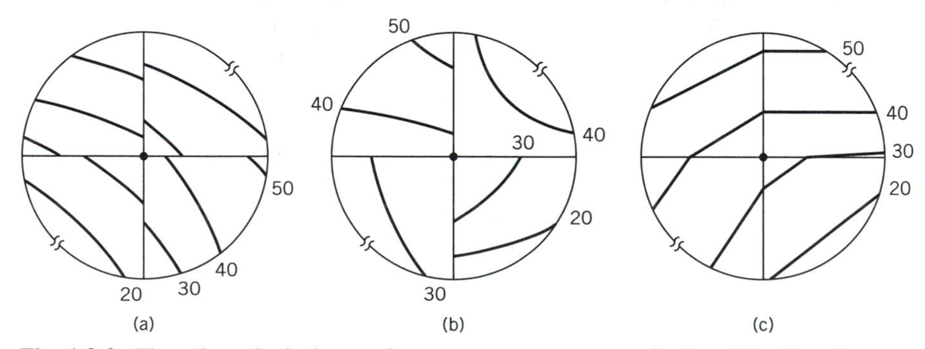

Fig. 1.3-2. Three hypothetical sets of stress contours near a node shared by four elements. (a) Without nodal averaging: imperfect but adequate continuity. (b) Without nodal averaging: inadequate continuity. (c) After nodal averaging: continuity, but difficult to say whether the raw data were good or bad.

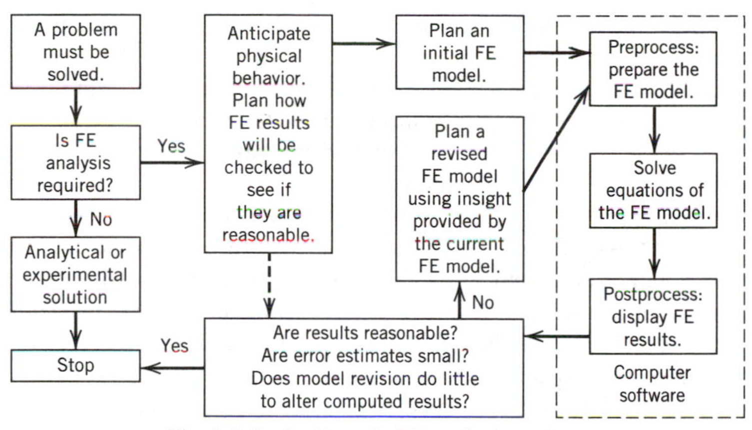

Fig. 1.3-3. Outline of a FE analysis project.

phase of verification is done *after* doing FE analysis, there will be a tendency, perhaps unconscious, to obtain analytical results that agree with FE results already obtained. We tend to find what we expect, whether it is there or not. Therefore some approximate results should be in hand *before* undertaking FE analysis. Figure 1.3-3 summarizes the procedure for FE analysis that is advocated in this book.

Organized and careful work will take less total time than a hurried approach that produces and propagates errors that must be discovered later and corrected. *Festina lente.*

1.4 DISCRETIZATION AND OTHER APPROXIMATIONS

Whatever the analysis method, we do not analyze the actual physical problem; rather, we analyze a mathematical model of it. Thus we introduce *modeling error.* For example, in elementary beam theory we represent a beam by a line (its axis) and typically ignore deformations associated with transverse shear. This is an excellent approximation for slender beams but not for very short beams. Or, for the axial-load problem of Fig. 1.4-1a, we would probably assume that a state of uniaxial stress prevails throughout the bar, which is proper if taper is slight but improper if taper is pronounced. Real structures are not so easily classified, as they are often built of parts that would be idealized mathematically in different ways and have cutouts, stiffeners, and connectors whose behavior is uncertain.

The foregoing considerations must be addressed in order to decide what types of elements to use and how many of them. If a beam is deep, transverse shear deformation may become important and should be included in beam elements. If a beam is *very* deep, two- or three-dimensional elements are more appropriate than beam elements. If a beam has a

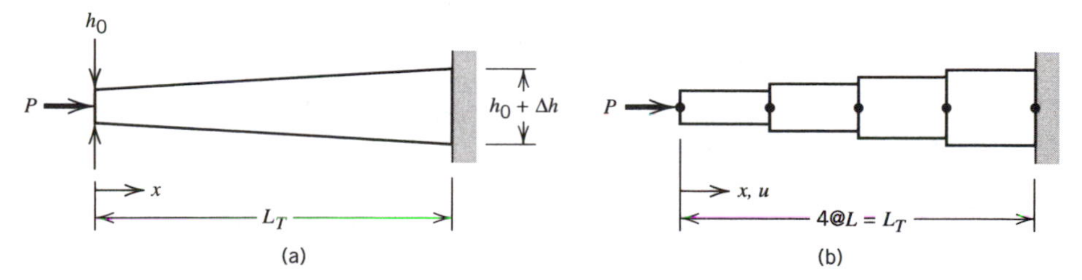

Fig. 1.4-1. (a) A tapered bar loaded by axial force P. (b) Discretization of the bar into four uniform two-node elements of equal length.

wide cross section, plate theory may be more appropriate than beam theory (then, of course, choose plate elements rather than beam elements). If an axisymmetric pressure vessel has a thick wall one should regard it as a solid of revolution rather than a shell of revolution and choose axisymmetric solid elements rather than axisymmetric shell elements.

Let us consider the axially loaded tapered bar of Fig. 1.4-1a in more detail and describe how the FE method implements the mathematical model. We will assume that a satisfactory mathematical model is based on a state of uniaxial stress. An analytical solution is then rather easy, but we pretend not to know it and ask for a FE solution instead. We *discretize* the mathematical model by dividing it into two-node elements of constant cross section, as shown in Fig. 1.4-1b. Each element has length L, accounts only for a constant uniaxial stress along its length, and has an axial deformation given by the elementary formula PL/AE. For each element, A may be taken as constant and equal to the cross-sectional area of the tapered bar at an x coordinate corresponding to the element center. The displacement of load P is equal to the sum of the element deformations. Intuitively, we expect that the exact displacement is approached as more and more elements are used to span the total length L_T. However, even if a great many elements are used there is an error, known as *discretization error*, which exists because the physical structure *and* the mathematical model each have infinitely many d.o.f. (namely, the displacements of infinitely many points) while the FE model has a finite number of d.o.f. (the axial displacements of its nodes).

How many elements are enough? Imagine that we carry out two FE analyses, the second time using a more refined mesh. The second FE model will have less discretization error than the first, and will also represent the geometry better if the physical object has curved surfaces. If the two analyses yield similar solutions, we suspect that results are not much in error. Or, we might establish a sequence of solutions by solving the problem more than twice, using a finer mesh each time. By study of how the sequence converges we may be able to state with some confidence that results from the finest mesh are in error by less than (say) 5%.

After the analyst has introduced modeling error and discretization error, the computer introduces *numerical error* by rounding or truncating numbers as it builds matrices and solves equations. Usually numerical error is small, but some modeling practices can greatly increase it.

Finally, it must be admitted that the software almost certainly contains errors [5.6]. Commercial software packages are large, versatile, and under continual revision. It is practically impossible to get everything right. Many errors either make a software feature inoperable or cause the program to crash, but some can lead to erroneous results. It is tempting to blame all strange results on the software, but it is *far* more often the case that we have blundered in modeling or in describing the model to the software. Strange results are obtained so often that (to repeat) it is vital that the analyst be able to *recognize* that results are strange.

1.5 RESPONSIBILITY OF THE USER

FE computer programs have become widely available, easier to use, and can display results with attractive graphics. Even an inept user can produce some kind of answer. It is hard to disbelieve FE results because of the effort needed to get them and the polish of their presentation. But smooth and colorful stress contours can be produced by *any* model, good or bad. It is possible that most FE analyses are so flawed that they cannot be trusted. Even a poor mesh, inappropriate element types, incorrect loads, or improper supports may produce results that appear reasonable on casual inspection. A poor model may have defects that are not removed by refinement of the mesh.

A responsible user *must* understand the physical nature of the problem and the behavior of finite elements well enough to prepare a suitable model and evaluate the quality of the results. Competence in using FE for stress analysis does not imply competence in using FE for (say) magnetic field problems. Responsibility for results produced is taken by the engineer who uses the software, not the software vendor, even if results are affected by errors in the software.

Figure 1.5-1 is an example of discrepancies that may appear [1.6]. A pressure pulse is applied to a straight beam with hinge supports. The loading causes the material to yield and the beam to vibrate. Analysis seeks to track the lateral displacement of the midpoint as a function of time. The results plotted come from ten reputable analysis codes and were obtained by users regarded as expert. Yet if any of the curves is correct we cannot tell which one it is. Admittedly, the problem is difficult. The results indicate "strong sensitivities of both physical and computational nature" [1.6]. This example reminds us that any analysis program is based on theory and approximation, and that a user may push the program beyond its range of validity [1.7].

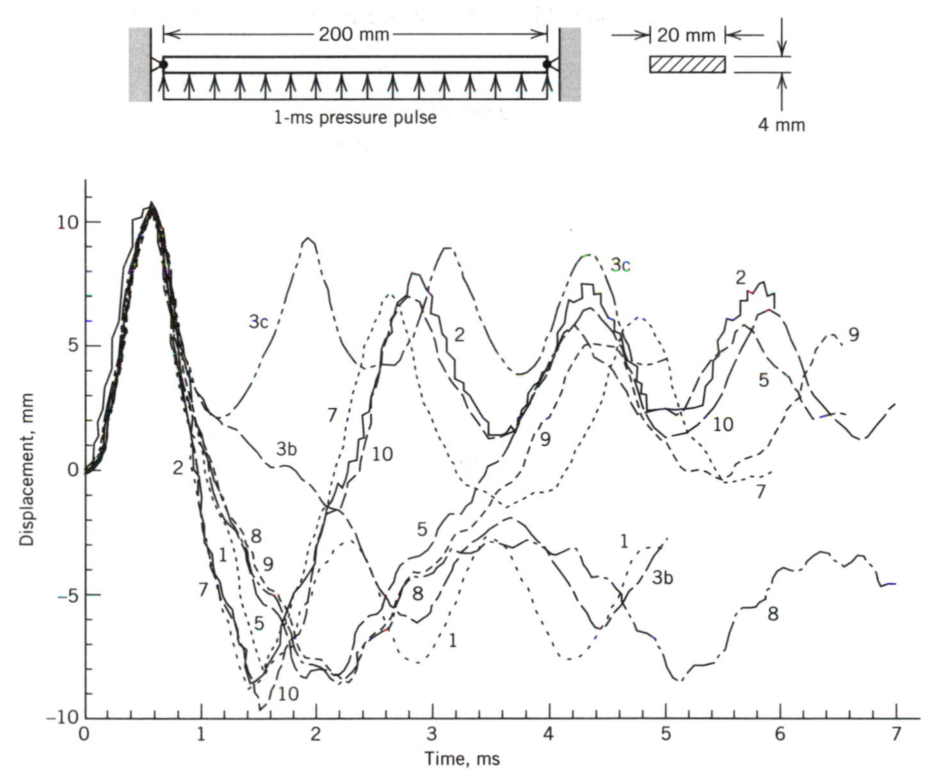

Fig. 1.5-1. Lateral midpoint displacement versus time for a beam loaded by a pressure pulse [1.6] The material is elastic–perfectly plastic. Plots were generated by various users and various codes.

1.6 ELEMENTARY MATRIX ALGEBRA

One need not understand matrix algebra in order to use FE software. However, our explanations of algorithms and of how elements behave are conveniently stated in matrix format. The following matrix theory is used in this book.

A matrix contains numbers, and/or symbols that represent numbers, arrayed in rows and columns. A matrix may be denoted by boldface type, **A**, or by use of brackets, $[A]$.

$$\mathbf{A} = \begin{bmatrix} A_{11} & A_{12} & \cdots & A_{1n} \\ A_{21} & A_{22} & & A_{2n} \\ \vdots & \vdots & & \vdots \\ A_{m1} & A_{m2} & \cdots & A_{mn} \end{bmatrix} \tag{1.6-1}$$

Matrix **A** has m rows and n columns, where m and n are positive integers of any magnitude. If $m = n$, **A** is called square, and n is its *order*. Coefficients with like subscripts (A_{11}, A_{22}, etc.) lie on the *diagonal* of a square matrix. If $m = 1$, **A** is a *row matrix* (also called a *row vector*); if $n = 1$, **A** is a *column matrix* (also called a *column vector*). Braces are often used to indicate a column vector; for example, $\{A\}$ means that **A** has only one column. If $m = n = 1$, **A** is the scalar $\mathbf{A} = A$.

If two matrices **A** and **B** have the same m and the same n, they may be added or subtracted term by term; for example, in $\mathbf{C} = \mathbf{A} + \mathbf{B}$, $C_{ij} = A_{ij} + B_{ij}$. A scalar multiplier of **A** acts on every term of **A**; for example, $\lambda\mathbf{A}$ contains λA_{11}, λA_{12}, and so on. The integral (or derivative) of a matrix with respect to a scalar parameter, such as time, is a matrix that contains the integral (or derivative) of every term.

The *transpose* of **A** is **A** but with rows and columns interchanged. Thus

$$\mathbf{A}^T = \begin{bmatrix} A_{11} & A_{21} & A_{31} & \cdots \\ A_{12} & A_{22} & A_{32} & \cdots \\ A_{13} & A_{23} & A_{33} & \cdots \\ \vdots & \vdots & \vdots & \end{bmatrix} \tag{1.6-2}$$

If $\mathbf{A}^T = \mathbf{A}$, the matrix is called *symmetric*, and $A_{12} = A_{21}$, $A_{13} = A_{31}$, and so on. A symmetric matrix must be square ($m = n$).

The product of two matrices is

$$\underset{\ell\times m}{\mathbf{A}}\ \underset{m\times n}{\mathbf{B}} = \underset{\ell\times n}{\mathbf{P}} \quad \text{where} \quad P_{ij} = \sum_{k=1}^{m} A_{ik}B_{kj} \tag{1.6-3}$$

for example, $P_{23} = A_{21}B_{13} + A_{22}B_{23} + A_{23}B_{33} + \cdots$. For multiplication, **A** and **B** must be *conformable*; that is, if **A** has m columns then **B** must have m rows. An example of multiplication is

$$\begin{bmatrix} 1 & 2 \\ 3 & 4 \end{bmatrix}\begin{bmatrix} 5 & 6 & 7 \\ 8 & 9 & 1 \end{bmatrix} = \begin{bmatrix} 21 & 24 & 9 \\ 47 & 54 & 25 \end{bmatrix} \tag{1.6-4}$$

In general, $\mathbf{AB} \neq \mathbf{BA}$. If **B** is square and symmetric, so is the product $\mathbf{P} = \mathbf{A}^T\mathbf{BA}$. If **A** is a column vector, then $\mathbf{A}^T\mathbf{BA}$ is a scalar. The transpose of a product is the product of the transposes in reverse order; that is, if $\mathbf{P} = \mathbf{AB}$, then $\mathbf{P}^T = \mathbf{B}^T\mathbf{A}^T$.

A *unit* matrix, or *identity* matrix, is denoted by **I**. It is a *diagonal* matrix of 1's; that is,

$$\mathbf{I} = \begin{bmatrix} 1 & 0 & \cdots & 0 \\ 0 & 1 & \cdots & 0 \\ \vdots & \vdots & & \vdots \\ 0 & 0 & \cdots & 1 \end{bmatrix} \tag{1.6-5}$$

The *inverse* of a square matrix **A** is denoted by $\mathbf{A}^{-1}$, where $\mathbf{A}^{-1}$ is constructed in such a way that $\mathbf{A}^{-1}\mathbf{A} = \mathbf{I}$. It is also true that $\mathbf{A}\mathbf{A}^{-1} = \mathbf{I}$. In this book we need to know what $\mathbf{A}^{-1}$ means but not how to construct $\mathbf{A}^{-1}$ from **A**. The inverse of a product is the product of the inverses in reverse order; that is, if $\mathbf{P} = \mathbf{AB}$, then $\mathbf{P}^{-1} = \mathbf{B}^{-1}\mathbf{A}^{-1}$.

A set of simultaneous linear algebraic equations may be symbolized as

$$\mathbf{KD} = \mathbf{R} \tag{1.6-6}$$

where **K** is a square matrix of known constants, **R** is a column vector of known constants, and **D** is a column vector of unknowns. Solution for **D** may be symbolized as

$$\mathbf{D} = \mathbf{K}^{-1}\mathbf{R} \tag{1.6-7}$$

In FE work, **K** is a "stiffness" matrix that is usually large and sparse. It would be wasteful of storage and time to invert it. Thus $\mathbf{D} = \mathbf{K}^{-1}\mathbf{R}$ usually means "solve for the unknowns," probably by some efficient form of Gauss elimination or perhaps by an iterative method. Solving Eq. 1.6-6 for **D** is a major part of FE calculations, but usually the user need not know how the software goes about it.

A square matrix is called *singular* if its determinant is zero. If **K** in Eq. 1.6-6 is singular, there is no unique solution vector **D**, and standard equation-solving subroutines will fail. As examples, the following matrices are singular.

$$\begin{bmatrix} 1 & 0 \\ 0 & 0 \end{bmatrix} \quad \begin{bmatrix} 1 & -1 \\ -1 & 1 \end{bmatrix} \quad \begin{bmatrix} 2 & 4 \\ 4 & 8 \end{bmatrix} \tag{1.6-8}$$

Let **K** be an n by n matrix and **D** an n by 1 column vector. Also let $\mathbf{D} \neq \mathbf{0}$, which means that at least one coefficient D_i is nonzero. Then, for all **D**,

$$\text{if } \mathbf{D}^T\mathbf{K}\mathbf{D} > 0, \quad \mathbf{K} \text{ is called positive definite} \tag{1.6-9a}$$

$$\text{if } \mathbf{D}^T\mathbf{K}\mathbf{D} \geq 0, \quad \mathbf{K} \text{ is called positive semidefinite} \tag{1.6-9b}$$

A positive definite matrix is nonsingular. In stress analysis, a stiffness matrix **K** is positive semidefinite (and singular) if supports of the FE structure do not prevent all possible rigid-body motions.

ANALYTICAL PROBLEMS

1.1 (a) Show that Eqs. 1.2-3 follow from Eqs. 1.2-1 and 1.2-2.

(b) Differentiate Eqs. 1.2-3 to obtain expressions for strains in terms of nodal displacements.

1.2 Show that sides 1–2 and 2–3 of the triangular element in Fig. 1.2-1b remain straight as the element is deformed.

1.3 Equations 1.2-1 may be applied to each of the elements shown, and conditions analogous to Eqs. 1.2-2 used to express displacements $u = u(x, y)$ and $v = v(x, y)$ in terms of nodal d.o.f. u_i and v_i. Carry out these operations. (As a partial check, note that the resulting expressions must yield $u = u_i$ and $v = v_i$ when $x = x_i$ and $y = y_i$, where i is 1, 2, or 3.)

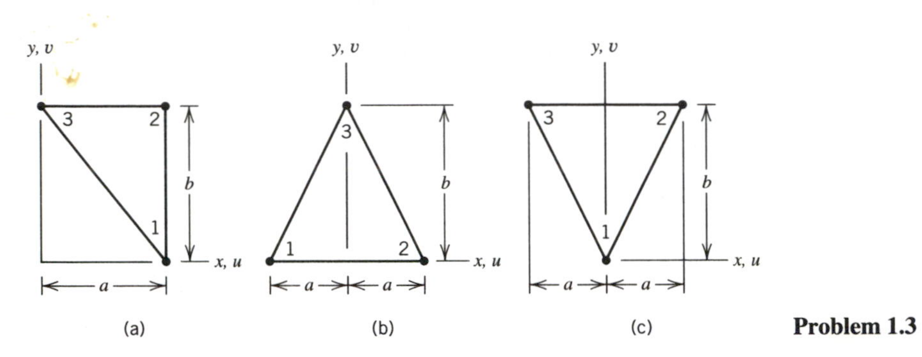

Problem 1.3

1.4 Show that side 1–3 of each of the elements in Problem 1.3 remains straight when the element is deformed.

1.5 (a) In terms of the β_i and x and y, evaluate the strains ε_x, ε_y, and γ_{xy} associated with the displacement field of a six-node triangle, Eq. 1.2-5.

(b) Pure moment loading is applied to a cantilever beam built of these elements, as shown. Exact values of computed stresses are desired, if possible. Why are roller supports placed at A and at B, rather than pin supports as shown at C?

(c) Sketch an alternative arrangement of supports at the left end of the beam that would work just as well.

(d) Which of the β_i in Eq. 1.2-5 will be zero for this particular cantilever beam problem? Which of the other β_i are related to one another, and how? Consider the strains calculated in part (a) to answer these questions.

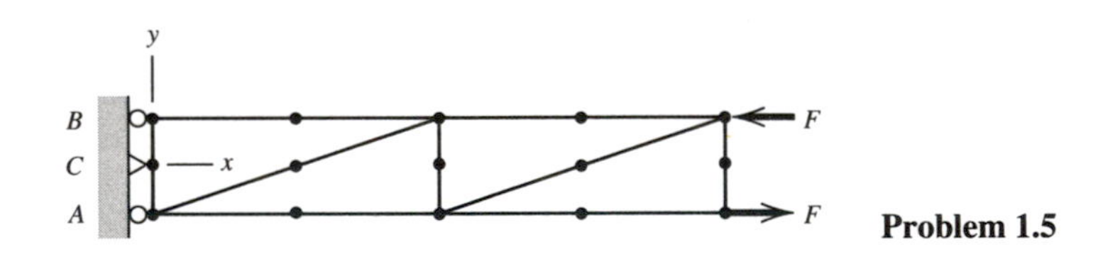

Problem 1.5

1.6 Elaborate on the list of stress predictions in the latter part of Section 1.3. For example, in Fig. 1.3-1a, where do you expect that stresses σ_x, σ_y, or τ_{xy} should approach zero? Also, where do you expect that these stresses may have large magnitudes, and of what algebraic signs?

1.7 Let dimensions in Fig. 1.3-1a be $g = 16$ mm, $h = 18$ mm, $\ell = 28$ mm, and $r = 6$ mm. Also let the thickness be t and the elastic modulus be E. Place roller supports along AE, as in Fig. 1.3-1. Use mechanics of materials analysis to estimate the following in terms of E, t, and load P.

(a) Stress σ_y at A and at E. Use straight beam theory.

(b) Stress σ_y at A and at E. Use curved beam theory.

(c) The σ_x of largest magnitude along AE.
(d) The vertical displacement component of load P.

1.8 A bar element used for the FE model in Fig. 1.4-1b has two d.o.f., namely, the axially directed displacement at each end. Express the element displacement field in the form of Eq. 1.2-1 and in the form of Eq. 1.2-3. Let $x = 0$ at the left end of the element.

1.9 Assume that the bar in Fig. 1.4-1a has a rectangular cross section. Let $\Delta h = 3h_0$ and let thickness t (perpendicular to the paper) be constant. Evaluate the following in terms of P, L_T, h_0, and t.
(a) The exact displacement of load P.
(b) The displacement of P using one, two, and then four uniform elements of equal length. What is the percentage error in each case?
(c) On a set of axes showing x (abscissa) and $\sigma_x/(P/th_0)$ (ordinate), plot the exact axial stress and the axial stress prediction of the four-element model. By approximately what factor are stress errors reduced each time the number of elements is doubled?

COMPUTATIONAL PROBLEMS

No specific computational problems are suggested in this chapter. However, students may wish to get acquainted with the FE software chosen for the remainder of the course by using it to solve simple bar and beam problems for which tabulated solutions are readily available.

CHAPTER 2

Bars and Beams. Linear Static Analysis

Stiffness matrices are developed for bar elements and beam elements. The physical meaning of these matrices is explained. Also explained is how loads are treated, what support conditions are appropriate for a structure built of bar or beam elements, and how the formulation yields displacements and stresses. Finally, an example application shows how beam elements may be used in practice.

2.1 INTRODUCTION

Static analysis omits time as an independent variable and is appropriate if deflections are constant or vary only slowly. A structure forced to vibrate at a frequency less than about one-third of its lowest natural frequency is a case in point. Such "quasistatic" problems may include *steady* inertia loads, such as those due to spinning about an axis at constant speed. *Linear* static analysis excludes plastic action and deflections large enough to change the way loads are applied or resisted. Thus elements that fail, large rotations, and gaps that open or close are excluded.

After doing an approximate preliminary analysis, planning how to do the computational analysis, and perhaps sketching an initial FE model, the analyst turns to software. FE analysis requires that the following steps be taken:

1. Prepare the FE model. The analyst must
   a. discretize the structure or continuum by dividing it into finite elements,
   b. prescribe how the structure is loaded, and
   c. prescribe how the structure is supported.
2. Perform the calculations. The software must
   a. generate the stiffness matrix **k** of each element,
   b. connect elements together, that is, assemble the element **k** matrices to obtain the structure or "global" matrix **K**,
   c. assemble loads into a global load vector **R**,
   d. impose support conditions, and
   e. solve the global equations **KD** = **R** for the vector **D** of unknowns. In structural problems **D** contains displacement components of the nodes.
3. Postprocess the information contained in **D**. In stress analysis this means compute strains and stresses.

Step 1a requires that the analyst exercise judgment about what types of element to use and how coarse or refined the mesh should be in different regions of the model. Steps 1b and 1c are often more straightforward than step 1a but it is easy to be inattentive and do them improperly. The work in step 1 is greatly assisted by the preprocessor portion of the software. Nevertheless, this phase of the analysis will probably take considerable time. Step 2 is carried out automatically by the software. Similarly, step 3 is automatic, although the analyst must instruct the program as to which results to present and the format of their presentation. The displaced shape and various stress contours are usually plotted.

Except for discretization and the plotting of stress contours, the foregoing FE procedure is also applied to the numerical analysis of trusses and frames. These structures are inherently discretized, in the sense that their members are already separate elements. In our terminology, truss elements are hinged at connection points and resist only axial force; frame elements are welded together at connection points and resist axial and transverse forces and bending moments. All these members can be regarded as special cases of what we will call a *3D beam element*, which resists axial force, transverse shear force in each of two directions, bending about each principal axis of the cross section, and torque about the longitudinal axis of the member. The response of the member to these loads can be formulated exactly, or at least quite accurately, using only the tools of mechanics of materials. In this chapter we examine beam elements, beginning with plane bar and plane beam elements as special cases, in order to explain the nature of a stiffness matrix, how loads and supports are treated, and how stresses are extracted from displacements.

A crude initial model of a complicated structure is sometimes built of bars and beams because the effort is comparatively small and information useful in subsequent FE modeling may appear.

2.2 STIFFNESS MATRIX FORMULATION: BAR ELEMENT

Direct Method. Consider a uniform prismatic elastic bar of length L, Fig. 2.2-1, with elastic modulus E and cross-sectional area A. A node is located at each end. For now we allow only axially directed displacements. We displace first one node and then the other and, in each case, calculate forces that must be applied to nodes in order to maintain the displacement state. These forces are easily calculated from the elementary formula for stretching a bar an amount δ, namely, $\delta = FL/AE$, which gives force F as $F = (AE/L)\delta$. For the respective cases in Fig. 2.2-1, with $\delta = u_1$ and then $\delta = u_2$,

$$F_{11} = F_{21} = \frac{AE}{L}u_1 \quad \text{and} \quad F_{12} = F_{22} = \frac{AE}{L}u_2 \tag{2.2-1}$$

where F_{ij} is the force at node i ($i = 1, 2$) associated with displacement of node j ($j = 1, 2$). Next, these results are written in matrix format, allowing both nodes to displace simultaneously, and using the sign convention that *forces and displacements are positive in the same direction*. In the present case *positive is to the right*. Thus

$$\begin{bmatrix} F_{11} & -F_{12} \\ -F_{21} & F_{22} \end{bmatrix} \begin{Bmatrix} 1 \\ 1 \end{Bmatrix} = \begin{Bmatrix} F_1 \\ F_2 \end{Bmatrix} \quad \text{or} \quad \frac{AE}{L}\begin{bmatrix} 1 & -1 \\ -1 & 1 \end{bmatrix} \begin{Bmatrix} u_1 \\ u_2 \end{Bmatrix} = \begin{Bmatrix} F_1 \\ F_2 \end{Bmatrix} \tag{2.2-2}$$

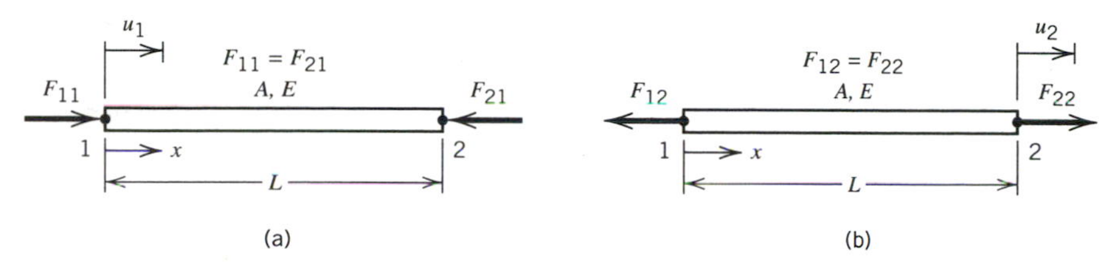

Fig. 2.2-1. Nodal forces associated with deformation of a two-node bar element. (a) Node 1 displaced u_1 units. (b) Node 2 displaced u_2 units.

where F_1 and F_2 are the resultant forces applied to the bar at nodes 1 and 2, $F_1 = F_{11} - F_{12}$ and $F_2 = -F_{21} + F_{22}$. The square matrix in the latter equation, including its scalar multiplier AE/L, is the element stiffness matrix **k**. Symbolically, we write Eq. 2.2-2 as $\mathbf{kd} = \mathbf{r}$, where $\mathbf{d} = [u_1 \quad u_2]^T$ for this element.

In Eq. 2.2-2 we see an instance of a general rule: *a column of* **k** *is a vector of nodal loads that must be applied to the element to sustain a deformation state in which the corresponding nodal d.o.f. has unit value and all other nodal d.o.f. are zero.* For example, with $u_1 = 1$ and $u_2 = 0$, the multiplication **kd** in Eq. 2.2-2 yields the first column of **k**:

$$\frac{AE}{L}\begin{bmatrix} 1 & -1 \\ -1 & 1 \end{bmatrix}\begin{Bmatrix} 1 \\ 0 \end{Bmatrix} = \begin{Bmatrix} F_1 \\ F_2 \end{Bmatrix} \quad \text{hence} \quad \begin{Bmatrix} F_1 \\ F_2 \end{Bmatrix} = \frac{AE}{L}\begin{Bmatrix} 1 \\ -1 \end{Bmatrix}u_1 = \begin{Bmatrix} F_{11} \\ -F_{21} \end{Bmatrix} \tag{2.2-3}$$

where F_{11} and F_{21} are shown in Fig. 2.2-1a.

Formal Procedure. The foregoing "direct method" can produce a stiffness matrix only for simple elements, where formulas from mechanics of materials provide relations between nodal displacements and associated nodal loads. For most elements a general formula for **k** must be used instead. We now take a first look at this formula, and the manipulations it requires, by applying it to the bar element, which is the simplest special case. The general formula is

$$\mathbf{k} = \int \mathbf{B}^T \mathbf{E} \mathbf{B} \, dV \tag{2.2-4}$$

where **B** is the *strain-displacement matrix*, **E** is the *material property matrix* (it may also be called the *constitutive matrix*), and dV is an increment of the element volume V. Equation 2.2-4 can be derived by stating that work is done by nodal loads that are applied to create nodal displacements, and that this work is stored in the element as elastic strain energy. (See Eqs. 3.1-9 and 3.1-10 for a more complete explanation.) To obtain **B** for the bar element we begin by writing an expression for axial displacement u of an arbitrary point on the bar. As shown in Fig. 2.2-2, linear interpolation of u between its nodal values u_1 and u_2 yields

$$u = \begin{bmatrix} \frac{L-x}{L} & \frac{x}{L} \end{bmatrix}\begin{Bmatrix} u_1 \\ u_2 \end{Bmatrix} \quad \text{or} \quad u = \mathbf{N}\mathbf{d} \tag{2.2-5}$$

where **N** is called the shape function matrix and **d** is the vector of element nodal d.o.f. In the present example **N** contains the two individual shape functions $N_1 = (L - x)/L$ and $N_2 = x/L$. Each shape function N_i describes how u varies with x when the corresponding

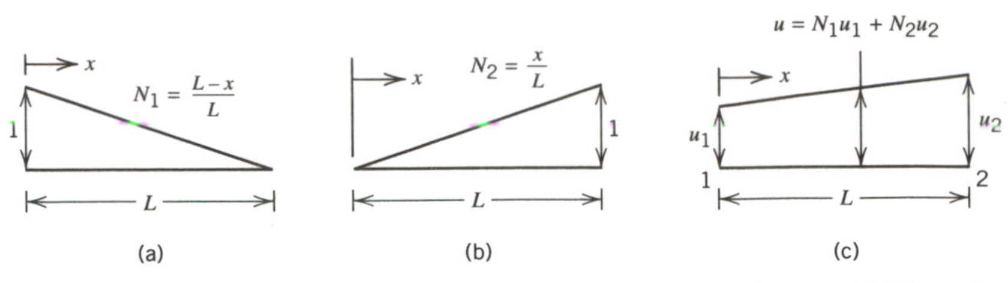

Fig. 2.2-2. (a,b) Shape functions N_1 and N_2 of a two-node bar element. (c) Linear interpolation of axial displacement between node 1 and node 2.

d.o.f. u_i is unity while the other is zero. Axial strain ε_x is the gradient of axial displacement:

$$\varepsilon_x = \frac{du}{dx} = \left[\frac{d}{dx}\mathbf{N}\right]\mathbf{d} = \mathbf{Bd} \quad \text{where} \quad \mathbf{B} = \left[-\frac{1}{L} \quad \frac{1}{L}\right] \tag{2.2-6}$$

Thus $\varepsilon_x = (u_2 - u_1)/L$, which is the basic definition of strain as change in length divided by original length. Finally, for the bar problem, matrix **E** is simply the elastic modulus E, a scalar, and dV is $A\ dx$. Equation 2.2-4 becomes

$$\mathbf{k} = \int_0^L \begin{Bmatrix} -1/L \\ 1/L \end{Bmatrix} E \left[-\frac{1}{L} \quad \frac{1}{L}\right] A\, dx = \frac{AE}{L}\begin{bmatrix} 1 & -1 \\ -1 & 1 \end{bmatrix} \tag{2.2-7}$$

which agrees with the stiffness matrix in Eq. 2.2-2. Note that the form of Eq. 2.2-4 guarantees that **k** will be a *symmetric* matrix.

Limitation. The displacement and strain fields of the element, Eqs. 2.2-5 and 2.2-6, clearly show a limitation of the two-node bar element: it can represent only a *constant* state of strain. Linear and higher order strain variations are not represented. Accordingly, if axial forces are applied only at nodes, the element agrees exactly with a mathematical model that represents the bar as a straight line having constant A and E between locations where axial forces are applied. If axial forces are instead distributed along all or part of the length, or if the bar is tapered, then the element is only approximate. Distributed load can still be applied, in the form of equivalent forces applied to nodes of a bar built of several elements; then exact results are *approached* as more and more elements are used to model the bar. Similarly, if a bar is tapered, so that its axial strain varies continuously, a stepwise-constant FE model becomes more and more accurate as more and more elements are used (Fig. 1.4-1).

That the two-node element is limited to a constant strain state can also be seen when the displacement field is written in terms of generalized coordinates β_i; that is, as $u = \beta_1 + \beta_2 x$. By differentiation, $\varepsilon_x = du/dx = \beta_2$, a constant. In Eq. 2.2-5 the two β_i have been replaced by the two nodal d.o.f. u_i.

2.3 STIFFNESS MATRIX FORMULATION: BEAM ELEMENT

We begin with a plane beam element that can resist only in-plane bending and transverse shear force. This element requires only four d.o.f. and will be called a "simple" plane beam element. A plane beam element that also resists axial force requires two additional d.o.f. It will be described later in this section and will be called a 2D beam element. Finally, a beam element in space that resists *all* components of nodal force and moment requires six d.o.f. per node and will be called a 3D beam element. This element will be discussed last.

Direct Method, Simple Plane Beam Element. Figure 2.3-1a shows a simple plane beam element. The element is prismatic, with elastic modulus E and centroidal moment of inertia I of its cross-sectional area. The beam centerline has lateral displacement $v = v(x)$. According to elementary beam theory, $v = v(x)$ is cubic in x for a uniform prismatic beam loaded only at its ends; that is, by the nodal forces and moments in Fig. 2.3-1b. Nodal d.o.f. consist of lateral translations v_1 and v_2 and rotations θ_{z1} and θ_{z2} about the z axis (normal to the paper). We will ignore transverse shear deformation in our explanations, although commercial software usually accounts for it.

The element stiffness matrix **k** can be constructed column by column, according to the general rule stated below Eq. 2.2-2. To obtain terms in a column we must solve a statically indeterminate beam problem, but this requires only elementary methods. Consider column 1 of **k**. Figure 2.3-1c shows nodal forces and moments that must be applied to sustain a deformation state in which the first d.o.f. has unit value and all other d.o.f. are zero. Nodal loads in Fig. 2.3-1 are labeled according to their position in **k** and with proper algebraic sign: positive directions are upward for translation and force and counterclockwise for rotation and moment. Clearly, not all numerical values of the k_{ij} can be positive; for example, in Fig. 2.3-1c, forces k_{11} and k_{31} must be of opposite sign to pre-

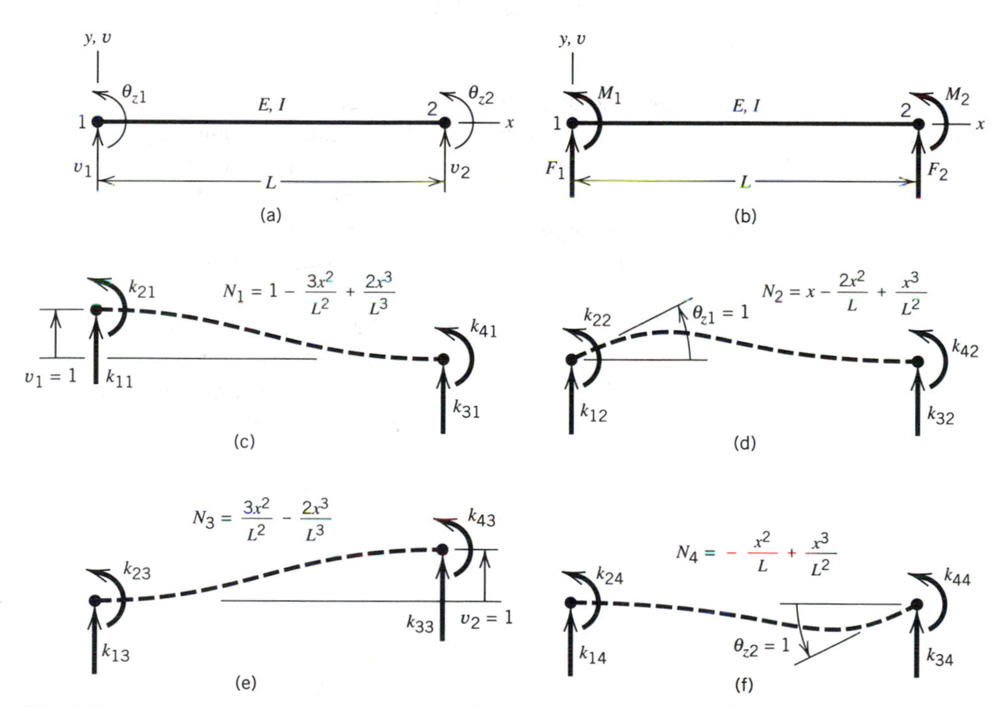

Fig. 2.3-1. (a) Simple plane beam element and its nodal d.o.f. (b) Nodal loads associated with the d.o.f. (c–f) Deflected shapes and shape functions associated with activation of each d.o.f. in turn. Nodal loads are labeled according to their position in **k**. (Reprinted from [2.2] by permission of John Wiley & Sons, Inc.)

serve equilibrium of vertical forces. To solve for the first column of **k**, that is, for the column vector $[k_{11} \quad k_{21} \quad k_{31} \quad k_{41}]^T$, the following conditions are used:

$$v_1 = 1 \quad \text{at node 1} \qquad 1 = \frac{k_{11}L^3}{3EI} - \frac{k_{21}L^2}{2EI} \tag{2.3-1a}$$

$$\theta_{z1} = 0 \quad \text{at node 1} \qquad 0 = -\frac{k_{11}L^2}{2EI} + \frac{k_{21}L}{EI} \tag{2.3-1b}$$

$$\sum(\text{forces})_y = 0 \qquad 0 = k_{11} + k_{31} \tag{2.3-1c}$$

$$\sum(\text{moments})_{\text{node 2}} = 0 \qquad 0 = k_{21} + k_{41} - k_{11}L \tag{2.3-1d}$$

The first two of these equations use standard beam deflection formulas and state that end deflections and rotations produced by force k_{11} and moment k_{21} are superposed to produce unit deflection and zero rotation at node 1. These two equations yield k_{11} and k_{21} in terms of E, I, and L. The latter two equations use statics and yield k_{31} and k_{41} when k_{11} and k_{21} are known. Similar arguments produce an analogous set of four equations for each of the remaining three deformation states. Each deformation state yields terms in one column of **k**. The result of this process is the element stiffness matrix,

$$\mathbf{k} = \begin{bmatrix} 12EI/L^3 & 6EI/L^2 & -12EI/L^3 & 6EI/L^2 \\ 6EI/L^2 & 4EI/L & -6EI/L^2 & 2EI/L \\ -12EI/L^3 & -6EI/L^2 & 12EI/L^3 & -6EI/L^2 \\ 6EI/L^2 & 2EI/L & -6EI/L^2 & 4EI/L \end{bmatrix} \tag{2.3-2}$$

which operates on the vector of nodal d.o.f. $\mathbf{d} = [v_1 \quad \theta_{z1} \quad v_2 \quad \theta_{z2}]^T$.

Formal Procedure, Simple Plane Beam Element. The special form of Eq. 2.2-4 applicable to a beam element is

$$\mathbf{k} = \int_0^L \mathbf{B}^T EI\, \mathbf{B}\, dx \tag{2.3-3}$$

where **B** is now a matrix that yields curvature d^2v/dx^2 of the beam element from the product **Bd**. The commonality of all forms of Eq. 2.2-4 is that in each case the expression $\mathbf{d}^T\mathbf{kd}/2$ represents strain energy in an element under nodal displacements **d**. In bars, strain energy depends on axial strain; in beams, strain energy depends on curvature. Energy principles are matters of theory that are not essential to an understanding of how elements behave (see Eq. 3.1-9 for a brief explanation related to Eq. 2.2-4). In terms of generalized coordinates β_i the lateral displacement $v = v(x)$ of a plane beam element is the following cubic in x:

$$v = \beta_1 + \beta_2 x + \beta_3 x^2 + \beta_4 x^3 \tag{2.3-4}$$

The β_i can be stated in terms of nodal d.o.f. by making substitutions similar to those used in Eq. 1.2-2, for example, at $x = 0$, $v = v_1$ and $\theta_z = \theta_{z1}$, where $\theta_z = dv/dx$. Thus an alternative form of Eq. 2.3-4 uses shape functions N_i to interpolate the lateral displacement $v = v(x)$ of the beam from its nodal d.o.f. **d**.

$$v = [N_1 \quad N_2 \quad N_3 \quad N_4]\begin{Bmatrix} v_1 \\ \theta_{z1} \\ v_2 \\ \theta_{z2} \end{Bmatrix} = \mathbf{N}\mathbf{d} \tag{2.3-5}$$

The separate N_i are stated in Fig. 2.3-1. They may also be found in a comprehensive tabulation of beam deflection formulas, as each N_i states the deflected shape associated with a particular end translation or rotation. Curvature of the beam element is

$$\frac{d^2v}{dx^2} = \left[\frac{d^2}{dx^2}\mathbf{N}\right]\mathbf{d} = \mathbf{B}\mathbf{d} \tag{2.3-6}$$

where strain-displacement matrix **B** is the 1 by 4 row vector

$$\mathbf{B} = \begin{bmatrix} -\frac{6}{L^2}+\frac{12x}{L^3} & -\frac{4}{L}+\frac{6x}{L^2} & \frac{6}{L^2}-\frac{12x}{L^3} & -\frac{2}{L}+\frac{6x}{L^2} \end{bmatrix} \tag{2.3-7}$$

After substitution of Eq. 2.3-7 into Eq. 2.3-3, and rather tedious multiplication and integration, Eq. 2.3-2 again results.

Limitation. Subject to the usual restrictions—that the beam is initially straight, linearly elastic, without taper, and so on—a beam loaded by end forces and end moments has a deflected shape $v = v(x)$ that is cubic in x, just as described by Eq. 2.3-4 and the N_i of Fig. 2.3-1. Therefore an FE model built of beam elements provides an exact solution when force and/or moment loads are applied to its nodes. A uniformly distributed load produces a beam deflection v that is fourth degree in x. Accordingly, beam elements are inexact under distributed load, but exact results are approached as more and more elements are used in the FE model.

Stress. Flexural stress is computed as $\sigma_x = My/I$, and bending moment M is computed from curvature d^2v/dx^2, which in turn depends on nodal d.o.f. **d**.

$$M = EI\frac{d^2v}{dx^2} = EI\,\mathbf{B}\mathbf{d} \tag{2.3-8}$$

Equations 2.3-4 and 2.3-8 show that M caused by **d** varies linearly with x in each element.

2D Beam Element. A 2D beam element might also be called a plane frame element. It is a combination of a bar element and a simple plane beam element. It resists axial stretching, transverse shear force, and bending in one plane. By combination of Eqs. 2.2-7 and

2.3-2, the stiffness matrix of a 2D beam element that lies along the x axis is

$$\mathbf{k} = \begin{bmatrix} AE/L & 0 & 0 & -AE/L & 0 & 0 \\ 0 & 12EI/L^3 & 6EI/L^2 & 0 & -12EI/L^3 & 6EI/L^2 \\ 0 & 6EI/L^2 & 4EI/L & 0 & -6EI/L^2 & 2EI/L \\ -AE/L & 0 & 0 & AE/L & 0 & 0 \\ 0 & -12EI/L^3 & -6EI/L^2 & 0 & 12EI/L^3 & -6EI/L^2 \\ 0 & 6EI/L^2 & 2EI/L & 0 & -6EI/L^2 & 4EI/L \end{bmatrix} \begin{matrix} u_1 \\ v_1 \\ \theta_{z1} \\ u_2 \\ v_2 \\ \theta_{z2} \end{matrix} \qquad (2.3\text{-}9)$$

where the symbols on the right are appended to show the d.o.f. on which $\mathbf{k}$ operates.

3D Beam Element. A beam element in a general-purpose FE program has three-dimensional capability and may also be called a "space beam" element. For explanation, we introduce "global" coordinate axes XYZ and let the element lie along a "local" x axis (Fig. 2.3-2). Local coordinate axes xyz may arbitrarily be oriented in global XYZ space. The x axis is defined by the coordinates of nodes 1 and 2. The web of the beam lies in the xy plane, which contains nodes 1, 2, and 3. Node 3 is either an extra node or another node of the structure, whose coordinates serve to orient the xy plane in XYZ space. No d.o.f. of the element are associated with node 3. At node 1 and at node 2 the element has six d.o.f., namely, three displacements and three rotations, for a total of 12 d.o.f. per element. In the software, $\mathbf{k}$ of this element is formulated using d.o.f. in local coordinates; then $\mathbf{k}$ is transformed so that global d.o.f. replace local d.o.f. at each node, in preparation for attaching the element to adjacent elements that use the same global d.o.f. The element resists force in any direction and moment about any axis. The following data are needed by the program: nodal coordinates, elastic modulus E, shear modulus G, cross-sectional area A, principal moments of inertia I_{yy} and I_{zz} of A, torsional constant J, and transverse shear deformation factors f_y and f_z. Additional data are needed for stress computation, such as the appropriate y distance in the flexure formula $\sigma_x = M_z y/I_{zz}$. Note that if the cross section is noncircular, J is *not* the polar moment of the cross-sectional area A. J is a property of the cross section, such that the correct relative rotation of nodes 1 and 2 under torque T is given by TL/GJ. Often this J is much less than the polar moment of A. We will not pursue further details of this element but urge careful study of beam bending and twisting theory [2.1] as well as the software documentation.

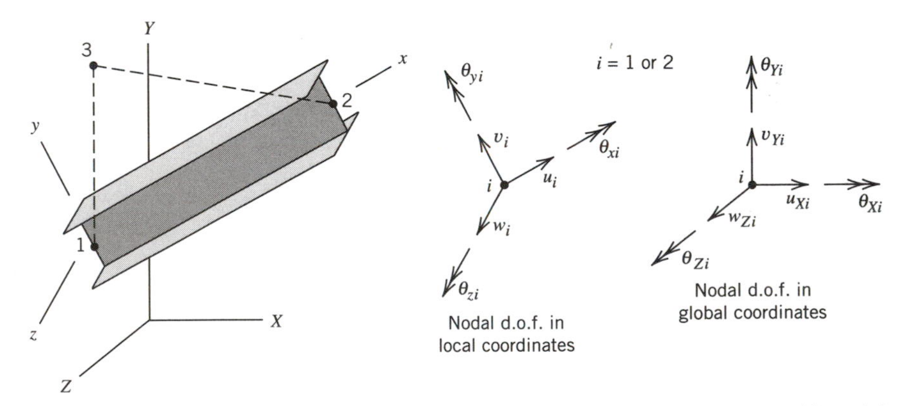

Fig. 2.3-2. 3D beam element arbitrarily oriented in global coordinates XYZ, with nodal d.o.f. in local and global coordinate systems.

A few words about support conditions: if a 3D beam element is used to model a cantilever beam along the x axis, what d.o.f. must be suppressed at the support? All six of them, including θ_{xi} to prevent the beam from being free to spin about its axis (even if no torque is applied).

Release. To model a continuous beam, either straight or curved (as in Fig. 2.5-3), adjacent elements are rigidly connected together at shared nodes. This means that adjacent elements have the same d.o.f. at nodes they share. Software usually allows the user to activate a "release" of one or more d.o.f. at a node, so that specified d.o.f. are not connected. For example, let two 3D beam elements be connected end to end. Release of all three rotational d.o.f. at the shared node makes the node a ball and socket joint. In effect, d.o.f. are not "released" but simply left unconnected when adjacent elements are put together.

Global and Local Coordinate Systems. The user defines the geometry of a FE model in a global coordinate system *XYZ*. Software typically generates an element stiffness matrix in a local coordinate system *xyz*, then automatically converts to the global system for assembly of elements. Global and local systems may be parallel or even coincident, in which case nodal displacement components are the same in both systems and the distinction between systems largely disappears. In our discussions we will use global coordinates *XYZ* only when it is desirable to distinguish between global and local systems.

2.4 PROPERTIES OF k AND K. AVOIDING SINGULARITY

Stiffness matrices **k** (element) and **K** (structure; global) are symmetric. This is true of any element or structure when there is a linear relationship between applied loads and the resulting deformations.

Each diagonal coefficient of **k** (and of **K**) is positive. We argue as follows. Imagine that a certain d.o.f. d_i is the only nonzero d.o.f., so that the load associated with d_i is $r_i = k_{ii}d_i$. Since d_i and r_i are positive in the same direction, a negative diagonal coefficient k_{ii} would mean that a load and its displacement are oppositely directed, which is unreasonable.

A structure that is either unsupported or inadequately supported has a singular stiffness matrix **K**, and FE software will be unable to solve the equations **KD** = **R** for nodal d.o.f. **D**. To prevent singularity, supports must be sufficient to prevent all possible rigid-body motions. These are motions that produce no deformation of the structure. For example, consider a one-element structure, namely, a single bar element. The formulation in Section 2.2 allows only axial translation. In effect, all nodal d.o.f. but u_1 and u_2 have already been suppressed. If unsupported, the bar element can have the rigid-body motion $u_1 = u_2 = c$, a constant. This motion is prevented by prescribing u_1 or u_2 as either zero or a nonzero value. Similarly, consider a one-element beam structure. If the four-d.o.f. simple plane beam element of Section 2.3 is unsupported, it can have two rigid-body motions in the xy plane, namely, lateral translation and rotation about a point, neither of which causes the element to bend. In terms of nodal d.o.f., these motions of the beam element can be written

$$\mathbf{d}_A = [c_1 \quad 0 \quad c_1 \quad 0]^T \quad \text{and} \quad \mathbf{d}_B = [0 \quad c_2 \quad c_2 L \quad c_2]^T \tag{2.4-1}$$

where $\mathbf{d}_A$ represents the rigid body translation $v = c_1$ and $\mathbf{d}_B$ represents rigid-body rotation through a small angle c_2 about node 1. The simple plane beam element is adequately supported if any two of its four d.o.f. are prescribed (except for prescription of only θ_{z1} and θ_{z2}, which would allow rigid-body translation). Essentially the same remarks apply to a multi-element structure. Usually it is not difficult to devise supports adequate to prevent rigid-body motion. (Support reactions can then be determined by writing only the equations of statics.) Adequate supports make a structure statically determinate. *More* than adequate supports make the support reactions statically indeterminate. This is perfectly acceptable and does not complicate the FE process in any way.

At each node of an assembled structure, general-purpose software makes three displacements and three rotations available for use as d.o.f. These are the structure or "global" d.o.f. If at any node a global d.o.f. causes no strain in any element attached to that node, the d.o.f. will not be resisted and the structure stiffness matrix will be singular . . . *unless* the offending global d.o.f. is restrained. Accordingly, part of the task of prescribing support conditions is imposing zero as the value of each "unresisted" d.o.f. Thus in modeling a 2D or 3D truss by two-node bar elements, all rotational nodal d.o.f. must be restrained at all nodes of the structure. Rotational d.o.f. are absent from the element formulation and a bar element has no stiffness with which to resist them. This does not prevent bar elements from having a relative rotation between them at a node they share, nor does it prevent a bar element from rotating in space because of unequal lateral displacements at its two nodes. At a node to which 3D *beam* elements are connected, all six d.o.f. are resisted by all beam elements at that node. In the absence of a release, beam elements are rigidly connected to one another at a shared node, which means that they have no relative rotation between them *at* the node. No restraint of nodal rotation is needed except in the infrequent situation of wanting to prevent the entire joint from rotating. A frame, like a truss, can be adequately supported by prescribing translational d.o.f. only.

A structure may have a singular **K** because it contains a mechanism. Imagine a straight beam, attached to a rigid support at each end, and modeled by two beam elements. Thus there is a node at the middle of the beam, to which both elements are connected. This FE model is stable and has no mechanism. Now if the two beam elements are replaced by two *bar* elements the FE model contains a mechanism because two collinear bar elements cannot resist a lateral force applied to the node they share. It does not matter that no such force may be applied; **K** is singular regardless of the load vector. (Physically, such a load could be sustained, but only after some lateral displacement has taken place; that is, after nonlinearity is taken into account. The present linear analysis yields **K** only for the undisplaced configuration.)

General-purpose FE software may regard all structures as three dimensional unless the user directs otherwise. Thus by default three displacements and three rotations are active at every node. No stiffness is associated with any d.o.f. unless it contributes to strain in at least one element. The user must assign a numerical value to each zero-stiffness d.o.f. Usually the assigned value is zero, thus suppressing the d.o.f. A program may automatically suppress d.o.f. not included in element formulations. Thus all rotational d.o.f. would automatically be suppressed if only bar elements were used, as in modeling a truss. Nevertheless, to minimize mistakes and surprises, a wise user will remember that in general there are six d.o.f. per node and will determine what the program actually does rather than assume what it will do.

2.5 MECHANICAL LOADS. STRESSES

Loads. Load may be applied as a force or a moment at a point or as surface pressure. Line load is conceptually intermediate to point load and surface pressure. Line load is force or moment distributed along a line (dimensions [force/length] for a force line load). Another possible load is body force loading, which acts at every material point in the body rather than only on the surface of the body. Sources of body force loading include self-weight (gravity) loading and acceleration. Thermal loading comes from temperature changes and is considered in the next section.

A concentrated force or moment is applied directly to a node. Moment load can be applied to a node only if at least one element connected to the node has rotational d.o.f. in its stiffness formulation. Input data to software consists of the magnitude, direction, and node associated with the force or moment. Distributed loading, such as pressure in a tank or line loading along a beam, acts *between* nodes and must be converted to direct nodal loading that is in some way equivalent. Most software is able to accomplish the conversion, whether or not elements are collinear or of equal length. The user need only tell the software what the loading is and where it acts. We will not detail the theory of the process [2.2]. The following results are for uniform loading on bar and beam elements.

Let a uniformly distributed axial force q act on a bar element, Fig. 2.5-1a. The dimensions of q are [force/length]. Load q may be externally applied. Or, it may represent the weight, or resistance to acceleration, of the element itself. The total force on an element of length L is qL. Half of this total is applied to each node. If two collinear elements of lengths L_a and L_b are connected, the node they share receives a total force $qL_a/2 + qL_b/2$. The final set of nodal forces on a straight bar modeled by equal-length elements appears in Fig. 2.5-1d. Equivalent nodal forces for a distributed load q that varies linearly are discussed in Section 3.9.

Theory indicates that a uniformly distributed transverse force on a beam element is replaced by nodal loads that consist of forces *and moments* (Fig. 2.5-2). The reader may recognize these loads as support reactions for a uniform beam fixed at both ends and uniformly loaded, except that nodal loads are directed opposite to beam support reactions. If elements of equal length and equal distributed load are assembled, moment loads cancel at nodes shared by two elements. The final result for collinear elements of equal length appears in Fig. 2.5-2d.

Note that no loads are needed at the supported ends of the structures in Figs. 2.5-1 and 2.5-2. As a general rule, load applied to a restrained d.o.f. may be omitted because such a load is reacted directly by the support rather than acting to deform the structure.

Usually, concentrated load is not applied at a non-nodal location because this circumstance is awkward to treat in the software. Instead, one simply arranges the FE mesh so that a node appears where the concentrated load must be applied. Beam elements may be an exception to this rule: beams are comparatively simple to treat and so often analyzed that some software is specially coded to accommodate a variety of non-nodal loadings.

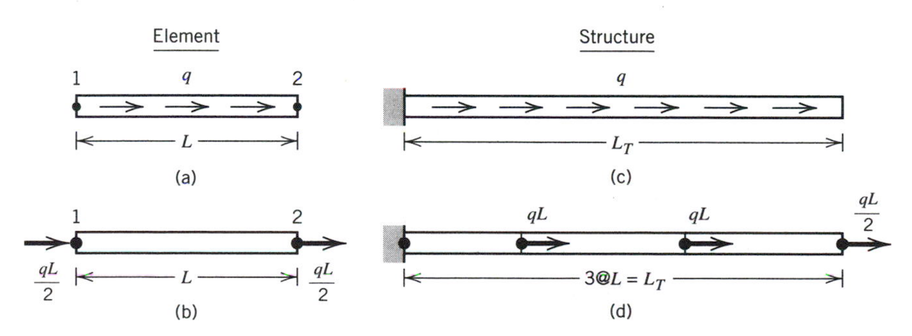

Fig. 2.5-1. Uniformly distributed axial force q on a two-node bar element and its conversion to equivalent nodal loads.

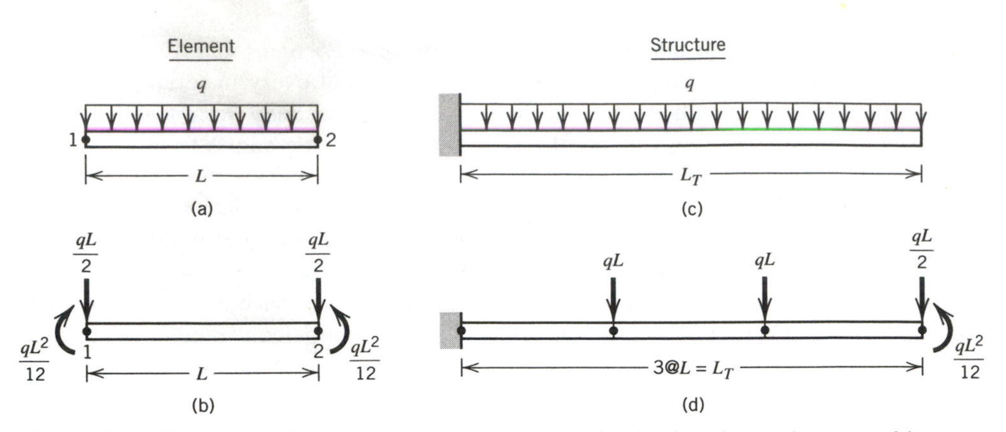

Fig. 2.5-2. Uniformly distributed transverse force q on a simple plane beam element and its conversion to equivalent nodal loads.

Remarks. Nodal loads that replace distributed loading in Figs. 2.5-1 and 2.5-2 come from FE theory and are called "kinematically equivalent" or "work-equivalent" nodal loads for the following reason. Let any one of the nodal d.o.f. d_i be nonzero, and use Eq. 2.2-5 or 2.3-5 to obtain the associated displacement field of the element. Compute work done by load q during this displacement, by integration of $uq\,dx$ (for a bar) or $vq\,dx$ (for a beam) over element length L. This work is equal to work done by the nodal load associated with d_i in acting through displacement d_i. Work-equivalent nodal loads are also *statically equivalent*, meaning that they have the same resultant force and the same moment about an arbitrarily chosen point as does the original distributed loading.

Nodal loading that is not work-equivalent is often called *lumped.* Lumped loading typically omits the nodal moments of work-equivalent loading. Lumped loading is often preferred for elements that have rotational d.o.f. Specifically, lumped loading is usually preferable for arches and shells and is often preferable for beams and plates. An example appears in Fig. 2.5-3. Nodal moments would clearly be spurious at the support nodes A and B. Also, if elements were of unequal length, work-equivalent loading would produce net moments at other nodes as well, but these moments would not be beneficial to accuracy. Work-equivalent loading and lumped loading both provide convergence toward exact results as the mesh is refined.

In general, computed nodal d.o.f. are not exact. But a uniform bar or beam represented by a FE model with work-equivalent nodal loads is an exception: computed nodal d.o.f. are exact. This does not mean that displacements are exact *between* nodes or that element stresses are exact. If the nodal loading is lumped, exact nodal d.o.f. will not be computed, but, depending on the situation, stresses at an arbitrary point may be *more* exact than obtained from work-equivalent loading. A user who wishes to know if software includes the nodal moment loads of Fig. 2.5-2 can find out by study of a one-element test case, in which computed displacements and bending moments are compared with those obtained by elementary beam theory.

Stresses. A FE program solves for nodal d.o.f. first, then (in present software) uses them to compute stresses. In a bar element, axial stress is $\sigma_x = E\varepsilon_x$, where E is the elastic modulus. Axial strain ε_x is given by $\varepsilon_x = \mathbf{Bd}$, where $\mathbf{B}$ is stated in Eq. 2.2-6. An example problem in Section 2.6 provides details of this process. Here we discuss the nature of the results. The FE model of a bar in Fig. 2.5-1d yields the displacements and stresses shown in Fig. 2.5-4. We see that stresses are discontinuous between elements. Indeed, this is the

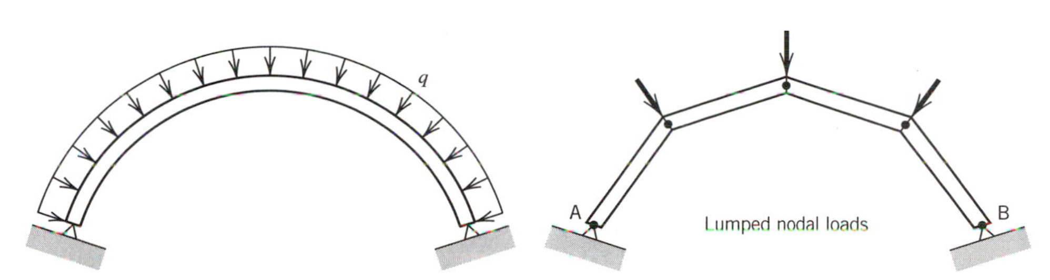

Fig. 2.5-3. Hinged arch with distributed normal loading q, and a coarse-mesh FE model with statically equivalent lumped loading.

way most types of element behave. Also, we see that displacements are more accurate than stresses. *This is usually the case*, because stresses are proportional to strains and strains are derivatives of displacement. Differentiation brings out differences between functions. For example, the two functions and $y = e^x$ and $y = 1 + x$ look very similar over the range $0 < x < 0.2$, but the first derivatives are $y' = e^x$ and $y' = 1$ (rather different) and the second derivatives are $y'' = e^x$ and $y'' = 0$ (very different).

In Fig. 2.5-4b we see that the most accurate values of stress are element center stresses and nodal average stresses. Unfortunately, the highest stress, σ_x at $x = 0$, is not as accurate. This is typical of FE models. Stresses of greatest interest usually appear at boundaries, but this is not where stresses are most accurately computed.

In a beam element we solve first for bending moment M rather than solving directly for stress. When nodal d.o.f. $\mathbf{d}$ are known, Eqs. 2.3-6 and 2.3-7 yield the curvature d^2v/dx^2, from which we obtain the bending moment $M = EI(d^2v/dx^2)$ and finally the flexural stress $\sigma_x = My/I$. Note that $\mathbf{B}$ is a function of x for this element, so we must decide where in the element to calculate the curvature. If software allows non-nodal loading on beam elements, the computed bending moment in each element is an algebraic sum: the foregoing M produced by nodal displacements and rotations, *plus* bending moment produced by the non-nodal loading with the element regarded as a fixed–fixed beam. Thus, for example, a uniformly loaded beam element that happens to undergo rigid-body motion will still display the end moments $qL^2/12$ seen in Fig. 2.5-2. This computation is in-

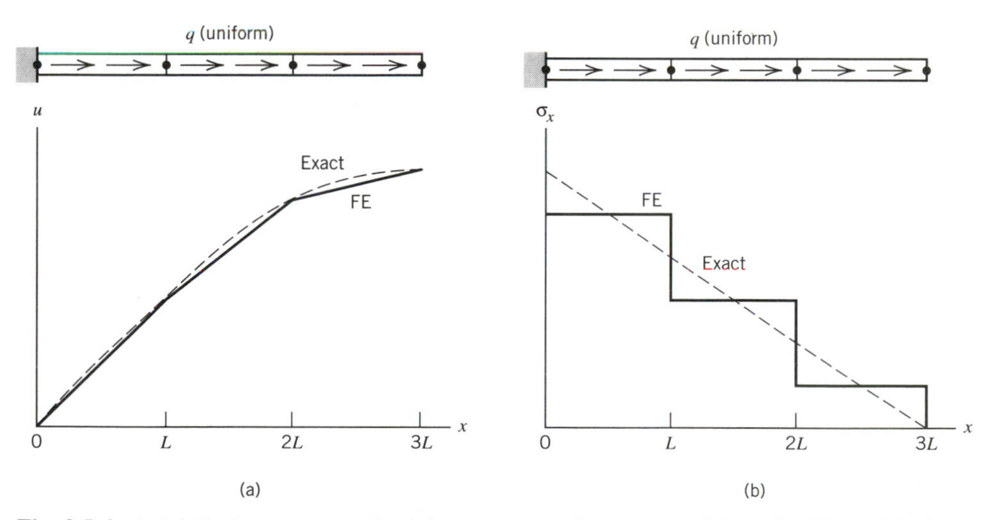

Fig. 2.5-4. Axial displacement u and axial stress σ_x in a bar, computed from the FE model of Fig. 2.5-1d.

dependent of whether loads are lumped for the purpose of constructing load vector **R** in the equations **KD** = **R**. "Fixed–fixed moments" can improve results, even providing exact bending moments in some situations, but in other situations may produce moments where none are expected (e.g., at A and B in Fig. 2.5-3).

3D beam elements may carry axial, bending, or torsional loads. A beam element sustains axial stress $\sigma_x = P/A$ and flexural stresses $\sigma_x = M_z y/I_{zz}$ and $\sigma_x = M_y z/I_{yy}$. These contributions to the resultant σ_x combine algebraically. Torque T about the axis of the beam creates shear stress $\tau = Tc/J$, where c and J must be *appropriate to the shape of the cross section* [1.5, 2.1]. Typical software may report the following at each end of an element: P, M_y, M_z, T, transverse shear forces V_y and V_z, the resultant σ_x values on y-parallel and z-parallel sides of the cross section, and shear stress associated with T.

2.6 THERMAL LOADS. STRESSES

If a homogeneous and isotropic elastic body is uniformly changed in temperature or has a temperature field that is linear in Cartesian coordinates, and the body is unrestrained by external supports, then its state of stress is unchanged. Thus if the body is initially unstressed it remains unstressed, although it does deform. More often, temperature gradients are more complicated, and thermal stresses arise with or without external supports.

FE thermal stresses are calculated by the following procedure, which applies to *any* kind of FE model: bar, solid, shell, and so on. These steps are carried out automatically by the software; the user need not take special action to activate them.

1. For each element, restrain all nodal d.o.f. and compute loads applied by the element to its nodes owing to temperature change. (We will not detail the theory, which is similar to the theory that yields work-equivalent nodal loads.)
2. Assemble the elements and element loads calculated in step 1. The result is a FE structure, as yet undeformed, whose nodal loads are produced by temperature changes.
3. Solve for nodal d.o.f., next compute element strains produced by these d.o.f., and then compute stresses from strains. These calculations are exactly the same as those used to calculate stresses produced by mechanical loads.
4. Superpose on stresses from step 3 the "initial" stresses, which are stresses that appear in step 1 when all d.o.f. are restrained and temperature change is applied.

Mechanical loads may be superposed on thermal loads. This is done in the assembly process, step 2.

Example 1. A simple bar problem illustrates the assembly of elements, application of loads and supports, and solution for stresses. These processes are carried out automatically by FE software. Software uses numbers but we will use symbols for the sake of clarity. Recall that software allows six d.o.f. per node, which means that in the present example all d.o.f. but axial displacements u_i must be suppressed at every node to prevent rigid-body motions and singularity of **K**. To save space we will assume that this has already been done, so that only axial displacement d.o.f. u_i are represented in what follows.

Let the bar in Fig 2.6-1 be uniform, with cross-sectional area A, elastic modulus E, and coefficient of thermal expansion α. Mechanical loading consists of forces P as shown. Thermal loading consists of uniform heating an amount ΔT. In this problem both elements have the same stiffness matrix **k** (Eq. 2.2-7) and the same thermal load vector $\mathbf{r}_T$.

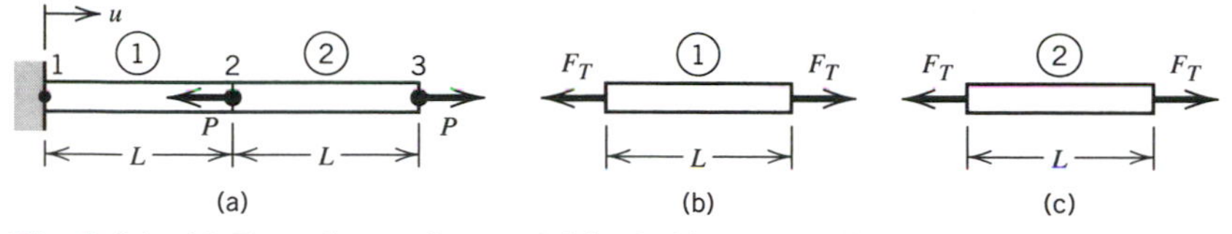

Fig. 2.6-1. (a) Two-element bar model loaded by externally applied forces P and by uniform heating an amount ΔT. (b) Loads $F_T = \alpha AE\,\Delta T$ associated with heating, applied by elements to nodes.

Specifically, for elements 1 and 2, respectively,

$$\mathbf{k}_1 = \mathbf{k}_2 = \frac{AE}{L}\begin{bmatrix} 1 & -1 \\ -1 & 1 \end{bmatrix} \qquad \mathbf{r}_{T1} = \mathbf{r}_{T2} = \begin{Bmatrix} -F_T \\ F_T \end{Bmatrix} \tag{2.6-1}$$

where $F_T = \alpha AE\,\Delta T$ is the force a bar would exert on confining walls when heated an amount ΔT, as is shown in elementary mechanics of materials. Assembly of elements yields the following set of global equations, whose unknowns are the axial displacement d.o.f. u_1, u_2, and u_3. (One may note that the physical meaning ascribed to columns of $\mathbf{k}_1$ and $\mathbf{k}_2$ is also true of the following structure stiffness matrix.)

$$\frac{AE}{L}\begin{bmatrix} 1 & -1 & 0 \\ -1 & 2 & -1 \\ 0 & -1 & 1 \end{bmatrix}\begin{Bmatrix} u_1 \\ u_2 \\ u_3 \end{Bmatrix} = \begin{Bmatrix} -F_T \quad + R_1 \\ F_T \quad -F_T \quad - P \\ F_T \quad + P \end{Bmatrix} \tag{2.6-2}$$

where R_1 is the force applied to node 1 by the support. This force is regarded as an unknown. Known mechanical loads P have been added as part of the assembly process. Note that d.o.f. u_2 receives stiffness and nodal load contributions from both elements to which it is connected. At node 2 the loads F_T are equal but oppositely directed and therefore cancel, leaving only load $-P$ at node 2. Load $-F_T$ at node 1 is reacted by the support and is discarded in the step we take next.

The stiffness matrix in Eq. 2.6-2 is singular because rigid-body axial translation is possible. Singularity is removed by the support condition, which is imposed by substituting $u_1 = 0$ into Eqs. 2.6-2. This leaves only u_2 and u_3 as unknown d.o.f. They are obtained by solving the equation

$$\frac{AE}{L}\begin{bmatrix} 2 & -1 \\ -1 & 1 \end{bmatrix}\begin{Bmatrix} u_2 \\ u_3 \end{Bmatrix} = \begin{Bmatrix} -P \\ F_T + P \end{Bmatrix} \tag{2.6-3}$$

By suppressing the first d.o.f. we have, in effect, discarded row 1 and column 1 from Eq. 2.6-2 to obtain Eq. 2.6-3. After solving Eq. 2.6-3, with $F_T = \alpha AE\,\Delta T$, we know the values of all d.o.f.

$$u_1 = 0 \qquad u_2 = \alpha L\,\Delta T \qquad u_3 = 2\alpha L\,\Delta T + \frac{PL}{AE} \tag{2.6-4}$$

Axial stress σ_x in each element is calculated from the formula

$$\sigma_x = E\varepsilon_x + \sigma_{x0} \tag{2.6-5}$$

where E is the elastic modulus, ε_x is the strain computed from nodal d.o.f., and σ_{x0} is the initial stress the element, here caused by temperature change with all element d.o.f. restrained. For the respective elements, Eq. 2.6-5 is

$$\sigma_{x1} = E\frac{u_2 - u_1}{L} + (-E\alpha\ \Delta T) = E\alpha\ \Delta T - E\alpha\ \Delta T = 0 \tag{2.6-6a}$$

$$\sigma_{x2} = E\frac{u_3 - u_2}{L} + (-E\alpha\ \Delta T) = E\alpha\ \Delta T + \frac{P}{A} - E\alpha\ \Delta T = \frac{P}{A} \tag{2.6-6b}$$

Heating has caused the bar to expand but has produced no stress because thermal expansion is unrestrained in the present example.

Example 2. If the foregoing example is now altered by fixing *both* ends of the bar, Eq. 2.6-2 reduces to a single equation. This equation and the resulting d.o.f. are

$$\frac{2AE}{L}u_2 = -P \quad \text{hence} \quad u_2 = -\frac{PL}{2AE}, \quad u_1 = u_3 = 0 \tag{2.6-7}$$

and Eq. 2.6-5 yields the element stresses

$$\sigma_{x1} = E\frac{u_2 - u_1}{L} + (-E\alpha\ \Delta T) = -\frac{P}{2A} - E\alpha\ \Delta T \tag{2.6-8a}$$

$$\sigma_{x2} = E\frac{u_3 - u_2}{L} + (-E\alpha\ \Delta T) = \frac{P}{2A} - E\alpha\ \Delta T \tag{2.6-8b}$$

All results for displacement and stress are exact in the foregoing examples because the loadings have not demanded a displacement field or a stress field more complicated than the elements can represent.

Spurious Stresses. The following example shows how thermal stresses may be incorrectly computed. Let a two-node bar element of length L be supported at only its left end and have a linear variation of temperature along its length, say

$$\Delta T = cx \tag{2.6-9}$$

where c is a constant and $x = 0$ at node 1. This ΔT causes the element to expand an amount $\alpha cL^2/2$, so that nodal d.o.f. are $u_1 = 0$ and $u_2 = \alpha cL^2/2$. Thus $\varepsilon_x = (u_2 - u_1)/L = \alpha cL/2$, and Eq. 2.6-5 yields

$$\sigma_x = E\varepsilon_x + (-E\alpha\ \Delta T) = E\alpha c\left(\frac{L}{2} - x\right) \tag{2.6-10}$$

However, this σ_x is incorrect. We know that the bar should be stress-free. The spurious σ_x arises because of a mismatch: the temperature field is linear in x but the strain field is constant. The correct stress, $\sigma_x = 0$, would be computed if ΔT were evaluated at the element center, $x = L/2$, and this ΔT taken as constant over the element, just as ε_x in Eq. 2.2-6 is constant over the element. Regardless of the element type (bar, plane, solid, etc.), it is usually best if the temperature gradient over an element is represented by a field of degree no higher than that of the strain field produced by nodal d.o.f. **d**. This usually requires that the actual ΔT variation be simplified for use in stress calculation. The user need not worry about this if the software has been suitably coded. (See also the discussion in Section 3.10.)

2.7 AN APPLICATION

We present an example problem, with emphasis on modeling and checking results rather than on matrices and manipulations. The structure is a flat oval bar loaded in its own plane, as shown in Fig. 2.7-1. Stresses and deflections of greatest magnitude are desired. The solution strategy suggested in Section 1.3 is used in the following analysis.

Preliminary Analysis. The structure is roughly circular. Therefore a crude analytical model of the problem is that of a circular ring having the same perimeter as the actual oval and loaded by concentrated forces, as shown in Fig. 2.7-2. Data in Fig. 2.7-1b are such that the radius of the substitute circular ring is $r = 78.2$ mm, and the pressure load produces the force $F = pt(b + c) = 300$ N. Handbook formulas [1.5] state deflections and bending moments in a circular ring loaded by two diametrically opposing forces. By superposing two such cases, one with inward forces and the other with outward forces, we obtain

$$\delta = 0.143 \frac{Fr^3}{EI} = 0.338 \text{ mm} \quad \text{and} \quad M = 0.5\,Fr = 11{,}730 \text{ N} \cdot \text{mm} \tag{2.7-1}$$

as the magnitudes of radial deflection and bending moment at loaded points. Hence direct axial stress and flexural stress in the circular ring have magnitudes

$$\sigma_a = \frac{F/2}{ht} = 3.33 \text{ MPa} \quad \text{and} \quad \sigma_b = \frac{M(h/2)}{I} = 174 \text{ MPa} \tag{2.7-2}$$

These results will be compared with FE results. This simple analysis is done before FE analysis to avoid the natural tendency to calculate a result that agrees with what FE analysis has led us to expect. Also, by having an approximate solution in hand, we will be

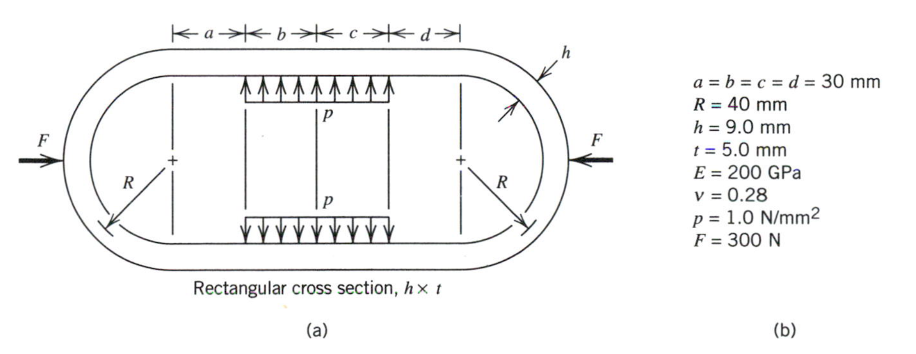

Fig. 2.7-1. (a) Plane structure under mechanical loading. (b) Data used in the numerical example.

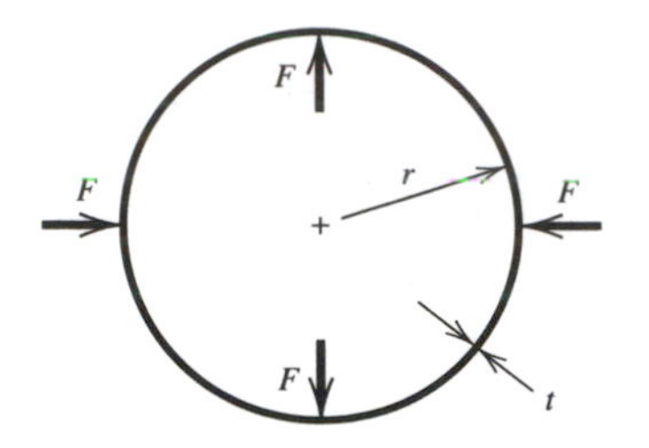

Fig. 2.7-2. Simple analytical model for approximate analysis, with $r = R + (a + b + c + d)/\pi$.

able to immediately detect FE results that happen to be greatly in error, owing perhaps to a blunder in data input.

FE Model and Analysis. There is symmetry of geometry, loading, and elastic properties about horizontal and vertical centerlines. Therefore only one quadrant need be analyzed, Fig. 2.7-3a. (Symmetry is discussed more fully in Section 4.12.) Supports shown are consistent with horizontal displacement allowed at end A, vertical displacement allowed at end D, and neither A nor D allowed to rotate. Between A and D, points will displace in both x and y directions and cross sections will rotate.

Beam elements are appropriate for this problem. A coarse-mesh model is shown in Fig. 2.7-3b. Portion BD is modeled by two elements, BC and CD, so that nodal force $qL/2$ and nodal moment $qL^2/12$ of Fig. 2.5-2 can conveniently be applied at C.* No moment load is needed at D because it would be reacted directly by the support. Support conditions indicated in Fig. 2.7-3c allow only translation u at A and only translation v at D. One need not also set $w = \theta_x = \theta_y = 0$ at nodes B and C; however, this excessive fixity would do no harm in the present problem. Similarly, the amount of fixity at A and D is excessive; for example, we need not restrain w, θ_x, or θ_y at D because similar fixity at A will avoid the possibilities of translation in the z direction and rigid-body rotations about x and y axes. $\mathbf{K}$ is rendered nonsingular by the support conditions.

Critique of FE Results. How good are the answers? Before comparing computed numbers with analytical approximations, and before looking at stress plots produced by the FE method, we look at the displaced shape. We sketch an intuitive approximation, shown dashed in Fig. 2.7-4a. Software will plot the computed displaced shape, scaled up to be easily visible, and animated so the model can be seen to move back and forth between its original shape and its deformed shape. We should see reasonable agreement between approximate and computed shapes. In particular, point A should move only to the right, point D should move only upward, and the FE model should not rotate at either point. Thus we check that the intended support conditions have indeed been imposed. Upon examining a list of computed numerical values of nodal d.o.f., we should see $w = \theta_x = \theta_y = 0$ at all nodes. (*Note*: Software typically plots only straight lines between nodes. Accordingly, a deformed beam element appears straight; its actual cubic curve is not displayed.)

If displacement results appear satisfactory after the foregoing inspection and comparison with analytical approximation, we proceed to examine stresses. We should find that the direct axial component of stress is tensile at A and compressive at D, while the flexural component of stress is tensile on the inside at A and tensile on the outside at D.

*Lumped loading omits this moment. The reader may wish to repeat this example, using lumped loading, to see how results are changed.

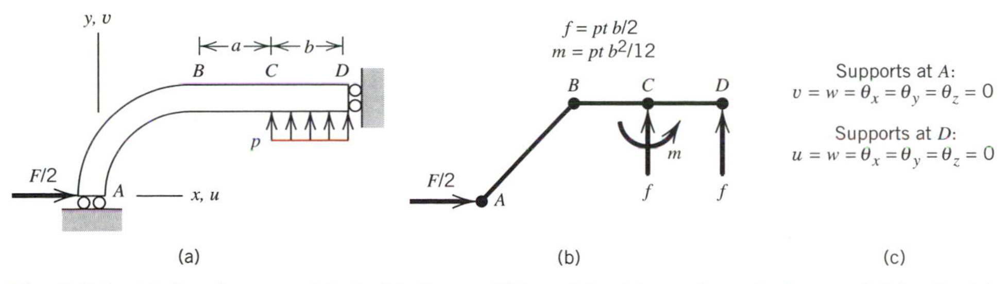

Fig. 2.7-3. (a) Quadrant modeled. (b) Coarse FE model with work-equivalent nodal loads. (c) Support conditions.

A summary of computed results for the FE model of Fig. 2.7-3b is as follows.

$$\begin{aligned} u_A &= 0.135 \text{ mm} & v_D &= 0.316 \text{ mm} \\ (\sigma_a)_A &= 0 & (\sigma_a)_D &= -3.33 \text{ MPa} \\ (\sigma_b)_A &= \pm 166 \text{ MPa} & (\sigma_b)_D &= \pm 117 \text{ MPa} \end{aligned} \tag{2.7-3}$$

where σ_a refers to the direct axial stress, $\sigma_a = P/A$, and σ_b refers to the bending component of stress, $\sigma_b = Mc/I$. (Some software may report the algebraic sum of axial and bending stresses at top, middle, and bottom surfaces of an element.) The value of σ_a at A at first looks wrong; why is it zero? Elementary statics, Fig. 2.7-4b, shows that member AB carries a transverse shear force but no axial force because of its 45° orientation. Accordingly, $(\sigma_a)_A = 0$ at A is correct for this particular FE model.

A finer mesh for the same problem is shown in Fig. 2.7-4c. Now arc AB is modeled by two chords rather than one, which is the most significant improvement of this mesh. Portion BC is not refined because doing so would make no difference: this part is straight and no loads are applied between B and C. Computed results for the finer-mesh FE model are

$$\begin{aligned} u_A &= 0.121 \text{ mm} & v_D &= 0.349 \text{ mm} \\ (\sigma_a)_A &= 1.80 \text{ MPa} & (\sigma_a)_D &= -3.33 \text{ MPa} \\ (\sigma_b)_A &= \pm 163 \text{ MPa} & (\sigma_b)_D &= \pm 116 \text{ MPa} \end{aligned} \tag{2.7-4}$$

These values are in reasonable agreement with results from the crude analytical approximation, Eqs. 2.7-1 and 2.7-2. Also, there is good agreement between Eqs. 2.7-3 and 2.7-4

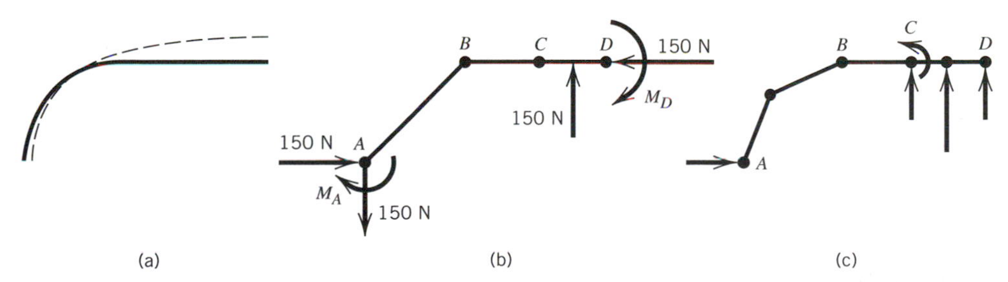

Fig. 2.7-4. (a) Original centerline (solid) and deformed centerline (dashed) of the quadrant modeled. Displacements are greatly exaggerated. (b) Free-body diagram, showing loads applied to the quadrant modeled. (c) A refined FE model with work-equivalent nodal loads.

except for σ_a at A, which is a small stress whose error has been satisfactorily explained. We conclude that the FE results are probably reliable, although one more mesh refinement and analysis would make us more comfortable with the results. Indeed, after analyses based on three or more successive mesh refinements, one might plot a result such as $(\sigma_b)_A$ versus element size. Extrapolation to zero element size would yield a prediction of $(\sigma_b)_A$ for infinite mesh refinement (see Section 5.15). The value of $(\sigma_b)_A$ obtained by extrapolation could be used to estimate the percentage error of $(\sigma_b)_A$ in each of the meshes actually used.

Some questions about the correctness of the model remain. Truly concentrated loads are not possible, therefore horizontal loads F are an idealization whose actual manner of application may have to be represented more precisely. The ratio h/R is probably small enough that transverse shear deformation is unimportant, but it does no harm to use beam elements that include it. If h becomes comparable to R, beam elements should be replaced by two-dimensional elements (Chapter 3). We have assumed from the outset that the material is linearly elastic. Stresses are not high and the elastic modulus suggests that the material is steel, so the assumption of linearity appears reasonable. The problem would be *much* more complicated if there were yielding.

The problem would also be much more complicated if R/h were so large that displacements became large. A linear analysis uses equilibrium equations written with respect to the initial (unloaded) geometry, while strictly they should be written with respect to the final (loaded) geometry. Usually the distinction is negligible because displacements are so small that the initial geometry is substantially unaltered by applied load. If displacements are *not* small the problem is much more complicated because the final geometry is not known in advance. Such a problem is called "nonlinear" because displacements and stresses are not directly proportional to applied loads.

ANALYTICAL PROBLEMS

2.1 Consider a two-d.o.f. bar element, as in Fig. 2.2-1, but let the cross-sectional area vary linearly with x from A_0 at $x = 0$ to $2A_0$ at $x = L$.

(a) Use the direct method to generate the element stiffness matrix. Suggestion: first compute the elongation produced by an axial force P.

(b) Use the formal procedure to generate the element stiffness matrix. Suggestion: use Eq. 2.2-6.

(c) The stiffness matrices of parts (a) and (b) do not agree. Why?

2.2 Consider a cable element of length L under constant tension T, as shown. Assume that lateral deflection v is linear in x and that $v \ll L$. Use the direct method to generate a 2 by 2 stiffness matrix that operates on d.o.f. v_1 and v_2. Suggestions: **k** depends only on T and L; consider $v_1 = 1$ and $v_2 = 0$, then $v_1 = 0$ and $v_2 = 1$, while T maintains constant direction.

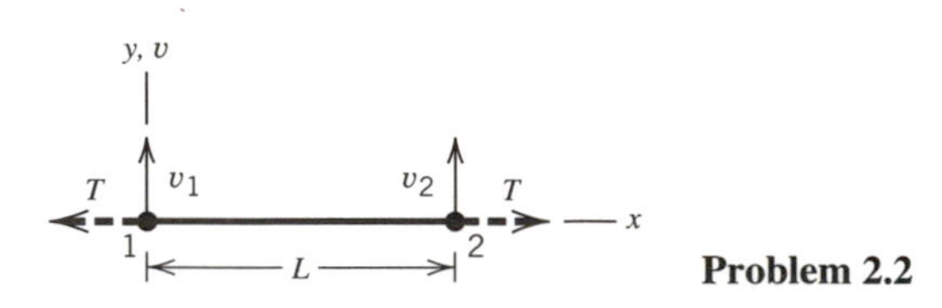

Problem 2.2

2.3 Use the direct method to generate the element stiffness matrix of a prismatic shaft element under torque loading. The d.o.f. are θ_1 and θ_2, where θ is an angle of rotation about the axis of the shaft.

2.4 (a) Complete the derivation begun in Eqs. 2.3-1; that is, generate column 1 of **k** for a simple plane beam element.
(b,c,d) Similarly, generate columns 2, 3, and 4 of **k**.

2.5 Use Eqs. 2.3-3 and 2.3-7 to generate the stiffness matrix of a simple plane beam element.

2.6 (a) In each of the simple plane beam elements shown, two d.o.f. are restrained and only the two d.o.f. labeled remain unrestrained. In each case, generate the 2 by 2 stiffness matrix that operates on the unrestrained d.o.f. Use the direct method.
(b) Repeat part (a), but use the formal procedure (Eq. 2.3-3, with **B** a 1 by 2 row matrix).

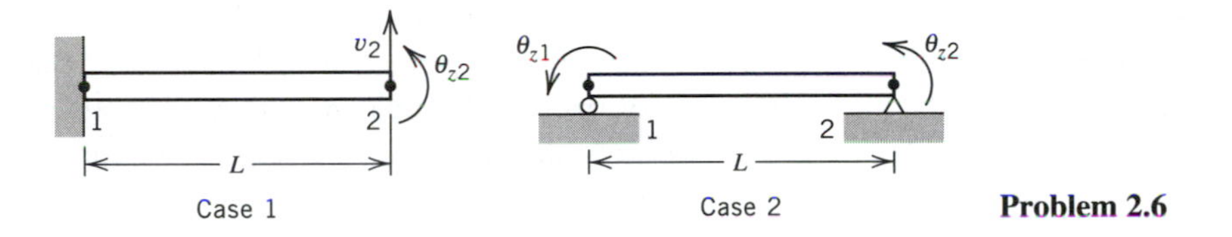

Problem 2.6

2.7 Imagine that by experiment it is known that end force and end moment as shown in the sketch are required in order to elevate the left end of the beam 1.0 mm without rotating this end. Fill in as many numerical values as you can in an element stiffness matrix that operates on nodal d.o.f. $[v_1 \quad \theta_{z1} \quad v_2 \quad \theta_{z2}]^T$, where v_1 and v_2 are measured in millimeters. To do so, use the given data, physical argument, statics, and symmetry considerations, but not beam deflection formulas or Eq. 2.3-2.

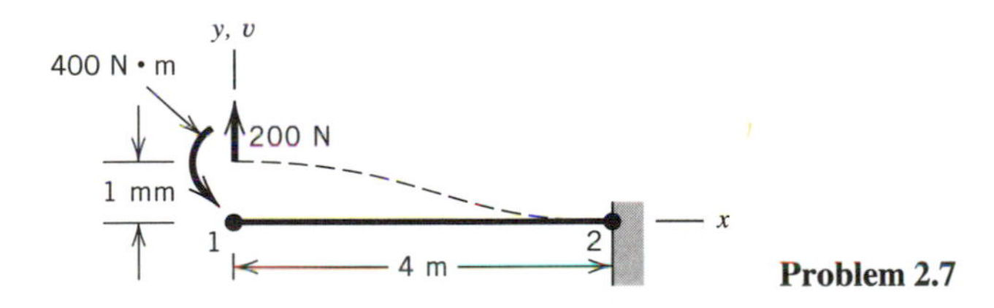

Problem 2.7

2.8 A simply supported beam of length L under a half sine wave of distributed transverse loading, $q = q_0 \sin(\pi x/L)$, has the deflected shape $v = v_{0e} \sin(\pi x/L)$, where v_{0e} is the center deflection. An approximate deflected shape is $v = 4v_{0a}x(L - x)/L^2$. If $v_{0e} = v_{0a}$, what are the percentage errors associated with the approximation? Examine deflection at quarter points, rotation at supports, bending moment at midspan and at supports, and transverse shear force at supports.

2.9 A uniformly distributed axial force q acts on a uniform bar, as shown. Let the FE model consist of n two-node elements, each of length L_T/n. For $n = 1$, $n = 2$, and $n = 4$, what are the percentage errors of displacement at the right end and axial stress at the left end?

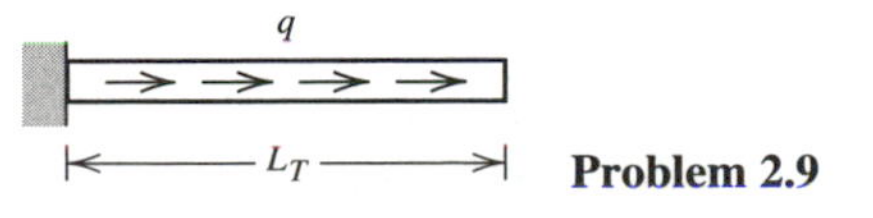

Problem 2.9

2.10 For the beam problem shown, what is the coarsest FE mesh that gives a nontrivial result? What are the percentage errors in center deflection and end bending moment of such a FE model? Obtain results needed to answer this question by use of
(a) elementary beam theory (applied to the FE model);
(b) matrices in Section 2.3.

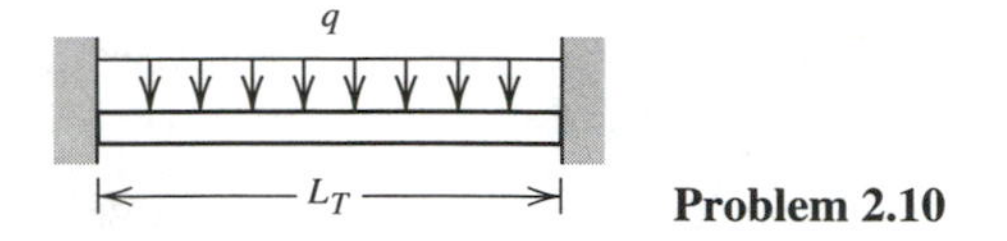

Problem 2.10

2.11 Activate d.o.f. of a simple plane beam element one at a time; that is, $v_1 > 0$ while $\theta_{z1} = v_2 = \theta_{z2} = 0$, and so on. Thus there are four cases. In each case, show that nodal loads in Fig. 2.5-2b are work-equivalent.

2.12 Distributed lateral force q and the cantilever beam are both uniform (see sketch). Compute the tip deflection and root bending moment using work-equivalent nodal loading (Fig. 2.5-2b). Then repeat the calculation, this time with lumped loading (i.e., omit the nodal moment portion of the work-equivalent loading). Compute percentage errors in each case. Suggestion: deflections of the FE model can be obtained by use of standard beam formulas.
(a) Use a single element.
(b) Use two elements, each of length $L_T/2$.

2.13 The bar shown is confined between rigid walls. Cross-sectional area A varies linearly from A_0 to $1.6A_0$. The bar is initially stress-free, then is uniformly heated an amount ΔT. Compute stresses in a FE model that contains three elements, each of length $L_T/3$ and having the respective cross-sectional areas $1.1A_0$, $1.3A_0$, and $1.5A_0$. On axes x (abscissa) and $\alpha E\ \Delta T$ (ordinate), plot the exact stress field and the FE stress field.

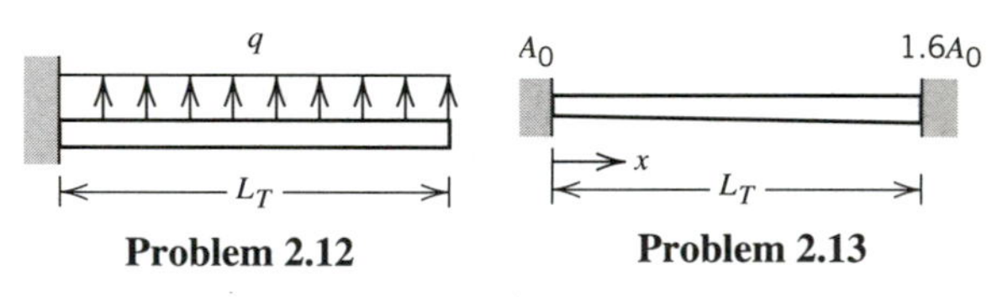

Problem 2.12 **Problem 2.13**

COMPUTATIONAL PROBLEMS

In the following problems compute peak values of displacement and stress or bending moment. Exploit symmetry if possible. When mesh refinement is used, estimate the maximum percentage error of FE results in the finest mesh. Where material properties are needed but not stated, use those of steel.

A FE analysis should be preceded by an alternative analysis, probably based on statics and mechanics of materials, and oversimplified if necessary. If these results and FE results have substantial disagreement we are warned of trouble somewhere.

2.14 (a) Does the software you use include transverse shear deformation in its beam elements? Disregard the software documentation; instead, devise and run test cases to find out directly.
(b) Similarly, use test cases to discover if nodal moments due to distributed load are applied to beam elements. (These moments may appear in **R** and/or in stress computation.)

2.15 Modify the problem of Fig. 2.7-1 by changing the following dimensions to the values indicated:

(a) $c = d = 0$, others as in Fig. 2.7-1.

(b) $a = d = 70$ mm, others as in Fig. 2.7-1.

(c) $b = c = 0$, others as in Fig. 2.7-1; also add a central vertical member of cross section h by t.

2.16 The beam shown is uniformly tapered in width. Let Poisson's ratio be zero. Create FE models by using elements of constant cross section, in the manner of Fig. 1.4-1, but with beam elements. Use one, then two, then four, and so on, elements of equal length. Choose convenient numbers for length L and force P. The problem may be repeated using a tip moment rather than a tip force, or using a distributed load.

2.17 Members of the plane structure shown may be bars pinned together at joints to create a truss or beams rigidly connected together at joints to create a frame. For the frame model one may assume that rotations at the wall are either permitted or prohibited. Investigate how much difference there is between the truss model and the frame model. Assume that all members have a square cross section, b units on a side. Let $H = 120$ mm, $L = 160$ mm, and $P = 1.0$ N. Consider the cases $b = 5$ mm, $b = 15$ mm, and $b = 30$ mm.

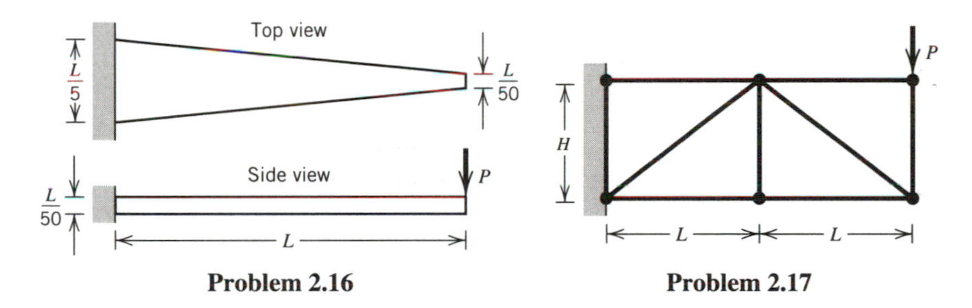

Problem 2.16 **Problem 2.17**

2.18 The beam structure shown has unit thickness normal to the paper. Depth h of the cross section varies linearly in the axial direction. Confine displacements to the plane of the paper.

(a) $L_1 = L_2 = s = 200$ mm, $h_1 = h_2 = 40$ mm, $h_c = 20$ mm.

(b) $L_1 = L_2 = s = 200$ mm, $h_1 = h_2 = 20$ mm, $h_c = 40$ mm.

(c) $L_1 = 200$ mm, $L_2 = 0$, $s = 80$ mm, $h_1 = 40$ mm, $h_2 = h_c = 20$ mm.

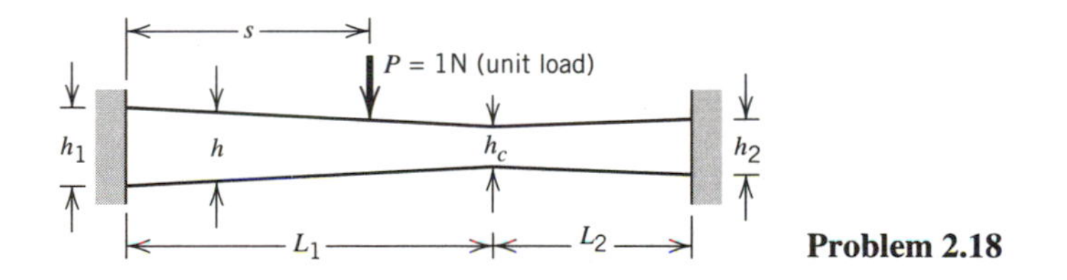

Problem 2.18

2.19 Problem 2.18 can be repeated with one end hinged.

2.20 Consider the problem of a beam on an elastic foundation. The foundation can be modeled by bar elements that connect nodes of the beam to a fixed support and act as linear springs of stiffness $k = AE/L$ (see sketch). This is not the best foundation model, but it is instructive to see how displacements and bending moments in the beam converge toward exact results as the mesh is refined. Analytical solutions for various kinds of loading and support on a uniform beam are available [1.5, 2.1]. Beams of finite length or with step changes in cross section may also be considered.

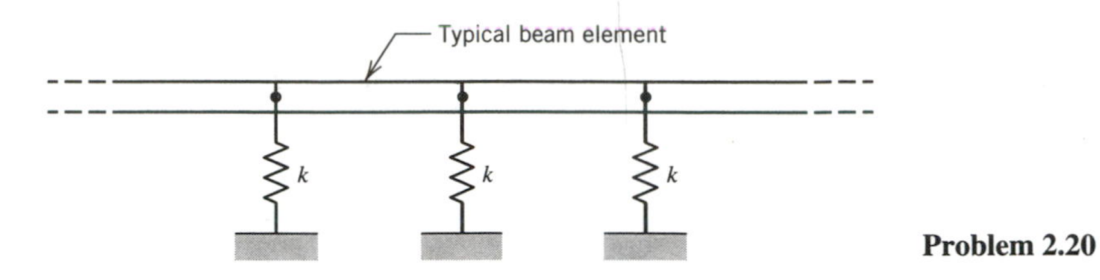

Problem 2.20

2.21 The structure shown has unit thickness normal to the paper. Depth h of the cross section varies linearly in the circumferential direction. Confine displacements to the plane of the paper. Some possible choices of geometry are as follows:
(a) $\phi_1 = 45°$, $\phi_2 = 90°$, $R = 500$ mm, $h_1 = h_2 = 30$ mm.
(b) $\phi_1 = 20°$, $\phi_2 = 40°$, $R = 500$ mm, $h_1 = h_2 = 30$ mm.
(c) $\phi_1 = 20°$, $\phi_2 = 90°$, $R = 500$ mm, $h_1 = h_2 = 30$ mm.
(d) $\phi_1 = 20°$, $\phi_2 = 90°$, $R = 500$ mm, $h_1 = 30$ mm, $h_2 = 10$ mm.

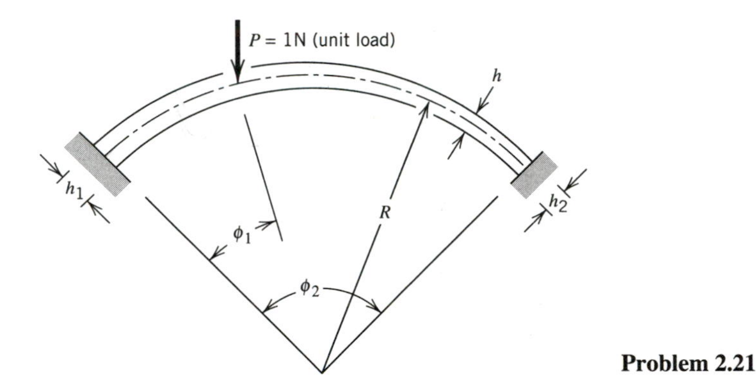

Problem 2.21

2.22 Problem 2.21 can be repeated with one or both ends hinged. Also, uniform or nonuniform heating can be applied. Yet another set of problems is generated by prescribing nonzero values of translational and/or rotational d.o.f. at one end.

2.23 Use the geometry shown for Problem 2.21, but orient load P so that it acts *normal* to the plane of the paper, and let h represent the diameter of a circular cross section. Thus the structure becomes a balcony beam, which has both bending and twisting deformations. The specific configurations in Problem 2.21 can be analyzed. Or, prescribed nonzero d.o.f. may now include twist.

2.24 Idealize a bicycle wheel as a planar structure having 36 radial spokes. Properties are as follows [2.3]. Spokes: diameter = 2.1 mm, $E = 210$ GPa, length = 309.4 mm from the center of the wheel to the centroidal axis of the rim cross section. Rim: $A = 138.4\ \text{mm}^2$, $E = 70$ GPa, $\nu = 0.33$, centroidal I of $A = 1469\ \text{mm}^4$, $I/c = 176\ \text{mm}^3$ (for stress calculation). Assume that initial tension in the spokes is sufficient to maintain tension in every spoke when load is applied. Consider the following loadings.
(a) A vertical force of 490 N applied by the road.
(b) A force of 100 N applied tangentially by caliper brakes at the top of the wheel. Neglect the mass of the wheel.

CHAPTER 3

Plane Problems

Necessary preliminaries from solid mechanics theory are reviewed. Next, plane elements of several types are discussed, with particular attention to element displacement fields and what they portend for element behavior. Treatment of loads and calculation of stresses are discussed. An example application closes the chapter.

3.1 INTRODUCTION

Stress–Strain–Temperature Relations. By definition, a plane body is flat and of constant thickness. Let Cartesian coordinates *xy* lie in the plane of the body. As explained in elementary mechanics of materials, the plane stress–strain relation (or constitutive relation) of a *linearly elastic and isotropic material* is

$$\begin{Bmatrix} \varepsilon_x \\ \varepsilon_y \\ \gamma_{xy} \end{Bmatrix} = \begin{bmatrix} 1/E & -\nu/E & 0 \\ -\nu/E & 1/E & 0 \\ 0 & 0 & 1/G \end{bmatrix} \begin{Bmatrix} \sigma_x \\ \sigma_y \\ \tau_{xy} \end{Bmatrix} + \begin{Bmatrix} \varepsilon_{x0} \\ \varepsilon_{y0} \\ \gamma_{xy0} \end{Bmatrix} \tag{3.1-1}$$

where E is the elastic modulus, ν is Poisson's ratio, and G is the shear modulus, $G = 0.5\, E/(1 + \nu)$. The last column vector in Eq. 3.1-1 contains initial strains (described below). Abbreviated, Eq. 3.1-1 is written $\boldsymbol{\varepsilon} = \mathbf{E}^{-1}\boldsymbol{\sigma} + \boldsymbol{\varepsilon}_0$. If this equation is solved for the stress vector $\boldsymbol{\sigma}$, we have

$$\boldsymbol{\sigma} = \mathbf{E}\boldsymbol{\varepsilon} + \boldsymbol{\sigma}_0 \tag{3.1-2}$$

in which $\boldsymbol{\sigma}_0 = -\mathbf{E}\boldsymbol{\varepsilon}_0$ and

$$\mathbf{E} = \frac{E}{1-\nu^2}\begin{bmatrix} 1 & \nu & 0 \\ \nu & 1 & 0 \\ 0 & 0 & (1-\nu)/2 \end{bmatrix} \quad \text{for plane stress} \tag{3.1-3}$$

Equations 3.1-1 and 3.1-3 pertain to a *plane stress* condition, in which $\sigma_z = \tau_{yz} = \tau_{zx} = 0$. Initial strains $\boldsymbol{\varepsilon}_0$ caused by temperature change ΔT are $\varepsilon_{x0} = \varepsilon_{y0} = \alpha\, \Delta T$ and $\gamma_{xy0} = 0$, where α is the coefficient of thermal expansion. The thickness is free to increase or decrease in response to stresses in the *xy* plane. In a *plane strain* condition thickness change is prevented. Equation 3.1-2 still states the stress–strain relation, but **E** is

$$\mathbf{E} = \frac{E}{(1+\nu)(1-2\nu)} \begin{bmatrix} (1-\nu) & \nu & 0 \\ \nu & (1-\nu) & 0 \\ 0 & 0 & (1-2\nu)/2 \end{bmatrix} \quad \text{for plane strain} \qquad (3.1\text{-}4)$$

and initial strains due to temperature change ΔT are $\varepsilon_{x0} = \varepsilon_{y0} = (1+\nu)\alpha\,\Delta T$ and $\gamma_{xy0} = 0$. As an example of plane strain, if a flat plate is bent so as to become a cylinder with z its axis, cross sections in planes normal to the z axis are in a state of plane strain (except very near ends of the cylinder). Stresses are independent of z in plane stress and in plane strain conditions. As the thickness of a plane body increases, from much less to greater than in-plane dimensions of the body, there is a transition of behavior from plane stress toward plane strain.

If $\nu = 0.5$ the material is incompressible. If ν approaches 0.5 and plane *strain* conditions prevail, Eq. 3.1-4 shows that strains ε_x and ε_y are associated with very large stresses σ_x and σ_y. This circumstance may cause trouble in FE analyses because of numerical ill-conditioning.

Equation 3.1-2 need not be restricted to isotropy. In the most general case of anisotropy, **E** is a full matrix, and for a plane problem it contains six independent elastic constants. The theory and computational processes of FE are not made more complicated by anisotropy. However, there is often practical difficulty in obtaining numerical values of elastic constants. Also, it is harder to judge the validity of results because response to loads is not as easily visualized and approximate calculations become more difficult.

Stresses in plane stress problems may be called *membrane stresses.* They are constant through the z-direction thickness. In contrast, the bending stresses that appear in plates and shells vary from tension to compression through the thickness and by definition are absent if the problem is plane. One should bear in mind that all physical structures are three-dimensional, so that regarding a problem as plane (or as a bar, beam, plate, or shell) implies that at least a small amount of idealization has already taken place.

Strain–Displacement Relations. FE theory makes extensive use of *strain–displacement relations*. They are used to obtain a strain field from a displacement field. Recall that normal strain is defined as change in length divided by original length and that shear strain is defined as the amount of change in a right angle. Thus in Fig. 3.1-1 we have $\varepsilon_x = \Delta u/\Delta x$, $\varepsilon_y = \Delta v/\Delta y$, and $\gamma_{xy} = \Delta u/\Delta y + \Delta v/\Delta x$. However, in general the x-direction displacement

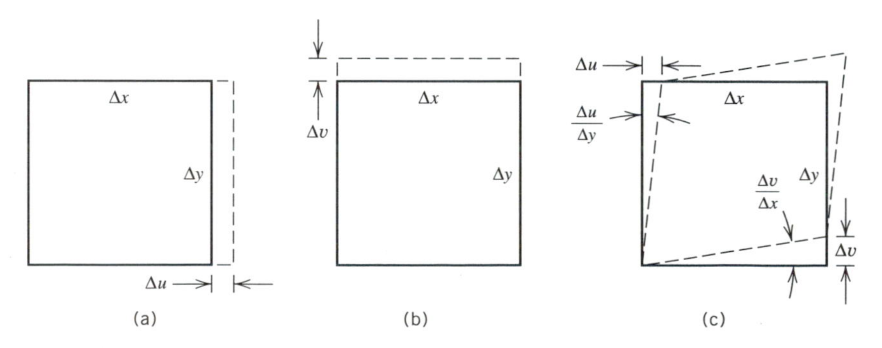

Fig. 3.1-1. A rectangle of incremental size, subjected to (a) x-direction strain, (b) y-direction strain, and (c) shear strain.

u and the y-direction displacement v are both functions of the coordinates, $u = u(x, y)$ and $v = v(x, y)$ in a plane problem. Therefore we must use partial derivatives. Doing so, and passing to the limit, we write

$$\varepsilon_x = \frac{\partial u}{\partial x} \qquad \varepsilon_y = \frac{\partial v}{\partial y} \qquad \gamma_{xy} = \frac{\partial u}{\partial y} + \frac{\partial v}{\partial x} \tag{3.1-5}$$

or, in alternative matrix formats,

$$\begin{Bmatrix} \varepsilon_x \\ \varepsilon_y \\ \gamma_{xy} \end{Bmatrix} = \begin{bmatrix} \partial/\partial x & 0 \\ 0 & \partial/\partial y \\ \partial/\partial y & \partial/\partial x \end{bmatrix} \begin{Bmatrix} u \\ v \end{Bmatrix} \quad \text{or} \quad \boldsymbol{\varepsilon} = \boldsymbol{\partial}\mathbf{u} \tag{3.1-6}$$

These strain definitions are suitable if the material has *small strains* and *small rotations*. Otherwise the strain definitions must be more extensive [3.1].

Displacements in a plane finite element are interpolated from nodal displacements u_i and v_i as follows:

$$\begin{Bmatrix} u \\ v \end{Bmatrix} = \begin{bmatrix} N_1 & 0 & N_2 & 0 & \cdots \\ 0 & N_1 & 0 & N_2 & \cdots \end{bmatrix} \begin{Bmatrix} u_1 \\ v_1 \\ u_2 \\ v_2 \\ \vdots \end{Bmatrix} \quad \text{or} \quad \mathbf{u} = \mathbf{N}\mathbf{d} \tag{3.1-7}$$

where the N_i are separate shape (or interpolation) polynomials and $\mathbf{N}$ is called the *shape function matrix*. According to Eq. 3.1-7, u depends only on the u_i, v depends only on the v_i, and u and v use the same interpolation polynomials. This is a common arrangement but it is not mandatory. An instance of Eq. 3.1-7, for a particular triangular element, appears in Eq. 1.2-3. From Eqs. 3.1-6 and 3.1-7 we obtain

$$\boldsymbol{\varepsilon} = \boldsymbol{\partial}\mathbf{N}\mathbf{d} \quad \text{or} \quad \boldsymbol{\varepsilon} = \mathbf{B}\mathbf{d} \qquad \text{where} \quad \mathbf{B} = \boldsymbol{\partial}\mathbf{N} \tag{3.1-8}$$

Matrix $\mathbf{B}$ is called the *strain–displacement matrix*.

A General Formula for k. Several texts on mechanics of materials derive an expression for U_0, the strain energy per unit volume of an elastic material. In terms of strains and in matrix format, this expression is $U_0 = \boldsymbol{\varepsilon}^T\mathbf{E}\boldsymbol{\varepsilon}/2$. Upon integrating over element volume V and substituting from Eq. 3.1-8, we obtain element strain energy U as

$$U = \tfrac{1}{2}\int \boldsymbol{\varepsilon}^T\mathbf{E}\,\boldsymbol{\varepsilon}\,dV = \tfrac{1}{2}\mathbf{d}^T\int \mathbf{B}^T\mathbf{E}\,\mathbf{B}\,dV\,\mathbf{d} = \tfrac{1}{2}\mathbf{d}^T\mathbf{k}\,\mathbf{d} \tag{3.1-9}$$

One can interpret Eq. 3.1-9 as follows. Let any element d.o.f., say the ith d.o.f., be increased from zero to the value d_i. This is accomplished by applying to the d.o.f. a force that increases from zero to F_i. The work done is $F_i d_i/2$, just as it would be if stretching a linear spring an amount d_i. This work is stored as strain energy U. Equation 3.1-9 says that work $F_i d_i/2$ is equal to strain energy in the element when the element displacement field is that produced by d_i and the element shape functions. For example, if $d_i = u_1$, we see from Eq. 3.1-7 that the element displacement field is $u(x, y) = N_1 u_1$ and $v(x, y) = 0$.

In Eq. 3.1-9, the expression

$$\mathbf{k} = \int \mathbf{B}^T \mathbf{E} \mathbf{B} \, dV \tag{3.1-10}$$

is identified as an expression for the *element stiffness matrix*. Equation 3.1-10 is not restricted to plane problems; it is applicable to all displacement-based finite elements. The direct method of generating **k**, applied to bars and beams in Chapter 2, is not general because there are no formulas that relate nodal forces to nodal displacements for elements of arbitrary shape.

We see from Eq. 3.1-10 that for a given **E**, the nature of **k** depends entirely on **B**, which in turn is derived from **N** by prescribed differentiations. In other words, *the behavior of an element is governed by its shape functions.* In subsequent sections we will examine the displacement and strain fields of several elements and use the field information to predict how the element will behave and what its defects will be. There are practical reasons for this study. In FE modeling, one seeks a good match between behavior that the actual structure is *expected* to display and behavior that elements are *able* to display, and one chooses element types, shapes, and sizes accordingly. Also, an analyst who understands the limitations of element behavior will not have unrealistically high expectations of the capabilities and accuracy of the FE method.

Loads. Mechanical loads include surface tractions, body forces, and concentrated forces and moments. Surface traction is a distributed load applied to a boundary of the structure, that is to a boundary line in two-dimensional problems and to a boundary surface in three-dimensional problems. Pressure loading is called a traction even though it pushes rather than pulls on the boundary. Also, traction may act either normal or tangent to a boundary (Fig. 3.1-2).

Body forces act throughout the volume of a structure rather than only on its surface. Body forces are usually caused by acceleration and occasionally by a magnetic field. Typical accelerations are the centripetal acceleration in rotating machinery and the acceleration of gravity, which produces self-weight loading. In one-dimensional elements there is no distinction between body and surface loads because mathematically the element is a line. Similarly, the self-weight of a horizontal plate can be replaced by pressure acting normal to the plate midsurface, which represents the plate mathematically.

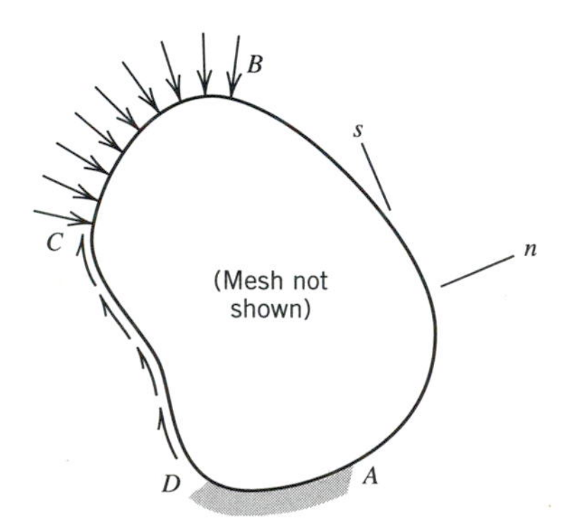

Fig. 3.1-2. Plane body with nonzero traction along BCD, zero traction along AB, and fixed support along DA. Axes n and s are, respectively, normal and tangent to the boundary.

Concentrated forces and moments come from prescribed loads and from the reactions of point supports. A concentrated load is a convenient substitute for a load of high intensity distributed over a small area.

Thermal loads are treated within the software in the manner described in Section 2.6.

Boundary Conditions. Boundary conditions include prescribed displacements and prescribed surface tractions. Usually both appear in a given problem because part of the boundary is supported and another part is loaded. In Fig. 3.1-2, for example, displacements are prescribed along *DA* and stresses are prescribed along *ABCD*. In a FE model, nodes along *AB* would receive no loads, nodes along *BCD* would be loaded by forces from the normal and tangential tractions, and all nodes along *ABCD* would be free to displace. Nodes along *DA* would have their d.o.f. set to zero and would not be loaded by prescribed forces. All boundary and internal nodes of the model except nodes along *DA* might be loaded by forces associated with self-weight loading and would be free to displace. In structural mechanics, the term "support condition" is used as a synonym for a displacement boundary condition.

Nature of the FE Approximation. We must preface our brief discussion by writing equilibrium equations and defining "compatibility." Stresses are in general functions of the coordinates, so that each stress has a rate of change with respect to x and y. In a plane problem the rates of change satisfy the equilibrium equations [6.1]

$$\frac{\partial \sigma_x}{\partial x} + \frac{\partial \tau_{xy}}{\partial y} + F_x = 0 \quad \text{and} \quad \frac{\partial \tau_{xy}}{\partial x} + \frac{\partial \sigma_y}{\partial y} + F_y = 0 \tag{3.1-11}$$

where F_x and F_y are body forces per unit volume. If Eqs. 3.1-11 are satisfied throughout a plane body, every differential element and the body itself are in static equilibrium. As for deformations, they are called *compatible* if displacement boundary conditions are met and the material does not crack apart or overlap itself.

Elasticity theory shows that if displacement and stress fields simultaneously satisfy equilibrium equations, compatibility, and boundary conditions on stress, then the solution obtained is *exact*. How is an exact solution approached by a FE approximation? Let elements be based on polynomial displacement fields, as in this book and indeed as for most elements in common use. Then the compatibility requirement is satisfied exactly within elements. Equilibrium equations and boundary conditions on stress are not satisfied: that is, Eqs. 3.1-11 are not satisfied at most points within the FE model, and stress boundary conditions are not satisfied at most points on the boundary (e.g., in Fig. 3.1-2 the computed stresses σ_n and τ_{ns} will not be precisely zero all along the unloaded boundary *AB*). Stress boundary conditions and equilibrium equations are satisfied in an *average* sense: it can be shown that integrals of the left-hand sides of Eqs. 3.1-11 vanish over each element. As a mesh is repeatedly refined, pointwise satisfaction of stress boundary conditions and equilibrium equations is approached more and more closely.

The foregoing discussion also applies to three-dimensional elastic problems, for which Eqs. 3.1-11 must be expanded to include all six stresses. The nature of a FE solution is further discussed in Section 4.8.

3.2 CONSTANT STRAIN TRIANGLE (CST)

The CST element, Fig. 3.2-1, is perhaps the earliest and simplest finite element. In terms of generalized coordinates β_i its displacement field is

$$u = \beta_1 + \beta_2 x + \beta_3 y$$
$$v = \beta_4 + \beta_5 x + \beta_6 y \tag{3.2-1}$$

and, from Eqs. 3.1-5 and 3.2-1, the resulting strain field is

$$\varepsilon_x = \beta_2 \qquad \varepsilon_y = \beta_6 \qquad \gamma_{xy} = \beta_3 + \beta_5 \tag{3.2-2}$$

We see that strains do not vary within the element; hence the name "constant strain triangle" (CST for short). The element may also be called a "linear triangle" because its *displacement* field is linear in x and y. Element sides remain straight as the element deforms. (Element sides have the appearance of bar elements discussed in Chapter 2, but a plane FE is not an assemblage of bars; a plane FE is the region bounded by its sides.)

We omit the algebra needed to recast Eqs. 3.2-1 in the shape-function form of Eq. 3.1-7. The algebra is tedious [2.2] and does not help us understand how the element behaves. The strain field obtained from the shape functions, in the form $\boldsymbol{\varepsilon} = \mathbf{Bd}$, is

$$\begin{Bmatrix} \varepsilon_x \\ \varepsilon_y \\ \gamma_{xy} \end{Bmatrix} = \frac{1}{2A} \begin{bmatrix} y_{23} & 0 & y_{31} & 0 & y_{12} & 0 \\ 0 & x_{32} & 0 & x_{13} & 0 & x_{21} \\ x_{32} & y_{23} & x_{13} & y_{31} & x_{21} & y_{12} \end{bmatrix} \begin{Bmatrix} u_1 \\ v_1 \\ u_2 \\ v_2 \\ u_3 \\ v_3 \end{Bmatrix} \tag{3.2-3}$$

where x_i and y_i are nodal coordinates (i = 1, 2, 3), $x_{ij} = x_i - x_j$ and $y_{ij} = y_i - y_j$ (i, j = 1, 2, 3), and $2A$ is twice the area of the triangle, $2A = x_{21}y_{31} - x_{31}y_{21}$. Node numbers are arbitrary except that the sequence 123 must go counterclockwise around the element if A is to be positive. Again we see that strains do not vary within the element. Applying Eq. 3.1-10 we obtain the element stiffness matrix:

$$\mathbf{k} = \mathbf{B}^T\mathbf{EB}tA \tag{3.2-4}$$

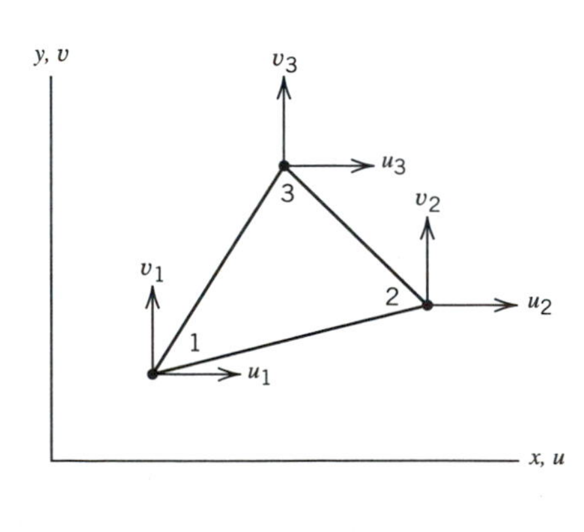

Fig. 3.2-1. A constant strain triangle. Its six nodal d.o.f. are shown.

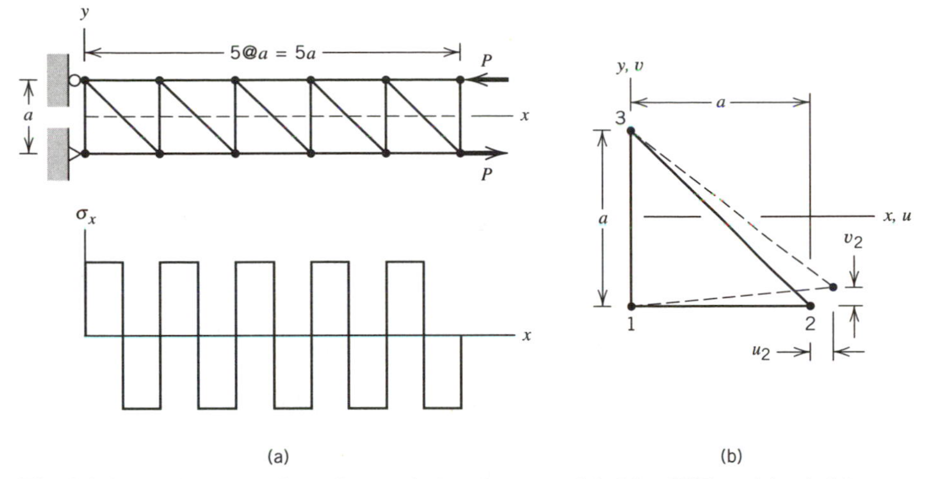

Fig. 3.2-2. (a) Stress σ_x along the x axis in a beam modeled by CSTs and loaded in pure bending. (b) Deformation of the lower-left CST in the model.

where t is the element thickness, assumed constant, and **E** comes from Eq. 3.1-3 if plane stress conditions prevail. Integration in Eq. 3.1-10 is trivial because **B** and **E** contain only constants.

The CST gives good results in a region of the FE model where there is little strain gradient. Otherwise the CST does not work well. This is evident if we ask the CST to model pure bending (Fig. 3.2-2). The x axis should be stress-free because it is the neutral axis of the beam. Instead, the FE model displays σ_x as a square wave pattern. Also, the FE model predicts deflections and σ_x stresses that are only about one-quarter of the correct values. The inability of the CST to represent an ε_x that varies linearly with y is partly to blame for this poor result. But the CST also develops a spurious shear stress when bent. This is seen in Fig. 3.2-2b. It is proper that u_2 and v_2 appear as shown, but v_2 creates a shear stress that should not be present. An expression for the shear stress can be obtained by use of Eqs. 3.1-2, 3.1-3, and 3.2-3. Despite defects of the CST, correct results are approached as a mesh of CST elements is repeatedly refined.

3.3 LINEAR STRAIN TRIANGLE (LST)

The LST element is shown in Fig. 3.3-1. It has midside nodes in addition to vertex nodes. The d.o.f. are u_i and v_i at each node i, $i = 1, 2, \ldots, 6$, for a total of 12 d.o.f. In terms of generalized coordinates β_i its displacement field is

$$
\begin{aligned}
u &= \beta_1 + \beta_2 x + \beta_3 y + \beta_4 x^2 + \beta_5 xy + \beta_6 y^2 \\
v &= \beta_7 + \beta_8 x + \beta_9 y + \beta_{10} x^2 + \beta_{11} xy + \beta_{12} y^2
\end{aligned}
\tag{3.3-1}
$$

and, from Eqs. 3.1-5 and 3.3-1, the resulting strain field is

$$
\begin{aligned}
\varepsilon_x &= \beta_2 + 2\beta_4 x + \beta_5 y \\
\varepsilon_y &= \beta_9 + \beta_{11} x + 2\beta_{12} y \\
\gamma_{xy} &= (\beta_3 + \beta_8) + (\beta_5 + 2\beta_{10})x + (2\beta_6 + \beta_{11})y
\end{aligned}
\tag{3.3-2}
$$

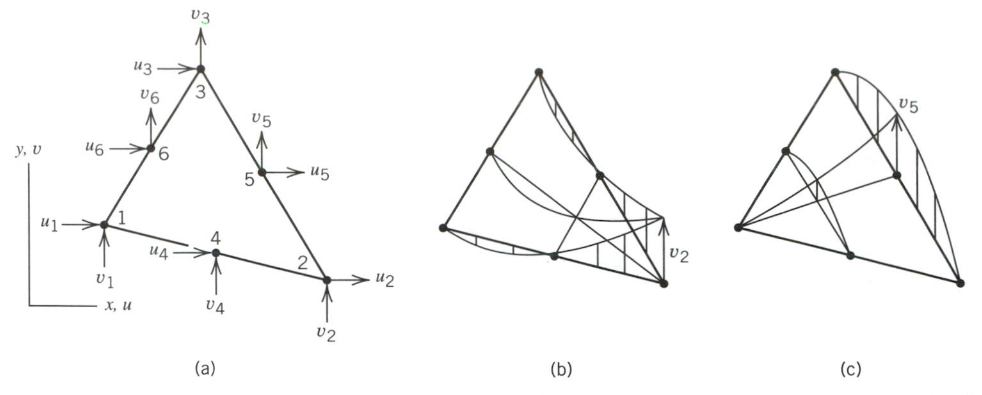

Fig. 3.3-1. (a) A linear strain triangle and its six nodal d.o.f. (b) Displacement mode associated with nodal d.o.f. v_2. (c) Displacement mode associated with nodal d.o.f. v_5. (For visualization only, imagine that displacement occurs normal to the plane of the element.) (*b* and *c* reprinted from [2.2] by permission of John Wiley & Sons, Inc.)

The strain field can vary linearly with x and y within the element; hence the name "linear strain triangle" (LST for short). The element may also be called a "quadratic triangle" because its *displacement* field is quadratic in x and y. Element sides deform into quadratic curves when a single d.o.f. is activated, as shown in Figs. 3.3-1b and 3.3-1c. The LST has all the capabilities of the CST, few as they are, and more. For example, Eq. 3.3-2 shows that strain ε_x can vary linearly with y. If the pure bending problem of Fig. 3.2-2a is solved using LST elements, exact results for deflection and stress are obtained. Additional numerical examples appear in Section 3.11.

The element stiffness matrix is most easily generated using "area coordinates." The procedure does not help in understanding how the element behaves. Details may be found elsewhere [2.2]. We note only that the product $\mathbf{B}^T\mathbf{E}\mathbf{B}$ is quadratic in the coordinates and that integration required in Eq. 3.1-10 can be done either in closed form by special formulas or numerically (Section 4.5). Numerical integration is necessary if element sides are not straight but curved, that is, *initially* curved, when all nodal d.o.f. are zero.

3.4 BILINEAR QUADRILATERAL (Q4)

The Q4 element is a quadrilateral that has four nodes. Its nodal d.o.f. are shown in Fig. 3.4-1. In terms of generalized coordinates β_i, its displacement field is

$$\begin{aligned} u &= \beta_1 + \beta_2 x + \beta_3 y + \beta_4 xy \\ v &= \beta_5 + \beta_6 x + \beta_7 y + \beta_8 xy \end{aligned} \tag{3.4-1}$$

The name "bilinear" arises because the form of the expressions for u and v is the product of two linear polynomials, that is, $(c_1 + c_2 x)(c_3 + c_4 y)$, where the c_i are constants. There are four parameters in each displacement expansion: four β_i for u and four β_i for v in Eq. 3.4-1, or four shape functions N_i in Eq. 3.4-3 below. From Eqs. 3.1-5 and 3.4-1, the element strain field is

$$\begin{aligned} \varepsilon_x &= \beta_2 + \beta_4 y \\ \varepsilon_y &= \beta_7 + \beta_8 x \\ \gamma_{xy} &= (\beta_3 + \beta_6) + \beta_4 x + \beta_8 y \end{aligned} \tag{3.4-2}$$

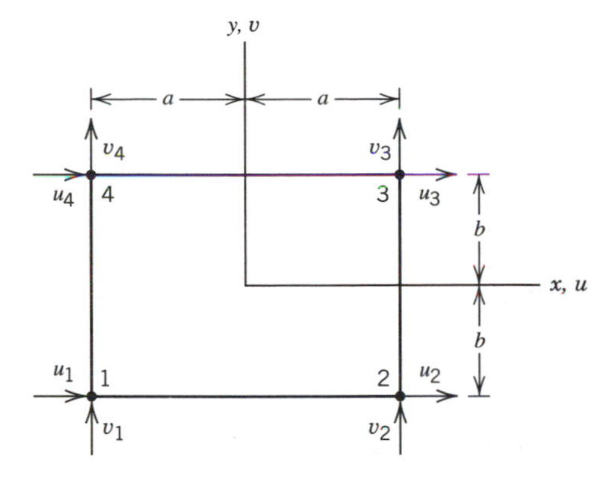

Fig. 3.4-1. A bilinear quadrilateral and its eight nodal d.o.f.

Important aspects of element behavior can be deduced from Eqs. 3.4-1 and 3.4-2. The strain field shows that ε_x is independent of x, which means that the Q4 element cannot exactly model a cantilever beam under transverse tip force (Fig. 3.4-2a), where axial strain varies linearly with x. Moreover, the Q4 element cannot exactly model a state of pure bending, despite its ability to represent an ε_x that varies linearly with y. Consider Fig. 3.4-2b, which shows a block of material loaded in pure bending. We know from beam theory that shear strain γ_{xy} is absent, that plane sections remain plane, and that top and bottom edges become arcs of practically the same radius of curvature, as shown by dashed lines in Fig. 3.4-2b. A Q4 element loaded in pure bending is shown in Fig. 3.4-2c. Sides rotate, as shown by dashed lines, but top and bottom edges remain straight. This result is dictated by Eqs. 3.4-1: along edges y = constant, displacement v is linear in x. Indeed *all* sides of a Q4 element deform as straight lines. Therefore right angles in the element are not preserved under pure moment loading and in consequence shear strain appears everywhere in the element except along the y axis. The same result can also be seen from Eqs. 3.4-2: the displacement mode of Fig. 3.4-2c requires that β_4 be nonzero so that ε_x will vary linearly with y, but β_4 also appears in the expression for γ_{xy}; therefore a Q4 element that bends also develops shear strain. (This trouble does not appear in the LST in pure bending: when β_5 in Eqs. 3.3-2 is nonzero, β_{10} assumes a value such that $\beta_5 + 2\beta_{10}$ is zero in the shear strain expression.)

Clearly, arguments of the preceding paragraph apply in similar fashion when bending moments are applied to top and bottom edges of the element instead of to sides. The physical consequence of these defects is that the Q4 element is too stiff in bending because an applied bending moment is resisted by spurious shear stress as well as by the ex-

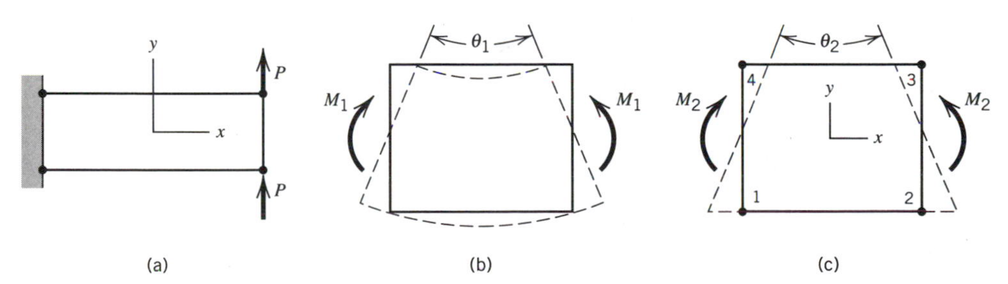

Fig. 3.4-2. (a) A one-element cantilever beam under transverse tip loading. (b) Correct deformation mode of a rectangular block in pure bending. (c) Deformation mode of the bilinear quadrilateral under bending load.

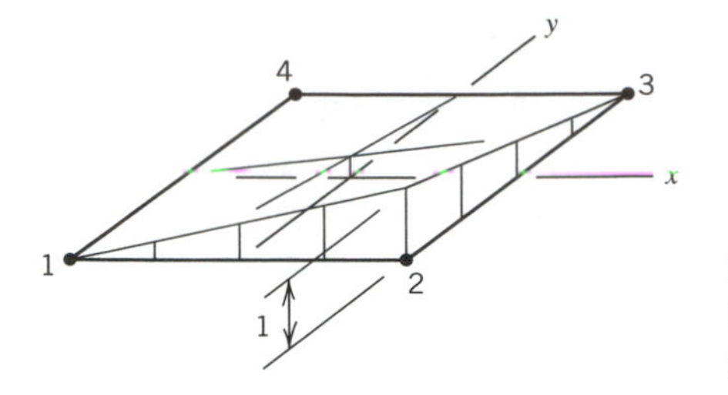

Fig. 3.4-3. Shape function N_2 of the bilinear quadrilateral. (For visualization only, imagine that displacement occurs normal to the xy plane.)

pected flexural stresses. Further discussion of this problem, as well as a remedy for it, appears in Section 3.6.

If the β_i in Eq. 3.4-1 are expressed in terms of nodal d.o.f., we obtain the displacement field in the form of Eq. 3.1-7, where

$$N_1 = \frac{(a-x)(b-y)}{4ab} \qquad N_2 = \frac{(a+x)(b-y)}{4ab}$$
$$N_3 = \frac{(a+x)(b+y)}{4ab} \qquad N_4 = \frac{(a-x)(b+y)}{4ab} \tag{3.4-3}$$

A representative shape function (N_2) is plotted in Fig. 3.4-3. Note that $N_2 = 1$ at node 2 and $N_2 = 0$ at every other node. *This is true of shape functions in general, for any element type*; that is, $N_i = 1$ at node i and $N_i = 0$ at node j where $j \neq i$. In the format of Eq. 3.1-8, the element strain field is

$$\begin{Bmatrix} \varepsilon_x \\ \varepsilon_y \\ \gamma_{xy} \end{Bmatrix} = \frac{1}{4ab} \begin{bmatrix} -(b-y) & 0 & (b-y) & 0 & \cdots \\ 0 & -(a-x) & 0 & -(a+x) & \cdots \\ -(a-x) & -(b-y) & -(a+x) & (b-y) & \cdots \end{bmatrix} \begin{Bmatrix} u_1 \\ v_1 \\ u_2 \\ v_2 \\ \vdots \\ v_4 \end{Bmatrix} \tag{3.4-4}$$

We can again deduce that the deformation mode of Fig. 3.4-2c contains spurious shear strain by substituting the nodal d.o.f. of this mode into Eq. 3.4-4.

Equilibrium (Eqs. 3.1-11) is not satisfied at every point in the Q4 element unless $\beta_4 = \beta_8 = 0$ in Eqs. 3.4-1, in which case a state of constant strain prevails. Despite this and other criticism of the Q4 element, it converges properly with mesh refinement and in most problems it works better than the CST element (which *always* satisfies Eqs. 3.1-11). Examples of element behavior appear in Section 3.11.

Equations stated in this section restrict the Q4 element to rectangular shape, but this restriction can be overcome; see Section 3.8.

3.5 QUADRATIC QUADRILATERAL (Q8)

The Q8 element is shown in Fig. 3.5-1. In terms of generalized coordinates β_i its displacement field is

$$u = \beta_1 + \beta_2 x + \beta_3 y + \beta_4 x^2 + \beta_5 xy + \beta_6 y^2 + \beta_7 x^2 y + \beta_8 xy^2$$
$$v = \beta_9 + \beta_{10} x + \beta_{11} y + \beta_{12} x^2 + \beta_{13} xy + \beta_{14} y^2 + \beta_{15} x^2 y + \beta_{16} xy^2 \tag{3.5-1}$$

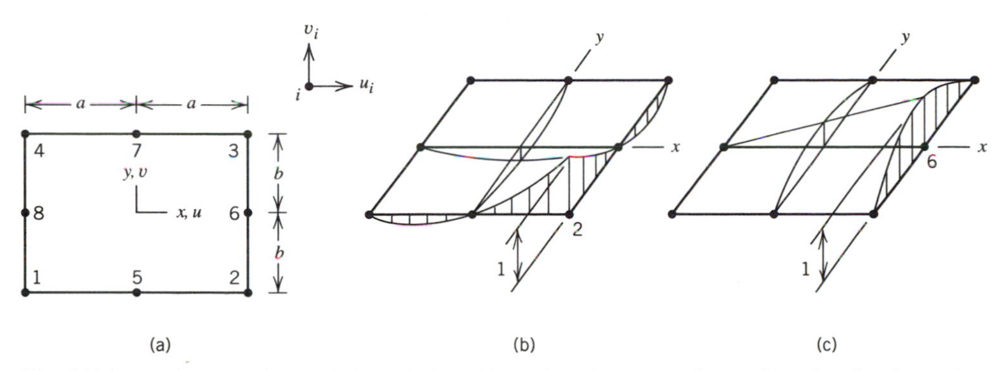

Fig. 3.5-1. (a) A quadratic quadrilateral. (b,c) Shape functions N_2 and N_6 . (For visualization only, imagine that displacement occurs normal to the xy plane.)

In more compact form than Eq. 3.1-7, the displacement field in terms of shape functions N_i is

$$u = \sum N_i \, u_i \qquad v = \sum N_i \, v_i \tag{3.5-2}$$

where index i runs from 1 to 8, which explains the "8" in the name Q8. As examples, two of the eight shape functions are

$$\begin{aligned} N_2 &= \tfrac{1}{4}(1+\xi)(1-\eta) - \tfrac{1}{4}(1-\xi^2)(1-\eta) - \tfrac{1}{4}(1+\xi)(1-\eta^2) \\ N_6 &= \tfrac{1}{2}(1+\xi)(1-\eta^2) \end{aligned} \tag{3.5-3}$$

where $\xi = x/a$ and $\eta = y/b$. By looking at a typical edge, for example, the edge $x = a$, we see from either Eqs. 3.5-1 or 3.5-3 that displacements are quadratic in y, which means that the edge deforms into a parabola when any single d.o.f. on that edge is nonzero. From Eqs. 3.1-5 and 3.5-1, the element strain field is

$$\begin{aligned} \varepsilon_x &= \beta_2 + 2\beta_4 x + \beta_5 y + 2\beta_7 xy + \beta_8 y^2 \\ \varepsilon_y &= \beta_{11} + \beta_{13} x + 2\beta_{14} y + \beta_{15} x^2 + 2\beta_{16} xy \\ \gamma_{xy} &= (\beta_3 + \beta_{10}) + (\beta_5 + 2\beta_{12})x + (2\beta_6 + \beta_{13})y \\ &\quad + \beta_7 x^2 + 2(\beta_8 + \beta_{15})xy + \beta_{16} y^2 \end{aligned} \tag{3.5-4}$$

Each of the three strains contains all linear terms and some quadratic terms (e.g., there is no x^2 term in the ε_x expression). The Q8 element can represent exactly all states of constant strain, and states of pure bending if it is rectangular. Nonrectangular shapes are permitted; see Section 3.8. Examples of element behavior appear in Section 3.11.

3.6 IMPROVED BILINEAR QUADRILATERAL (Q6)

The principal defect of the Q4 element is its overstiffness in bending, which can be illustrated by comparison of the bending moments in Figs. 3.4-2b and 3.4-2c. Let the rectangular block and the element have the same dimensions, elastic modulus E, and Poisson ratio ν. Then apply whatever bending moments M_1 and M_2 are necessary to make vertical

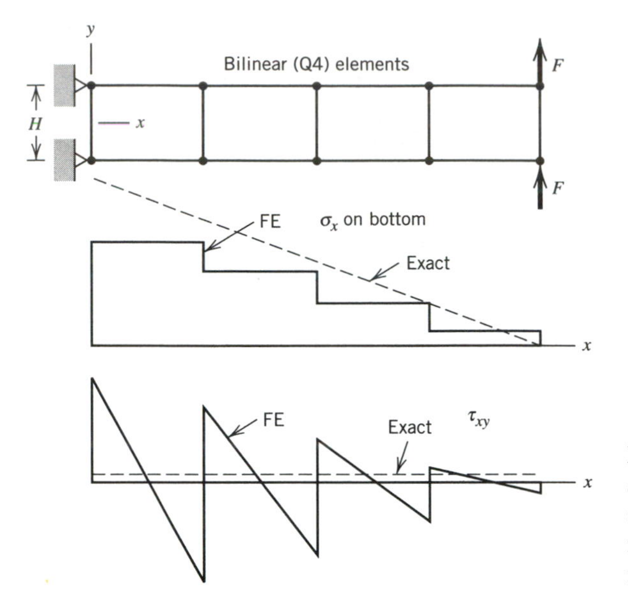

Fig. 3.6-1. Qualitative variation of axial stress and average transverse shear stress in a cantilever beam modeled by rectangular Q4 elements.

sides of the block and the element include the same angle, $\theta_1 = \theta_2$. Moment M_1 is the correct value. It can be shown that M_2 is

$$M_2 = \frac{1}{1+\nu}\left[\frac{1}{1-\nu} + \frac{1}{2}\left(\frac{a}{b}\right)^2\right]M_1 \tag{3.6-1}$$

where a and b are dimensions shown in Fig. 3.4-1. If aspect ratio a/b increases without limit, so does M_2, which means that the Q4 element becomes infinitely stiff in bending. This phenomenon is called "locking" [3.2]. In practice we avoid elements of large aspect ratio, and a FE mesh does not "lock" but rather is overly stiff when bent, as explained in Section 3.4. Qualitative results appear in Fig. 3.6-1. Deflections and axial stress in the FE model are smaller than the exact values, and transverse shear stress is greatly in error except along the y-parallel centerline of each element.

A remedy for the trouble is fairly simple and produces an element sometimes called the Q6 element [2.2, 3.3]. Its displacement expansions for u and v each contain six shape functions; that is,

$$\begin{aligned} u &= \sum_{i=1}^{4} N_i u_i + (1-\xi^2)g_1 + (1-\eta^2)g_2 \\ v &= \sum_{i=1}^{4} N_i v_i + (1-\xi^2)g_3 + (1-\eta^2)g_4 \end{aligned} \tag{3.6-2}$$

where $\xi = x/a$, $\eta = y/b$, and the N_i in the summations are shape functions of the Q4 element, Eq. 3.4-3. In Eq. 3.6-2 we have simply augmented the displacement field of the Q4 element by modes that describe a state of constant curvature. This is easy to see for modes associated with d.o.f. g_2 and g_3: as shown in Fig. 3.6-2a, they allow edges of the element to become curved. Accordingly, the Q6 element can model bending with either an x-parallel neutral axis or a y-parallel neutral axis; indeed g_2 and g_3 can be nonzero si-

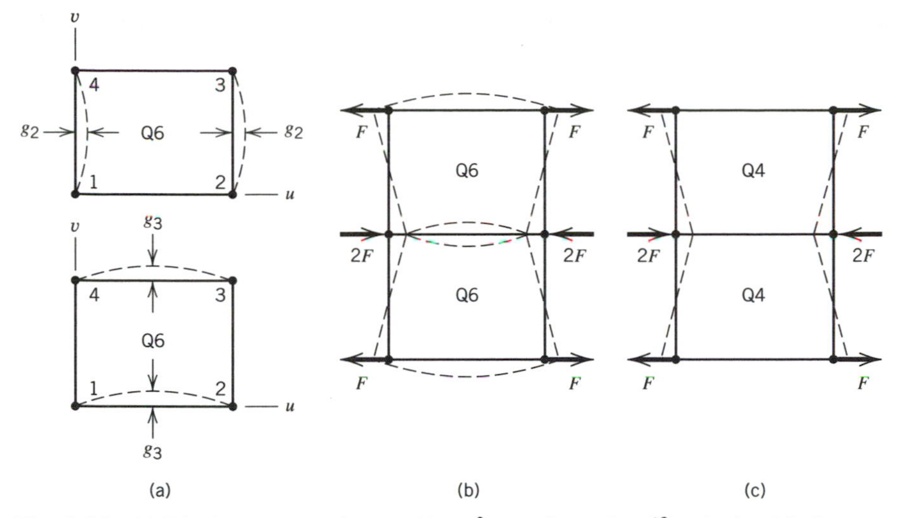

Fig. 3.6-2. (a) Displacement modes $u = (1 - \eta^2)g_2$ and $v = (1 - \xi^2)g_3$ in the Q6 element. (b) Incompatibility between adjacent Q6 elements. (c) No incompatibility between adjacent Q4 elements.

multaneously. Modes associated with d.o.f. g_1 and g_4 allow the existence of strains normal to a beam axis that appear because of the Poisson effect. From Eqs. 3.1-5 and 3.6-2, shear strain in the Q6 element is

$$\gamma_{xy} = \sum_{i=1}^{4} \frac{\partial N_i}{\partial y} u_i + \sum_{i=1}^{4} \frac{\partial N_i}{\partial x} v_i - \frac{2y}{b^2} g_2 - \frac{2x}{a^2} g_3 \tag{3.6-3}$$

In pure bending, the negative terms $(2y/b^2)g_2$ and $(2x/a^2)g_3$ are equal in magnitude to positive terms produced by the summations, thus permitting shear strain to vanish, as is proper. The Q6 element can represent pure bending exactly, but only if the element is rectangular. (This point is discussed further in connection with Fig. 3.8-2.) Qualitative results appear in Fig. 3.6-3. Axial stress is exact along the y-parallel centerline of each element, and average transverse shear stress is exact everywhere. Further examples of element behavior appear in Section 3.11. The Q6 element, or a differently formulated

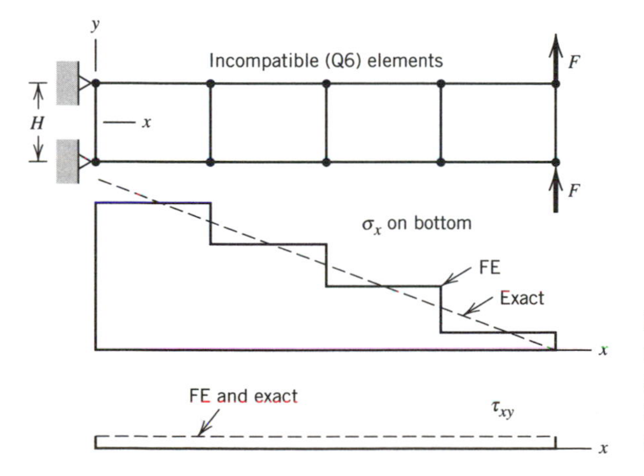

Fig. 3.6-3. Qualitative variation of axial stress and average transverse shear stress $2F/A = 2F/Ht$ in a cantilever beam modeled by rectangular Q6 elements.

element of comparable behavior, is usually the default option for a four-node quadrilateral in commercial software.

The d.o.f. g_1 through g_4 are *internal* d.o.f. Unlike nodal d.o.f. u_i and v_i , they are not connected to corresponding d.o.f. in adjacent elements. Modes associated with d.o.f. g_i are *incompatible*. That is, under some (but not all) loadings, an overlap or a gap may appear between adjacent elements. This point is made in Fig. 3.6-2b. No gap appears with Q4 elements under similar loading (Fig. 3.6-2c). Indeed no gaps or overlaps appear in a physical continuum; why then is the Q6 element acceptable? It is because elements approach a state of constant strain as a mesh is repeatedly refined. In a state of constant strain all initially straight lines, including element edges, remain straight after deformation. Then there is no incompatibility between elements. Thus mesh refinement produces convergence toward correct results.

3.7 ELEMENTS WITH "DRILLING" D.O.F.

A "drilling" d.o.f. is a rotational d.o.f. whose vector is normal to the plane of an element. Thus θ_{zi} is a drilling d.o.f. at node i for an element in the xy plane. Elements with drilling d.o.f. are not yet in common use, so we discuss them only briefly.

An element edge that has a midside node can deform into either a straight or a parabolic shape. As shown below, translational d.o.f. at midside can be expressed in terms of translational and drilling d.o.f. at corners. This permits an exchange: translational d.o.f. at midsides are traded for drilling d.o.f. at corners. Consider the LST element in Fig. 3.7-1a. Let u_n represent displacement normal to side 2–3. Normal displacement at node 5 is written

$$u_{n5} = \tfrac{1}{2}(u_{n2} + u_{n3}) + \tfrac{1}{8}(\theta_{z3} - \theta_{z2})L_{23} \tag{3.7-1}$$

where θ_{z2} and θ_{z3} are drilling d.o.f. at nodes 2 and 3 and L_{23} is the length of side 2–3. Tangential displacement at midside is written

$$u_{s5} = \tfrac{1}{2}(u_{s2} + u_{s3}) \tag{3.7-2}$$

These equations mean, for example, that if $u_{n2} = u_{n3} = 0$ and $\theta_{z3} = -\theta_{z2} = \bar{\theta}$, side 2–3 displaces into a parabola with end rotations $\bar{\theta}$ and center displacement $u_{n5} = \bar{\theta}L_{23}/4$. If $\theta_{z2} =$

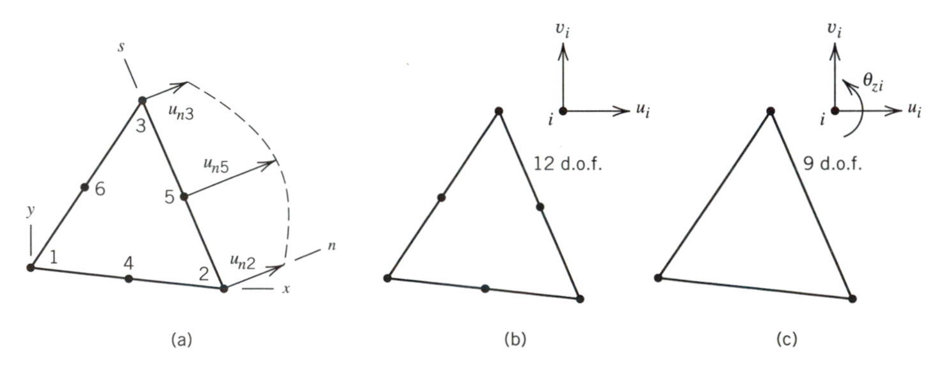

Fig. 3.7-1. (a) Displacements normal to one side of a LST element. (b) The d.o.f. in a LST element. (c) The d.o.f. in a triangular element with drilling d.o.f.

θ_{z3}, the edge remains straight. Equation 3.7-2 constrains tangential strain ε_s to be constant along edge 2–3.

Equations analogous to Eqs. 3.7-1 and 3.7-2 are also written for the remaining two sides. When these six equations are substituted into the shape functions of the LST element, along with sine and cosine terms to relate directions n and s to directions x and y, midside nodes disappear and their d.o.f. are replaced by drilling d.o.f. at corner nodes. Because the three edge strains ε_s are constrained to be constant, there is a reduction of three d.o.f. in the process. Also, the drilling d.o.f. have a "do-nothing" aspect that makes them less effective: if they are all equal, they have no effect on element strains. Thus we do not have nine *useful* d.o.f. in Fig. 3.7-1c.

The Q8 element can be similarly treated. We then exchange a 16 d.o.f. element having corner and midside nodes, each with translational d.o.f., for a 12 d.o.f. element having only corner nodes, each with translational and drilling d.o.f. Again, if all drilling d.o.f. are equal, the element displays no strain. The 12 d.o.f. element behaves rather like the Q6 element but cannot model pure bending exactly if Poisson's ratio ν is nonzero. To see this, consider a one-element beam that lies along the x axis. Under pure bending we should have $\varepsilon_y = -\nu\varepsilon_x = -\nu(My/EI)$. But ε_y cannot be linear in y because constraints such as Eq. 3.7-2 make ε_y independent of y.

FE analysis of shells provides a motivation for the use of drilling d.o.f. A shell element combines membrane and bending actions and thus is analogous to a 3D beam element, which combines bar and beam actions. Three displacements and three rotations are active at each node of a shell, in order to accommodate 3D beam elements that serve as stiffeners, and also to allow shell elements to meet at an angle, as they do along a fold line. If flat elements are used, they model a shell as a faceted surface and accordingly all interelement boundaries become fold lines. Since three rotational d.o.f. per node are active at the global level it would be wasteful to omit their element-normal (drilling) component at the element level if its inclusion can provide an improved element.

3.8 ELEMENTS OF MORE GENERAL SHAPE

Equations in Sections 3.4, 3.5, and 3.6 have a form that limits quadrilateral elements to rectangular shape. However, commercial software allows quadrilateral elements to be of *general* shape. The limitation to rectangular shape is overcome by expressing displacements and strains in an auxiliary coordinate system. The theory is explained in Sections 4.4 to 4.7. For now the following brief description is sufficient.

Examples of nonrectangular elements appear in Fig. 3.8-1. Sides having side nodes may be curved, which allows a better geometric approximation of a curved boundary. A side node may also be shifted toward a corner. *But these geometric distortions are usu-*

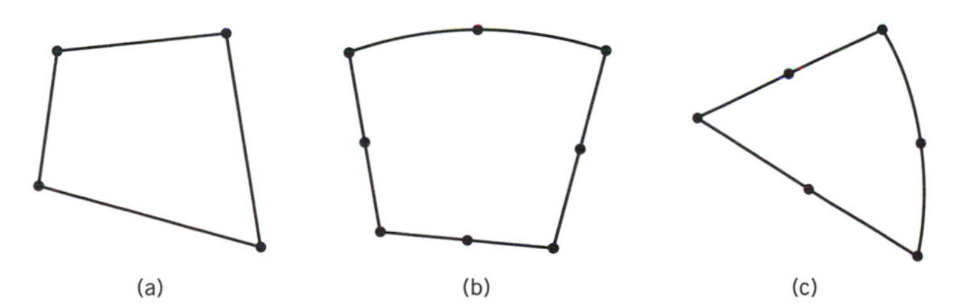

Fig. 3.8-1. (a,b) Nonrectangular quadrilaterals. (c) Triangular element with a curved side. (Note: curved sides are neither necessary nor advocated.)

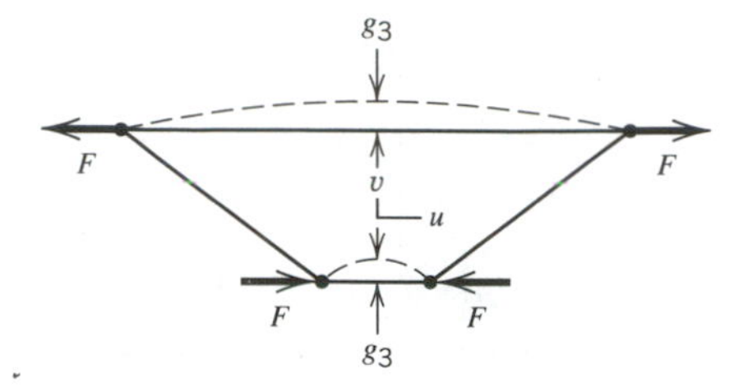

Fig. 3.8-2. Dashed lines show the displacement mode $v = (1 - \xi^2)g_3$ in a Q6 element of trapezoidal shape.

ally detrimental to accuracy. Of course we cannot model a structure of arbitrary shape using only rectangles; we must be able to fit the actual geometry reasonably well and be able to grade the mesh from coarse to fine near a region of interest. However, as a rule *elements behave best when they are of a compact regular shape.* Accordingly, it is usually best to keep corner angles approximately equal and avoid elongated elements. LST and Q8 elements should usually have straight sides and side nodes at midside. It is proper to use a curved side to fit the shape of a hole or a fillet, but all element sides *internal* to the mesh should be straight. Whether well shaped or badly shaped, all elements discussed thus far can represent exactly any state of constant strain and will provide convergence toward correct results as a mesh is repeatedly refined.

For one type of element in particular it is not hard to see that accuracy must decline with increasing shape distortion. Under pure bending, top and bottom surfaces of a beam have practically the same radius of curvature. Let a pure bending load be applied to the distorted Q6 element shown in Fig. 3.8-2. Mode $v = (1 - \xi^2)g_3$ is activated so that top and bottom edges become arcs, as shown by dashed lines. However, the arcs have much different radii. The discrepancy increases as the amount of shape distortion increases.

Remember: quadrilaterals of general shape behave much like the rectangular elements described in Sections 3.4, 3.5, and 3.6, provided that elements are well shaped as explained above.

3.9 LOADS

Mechanical loads consist of concentrated loads at nodes, surface traction, and body force. Traction and body force loads cannot be applied directly to a FE model. Instead, they must be converted to equivalent nodal loads in the manner now described. Here we consider plane elements that have translational d.o.f. only.

In a plane problem, surface traction may act on internal and/or external boundaries of

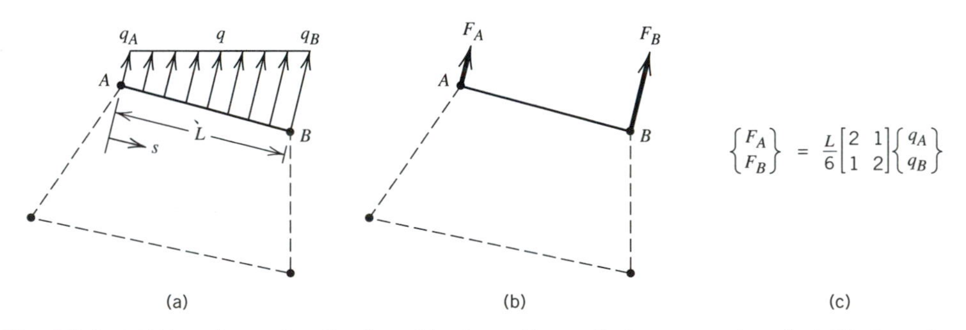

Fig. 3.9-1. (a) Linearly varying distributed load on a linear-displacement edge. (b,c) Work-equivalent nodal loads.

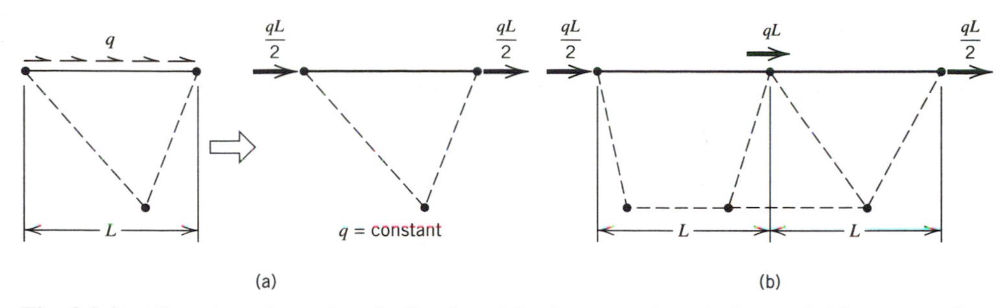

Fig. 3.9-2. Allocation of a uniformly distributed load q as work-equivalent nodal forces on edges having linear displacement variation.

a FE mesh. A traction has arbitrary orientation with respect to a boundary but usually is expressed in terms of components normal and tangent to the boundary. A commonly used traction is pressure p, which acts normal to a boundary in the compressive sense. Boundary pressure p is equivalent to a line load $q = pt$, where t is the thickness of the model. The dimensions of q are [force/length].

In Fig. 3.9-1, traction q and the edge displacement in the direction of q both vary linearly with the edge-tangent coordinate s. Nodal loads F_A and F_B are applied to the FE model instead of q. These loads are "work-equivalent," meaning that an edge displacement produced by displacements of nodes A and B causes distributed load q to do the same work as is done by nodal loads F_A and F_B in moving through the nodal displacements. That is, if $v' = v'(s)$ is the component of displacement normal to edge AB in Fig. 3.9-1, work-equivalency requires

$$F_A v'_A + F_B v'_B = \int_0^L v'(q\, ds) \tag{3.9-1}$$

where $v' = (L - s)v'_A/L + sv'_B/L$, for all values of v'_A and v'_B. The mesh layout and displacement field *within* the element do not matter: if q and the *edge* displacement are linear, then work-equivalent loads are as stated in Fig. 3.9-1c. Nodal loads combine at shared nodes. This point is made in Fig. 3.9-2, where a uniform edge-tangent traction is shown acting on elements whose edges are collinear and of the same length.

Similar results appear in Figs. 3.9-3 and 3.9-4. Traction q in Fig. 3.9-3 varies quadratically with s, as does edge displacement in the direction of q. If edge ABC is straight and node B is at midedge, then work-equivalent nodal loads are as stated in Fig. 3.9-3c.

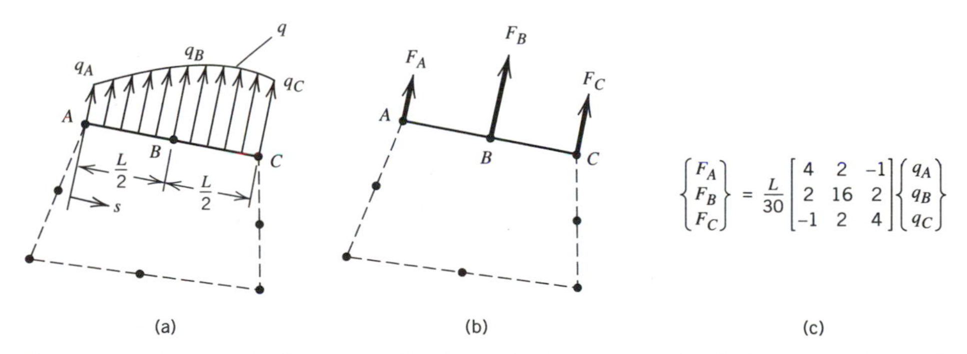

Fig. 3.9-3. (a) Quadratically varying distributed load on a quadratic-displacement edge. (b,c) Work-equivalent nodal forces.

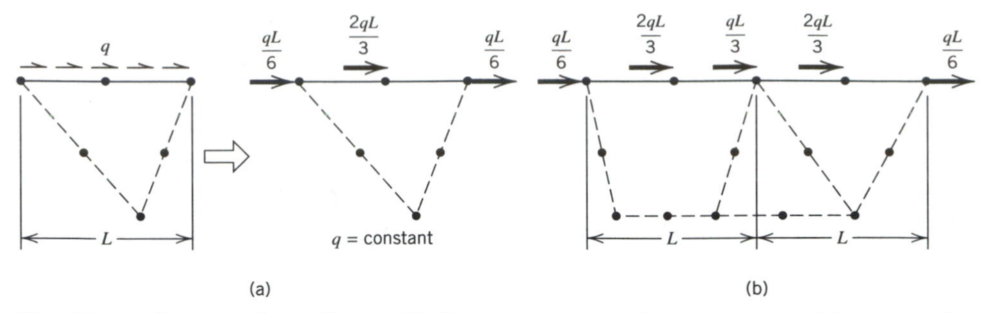

Fig. 3.9-4. Allocation of a uniformly distributed load q as work-equivalent nodal forces on edges having quadratic displacement variation and midside nodes. (See data in Fig. 3.9-3c.)

Accordingly, a *uniform* q produces the nodal loads shown in Fig. 3.9-4. Note that edge nodes carry more force than corner nodes. Work-equivalent nodal loads usually provide greater accuracy than a "lumping" in which all nodes carry the same force, although in either case exact results will be approached as the mesh is repeatedly refined.

Figure 3.9-5 shows work-equivalent loads for a uniform body force in the negative y direction. The total force is weight W of the element in each case. The orientation of an element in the xy plane does not matter, but quadrilaterals must be rectangular if the nodal loads shown are to be work-equivalent. In Fig. 3.9-5c and 3.9-5d we see some surprises. Vertex nodes of the LST element are not loaded. Corner nodes of the Q8 element carry *upward* loads, but the sum of all eight nodal loads is W, acting downward, as must be the case.

Most software is capable of automatically calculating equivalent nodal loads of proper magnitude and direction and combining them at shared nodes. The user need only prescribe the direction and intensity of the distributed loading. Software does not require that edges be collinear or of the same length.

A concentrated moment cannot be applied to a node of CST, LST, Q4, Q8, or Q6 elements because these elements use only translational d.o.f. This means that if a 2D beam element is attached to plane elements in the manner shown in Fig. 3.9-6a there will be a hinge connection at A, which transmits only force. This arrangement is a mechanism and $\mathbf{K}$ will be singular. An ad-hoc arrangement that transmits both force and moment is shown in Fig. 3.9-6b. The beam has been extended into the plane body by adding two beam elements, AB and BC. (Adding only one beam element, AB, is an equally plausible alternative [3.4]). Translational d.o.f. of beam elements and plane elements are connected at nodes A, B, and C. Nodal d.o.f. θ_{zi} at these nodes are associated with only the beam ele-

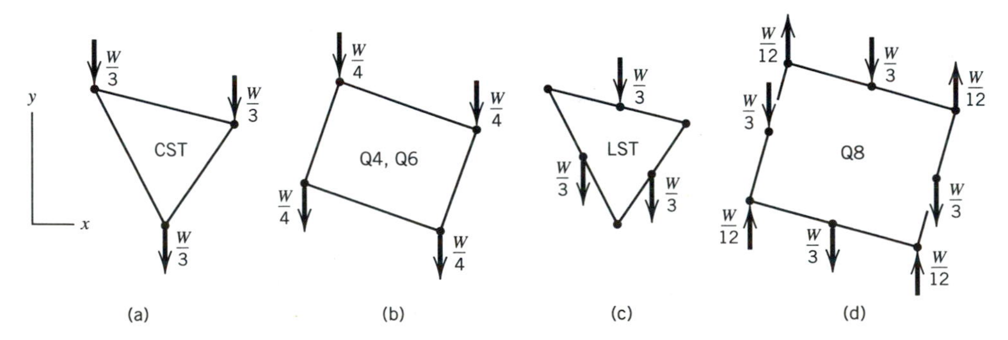

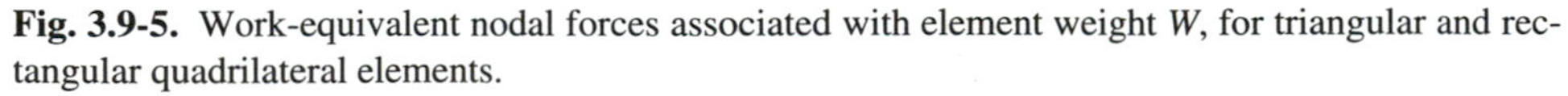

Fig. 3.9-5. Work-equivalent nodal forces associated with element weight W, for triangular and rectangular quadrilateral elements.

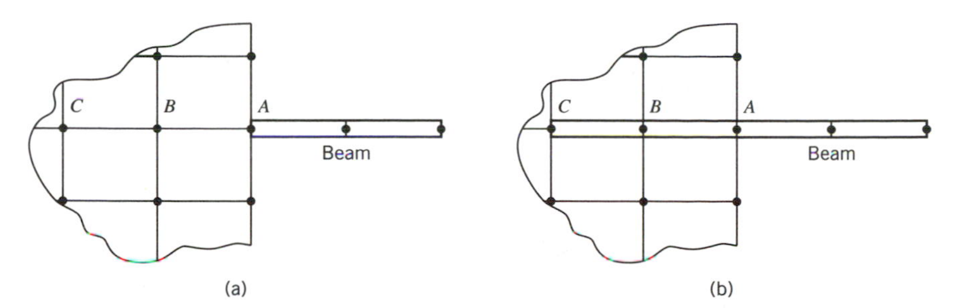

Fig. 3.9-6. Connecting a 2D beam element to plane elements. (a) No moment is transferred. (b) Moment is transferred.

ments. If plane elements have drilling d.o.f. (Section 3.7) the connection of Fig. 3.9-6a transfers moment, although such a connection is not recommended. In any case one should not expect that stresses in the plane body will be accurately calculated near node *A*. An alternative way to connect beam elements and plane elements is noted in Section 4.13.

3.10 STRESS CALCULATION. OTHER REMARKS

Stresses. After nodal d.o.f. have been computed, conventional software calculates stresses by means of Eqs. 3.1-2 and 3.1-8, that is, $\boldsymbol{\sigma} = \mathbf{EBd} + \boldsymbol{\sigma}_0$. This equation is applied element by element, not globally. In general, $\mathbf{B}$ is a function of the coordinates, so the user (or the author of the software) must decide where in the element stresses should be computed. Stresses tend to be more accurate within an element than on its boundary (see Fig. 2.5-4). This is unfortunate because stresses are usually largest at boundaries of the structure, which of course are also boundaries of some elements. Usually it is best to calculate stresses at certain points within an element, then extrapolate from these values to obtain element boundary stresses. This matter is discussed further in Section 4.7.

An alternative method of stress calculation has been devised [3.5], but at the present writing it is not available in commercial software. The alternative method does not use the conventional calculation $\boldsymbol{\varepsilon} = \mathbf{Bd}$. Instead, it computes element nodal forces $\mathbf{r} = \mathbf{kd}$, where element nodal d.o.f. $\mathbf{d}$ are available from $\mathbf{D}$ after solving the global equations $\mathbf{KD} = \mathbf{R}$. Then a least-squares process is used to compute an element stress field that equilibrates $\mathbf{r}$. The alternative method is more complicated than the conventional method but has three important benefits. First, there is better accuracy when loads (rather than nonzero displacements) are prescribed. In contrast to the conventional method, stresses may be at least as accurate as displacements. This happens because nodal forces equilibrate loads applied to the structure and therefore may be exact or nearly so even when displacements are underestimated due to overly stiff elements (e.g., as in Figs. 3.6-1 and 3.11-1c). Second, the method avoids difficulties associated with matching strain fields and temperature fields [3.6], which is described in a simple context at the end of Section 2.6. Third, the method is relatively insensitive to element shape distortions.

Software may report stresses in either local or global coordinates. For example, flexural stress in a beam is reported in local coordinates because by definition flexural stress is a normal stress in the beam's axial direction. As another example, if a plane element is arbitrarily oriented in space its membrane stresses will be reported with reference to local axes *xy* in the plane of the element. The user of software must study the documentation to understand how stresses are presented and what options are available.

Some useful stress quantities are *invariant*; that is, they have the same numerical value in any coordinate system. One such quantity is the von Mises, or "effective," or "equivalent" stress

$$\sigma_e = \frac{1}{\sqrt{2}}\left[(\sigma_1 - \sigma_2)^2 + (\sigma_2 - \sigma_3)^2 + (\sigma_3 - \sigma_1)^2\right]^{1/2} \tag{3.10-1}$$

where σ_1, σ_2, and σ_3 are the three principal stresses at the point in question, with σ_1 the algebraically largest and σ_3 the algebraically smallest. Equation 3.10-1 reduces to $\sigma_e = \sigma_1$ if σ_1 is uniaxial, that is, if $\sigma_2 = \sigma_3 = 0$. Note that σ_e may exceed the magnitude of σ_1 as, for example, when $\sigma_1 = -\sigma_3$. An alternative form of Eq. 3.10-1, which provides the same value of σ_e, can be written in terms of all six nonprincipal stresses (three normal and three shear). Another invariant stress is the "stress intensity" SI,

$$\text{SI} = \sigma_1 - \sigma_3 \tag{3.10-2}$$

which is twice the maximum shear stress. Note that SI is *not* the stress intensity *factor* used in fracture mechanics. In general, one does not associate a direction with σ_e. The planes on which SI acts can be determined, but one usually does not care what they are. Both σ_e and SI are used in failure theories, which state that yielding begins when σ_e or SI (depending on the theory) reaches a limiting value.

Because σ_e represents the *entire* state of stress, contours of σ_e are often plotted and examined for their interelement continuity, as a way to visually estimate the discretization error of computed stresses. Contours of (say) σ_x might be similarly informative in one part of a FE model but not in another part because a stress other than σ_x is dominant there. Symmetry of the FE model and its loads and supports provides symmetry of σ_e contours but may not provide symmetry of contours of any particular stress that contributes to σ_e.

As an option in most software, stresses may be averaged at nodes. Thus if n elements meet at a node, the n values of (say) σ_x are added and the sum divided by n. Sometimes contributions to the sum are weighted, by element volume, proximity of the element centroid to the node, or some other factor. At nodes interior to the mesh, the average stress may be the most accurate stress that the current discretization can provide. At nodes on the boundary of the mesh, greatest stress accuracy is usually provided by extrapolation, using a polynomial field fitted to stress values at several nearby points, including interior nodes and/or points within elements.

However, there are good reasons *not* to average stresses at nodes. Two parts joined by a shrink fit have different normal stresses in directions tangent to the interface. An average stress would not represent the actual stress on either side of the discontinuity. A discontinuity of thickness or modulus also causes a discontinuity in stress. As examples, in Fig. 3.10-1a, σ_x would be discontinuous at $x = 0$ because an x-direction force is applied to different cross-sectional areas. In Fig. 3.10-1b, σ_y would be discontinuous at $x = 0$ because both parts have the same ε_y but $E_1\varepsilon_y \neq E_2\varepsilon_y$. In Fig. 3.10-1c, different coordinate systems are used for stress computation, and an average such as $(\sigma_x + \sigma_n)/2$ would make no sense for a node on interelement boundary AB. Finally, stress contours based on nodal average stresses are interelement-continuous and are thus deprived of error information. Stress contours plotted from *un*averaged stresses have discontinuities at interelement boundaries. The amount of discontinuity is a qualitative measure of whether or not mesh refinement is adequate (see Fig. 1.3-2).

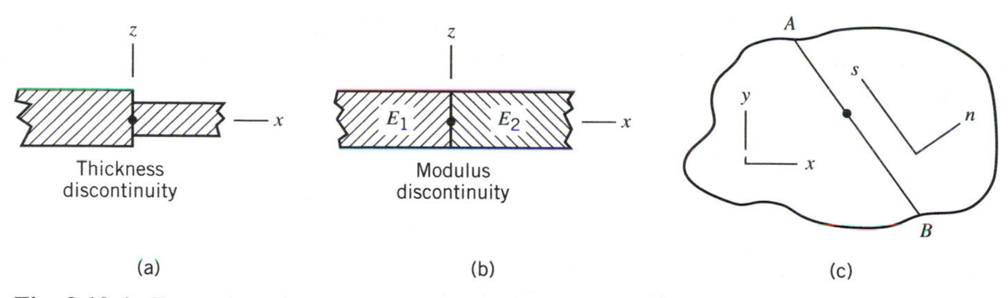

Fig. 3.10-1. Examples of situations in which stresses should *not* be averaged at a node. (a,b) Plane elements seen in cross section, with Cartesian coordinates *xyz*. (c) Plane elements seen in plan view, with interelement boundary *AB*.

Element Connections. Elements of different types can be connected to one another, but not in completely arbitrary fashion. Some unacceptable connections appear in Fig. 3.10-2. The "connection" at *A* is in fact no connection at all because a CST element has no side node. Connections between a two-node edge (CST and Q4 elements) and a three-node edge (LST and Q8 elements) should be avoided, because the side node is left unconnected, and clearly there is a mismatch in displaced shapes of adjacent edges: one is straight, the other parabolic (line *BC* in Fig. 3.10-2). Two two-node edges should not be connected to one three-node edge because two straight edges do not match a parabolic edge (lines *CD* and *EF*). Also, three-node edges should not be connected so that side nodes are joined to corner nodes: both edges deform as parabolas, but in general they are *different* parabolas (along line *GH*). The Q6 element is by nature incompatible but becomes compatible with mesh refinement. Connections like those in Fig. 3.10-2 are not fatal. They cause poor results locally but the effect dies away with distance, in accord with Saint-Venant's principle. Nevertheless, there is the danger that artificial local stress disturbances will be mistaken for actual physical behavior. One could make most connections in Fig. 3.10-2 "legal" by constraining three-node edges to remain straight, but then d.o.f. of the side node would be rendered useless. (Most software allows the user to impose such constraints.)

Elements with side nodes can be formulated in a way that allows any number of side nodes to be deleted. In Fig. 3.10-2, for example, the side node along *BC* could be deleted from the LST formulation, so that adjacent element sides along *BC* would both be two-node sides and would deform as straight lines. Accordingly, if done properly, elements of many differing types can be connected (e.g., Fig. 1.2-1).

Supports. Plane elements have no resistance to forces normal to their plane and no resistance to nodal moment loads (unless the elements have drilling d.o.f.). Accordingly, out-of-plane translation and all rotations must be suppressed at all nodes of a plane FE model. A moment load, if present, must be applied as equivalent couple-forces on a pair of

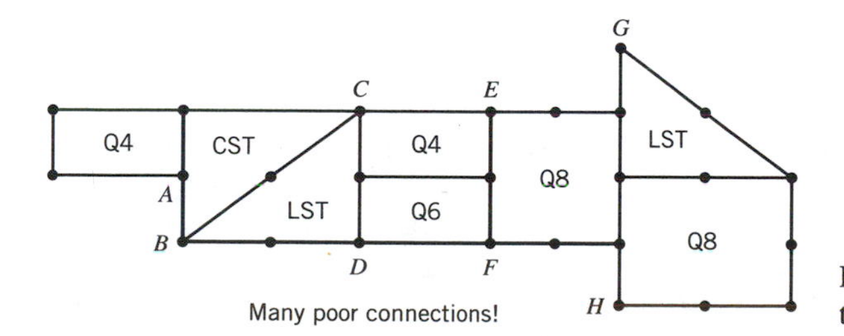

Fig. 3.10-2. Examples of how *not* to connect elements.

nodes. Elements remain able to rotate in the analysis plane. Also, nodes lie in the analysis plane and do nothing to inhibit transverse normal strain associated with the Poisson effect in a plane stress problem.

Other Element Types. Thus far we have used as nodal d.o.f. only "low-order" quantities, such as displacement d.o.f. in bar, plane, and solid elements, and displacement and rotation d.o.f. in beam, plate, and shell elements. Theory permits any number of d.o.f. per node. For example, we could formulate a bar element that uses axial displacement u and axial strain du/dx as d.o.f. at each node. Elements having such "extra" d.o.f. are not usually found in commercial software. Most current software is not structured to accommodate more than the essential number of d.o.f. per node. Also, interelement continuity of derivatives is not always proper (consider du/dx in Fig. 3.10-1a). Finally, stress boundary conditions may dictate relations among the extra d.o.f. at a boundary node but not their numerical values, which is awkward.

Elements discussed thus far, and indeed most elements in common use, are based on displacement fields such as Eqs. 3.2-1 and 3.3-1. There are other formulation methods, many based on simultaneous use of separate fields for displacement and stress. Such elements have displacement d.o.f. and the user may be unaware of the nature of the element. In any case the user should study the software documentation and try some simple test problems in order to understand how an element behaves before using it in applications.

3.11 COMPARATIVE EXAMPLES

Plane elements of different types can be compared by using them to solve a particular problem. We caution that a single problem does not tell all: an element type best in one problem may not be best in another. Also, different software may contain implementations of a given element that differ because of minor adjustments (that we have not discussed) whose purpose is to enhance element behavior. The reader is encouraged to try additional meshes for the following problem, and to try other problems as well. A particular suggestion is that FE be used to solve stress concentration problems, for which almost-exact results are widely available. Such problems illustrate the effects of element type, size, aspect ratio, and mesh layout.

The test problem chosen here is that of Fig. 3.11-1a, a cantilever beam of unit thickness loaded by a transverse tip force. Loads, properties, and dimensions are assumed to be in a consistent set of units. Plane stress conditions prevail. In the calculation of tip deflection, 6/5 is the standard transverse shear deformation factor for a rectangular cross section. In Fig. 3.11-1b, nodal loads on the quadratic edge come from Fig. 3.9-3c with $q_A = q_C = 0$, in accordance with the parabolic distribution of transverse shear stress that beam theory predicts. Support conditions are consistent with a fixed end but without restraint of y-direction deformations associated with the Poisson effect. Stresses are calculated in the conventional way, using Eqs. 3.1-2 and 3.1-8.

With only two nonzero d.o.f., the simple plane beam element of Section 2.3 solves the problem exactly when transverse shear deformation is included in its formulation. As expected, CST elements perform poorly. Q4 elements are better but not good. LST elements give an accurate deflection but a disappointing stress. Q6 and Q8 elements are the best performers. In the rectangular Q6 element, the stress σ_x at $x = 1$ (midway between nodes) is exact, but since σ_x is independent of x in this element, the same σ_x is reported at node B. In most of these FE models, distortion and elongation of elements are seen to reduce accuracy. In the latter Q8 example the amount of distortion is sufficient to provoke a

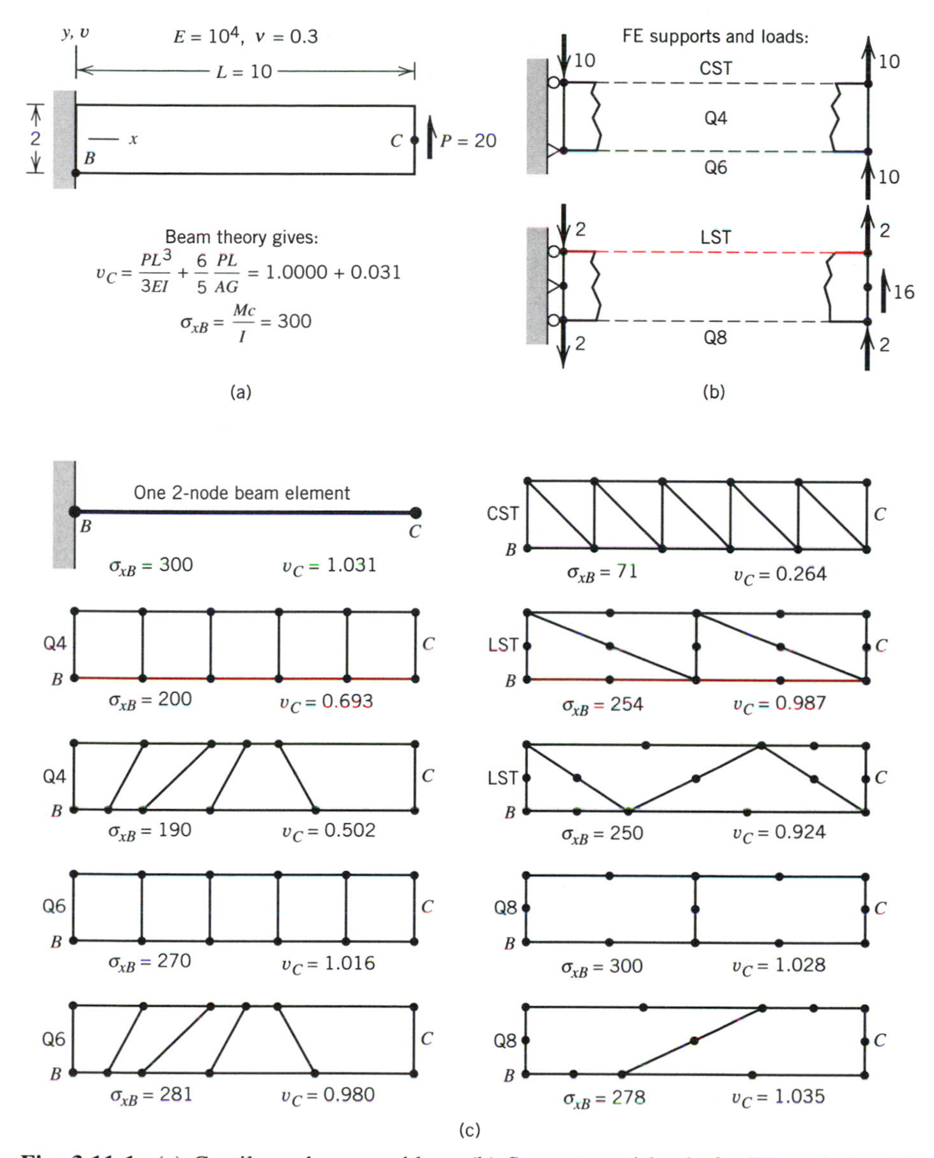

Fig. 3.11-1. (a) Cantilever beam problem. (b) Supports and loads for FE analysis. (c) Results from models built of various types of plane elements.

warning from the software. Further distortion of elements may cause accuracy to decline *precipitously*, not gradually as one might expect. Some arrangements of Q4 and Q6 elements are prone to locking [3.2, 3.7].

3.12 AN APPLICATION

A flat square plate contains a central circular hole, which is loaded by pressure p. The geometry and elastic properties are depicted in Fig. 3.12-1a. Plane stress conditions prevail. Magnitudes and locations of maximum principal stress are desired. The solution strategy suggested in Section 1.3 is used in the following analysis.

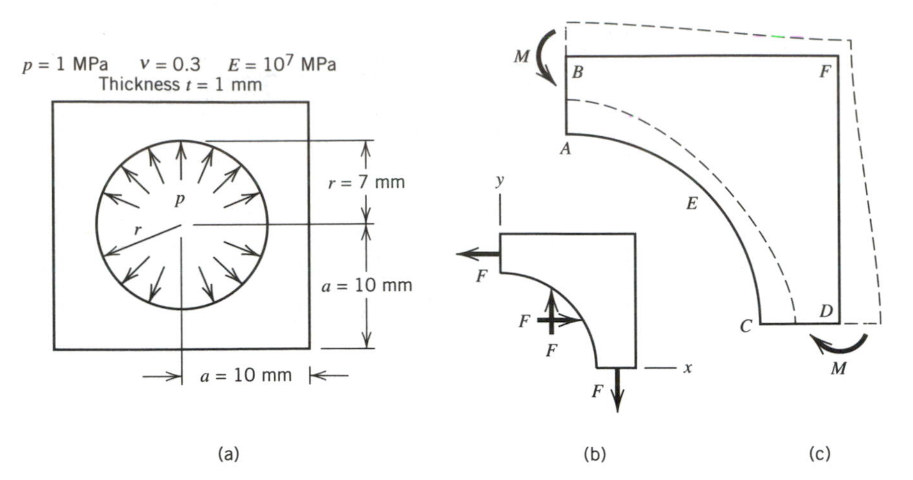

Fig. 3.12-1. (a) Flat plate with central hole loaded by internal pressure. (b) Forces that act on one quadrant. (c) The anticipated displaced shape, greatly exaggerated, is shown by dashed lines.

Preliminary Analysis. Before undertaking a FE analysis we examine the problem in a physical way and make simple calculations, in order to anticipate where stresses will be largest, prepare a good FE model, and obtain approximate results for subsequent comparison with FE results.

Structure, geometry, and loading are all symmetric with respect to horizontal and vertical centerlines. This means that deflections and stresses will have the same symmetries and we can consider a single quadrant (symmetry is discussed in Section 4.12). Forces F that act on a representative quadrant are shown in Fig. 3.12-1b. It is easy to calculate F *exactly* by statics. The average normal stress on horizontal and vertical cross sections then follows.

$$F = prt = 7\,\text{N} \quad \text{and} \quad \sigma_{\text{ave}} = \frac{F}{(a-r)t} = 2.3\ \text{MPa} \tag{3.12-1}$$

Deformations must be symmetric with respect to horizontal and vertical centerlines, and we expect that pressure will push the slender parts further outward than the more massive corners. Accordingly, we anticipate the deformed shape shown in Fig. 3.12-1c. We see that the slender parts have acquired an inward curvature, which must be associated with bending moments M in the directions shown. The associated flexural stresses will be tensile on the outside, compressive on the inside, and will add algebraically to stress σ_{ave} of Eq. 3.12-1. Therefore it appears that the maximum stress may appear at B and D rather than at A and C. But there is another possibility: because arc AEC bends outward there will be tensile flexural stress at E. Therefore point E is another candidate for the location of maximum stress.

FE Model and Analysis. We might choose to model only one octant, because there is symmetry with respect to diagonals as well as centerlines. However, we choose instead to model a quadrant because support conditions are more straightforward and computed results can be checked for anticipated symmetries about the diagonal. We arbitrarily elect to use Q6 elements (Section 3.6), formulated in a way that permits nonrectangular shapes. For the sake of illustration, we deliberately choose a *very coarse* mesh for the initial FE

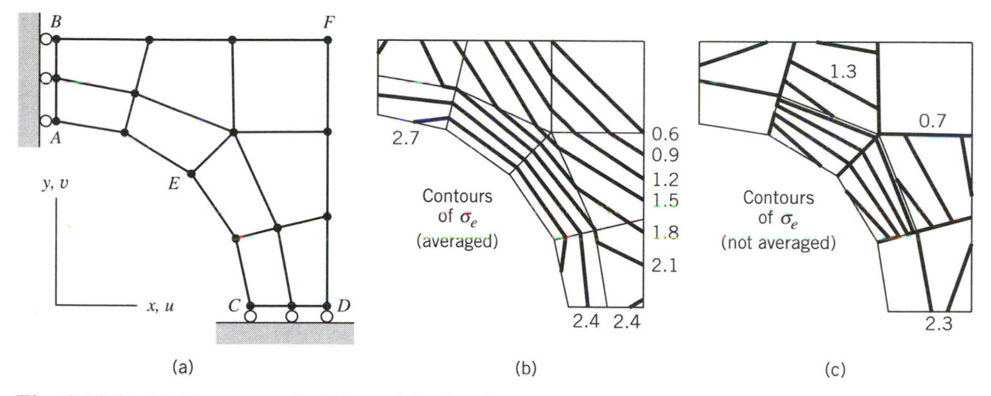

Fig. 3.12-2. (a) Coarse-mesh FE model, showing support conditions. (b) Contours of σ_e, from nodal average values. Stress units are megapascals (MPa). (c) Contours of σ_e, without nodal averaging, from individual elements.

model. The model and its support conditions are shown in Fig. 3.12-2a. The mesh is symmetric about diagonal *EF* and is coarsest near corner *F* where stresses are certain to be low because *F* is a free corner. With the software used, nodal loads associated with pressure along *AEC* are calculated automatically, and the only support conditions that the user need impose explicitly are $u_i = 0$ at nodes i along *AB* and $v_i = 0$ at nodes i along *CD*. Nodal translations w_i and all nodal rotation d.o.f. are automatically suppressed by the software used when it is told that the model is plane.

Critique of FE Results. Computed displacements are examined first, scaled up so as to be easily visible, and animated. Thus, on the computer screen, we see that nodes along *AB* have only *y*-direction displacement, nodes along *CD* have only *x*-direction displacement, all displacements are symmetric about diagonal *EF*, and the anticipated displaced shape indicated by dashed lines in Fig. 3.12-1c is indeed obtained. These results are in accord with the model we intended to describe to the software, so no blunder is yet in evidence. We postpone discussion of the maximum stress until after results from a finer mesh have been obtained. We qualitatively examine contour plots of the von Mises stress σ_e (defined in Eq. 3.10-1). Contours plotted from nodal average stresses and from element-by-element (unaveraged) stresses are shown in Figs. 3.12-2b and 3.12-2c. As expected, contours of σ_e are symmetric about diagonal *EF*. Aside from reflecting the coarseness of the mesh, averaged contours give little indication that results are unreliable. But unaveraged contours show *severe* interelement discontinuities. Interelement changes in stress are comparable in magnitude to the stresses themselves! It is now obvious that the coarse-mesh FE results are not to be trusted.

The quadrant is now modeled by a finer mesh, again using Q6 elements. The same support conditions as before are imposed on nodes along *AB* and *CD*. This time, just to see what happens, the mesh is made unsymmetric about the diagonal. Note that elements are smallest near points *A*, *B*, *C*, *D*, and *E*, where the largest stresses are expected, and elements are largest near *F*, where stresses are known to be low. Note also that elements near *E* are "squashed" in the radial direction because Fig. 3.12-2 suggests that stress gradients are much higher in the radial direction than in the circumferential direction. The displaced shape on the computer screen again appears satisfactory. Numerical values of nodal d.o.f. u_i and v_i are found to be not quite symmetric about the diagonal owing to asymmetry of the mesh. Averaged and unaveraged plots of von Mises stress σ_e are shown in Fig. 3.12-3. Results are greatly improved over the coarse mesh, but unaveraged con-

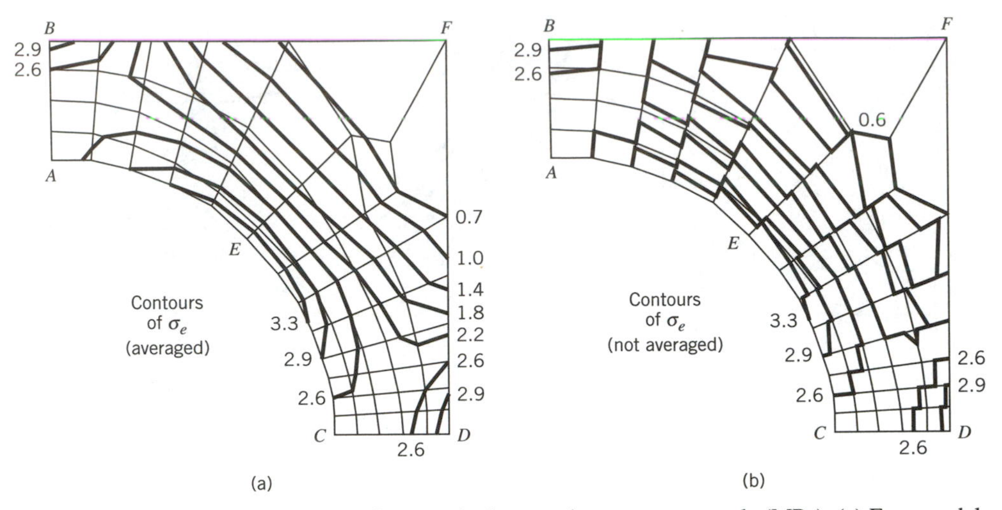

Fig. 3.12-3. Contours of σ_e from a finer mesh. Stress units are megapascals (MPa). (a) From nodal average values. (b) Without nodal averaging, from individual elements.

tours still have significant interelement discontinuity. Even averaged contours show appreciable changes in direction where they cross interelement boundaries near *AB* and *CD*. Also, contour lines lack symmetry about diagonal *EF* and do not intersect lines of symmetry *AB*, *CD*, and *EF* at 90° angles. All this suggests a need for even more mesh refinement.

Numerical results from both meshes are listed in Table 3.12-1. These numbers are obtained directly from output files, not by visual inspection of displacement plots and stress plots. Displacement results are reasonable; they show that *AB* and *CD* have shortened, as should be expected from the combination of compressive radial loading and the Poisson effect with circumferential tension. Computed displacements also show that the finer mesh is more flexible than the coarse mesh. Such is usually the case but cannot be guaranteed for the Q6 element because it is an incompatible element (see Section 4.8). At corner *F*, all stresses are zero according to theory. Computed values of σ_1 at *F* are small and decrease with mesh refinement, as expected. At a point such as *A*, σ_x (not shown in Table 3.12-1) is found to be *almost* equal to σ_1 at *A*. Theoretically, $\sigma_x = \sigma_1$ at *A*. The discrepancy is due to τ_{xy}, which is small but not quite zero as theory says it should be on an axis

TABLE 3.12-1. Selected displacements and maximum principal stress σ_1 in the FE models of Fig. 3.12-2 (coarse mesh) and Fig. 3.12-3 (finer mesh). Displacements are in mm. Stresses are in MPa.

Node	Coarse Mesh			Finer Mesh		
i	$10^6 u_i$	$10^6 v_i$	σ_1	$10^6 u_i$	$10^6 v_i$	σ_1
A	0	2.08	2.11	0	2.28	1.92
B	0	1.78	2.28	0	1.96	3.01
C	2.08	0	2.11	2.34	0	1.84
D	1.78	0	2.28	2.01	0	3.12
E	1.22	1.22	2.68	1.27	1.24	3.16
F	0.97	0.97	0.38	0.98	1.00	0.21

of symmetry. The average normal stress from points A and B or C and D is in satisfactory agreement with Eq. 3.12-1. The largest σ_1 anywhere in the structure is at A and C or at E (we cannot be sure which without more refinement) and has a numerical value of about 3.1 MPa.

In summary, computed results are reasonable but as yet we cannot trust them. An error measure for the stress field, discussed in Section 5.16, gives $\eta = 0.373$ for the coarse mesh and $\eta = 0.183$ for the finer mesh. These measures also indicate that results are not yet to be trusted. Another mesh refinement is called for. The next mesh should build on information in Fig. 3.12-3 by making elements smallest where stresses and stress gradients are largest. By plotting a particular stress or a particular displacement versus element size, as computed from three or more meshes, one could extrapolate to zero element size, and thus obtain a predicted result for infinite mesh refinement (see Section 5.15). Hence the percentage error of a result from a given mesh can be estimated.

We may now admit that a FE analysis probably is not needed: a solution of the problem appears in [3.8], where we find the experimentally determined values $\sigma_1 = 2.9$ MPa at B and D and $\sigma_1 = 2.7$ MPa at E. It is wise to ask at the outset if a FE analysis is really necessary, as it is not a trivial task.

Related Problems. If the problem is changed to one of plane strain rather than plane stress, computations eventually fail as Poisson's ratio ν approaches 0.5. By trial, it was found that coarse-mesh Q6 element results in plane strain were reasonable up to $\nu = 0.499999990$ but ridiculous when $\nu = 0.499999999$. In plane stress or plane strain, and for any value of ν, if ligament thickness $a - r$ becomes much less than a, the problem becomes inherently nonlinear because then stresses in ligaments are strongly influenced by the displaced shapes of ligaments, and the displaced shapes are not known in advance. A linear solution, as used in the foregoing example, presumes that displacements do nothing to alter the way load is carried.

ANALYTICAL PROBLEMS

3.1 (a) Over a distance dx, stress σ_x changes by the amount $(\partial\sigma_x/\partial x)dx$ as shown in the sketch. Stresses σ_y and τ_{xy} experience similar changes, over distances dx and dy. Force is stress times area, and thickness is constant. Take these remarks as suggestions and derive the plane equilibrium equations, Eqs. 3.1-11.

(b) Repeat part (a) but work in three dimensions. Thus there are three normal stresses and three shear stresses, and there are three equilibrium equations.

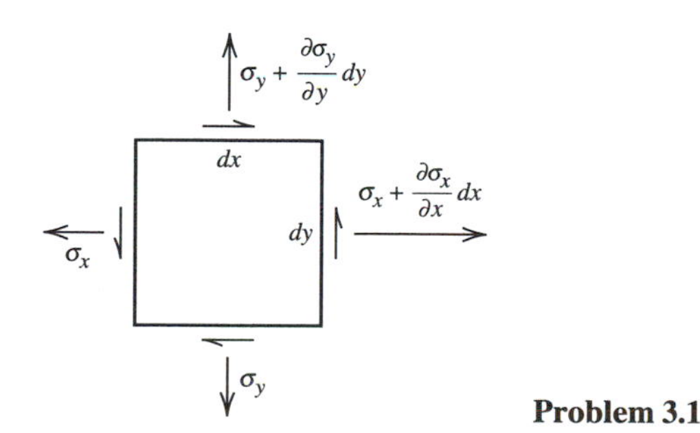

Problem 3.1

3.2 The cantilever beam shown is tip-loaded by a moment M. Assume that Poisson's ratio is zero. Use beam theory to compute the displacement components of points D, E, and F. Regard these results as nodal displacements, and use them to compute stresses in elements defined as follows.
(a) A CST element whose nodes are A, E, and C.
(b) A CST element whose nodes are B, D, and F.
(c) A Q4 element whose nodes are A, D, F, and C.
Express the stresses in terms of M, L, c, and thickness t.

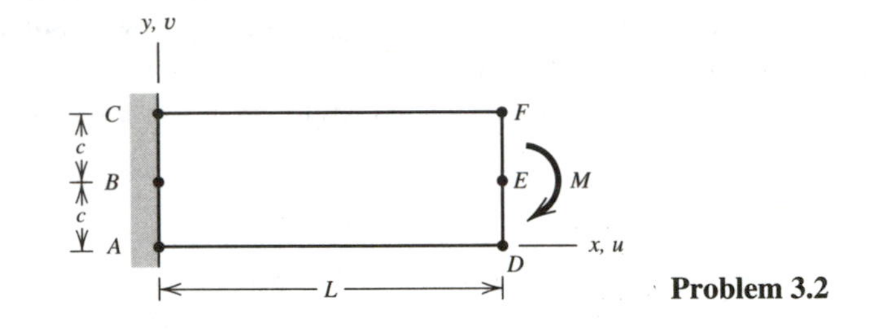

Problem 3.2

3.3 Repeat Problem 3.2 but replace moment M by a tip force P in the y direction. Neglect transverse shear deformation.

3.4 Evaluate the stresses in Fig. 3.2-2b, as suggested in the last sentence of Section 3.2. Let Poisson's ratio be zero. Suggestion: obtain u_2 and v_2 from beam theory. Will the FE model actually provide these deformations?

3.5 Let the cantilever beam of Fig. 3.2-2a be modeled by LST elements rather than CST elements. Apply a transverse tip force in the y direction. Will computed results be exact? Why or why not?

3.6 (a) For the element shown, determine shape function N_3 in terms of y and b (see Fig. 3.3-1 for a hint).
(b) Shape function N_4 for this element is $N_4 = 1 - (y/b) - (x/a)^2 + (y/2b)^2$. Show that N_4 is unity at node 4 and zero at the other five nodes.
(c) Let u_3, v_3, u_4, and v_4 be the only nonzero d.o.f. In terms of these d.o.f., x, y, a, and b, what are the element strains?

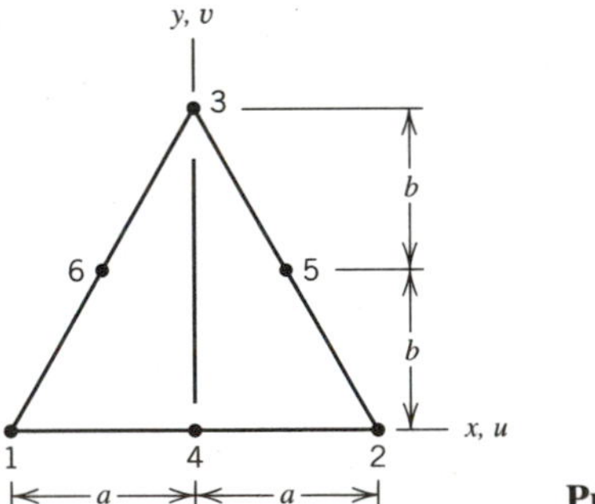

Problem 3.6

3.7 Generalized coordinates β_i can be expressed in terms of nodal d.o.f. by substitution: for example, in Eq. 3.2-1 we obtain $u_i = \beta_1 + \beta_2 x_i + \beta_3 y_i$, where $i = 1, 2, 3$. Thus we obtain a set of three equations that can be solved for β_1, β_2, and β_3.
(a) Write this set of equations in matrix format for the CST of Fig. 3.2-1. Do not bother to solve.

(b) Write an analogous set of four equations for the Q4 element of Fig. 3.4-1.
(c) By comparison of Eqs. 3.4-1 and 3.4-3, write the β_i in the first of Eqs. 3.4-1 in terms of nodal d.o.f. u_i.

3.8 A uniform beam is modeled by Q4 elements, as shown. Qualitatively, and without calculation, plot σ_x and τ_{xy} along the top edge from A to C, as predicted by the elements. Also plot the exact stresses according to beam theory. Consider each of the following loadings.
(a) $F_1 = 0, \quad F_2 = F_3$.
(b) $F_1 = 0, \quad F_2 = -F_3$.
(c) $F_1 > 0, \quad F_2 = F_3 = 0$.
(d) Repeat parts (a), (b), and (c) with Q6 elements.

3.9 Let axes x and y originate at node 1 of a Q4 element, as shown. For this choice of axes write appropriate shape functions N_1 through N_4, analogous to the shape functions in Eq. 3.4-3.

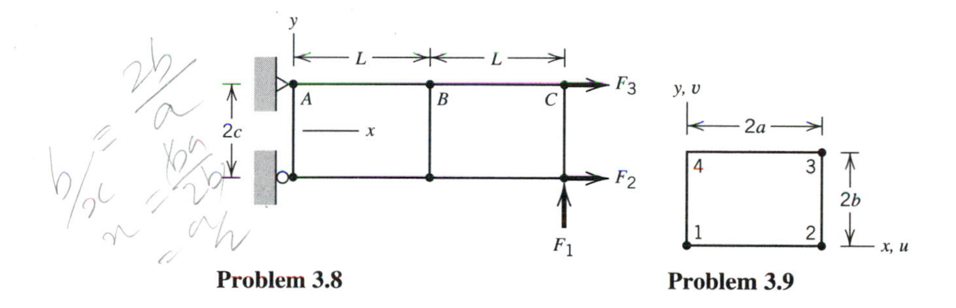

Problem 3.8 **Problem 3.9**

3.10 Imagine that nodal d.o.f. in Fig. 3.4-2c are $u_1 = u_3 = -c$, $u_2 = u_4 = c$, where c is a constant, and all $v_i = 0$. Use Eq. 3.4-4 to express element strains in terms of a, b, c, x, and y.

3.11 Let a Q8 element be a 2 by 2 square, so that $a = b = 1$ in Fig. 3.5-1. According to Eq. 3.5-3, what element strains are associated with nodal d.o.f. u_2, v_2, u_6, and v_6?

3.12 Imagine that nodal d.o.f. in a rectangular Q6 element are $v_1 = v_3 = c$, $v_2 = v_4 = -c$, where c is a constant, and all $u_i = 0$. If the element is a Q6 element, what is g_2 in Eq. 3.6-2?

3.13 For $0 \le x \le 2a$ in the sketch, lateral displacement v depends on v_1, v_2, and v_3. Shape functions for these d.o.f. are provided in the sketch. Obtain the four shape functions associated with d.o.f. v_1, v_3, θ_{z1} and θ_{z3}, where the two drilling d.o.f. replace v_2. Note: this is an exercise in manipulation, not a physical problem.

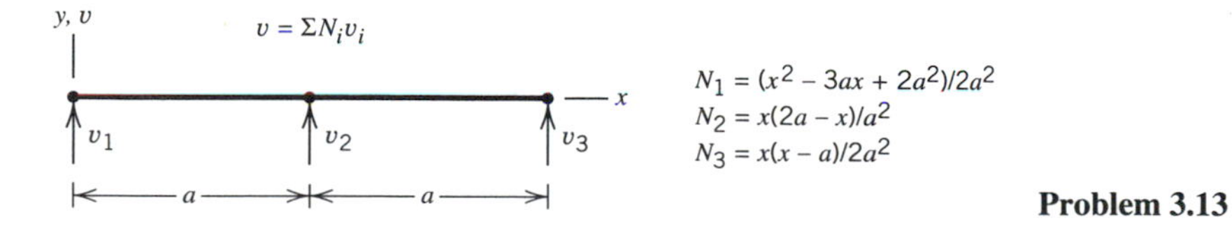

Problem 3.13

3.14 Let a plane element have three or more sides, side lengths L_i , midside normal displacements u_{ni} , and all corner translational d.o.f. set to zero. Show that use of drilling d.o.f. implies the constraint $\Sigma(u_{ni}/L_i) = 0$, where the summation includes all sides.

3.15 For rectangular elements, are Q4 and Q6 results in Fig. 3.11-1 in approximate agreement with Eq. 3.6-1? Suggestion: the coefficient of M_1 in Eq. 3.6-1 can be regarded as a factor that relates displacements in Q4 and Q6 elements subjected to the same bending moment.

3.16 Show that the nodal loads are work-equivalent in (a) Fig. 3.9-1, (b) Fig. 3.9-3, with $q_A = q_B = q_C$, (c) Fig. 3.9-5b, and (d) Fig. 3.9-5d.

√ **3.17** Show that nodal forces calculated according to Fig. 3.9-3c are statically equivalent to the following loadings on a plane body of unit thickness.

(a) $q_A = \sigma$, $q_B = 0$, and $q_C = -\sigma$ (corresponding to a flexural stress distribution with B on the neutral axis).

(b) $q_A = 0$, $q_B = \sigma/2$, $q_C = \sigma$ (corresponding to a flexural stress distribution with A on the neutral axis).

(c) $q_A = q_C = 0$, $q_B = \tau$, where τ acts tangent to ABC (corresponding to a shear stress distribution on a beam of rectangular cross section).

3.18 On a straight linear element edge, what $q = q(x)$ is equivalent to a concentrated nodal force? For example, set $F_A = 0$ in Fig. 3.9-1.

COMPUTATIONAL PROBLEMS

In the following plane problems compute significant values of stress and/or displacement, as appropriate. Exploit symmetry if possible. Choose convenient numbers and consistent units for material properties, dimensions, and loads. When mesh refinement is used, estimate the maximum percentage error of FE results in the finest mesh. Unless directed otherwise, assume unit thickness, plane stress conditions, and isotropic materials.

A FE analysis should be preceded by an alternative analysis, probably based on statics and mechanics of materials, and oversimplified if necessary. If these results and FE results have substantial disagreement we are warned of trouble somewhere.

3.19 The rectangular structure shown may be modeled by Q4, Q6, or Q8 elements. Use elements of approximately the shape shown. Space nodes uniformly along the right edge, where uniform pressure p is applied. Do computations twice: first represent p by equal nodal forces, then represent p by forces computed from Fig. 3.9-1 or Fig. 3.9-3, as appropriate.

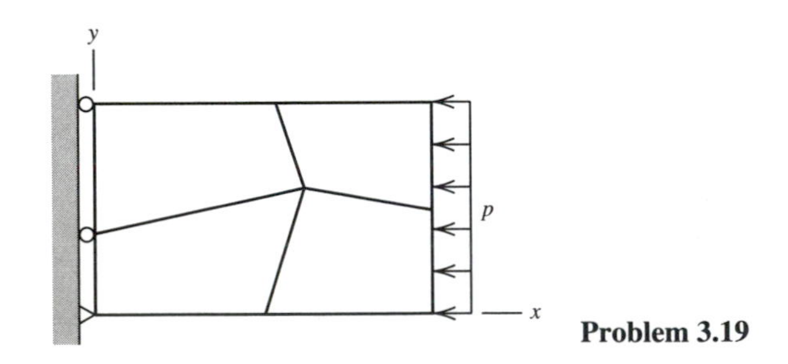

Problem 3.19

3.20 In Problem 3.19, let $p = 0$ and apply instead the temperature field $\Delta T = cx$, where c is a constant. Then repeat the calculations, using $\Delta T = cy$. Are results reasonable? If results differ for the two thermal loadings, explain why.

3.21 In Problem 3.19, use work-equivalent nodal loads but revise the mesh so as to include one or more of the "poor connections" of Fig. 3.10-2. Use additional elements as necessary but maintain the rectangular shape of the body. (With a fine mesh one can study the degree to which the effect of the connection is localized.)

3.22 One can undertake a systematic study of the effects of mesh distortion [3.7, 3.9]. For example, if the beams shown are modeled by Q4 or Q6 elements, one could vary ϕ (or ℓ/H) while keeping ℓ/H (or ϕ) constant. Using LST or Q8 elements, one could vary ϕ or s/H. Loading may be by tip moment or transverse tip force.

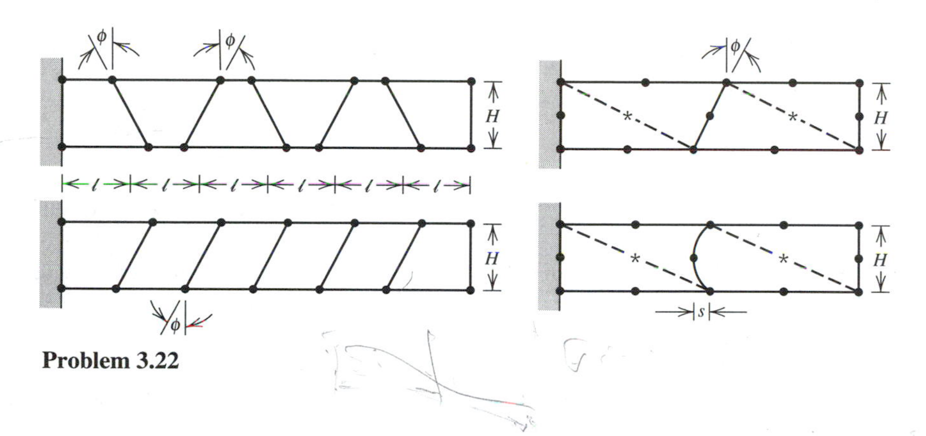

Problem 3.22

3.23 As a variant of Problem 3.22, use plane strain conditions, maintain a chosen mesh distortion, and investigate what happens as Poisson's ratio approaches 0.5.

3.24 Part (a) of the sketch shows a curved beam under pure bending load, as modeled by a single Q8 element having two curved sides. Alternative Q8 element models can have straight sides. The parallel sides may be tangent to arcs as in part (b), chords of arcs as in part (c), or something in between. Also, angle θ and the radius ratio r_0/r_i may be varied. By calculation, examine the relative merits of these FE models. For comparison, an analytical solution for circumferential and radial stresses is well known [1.5, 2.1].

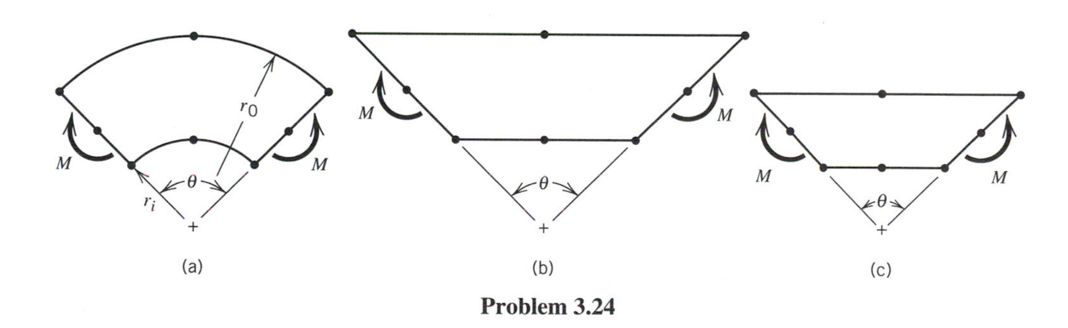

Problem 3.24

3.25 The rectangular plate shown contains a hole and is securely bonded to a rigid base on the bottom and to a rigid bar on the top. Peak values of σ_e (Eq. 3.10-1) in the neighborhood of the hole are desired. Load the model by translating the rigid bar, an amount (a) u_0 in the x direction, or (b) v_0 in the y direction.

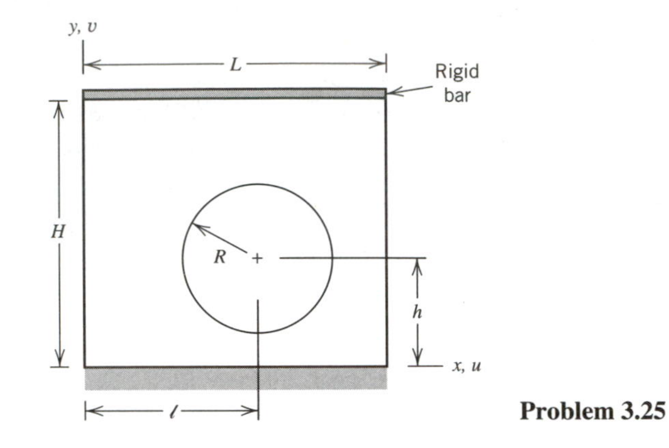

Problem 3.25

3.26 (a) As a variant of Problem 3.25, reinforce the hole by doubling the z-direction thickness of material from the edge of the hole out to a radius cR, where constant c might be 1.3 or so.

(b) As another variant of Problem 3.25, instead of doubling the thickness of material out to radius cR, uniformly raise the temperature of this material. Do not change the temperature of the remainder of the structure, but omit the rigid bar on top.

3.27 A centrally loaded beam is supported at both ends, as shown. Compute flexural stress σ_x at $x = 0$ on top and bottom surfaces. Compare FE results with the flexure formula $\sigma_x = My/I$.

(a) Choose numerical values such as $P = 1$, $L = 12$, and various values of H in the range $4 \leq H \leq 36$ [1.5]. Build the FE model using plane elements.

(b) Model the structure by a minimal number of 2D beam elements. Compare deflections with deflections obtained by use of plane elements in part (a).

(c) Repeat part (a) but make the right-hand support like the left, that is, impose both x- and y-direction restraint at both supports.

3.28 (a) The "fixed support" of a cantilever beam must in reality be elastic. Assume that the cantilever beam shown is attached to a very large plane body having the same thickness and elastic properties as the beam. By what amount is the tip rotation $\theta_z = ML/EI$ of the beam increased by deformation of its support [1.5, 3.10]?

(b) Investigate the beam-to-plane connection shown in Fig. 3.9-6b. How well does it model the elasticity of the beam's support?

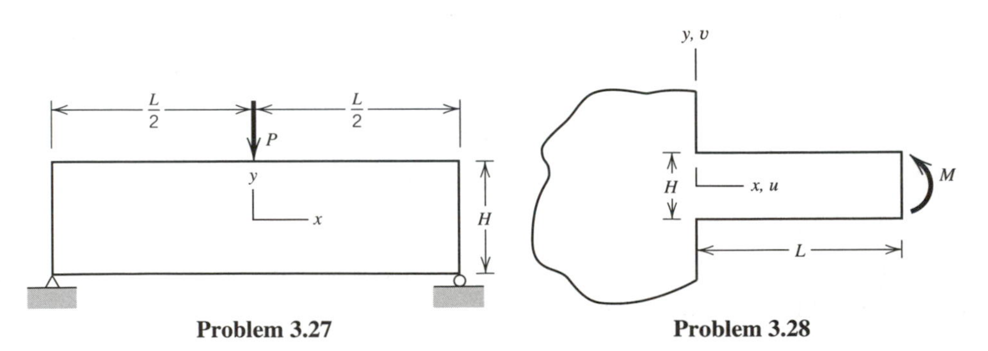

Problem 3.27 **Problem 3.28**

3.29 The two blocks of heights H_1 and H_2 shown are securely bonded together. Their material properties are identical except that their coefficients of thermal expansion α_1 and α_2 are different. Stresses due to uniform heating are desired. Supports (not shown) apply no force. Suggestions for checking: If H_1 and H_2 are much smaller than L, is it reasonable that stresses are independent of x except near ends? By inspection, what is τ_{xy} at $x = y = 0$ and at $x = L$, $y = 0$? What is a probable upper bound for the magnitude of any normal stress, for example, with $\nu = 0$ to make it simpler?

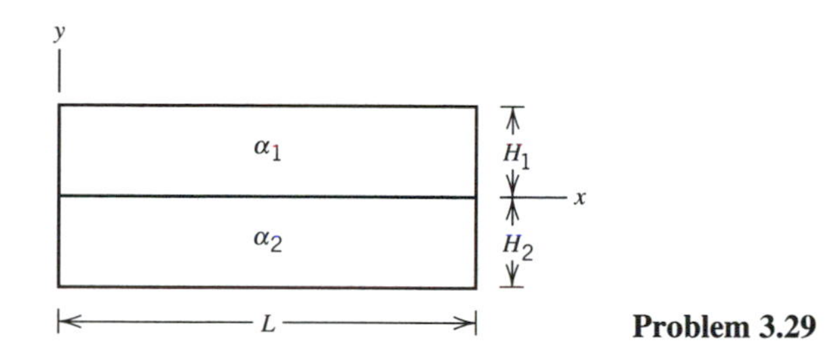

Problem 3.29

3.30 A horizontal elastic medium (e.g., soil or rock) is loaded by its own weight. Assume that the initial state of stress is hydrostatic, as in a fluid. Next, excavate a vertical cut of height H and/or a tunnel of radius R, as shown. What is the change in the state of stress and the final state of stress? Assume that plane strain conditions prevail. For an alternative initial state of stress, consider uniform compression in the x direction only.

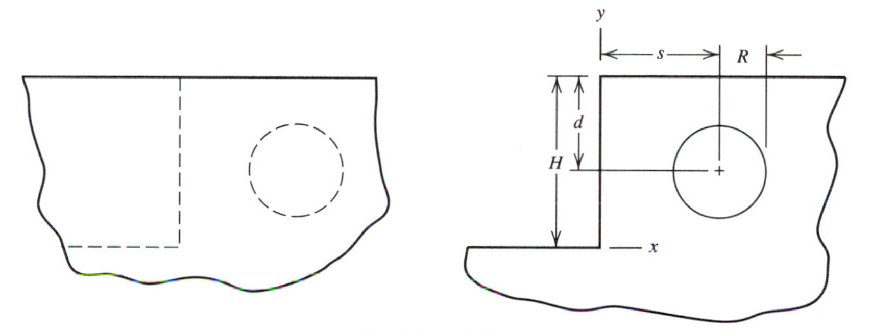

Problem 3.30

3.31 The structures shown consist of bars of square cross section (shown by double lines) securely connected to flat sheets of the same material. Centerlines of bars and midsurfaces of sheets lie in the same plane. Let FE models consist of plane elements of thickness t and bar elements of cross-sectional area A. In sketch (a), $H_1 + H_2 \approx 3L$, $L \approx 500t$, and $A \approx (H_1 + H_2)t$ are suggested. Is it reasonable to neglect the bending stiffness of the bars? Find out by repeating the analysis with bending stiffness included.

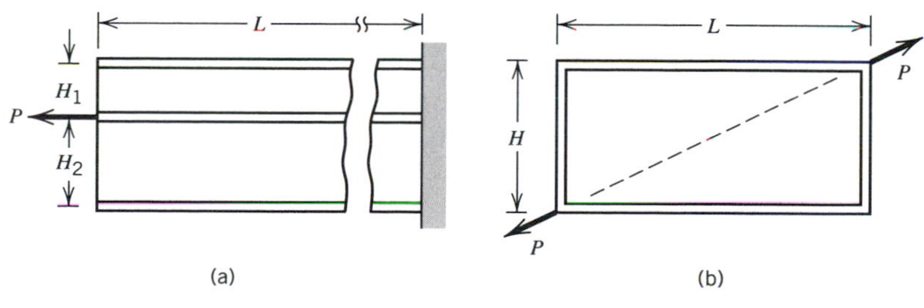

Problem 3.31

3.32 A thick-walled circular cylinder under internal or external pressure can be modeled by a row of plane elements, with two boundaries of the mesh constrained to displace only radially (see sketch). Should the mesh spacing be uniform in the radial direction, as shown? What is the effect of changing r_i/r_0? How many elements are needed for (say) 95% accuracy? What should angle θ be, and what happens if it is too large?

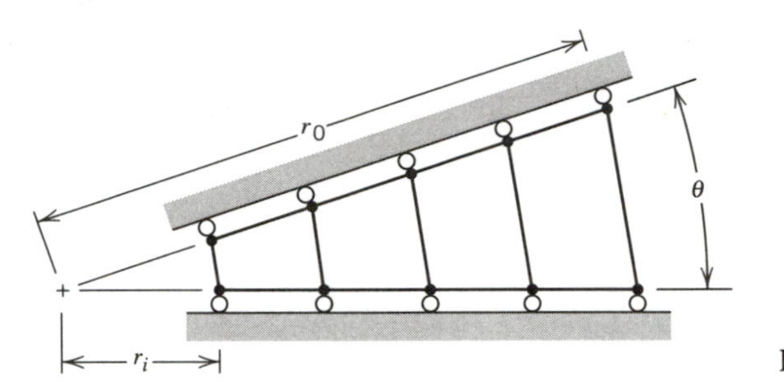

Problem 3.32

3.33 Analyze the structure depicted in Fig. 1.3-1.

3.34 Problems related to the example problem in Section 3.12 are as follows.

(a) Refine the mesh yet again and obtain reliable values of σ_1.

(b) Model an octant of the structure rather than a quadrant.

(c) Choose other values of r/a, or let the outline be a rectangle rather than a square.

3.35 Any of the preceding computational problems can be modified by making the material orthotropic. For a simple choice, with n and s orthogonal principal axes of the material, let $E_n = 8E_s$ and $G = 2E_s$, with zero Poisson ratios. Thus **E** becomes a diagonal matrix. Axes n and s may be oriented arbitrarily with respect to global axes.

3.36 Many stress concentration factors have been tabulated [1.5, 3.8], especially for plane bodies. Textbooks on mechanics of materials usually contain some of these results for circular holes and fillets in bars loaded in tension and in bending. A FE analysis can be undertaken, using progressive mesh refinement near the point of peak stress, until error is reduced to less than (say) 5%.

Isoparametric Elements and Solution Techniques

The chapter first discusses matrix sparsity, equation solving, and transformations. The associated manipulations are largely internal to software and typically the user has little control over them. The next four sections consider the popular isoparametric approach to FE formulation. The nature of the FE method and its convergence to correct results are then summarized. Final sections discuss infinite media, substructures, symmetry, and constraints. The latter topics are matters of element formulation and equation manipulation that are largely under the control of the FE user.

4.1 NODE NUMBERING AND MATRIX SPARSITY

Demands on computer storage and the speed of program execution are strongly influenced by the way in which global stiffness coefficients K_{ij} are stored. In turn, the storage format depends largely on how nodes and/or elements are numbered. Commercial software can be expected to contain an algorithm that chooses an effective numbering sequence for internal storage and processing, but the user may have to activate it by giving an appropriate command. In this section we summarize these considerations by means of simple examples.

Consider the five-element, six-node structure shown at the top of Fig. 4.1-1a. The nature of the physical problem is unimportant. In this example we assume that each element has two nodes and that there is only one d.o.f. per node. Element stiffness matrices have the forms

$$\mathbf{k}_{1-2} = \begin{bmatrix} a & b \\ b & c \end{bmatrix}, \quad \mathbf{k}_{2-4} = \begin{bmatrix} d & e \\ e & f \end{bmatrix}, \quad \mathbf{k}_{3-4} = \begin{bmatrix} g & h \\ h & i \end{bmatrix}, \quad \text{etc.} \tag{4.1-1}$$

In the software, structure stiffness matrix $\mathbf{K}$ is formed by assembling element stiffness matrices, taking care to place element stiffness coefficients in the proper rows and

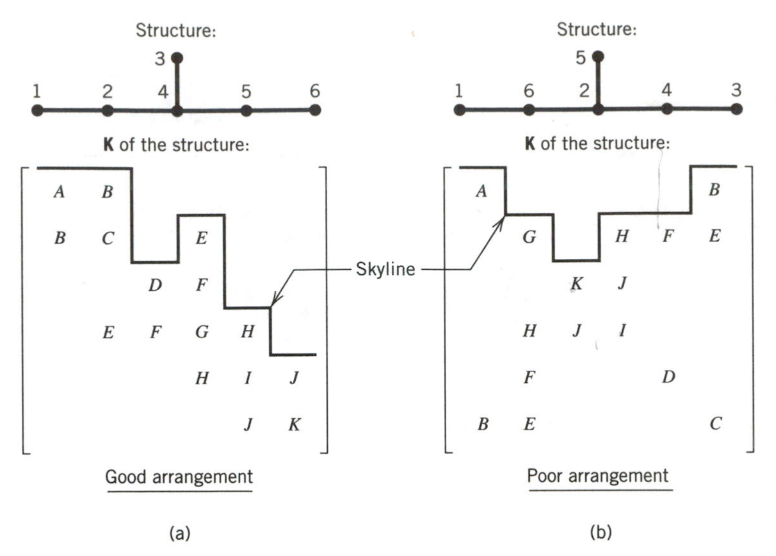

Fig. 4.1-1. Structures built of two-node elements having one d.o.f. per node and their symbolic stiffness matrices. Nonzero coefficients are represented by letters.

columns in the **K** array. Thus, for the six-d.o.f. structure in Fig. 4.1-1a, **K** is formed as the sum

$$\mathbf{K} = \begin{array}{c} \\ 1 \\ 2 \\ 3 \\ 4 \\ 5 \\ 6 \end{array} \begin{array}{c} \begin{array}{cccccc} 1 & 2 & 3 & 4 & 5 & 6 \end{array} \\ \begin{bmatrix} a & b & & & & \\ b & c & & & & \\ & & & & & \\ & & & & & \\ & & & & & \\ & & & & & \end{bmatrix} \end{array} + \begin{array}{c} \begin{array}{cccccc} 1 & 2 & 3 & 4 & 5 & 6 \end{array} \\ \begin{bmatrix} & & & & & \\ & d & & e & & \\ & & & & & \\ & e & & f & & \\ & & & & & \\ & & & & & \end{bmatrix} \end{array} + \begin{array}{c} \begin{array}{cccccc} 1 & 2 & 3 & 4 & 5 & 6 \end{array} \\ \begin{bmatrix} & & & & & \\ & & & & & \\ & & g & h & & \\ & & h & i & & \\ & & & & & \\ & & & & & \end{bmatrix} \end{array} + \cdots \qquad (4.1\text{-}2)$$

in which zeros are represented by blanks. Row and column numbers of nonzero coefficients are also the numbers of nodes to which elements are connected, for example, terms from element 2-4 appear in rows and columns 2 and 4. In Fig. 4.1-1a, the assembled **K** is represented in the same format used in Eq. 4.1-2, so that $A = a$, $B = b$, $C = c + d$, and so on. An alternative node numbering changes only the topology of **K**; that is, the same coefficients appear but in a different arrangement. An example is shown in Fig. 4.1-1b.

The assembly process illustrated by Eq. 4.1-2 can be explained as follows. From the formula (Eq. 3.1-9), strain energy in element 1-2 is $\mathbf{d}_{1-2}^T \mathbf{k}_{1-2} \mathbf{d}_{1-2}/2$, where $\mathbf{d}_{1-2}$ contains the d.o.f. of element 1-2. This is the same energy as $\mathbf{D}^T \mathbf{K}_{1-2} \mathbf{D}/2$, where **D** contains all d.o.f. of the structure and $\mathbf{K}_{1-2}$ contains only the stiffness coefficients of element 1-2 (in appropriate locations). Writing the strain energies of other elements in similar fashion and summing element energies to obtain total structure strain energy, we obtain $\mathbf{D}^T(\mathbf{K}_{1-2} + \mathbf{K}_{2-4} + \mathbf{K}_{3-4} + \cdots)\mathbf{D}/2 = \mathbf{D}^T\mathbf{K}\mathbf{D}/2$.

The global stiffness matrix **K** is *sparse*, meaning that it contains a great many zeros. A FE model having many d.o.f. may produce a **K** in which over 99% of the K_{ij} are zero. It would be wasteful to store and manipulate so many zeros. Accordingly, FE software uses sparse-matrix formats to store and process **K** (and other large matrices, such as the global mass matrix **M** used in dynamics). Examples of matrix topology appear in Fig. 4.1-1. In a good arrangement, Fig. 4.1-1a, the "bandwidth" is small: nonzero coefficients cluster in a narrow band along the diagonal. In general, there are some zero coefficients within the band, but there are *only* zeros outside it. The "skyline" bounds the top of the band, as shown in Fig. 4.1-1. Because **K** is symmetric and banded, we need store only its diagonal and coefficients between the diagonal and the skyline. It is common practice to store these K_{ij} in a one-dimensional array by working down columns of **K** from the skyline to the diagonal. Thus in Fig. 4.1-1a we consecutively store $A, B, C, D, E, F, G, H, I, J, K$. A separate "index" array records which of these are diagonal coefficients K_{ii}. In the "poor" arrangement of Fig. 4.1-1b there are more coefficients to store one-dimensionally because of zeros below the skyline: $A, G, K, H, J, I, F, 0, 0, D, B, E, 0, 0, 0, C$. Zeros below the skyline become nonzero as **K** is processed by a direct equation solver such as Gauss elimination. In a model with more nodes the differences between "good" and "poor" arrangements would be more striking.

The foregoing remarks presume that storage is governed by node numbering. Similar remarks can be made if information is stored and processed in a manner governed by element numbering. Then one speaks of "wavefront" rather than bandwidth. Numerical definitions of wavefront and bandwidth are available [4.1], but most FE users need only know that bandwidth and wavefront are similar measures of demands on storage space and computational effort and that smaller is better. Smaller bandwidth or wavefront usually results when consecutive node or element numbers run across the smaller dimension of the model. Fortunately, the user need not strive for small bandwidth or wavefront when preparing input data. Software can automatically revise node and element numbering so that internal processing is carried out compactly and efficiently, then convert back to the original numbering, so that results displayed by the postprocessor have the numbering used by the analyst in creating the model.

4.2 EQUATION SOLVING

Time-independent FE analysis requires that the global equations **KD** = **R** be solved for **D**. This may be done by a direct method or an iterative method. In a direct method—usually some form of Gauss elimination—the number of operations required is dictated by the number of d.o.f. and the topology of **K**. An iterative method requires an uncertain number of operations; calculations are halted when convergence criteria are satisfied or an iteration limit is reached.

Solution methods have been extensively studied over several years. For full discussion the reader is referred to [3.1], many numerical analysis textbooks, and current research papers. The following summary should be adequate for most users of FE software.

If a Gauss elimination solution is driven by node numbering, forward reduction proceeds in node number order and back substitution in reverse order, so that numerical values of d.o.f. at the first-numbered node are determined last. If the solution is driven by element numbering, assembly of element matrices may alternate with steps of forward reduction. Thus some eliminations are performed as soon as enough information has been assembled, then more assembly is carried out, then more eliminations, and so on, until all d.o.f. have been treated. Back substitution follows. The assembly-reduction process is like a "wave" that moves over the structure. A solver that works this way is called a "wavefront" or "frontal" equation solver. Wavefront is a measure of the number of coefficients being manipulated in one of the reduction steps.

The computation time of a direct solution is roughly proportional to nb^2, where n is the order of **K** and b is its bandwidth. For three-dimensional structures the computation time becomes large because b becomes large. Large b indicates high connectivity among d.o.f. Here an iterative solver may be faster because high connectivity speeds convergence. In contrast, a long slender structure has low b and low connectivity; an iterative solver would be slow to converge but a direct solver would be fast because b is small.

Frequently, a structure must be analyzed to determine the effects of several different load vectors **R**. This is done very effectively by a direct solver because most of its computational effort is expended on reduction of **K**. As long as the structure or FE model is not changed this need be done only once, regardless of the number of load vectors. In contrast, an iterative solver must treat each different load case as a new problem. Despite this disadvantage, iterative solvers may be best on parallel-processing computers. They may also be best for some nonlinear problems, in which **K** changes from load step i to load step $i + 1$, because solution $\mathbf{D}_i$ may be an excellent starting approximation for solution $\mathbf{D}_{i+1}$. An iterative solver can be coded so that operations are performed on separate elements and the results combined. Thus a global **K** need not be assembled, and storage requirements are reduced.

A direct solver works well for most problems. In most current software it is the only solution algorithm available and is used as a "black box." This situation is beginning to change.

4.3 TRANSFORMATIONS

Alternative Directions for D.O.F. The stiffness matrix of a finite element is most easily written in a local coordinate system. As examples, it is convenient to place a bar or beam element along the x axis and a plane element in the xy plane. But a FE model may require an element to be arbitrarily oriented in global coordinates XYZ. Rather than formulate element properties in global coordinates at the outset, it is easier to transform an element initially formulated in local coordinates. Transformation of this kind is carried out automatically by the software; the user is not obliged to activate it. The procedure used by the software is illustrated by the following example.

A two-node bar element is shown in Fig. 4.3-1a. Its stiffness matrix $\mathbf{k}'$ in local coordinates xy operates on d.o.f. u_1' and u_2' directed along the x axis. If the x axis lies in the XY plane, the relation between local d.o.f. and global d.o.f. is contained in the transformation

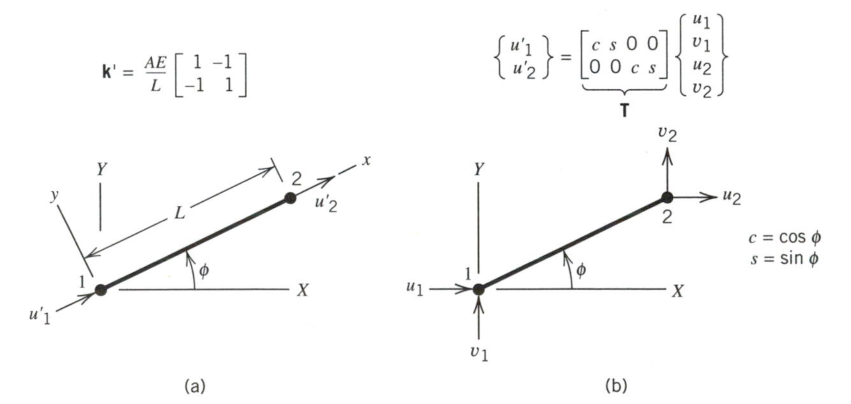

Fig. 4.3-1. (a) Stiffness matrix of a bar element in local coordinates xy. Local d.o.f. are u_1' and u_2'. (b) Transformation from local to global d.o.f. in the same plane.

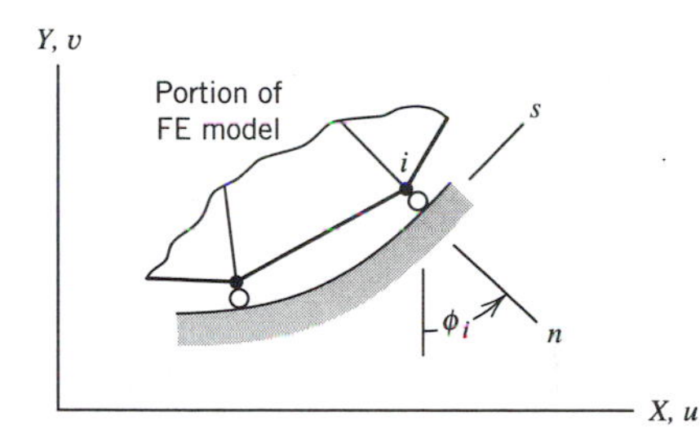

Fig. 4.3-2. Plane problem in which node i is allowed to displace only in a direction tangent to a rigid boundary.

matrix **T** of Fig. 4.3-1b. It can be shown [2.2] that the element stiffness matrix **k** that operates on global d.o.f. $[u_1 \quad v_1 \quad u_2 \quad v_2]^T$ is

$$\mathbf{k} = \mathbf{T}^T\mathbf{k}'\mathbf{T} = \frac{AE}{L}\begin{bmatrix} c^2 & cs & -c^2 & -cs \\ cs & s^2 & -cs & -s^2 \\ -c^2 & -cs & c^2 & cs \\ -cs & -s^2 & cs & s^2 \end{bmatrix} \tag{4.3-1}$$

If the x axis is not in the XY plane but is arbitrarily oriented with respect to XYZ axes, **T** becomes a 6 by 2 matrix containing direction cosines of the x axis and **k** becomes a 6 by 6 matrix. It seems preferable to regard such transformations as a change in the representation of element d.o.f. rather than as a change in orientation of the element.

A similar transformation is convenient in the example of Fig. 4.3-2. The support condition requires that d.o.f. at a typical node i have the relation $v_i = u_i \tan \phi_i$. Rather than impose this condition as a constraint (Section 4.13), we can replace u_i and v_i by u'_{in} and u'_{is}, which are d.o.f. in local coordinate directions, respectively, normal and tangent to the support. The support condition then becomes simply $u'_{in} = 0$. Other nodes along the support can be treated similarly, using the value of ϕ appropriate to each node. Indeed, the directions of nodal d.o.f. can be different at every node of the structure if the user wishes to establish a different local coordinate system at every node.

Offsets. In FE modeling we must often connect elements whose axes are parallel but not coincident. As an example, a floor slab is connected to a supporting beam on the lower surface of the slab. A FE model consists of plate elements and beam elements. We wish to connect beam nodes to plate nodes, but they are separated by a vertical distance. We can eliminate beam nodes by making them "slave" to plate nodes. The procedure, invoked by the user and carried out automatically by the software, is as follows [2.2, 4.2].

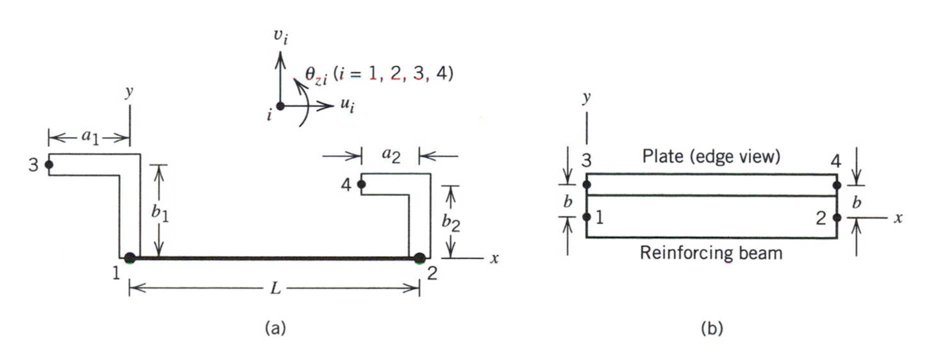

Fig. 4.3-3. (a) Nodes 1 and 2 are connected to nodes 3 and 4 by rigid links (not merely stiff, but *rigid* links). (b) An application.

In Fig. 4.3-3, let the element between nodes 1 and 2 resist both bending and axial deformation, and let imaginary rigid links connect nodes 1 and 2 to nodes 3 and 4, respectively. We allow arbitrary offsets a_i and b_i in x and y directions, but for simplicity omit offsets c_i in the z direction and displacements normal to the xy plane. Accordingly, the relation between d.o.f. at nodes 1 and 2 and d.o.f. at nodes 3 and 4 is

$$\begin{Bmatrix} u_1 \\ v_1 \\ \theta_{z1} \\ u_2 \\ v_2 \\ \theta_{z2} \end{Bmatrix} = \underbrace{\begin{bmatrix} 1 & & b_1 & & & \\ & 1 & a_1 & & & \\ & & 1 & & & \\ & & & 1 & & b_2 \\ & & & & 1 & a_2 \\ & & & & & 1 \end{bmatrix}}_{\mathbf{T}} \begin{Bmatrix} u_3 \\ v_3 \\ \theta_{z3} \\ u_4 \\ v_4 \\ \theta_{z4} \end{Bmatrix} \tag{4.3-2}$$

in which zeros are represented by blanks. In the special case of Fig. 4.3-3b, $a_1 = a_2 = 0$ and $b_1 = b_2 = b$. Let the element stiffness matrix that operates on d.o.f. at nodes 1 and 2 be called $\mathbf{k}'$. It is formulated in the usual way (see Eq. 2.3-9). To transform it to a matrix $\mathbf{k}$ that operates on d.o.f. at nodes 3 and 4 we carry out the transformation $\mathbf{k} = \mathbf{T}^T\mathbf{k}'\mathbf{T}$. If local axes xy are not parallel to global axes, another transformation analogous to Eq. 4.3-1 is performed. After assembly of $\mathbf{k}$ into the global $\mathbf{K}$, d.o.f. at nodes 1 and 2 do not appear in the global vector of d.o.f. $\mathbf{D}$, but these d.o.f. reappear during postprocessing to obtain stresses in the beam element.

4.4 ISOPARAMETRIC ELEMENTS: FORMULATION

The isoparametric formulation makes it possible to have nonrectangular elements, elements with curved sides, "infinite" elements for unbounded media, and singularity elements for fracture mechanics. Here we discuss only the four-node plane quadrilateral. Other isoparametric elements have more nodes and more shape functions but are very similar in that they use the same concepts and computational procedures.

An auxiliary coordinate system must be introduced in order that a quadrilateral may be

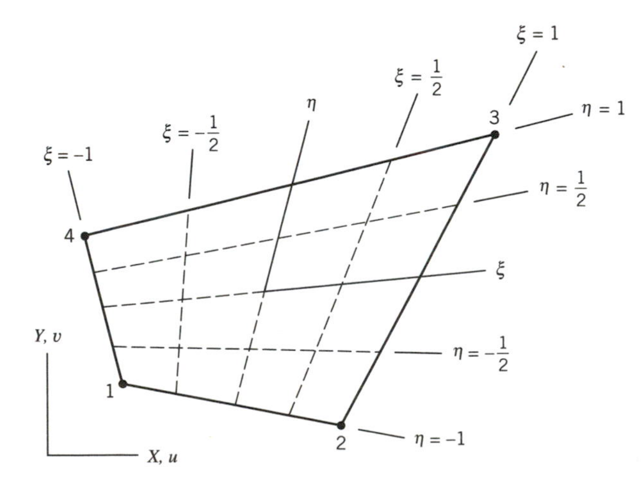

Fig. 4.4-1. Four-node plane isoparametric element. (Reprinted from [2.2] by permission of John Wiley & Sons, Inc.)

nonrectangular. This system, called $\xi\eta$ in Fig. 4.4-1, is a "natural" coordinate system. Its origin in global coordinates XY is at the average of the corner coordinates. In natural coordinates $\xi\eta$, element sides are always defined by $\xi = \pm 1$ and $\eta = \pm 1$, regardless of the shape or physical size of the element or its orientation in global coordinates XY. In general, axes ξ and η are not orthogonal and they have no particular orientation with respect to axes X and Y. Coordinates of a point within the element are defined by

$$X = \sum N_i X_i \qquad Y = \sum N_i Y_i \tag{4.4-1}$$

in which X_i and Y_i are coordinates of the corner nodes and the shape (or interpolation) functions N_i are

$$\begin{aligned} N_1 &= \tfrac{1}{4}(1-\xi)(1-\eta) \qquad & N_2 &= \tfrac{1}{4}(1+\xi)(1-\eta) \\ N_3 &= \tfrac{1}{4}(1+\xi)(1+\eta) \qquad & N_4 &= \tfrac{1}{4}(1-\xi)(1+\eta) \end{aligned} \tag{4.4-2}$$

These N_i are similar to the N_i in Eq. 3.4-3. Given ξ and η coordinates of a point we can use Eqs. 4.4-1 to calculate its X and Y coordinates. Displacements of a point are interpolated from nodal d.o.f. by use of the same shape functions:

$$u = \sum N_i u_i \qquad v = \sum N_i v_i \tag{4.4-3}$$

Displacements u and v are parallel to X and Y axes, not ξ and η axes. The name "isoparametric" derives from use of the same shape functions to interpolate both coordinates and displacements. A plane isoparametric element does not require a transformation of the type used in Fig. 4.3-1. (Global directions X and Y are used in the present section merely to avoid confusion with the local element-related directions x and y used in Section 4.3.)

In order to write the strain–displacement matrix $\mathbf{B}$ (Eq. 3.1-8) we must establish the relation between gradients in the two coordinate systems. Consider one of these gradients, the strain $\varepsilon_X = \partial u/\partial X$. We cannot immediately write the result because u is defined as a function of ξ and η rather than as a function of X and Y. We must start by differentiating with respect to ξ and η, and use the chain rule:

$$\begin{Bmatrix} \dfrac{\partial u}{\partial \xi} \\[2ex] \dfrac{\partial u}{\partial \eta} \end{Bmatrix} = \underbrace{\begin{bmatrix} \dfrac{\partial X}{\partial \xi} & \dfrac{\partial Y}{\partial \xi} \\[2ex] \dfrac{\partial X}{\partial \eta} & \dfrac{\partial Y}{\partial \eta} \end{bmatrix}}_{\mathbf{J}} \begin{Bmatrix} \dfrac{\partial u}{\partial X} \\[2ex] \dfrac{\partial u}{\partial Y} \end{Bmatrix} \tag{4.4-4}$$

where $\mathbf{J}$ is called the Jacobian matrix. Coefficients in $\mathbf{J}$ can be obtained from Eqs. 4.4-1.

$$\frac{\partial X}{\partial \xi} = \sum \frac{\partial N_i}{\partial \xi} X_i \,, \qquad \frac{\partial Y}{\partial \xi} = \sum \frac{\partial N_i}{\partial \xi} Y_i \,, \quad \text{etc.} \tag{4.4-5}$$

Equation 4.4-4 can be solved for the vector on the right-hand side. Hence strain ε_X becomes

$$\varepsilon_X = \frac{\partial u}{\partial X} = J_{11}^* \frac{\partial u}{\partial \xi} + J_{12}^* \frac{\partial u}{\partial \eta} \tag{4.4-6}$$

where J^*_{11} and J^*_{12} are coefficients in the first row of $\mathbf{J}^{-1}$ and

$$\frac{\partial u}{\partial \xi} = \sum \frac{\partial N_i}{\partial \xi} u_i \qquad \text{and} \qquad \frac{\partial u}{\partial \eta} = \sum \frac{\partial N_i}{\partial \eta} u_i \tag{4.4-7}$$

are obtained from Eqs. 4.4-3. The remaining strains ε_Y and γ_{XY} are formulated in similar fashion, and at last the strain–displacement matrix **B** can be written.

The element stiffness matrix is

$$\mathbf{k} = \int \mathbf{B}^T \mathbf{E} \mathbf{B} \, dV = \int_{-1}^{1} \int_{-1}^{1} \mathbf{B}^T \mathbf{E} \mathbf{B} \, t \, |\mathbf{J}| \, d\xi \, d\eta \tag{4.4-8}$$

where t is the element thickness and $|\mathbf{J}|$ is the determinant of **J** in Eq. 4.4-4. $|\mathbf{J}|$ can be regarded as a scale factor between areas; that is, $dX\,dY = |\mathbf{J}|\,d\xi\,d\eta$. In general, $|\mathbf{J}|$ is a function of the coordinates, but for a rectangle or a parallelogram it is constant and has the value $A/4$, where A is the area of the rectangle or parallelogram and the "4" is the area in $\xi\eta$ coordinates, where the element is always a square two units on a side.

Other plane isoparametric elements have more nodes; hence there are more shape functions N_i and more columns in **B**, but **J** is still 2 by 2 and there are still three rows in **B**. For 3D solid elements **J** is 3 by 3 and **B** has six rows.

4.5 GAUSS QUADRATURE AND ISOPARAMETRIC ELEMENTS

Integration in Eq. 4.4-8 may be done analytically by using closed-form formulas from a table of integrals. Alternatively, integration may be done numerically. Gauss quadrature is a commonly used form of numerical integration. It is better suited to numerical analysis than closed-form formulas. To begin our explanation of Gauss quadrature, we consider one-dimensional problems without particular reference to FE. Gauss quadrature evaluates the integral of a function as the sum of a finite number of terms:

$$I = \int_{-1}^{1} \phi \, d\xi \qquad \text{becomes} \qquad I \approx \sum_{i=1}^{n} W_i \, \phi_i \tag{4.5-1}$$

where W_i is a "weight" and ϕ_i is the value of $\phi = \phi(\xi)$ at a particular location often called a "Gauss point." Figure 4.5-1 shows examples of this process for Gauss rules of orders $n = 1$, $n = 2$, and $n = 3$. Gauss points are at $\xi = 0$, $\xi = \pm a$, and $\xi = 0, \pm b$ respectively. There exist tabulations of Gauss point locations and corresponding weights for values of n much larger than needed for FE work [4.3].

If $\phi = \phi(\xi)$ is a polynomial, n-point Gauss quadrature yields the exact integral if ϕ is of degree $2n - 1$ or less. Thus the form $\phi = c_1 + c_2\xi$ is exactly integrated by a one-point rule, the form $\phi = c_1 + c_2\xi + c_3\xi^2$ is exactly integrated by a two-point rule, and so on. Use of an excessive number of points, for example, a two-point rule for $\phi = c_1 + c_2\xi$, still yields the exact result. If ϕ is not a polynomial, but (say) the *ratio* of two polynomials, Gauss quadrature yields an approximate result. Accuracy improves as more Gauss points are used. Convergence toward the exact result may not be monotonic.

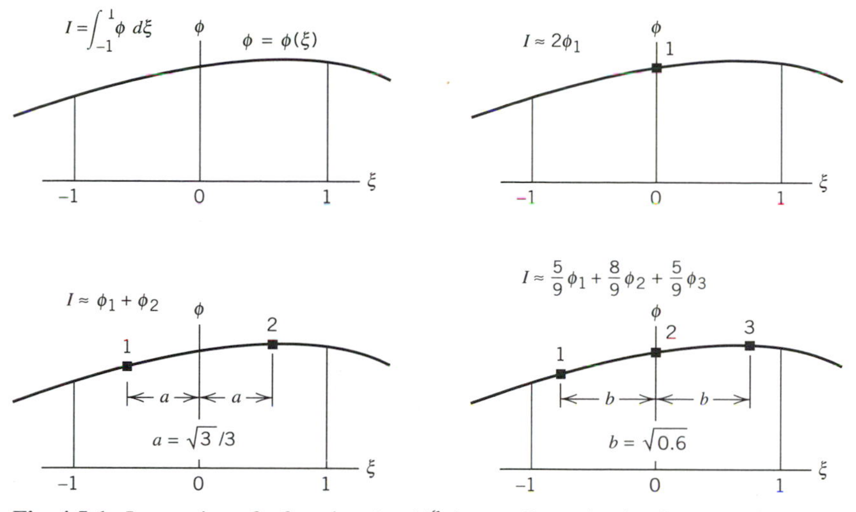

Fig. 4.5-1. Integration of a function $\phi = \phi(\xi)$ in one dimension by Gauss quadrature of orders 1, 2, and 3. Gauss points are numbered.

In two dimensions, integration is over a quadrilateral and a Gauss rule of order n uses n^2 points. The formula analogous to Eq. 4.5-1 is

$$I = \int_{-1}^{1}\int_{-1}^{1} \phi(\xi, \eta) \, d\xi \, d\eta \approx \sum_{i=1}^{n}\sum_{j=1}^{m} W_i W_j \, \phi(\xi_i, \eta_j) \tag{4.5-2}$$

where $W_i W_j$ is the product of one-dimensional weights. Usually $m = n$; that is, the same number of points are used in each direction. If $m = n = 1$, ϕ is evaluated at $\xi = \eta = 0$ and $I \approx 4\phi_1$. Gauss points for four-point and nine-point rules are shown in Fig. 4.5-2, and the corresponding integrals are

$$I \approx \phi_1 + \phi_2 + \phi_3 + \phi_4 \tag{4.5-3a}$$

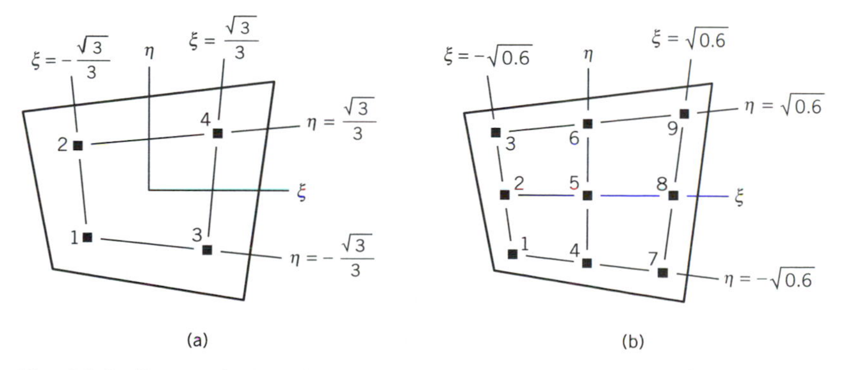

Fig. 4.5-2. Gauss point locations for integration of a function $\phi = \phi(\xi, \eta)$ in two dimensions, using orders 2 and 3. (Reprinted from [2.2] by permission of John Wiley & Sons, Inc.)

$$I \approx \frac{25}{81}(\phi_1+\phi_3+\phi_7+\phi_9) + \frac{40}{81}(\phi_2+\phi_4+\phi_6+\phi_8) + \frac{64}{81}\phi_5 \qquad (4.5\text{-}3b)$$

In three dimensions, Gauss quadrature of order n over a hexahedron involves n^3 points, three summations, and the product of three weight factors. Analogous numerical integration formulas are available for integration over triangles and tetrahedra [2.2].

Consider again the plane four-node element discussed in Section 4.4. Its stiffness matrix integrand $\mathbf{B}^T\mathbf{EB}t|\mathbf{J}|$ is an 8 by 8 matrix. Because it is a *symmetric* matrix, only 36 of its 64 coefficients are different from one another. Each of these coefficients has the form $\phi = \phi(\xi, \eta)$ and each must be integrated over the element area. In computer programming, a p-point integration rule requires p passes through an integration loop. Each pass requires evaluation of $\mathbf{B}$ and $|\mathbf{J}|$ at the coordinates of a Gauss point, computation of the product $\mathbf{B}^T\mathbf{EB}t|\mathbf{J}|$, and multiplication by weight factors. Each pass makes a contribution to $\mathbf{k}$, which is fully formed when all p passes have been completed. Clearly, there is considerable computation required in this process.

For an element of general shape, each coefficient in the matrix $\mathbf{B}^T\mathbf{EB}t|\mathbf{J}|$ is the ratio of two polynomials in ξ and η. The polynomial in the denominator comes from $\mathbf{J}^{-1}$: when $\mathbf{J}$ of Eq. 4.4-4 is inverted, $|\mathbf{J}|$ becomes the denominator of every coefficient in $\mathbf{J}^{-1}$ and hence appears in the denominator of every coefficient in $\mathbf{B}$. Analytical integration of $\mathbf{k}$ would require the use of cumbersome formulas. Numerical integration is simpler but in general it is not exact, so that $\mathbf{k}$ is only approximately integrated regardless of the number of integration points. Should we use very few points for low computational expense or very many points to improve the accuracy of integration? The answer is *neither*, for reasons explained in the next section.

4.6 CHOICE OF QUADRATURE RULE. INSTABILITIES

A FE model is usually inexact, and usually it errs by being too stiff (see Section 4.8). Overstiffness is usually made *worse* by using more Gauss points to integrate element stiffness matrices because additional points capture more higher-order terms in $\mathbf{k}$. These terms resist some deformation modes that lower-order terms do not, and therefore act to stiffen an element. Accordingly, *greater* accuracy in the integration of $\mathbf{k}$ usually produces *less* accuracy in the FE solution, in addition to requiring more computation.

On the other hand, use of too *few* Gauss points produces an even worse situation known by various names: instability, spurious singular mode, mechanism, kinematic mode, zero-energy mode, and hourglass mode. Instability (not of the buckling type) occurs if one or more deformation modes happen to display zero strain at all Gauss points. One must regard Gauss points as strain sensors. If Gauss points sense no strain under a certain deformation mode, the resulting $\mathbf{k}$ will have no resistance to that deformation mode.

A simple illustration of instabilities appears in Fig. 4.6-1. Four-node plane elements are integrated by a one-point Gauss rule. In the lower left element, with c a constant, the three instabilities shown have the respective forms (b) $u = cxy$, $v = 0$; (c) $u = 0$, $v = -cxy$; and (d) $u = cy(1-x)$, $v = cx(y-1)$. We easily check that each of these displacement fields produces strains $\varepsilon_x = \varepsilon_y = \gamma_{xy} = 0$ at the Gauss point, $x = y = 0$. *Non*rectangular elements behave in the same way. Even if the mesh had just enough supports to prevent rigid-body motion it could still display these modes, without strain at the Gauss points, and hence without strain energy. The FE model would have no resistance to loadings that

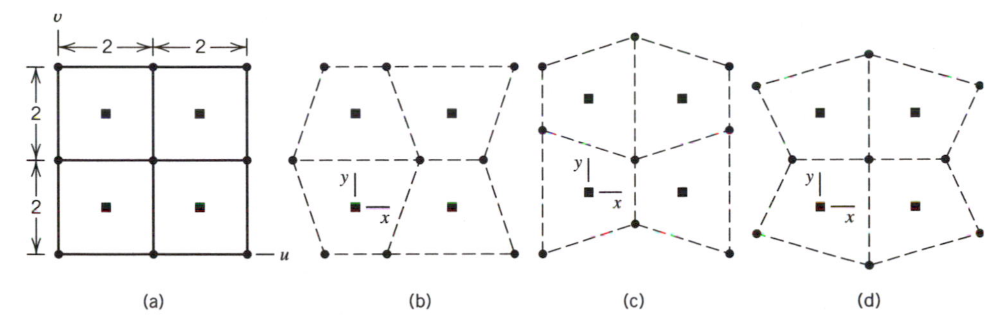

Fig. 4.6-1. (a) Undeformed plane 2 by 2 four-node square elements. Gauss points are shown by solid squares. (*b,c,d*) "Instability" displacement modes. (Reprinted from [2.2] by permission of John Wiley & Sons, Inc.)

would activate these modes. The global **K** would be singular *regardless* of how the structure is loaded. These spurious displacement modes rarely appear in isolation. Usually they are superposed on "legitimate" displacement modes, which makes them hard to identify.

When supports are adequate to make **K** nonsingular, there may yet be a near-instability that is troublesome. Consider Fig. 4.6-2a. All d.o.f. are fixed at the support and each element stiffness matrix is integrated with a single Gauss point. Restraint provided by the support is felt less and less with increasing distance from it. If L is several times H, the computed displacement of load P may be greater than the length of the bar! At the same time displacements and stresses at the Gauss points will have good accuracy unless the spurious displacements overwhelm the solution.

A plane eight-node element whose stiffness matrix is integrated with four Gauss points has the "hourglass" instability shown in Fig. 4.6-2b. This mode is of no concern because it is *noncommunicable*: there is no way that two adjacent elements can both display this mode while remaining connected, even if nodal d.o.f. are reversed from Fig. 4.6-2b in one of the two elements. Accordingly, a mesh of two or more such elements has no such instability. However, a near-instability, roughly analogous to that in Fig. 4.6-2a, is possible if adjacent elements have greatly different moduli and an edge-normal force is applied to the stiffer element at a midside node. In this case hourglassing is only lightly resisted by the softer element. Also, if a ninth node were added at the element center, two additional instabilities would be possible under four-point quadrature, both of them communicable [2.2, 3.2].

The default option in commercial software usually calls for the smallest number of

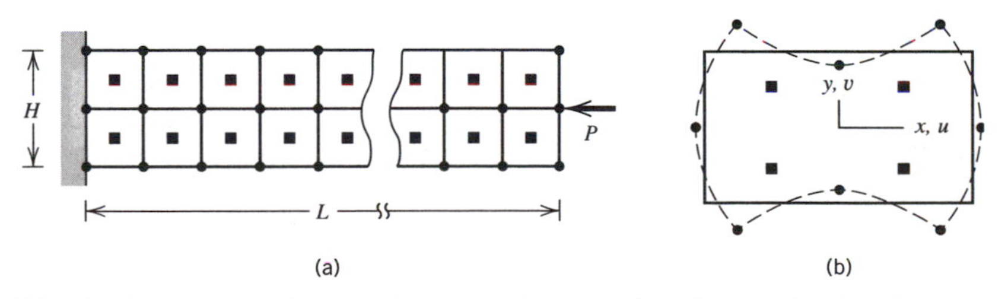

Fig. 4.6-2. (a) Mesh of four-node square elements with all nodes fixed at the support. Gauss points are shown by solid squares. (b) "Hourglass" instability displacement mode in a single eight-node element integrated by four Gauss points.

Gauss points that will make instability impossible. Occasionally, fewer points are used, but then "stabilization" devices are invoked that prevent instability. A user who chooses other quadrature rules or overrides a stabilization device must be aware of the possible difficulties and how to avoid them.

4.7 STRESS CALCULATION AND GAUSS POINTS

Calculated stresses $\boldsymbol{\sigma} = \mathbf{EBd}$ are often most accurate at Gauss points. This statement can be made plausible by returning to the problem of Fig. 4.6-2a. As shown in Fig. 4.7-1, a large and spurious bending deformation, associated with rotation of y-parallel element sides, is superposed on a constant strain state that is essentially correct. The spurious deformation has no effect on strains at the Gauss point. In more mundane situations, and with various element types, it is not hard to realize that strains are likely to vary over an element and are therefore likely to be more accurate at some locations than others. It happens that the locations of greatest accuracy are apt to be the same Gauss points that were used for integration of the stiffness matrix [3.2, 4.4]. Consider Fig. 4.7-2, for example, which shows a portion of a beam in which shear strain is γ_{xy} constant along the x axis. The shear strain calculated by FE displays a quadratic variation that is most accurate at x coordinates of the Gauss points.

In summary, it is common practice to use an order 2 Gauss rule (four points) to integrate $\mathbf{k}$ of four- and eight-node plane elements, and common practice to compute strains and stresses at these same points. Similarly, three-dimensional elements often use eight Gauss points for stiffness integration and stress calculation. Stresses at nodes or at other element locations are obtained by extrapolation or interpolation from Gauss point values. Thus the element stress field is represented as bi- or trilinear in isoparametric coordinates; for example, in a plane element σ_x is represented by the form $\sigma_x = c_1 + c_2\xi + c_3\eta + c_4\xi\eta$. For eight-node elements this is a polynomial of lower degree than contained in the $\mathbf{B}$ matrix and therefore some information has been discarded. Nevertheless, accuracy is usually greatest when the element stress field is a polynomial fitted to Gauss point values.

4.8 NATURE OF FINITE ELEMENT SOLUTION

The FE method is a form of the Rayleigh–Ritz method, which is a classic approximation technique originated by Lord Rayleigh in 1870 and generalized by W. Ritz in 1909. In the classical Rayleigh–Ritz method one begins with a displacement assumption in terms of generalized coordinates β_i, for example, Eqs. 3.2-1, 3.3-1, and 3.4-1. However, the as-

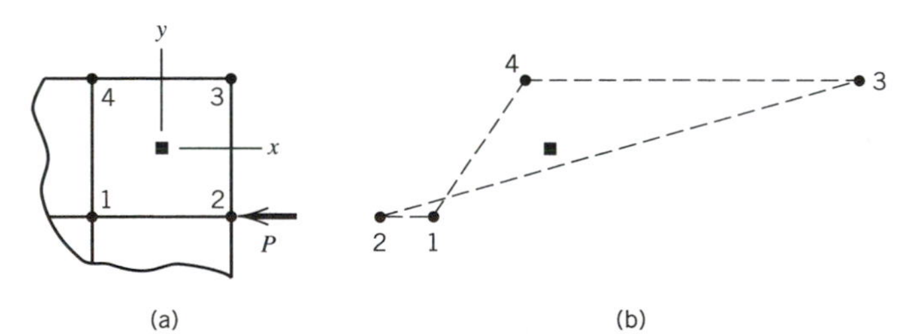

Fig. 4.7-1. (a) Upper right-hand element in Fig. 4.6-2a. (b) Possible displacement mode of this element.

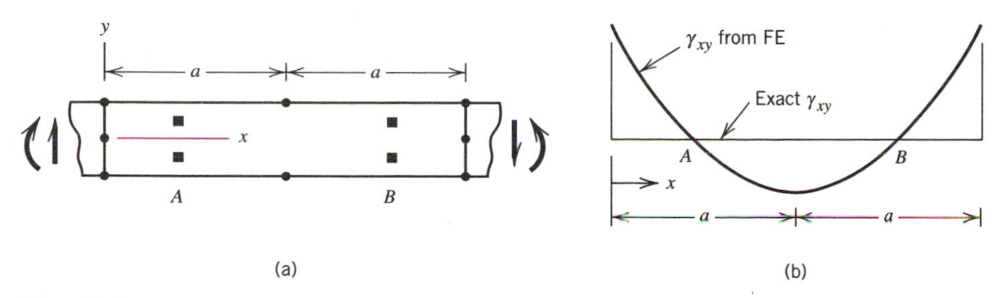

Fig. 4.7-2. (a) Portion of a beam modeled by a single layer of eight-node elements. (b) Shear strain along the x axis.

sumed field applies to the *entire body*, not element by element in piecewise fashion; indeed, there are no elements and no nodes. The assumed field must satisfy compatibility conditions within the body and displacement boundary conditions. For example, a lateral displacement field $v = v(x)$ for a cantilever beam without transverse shear deformation must be single-valued and must display $v = 0$ and $dv/dx = 0$ at the support. One forms an energy expression that includes strain energy of the body and work done by applied loads. Minimization of the total energy with respect to the β_i yields simultaneous algebraic equations that can be solved for the β_i. The FE method differs in that it uses a displacement field defined in piecewise fashion and uses nodal d.o.f. instead of the β_i. These modifications make it much easier to write a computer program to carry out the calculations. The FE method can be regarded as a modern way of arranging procedural details of the Rayleigh–Ritz method.

In the Rayleigh–Ritz method, the model can deform only into shapes contained in the assumed displacement field. For example, if we assume the field $v = \beta_1 x^2 + \beta_2 x^3$ for lateral displacement of a uniform cantilever beam fixed at $x = 0$, the model is constrained to deform into only the modes $v = x^2$ and $v = x^3$, with the respective amplitudes β_1 and β_2. If the loading happens to be distributed rather than concentrated at the free end, the correct field $v = v(x)$ is more complicated than this. Then the assumed field has, in effect, applied constraints that prevent the correct displacement field from appearing. Constraints have a stiffening effect. Accordingly, the Rayleigh–Ritz method yields "lower bound" displacements, that is, displacements that are either exact or too small as compared with an exact solution of the mathematical model (the mathematical model is beam theory in the foregoing example). This does not mean that displacements are too small at every point; it means that work done by applied loads is too small (recall that work is load times displacement increment, integrated over the structure volume). In other words, displacements are too small in an average sense. If the load consists of a *single* force or moment, we can say with certainty that the Rayleigh–Ritz method predicts a load-parallel displacement of the loaded point that is either exact or too small as compared with the mathematical model.

The foregoing remarks also apply to a FE solution using displacement-based elements (like those in Chapter 3), *provided that* (a) nodal loads from distributed loading are applied in work-equivalent fashion , (b) elements are compatible, and (c) elements are integrated exactly. These restrictions mean that if we use load lumping (Section 2.5), incompatible elements (Section 3.6), or (say) four Gauss points to integrate **k** of an eight-node plane rectangular element or **k** of a four-node nonrectangular plane element, then we cannot guarantee that computed displacements are a lower bound. Nevertheless, even if we violate these restrictions in solving a practical problem, it is more likely that the FE model will be too stiff than too flexible.

Additional remarks about the nature of a FE approximation appear in Section 3.1.

4.9 CONVERGENCE REQUIREMENTS. PATCH TEST

Imagine that a given problem is solved repeatedly, each time using a finer FE mesh. Will the sequence of solutions converge toward exact displacements, strains, and stresses? The answer is *yes* provided that the elements used pass a *patch test* [3.2].

To perform a patch test, one builds a simple FE model, that is, a "patch" of elements, such that at least one node is internal to the patch (rather than on its boundary), and having just enough supports to prevent rigid-body motion. Work-equivalent loads consistent with a constant stress state are applied. In Fig. 4.9-1a, the rectangular outline of the patch and uniform spacing of nodes along its left and right edges make it easy to assign loads F and $2F$, which are work-equivalent loads consistent with a uniform x-direction traction. No load is applied to the internal node. The correct response to this loading is constant stress $\sigma_x = 2F/Ht$, where t is the constant z-direction thickness. (We speak of stresses rather than strains only because most software reports stresses rather than strains.) One analyzes the "patch" model like any other FE model and examines the computed stresses. If the stress results are *exact*, that is, if $\sigma_x = 2F/Ht$ and all other stresses are zero at all stress calculation points, then the patch test for σ_x is passed. Other states of stress, that is, σ_y = constant and τ_{xy} = constant for plane elements, should also be patch-tested. If an element passes patch tests we can be sure that, when this type of element is used in the FE model of a practical problem, exact results will be approached as the mesh is repeatedly refined. Here "exact" means perfect agreement with the *mathematical* model on which the element is based; that is, beam theory, plate theory, or whatever. In other words, prior to convergence the FE model disagrees with its mathematical model because of *discretization error*, which tends to zero with mesh refinement if elements pass patch tests. Whether or not the mathematical model is a good representation of physical reality is another matter.

A successful patch test indicates that when an element is used in a mesh, rather than in isolation, it is able to display (a) a state of constant strain, (b) rigid-body motion without strain, and (c) compatibility with adjacent elements when a state of constant strain prevails. An element that meets these requirements may be called a *valid* element. It is not hard to see that these requirements must be met if there is to be convergence toward correct results with mesh refinement. Consider Fig. 4.9-1b. From A to B the exact strain ε_x varies linearly with x. This variation is approximated in stair-step fashion by constant strain elements between A and B and can be approximated arbitrarily closely by using more and more elements. As a counterexample, if we were to use an (invalid) element that could display only a linear *variation* of ε_x such as $\varepsilon_x = cx$, where c is a constant, we would see a sawtooth plot of ε_x that would remain inexact regardless of the number of elements used. As for rigid-body motion, from B to C in Fig. 4.9-1b elements must be able to display rigid-body motion without strain. Finally, the theoretical need for interelement compatibility was noted in Section 3.1. A valid incompatible element (e.g., element Q6 of Section 3.6) displays its incompatible mode only when there is a strain gradient. Its displacements are compatible when it is in a state of constant strain. With repeated mesh refinement, the change in a strain field over an element becomes negligible in comparison with the constant part of the strain field. Accordingly, as a mesh is repeatedly refined, elements must become compatible if they were not so already.

The foregoing arguments make it plausible that patch tests check that all convergence requirements are met in the limit of mesh refinement, when each element must approach a state of constant strain. This is all that is required for convergence. Passing patch tests says nothing about the *speed* of convergence; that is, passing patch tests shows that an element works, not that it works well.

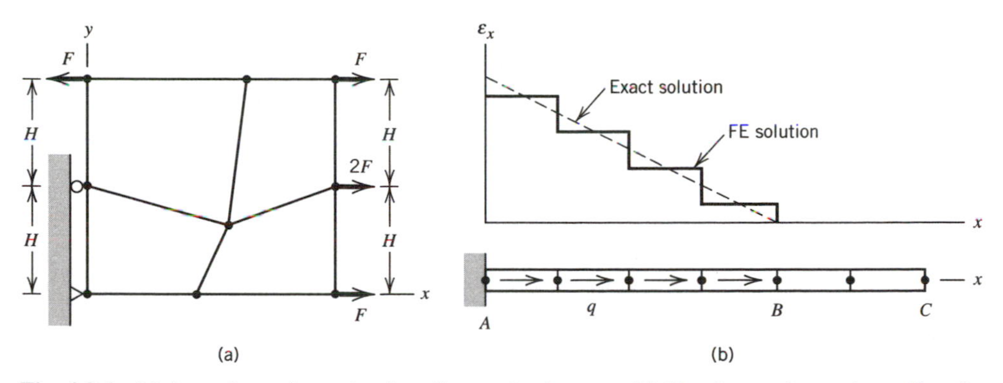

Fig. 4.9-1. (a) A patch test for σ_x in plane four-node elements. (b) Six-element bar under uniformly distributed load q over portion AB.

Plate elements carry load by bending and must display constant curvature states in order to pass patch tests. Applied loads must be consistent with constant states of $\partial^2 w/\partial x^2$, $\partial^2 w/\partial y^2$, and $\partial^2 w/\partial x\,\partial y$, where $w = w(x, y)$ is the displacement in the z direction, normal to the plate midsurface $z = 0$. Constant-curvature states produce constant strain states in z = constant layers of a plate.

One who uses FE rather than developing new elements will probably not use patch tests to study the validity of elements. Nevertheless, patch tests can be helpful in learning about FE and in learning how to use software because patch tests are simple, data are easy to prepare, and exact results are known.

4.10 INFINITE MEDIA AND INFINITE ELEMENTS

Occasionally, a region of interest is embedded in a medium so large that it can be considered unbounded. For example, Fig 4.10-1a represents a thick slab supported by soil. For the present discussion it does not matter whether the problem is two- or three-dimensional. Stresses in and near the slab are desired. A coarse-mesh FE model is shown in Fig. 4.10-1b. If arc CD is far enough from the slab, the FE model may be terminated

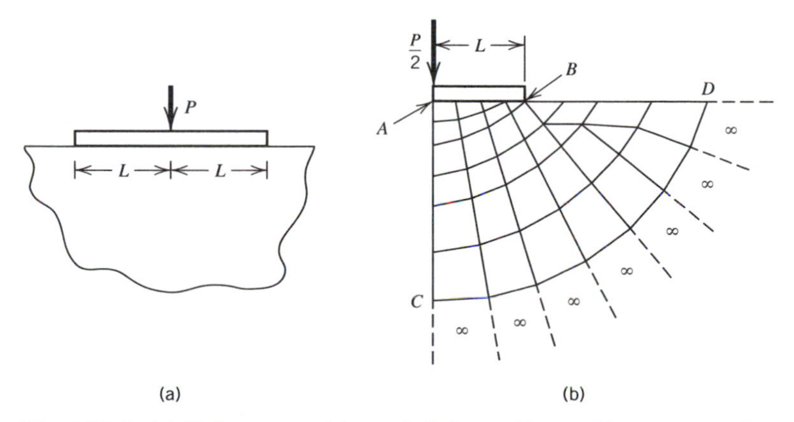

Fig. 4.10-1. (a) Slab supported by an infinite medium, with symmetry about the vertical centerline. (b) FE model of half the structure, with infinite elements denoted by ∞.

there by making *CD* a fixed support. But how far is far enough? Too close introduces error; too far produces a large and unwieldy FE model. And if the problem is dynamic, a fixed support will reflect waves regardless of the size of the FE model. Some way of representing the "far field" is desirable.

One way to treat the problem is by introducing "infinite" elements [4.5, 4.6]. Instead of fixing d.o.f. along *CD* in Fig. 4.10-1b, these d.o.f. are connected to a single layer of infinite elements. An infinite element is produced by using special shape functions in the isoparametric formulation, so that nodes on one side of an element are made to move off to infinity. In Fig. 4.10-1b these sides are downward and to the right of *CD* and do not appear. "Mapping to infinity" has the effect of making displacements decay toward zero with increasing distance from *CD*. Infinite elements can be used for various problems of continua, stress analysis being only one, and for either time-independent or harmonic wave problems. It appears that they cannot be used for transient problems such as shock waves.

Another way to treat the problem is by using boundary elements (BE) to model the medium that supports the slab. We make no attempt here to explain the workings of the BE method. Suffice it to say that FE and BE models can be connected, that BE models can easily represent infinite media, and that a BE model has nodes only on its boundary. Thus in Fig. 4.10-1b a BE model of the soil would contact the slab directly and would have no nodes within the region *ABCD*, which simplifies the task of data preparation. Despite the reduction in number of d.o.f., the computation time of BE may be greater than that of FE because global BE matrices are full and unsymmetric.

4.11 SUBSTRUCTURES

Substructuring is a process of analyzing a large FE model as a collection of component FE models. It will be easier to understand why this is done if we first describe how it is done, which is as follows.

1. Divide the FE model into two or more parts (substructures) by cutting along lines of nodes. Preferably, cuts are made across narrow parts of the model, so as to reduce both the number of d.o.f. on cutting lines and the interaction between substructures. For example, we choose cuts along hatched lines in Fig. 4.11-1a rather than cuts along the middle of wings or along the fuselage.

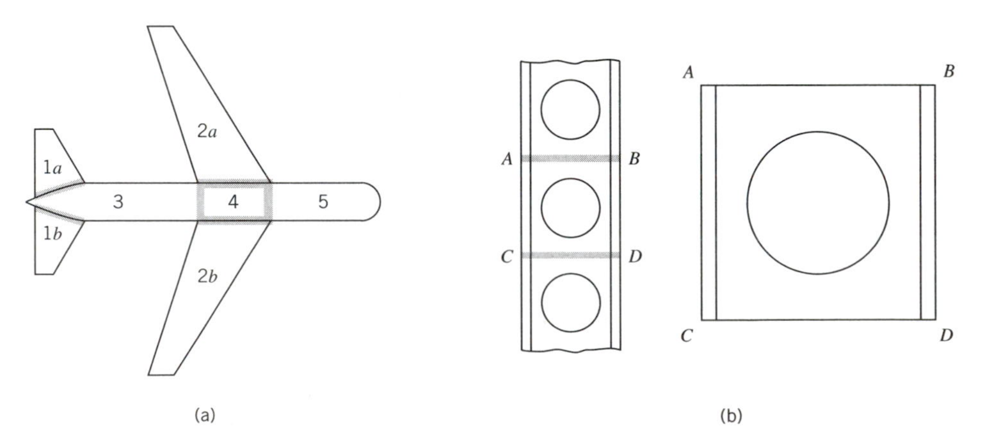

Fig. 4.11-1. (a) Possible substructures 1*a*, 1*b*, ..., 5 of a hypothetical aircraft. (b) Castellated beam, with typical repeating substructure *ABCD*. Elements of the substructures are not shown.

2. Create a FE model of each substructure and obtain a set of global equations $\mathbf{K}_s\mathbf{D}_s = \mathbf{R}_s$ for each substructure. Begin to solve these equation sets, for example, by Gauss elimination, until all d.o.f. *not* on cutting lines have been eliminated and only "attachment" d.o.f. $\mathbf{D}_a$ on the cutting lines remain. $\mathbf{D}_a$ is a small subset of $\mathbf{D}_s$. Symbolize the reduced equation set for a single substructure by $\mathbf{K}_a\mathbf{D}_a = \mathbf{R}_a$.

3. Assemble the reduced equation sets of all substructures, to obtain global equations $\mathbf{K}_A\mathbf{D}_A = \mathbf{R}_A$, where $\mathbf{D}_A$ contains all attachment d.o.f. $\mathbf{D}_a$ of all substructures. (This set of equations is the same reduced set that would result if *all* d.o.f. of the entire structure had been assembled to form global equations $\mathbf{KD} = \mathbf{R}$, then Gauss elimination applied until only attachment d.o.f. $\mathbf{D}_A$ remain.) *Note*: Attachment d.o.f. of mating substructures must match in number, placement, type, and orientation.

4. Solve the equations $\mathbf{K}_A\mathbf{D}_A = \mathbf{R}_A$ for $\mathbf{D}_A$. Thus attachment d.o.f. $\mathbf{D}_a$ become known for all substructures. Return to the substructure equations $\mathbf{K}_s\mathbf{D}_s = \mathbf{R}_s$ created and partially solved in step 2: now solve for the remaining d.o.f. in $\mathbf{D}_s$ by back substitution. Finally, postprocess to obtain stresses in elements.

The substructure assembly process, step 3, is the same process used to assemble individual elements of a standard FE model. In effect, a substructure is a large element that has internal d.o.f. as well as d.o.f. on its boundary. Indeed, substructures are sometimes called "superelements." Other terminology may refer to attachment d.o.f. as "masters" and other d.o.f. as "slaves." A capability for substructuring is included in large commercial software packages.

A substructuring approach becomes appropriate when the structure is large and can be cut into substructures that do not interact strongly. Then individual substructures can be repeatedly revised, in design or in FE modeling, always using the same attachment d.o.f. $\mathbf{D}_a$ originally calculated from the assembled substructures. Different design groups, even different companies, can work on different substructures. Indeed, the location of substructure boundaries may be dictated by binding agreements between subcontractors. Only occasionally, when it is felt necessary to update the values of the attachment d.o.f., are substructures assembled and the resulting global equations solved.

Another motivation for substructuring appears when nonlinearities such as plastic action are confined to a single part of the structure. The linear part, whose reduced stiffness matrix $\mathbf{K}_a$ does not change as loading increases, can be represented by a substructure. Its attachment d.o.f. $\mathbf{D}_a$ are shared by the nonlinear part, whose properties and matrices must be repeatedly revised as loading increases.

Finally, there is an advantage to substructuring if the FE model contains many repetitions of the same geometry (Fig. 4.11-1b). Then the same set of reduced substructure equations $\mathbf{K}_a\mathbf{D}_a = \mathbf{R}_a$ applies to each of the repeating substructures. Repeated assembly of the same $\mathbf{K}_a$ and $\mathbf{R}_a$ arrays, with appropriate node numbers, yields the global arrays $\mathbf{K}_A$ and $\mathbf{R}_A$ of the assembled substructures.

Substructuring in static stress analysis does not introduce any additional approximation. Nor does it reduce computational effort in the *very rare* situation of having no repeating parts, no nonlinearities, and no revisions in design or modeling. Substructuring increases the number of computer files needed to do an analysis. Clearly, it is possible to loose track of pieces of the puzzle. The analyst is advised to plan carefully and keep records.

4.12 SYMMETRY

Types of symmetry include reflective, skew, axial, and cyclic. If symmetry is recognized and exploited, the size of the FE model is reduced. Thus there is less input data to prepare and less computation to do.

A structure has *reflective* or *mirror* symmetry if there is symmetry of geometry, support conditions, and elastic properties with respect to a plane. Reflective symmetry of structure *and loads* is shown in Fig. 4.12-1a: if reflected by the plane $x = 0$, the left half yields the right half and vice versa. One could say that reflection brings the structure and its loads into "self-coincidence." Analysis of either half yields a complete solution because symmetric loading on a symmetric structure produces symmetric results.

If $P_x = P_y$ in Fig. 4.12-1a, the planes $x = 0$, $y = 0$, $x = y$, and $x = -y$ are all planes of reflective symmetry, and we need analyze only one octant of the structure, using $P_x/2$ as the load. Supports on a symmetry plane in Fig. 4.12-1a must allow only motion radially from the origin $x = y = 0$ (as in Fig. 3.12-2a). A similar example appears in Fig. 4.12-1b: analysis of the right (or left) half of the beam, with rotation θ_z prevented at $x = 0$, provides a complete solution of the problem. These examples are very simple, but one can see that if the structure were large and complicated it would be a waste of effort to ignore symmetry and prepare a model of the entire structure.

Note that loads as well as structure may be cut by a plane of symmetry. In Fig. 4.12-1a, if only half the structure is retained because plane $x = 0$ is used as a plane of reflective symmetry, loads P_y become $P_y/2$ on the half retained. Similarly, if a stiffening beam (as might be used beneath a floor slab) is longitudinally bisected by a plane of reflective symmetry, only half its stiffness is retained.

The problem shown in Fig. 4.12-1c is *antisymmetric* because of the loading. Reflection about the plane $x = 0$, followed by *reversal* of all loads, results in self-coincidence. Again, analysis of half the structure yields a complete solution. Note, however, that support conditions differ in Figs. 4.12-1b and 4.12-1c.

Rules that help in setting the correct support conditions for reflective symmetry are as follows. The conditions stated apply *only* to boundary nodes of the FE model that lie in a plane of reflective symmetry of the entire structure. If the problem is *symmetric:*

1. Translations have no component normal to a plane of symmetry.
2. Rotation vectors have no component parallel to a plane of symmetry.

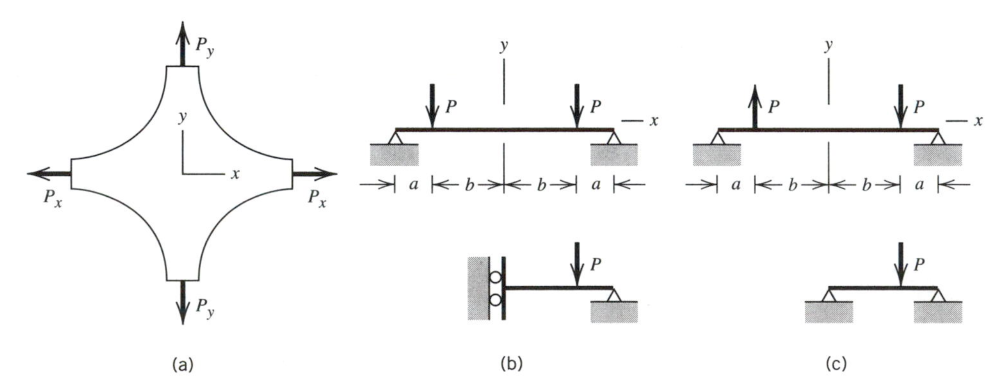

Fig. 4.12-1. (a) Plane structure having reflective symmetry about $x = 0$ and $y = 0$ planes. (b) Beam under symmetric load. (c) Beam under antisymmetric load.

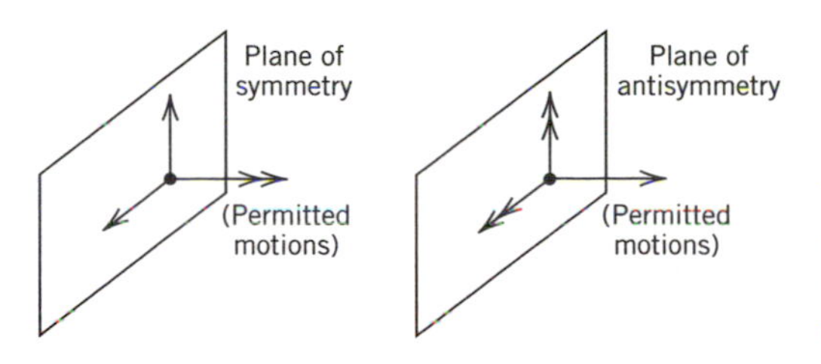

Fig. 4.12-2. The d.o.f. permitted (i.e., not restrained) at a node in a plane of symmetry or antisymmetry. A double-headed arrow represents a rotational d.o.f.

If the problem is *antisymmetric*, that is, symmetric except that loads must be reversed to achieve self-coincidence:

1. Translations have no component parallel to a plane of antisymmetry.
2. Rotation vectors have no component normal to a plane of antisymmetry.

Figure 4.12-2 depicts these rules in terms of d.o.f. *permitted* rather than d.o.f. restrained. The reader should verify that these rules hold for the special cases in Figs. 4.12-1b and 4.12-1c.

If one suspects the presence of symmetries but their nature is not clear, one may do a coarse-mesh analysis, either of the entire structure or a part of it that is obviously treatable by symmetry considerations. Computed results may confirm or refute the existence of the suspected symmetries.

Figure 4.12-3 is an example of how symmetry concepts might be applied even when obvious symmetries are not present [5.4]. By regarding the load as the sum of symmetric and antisymmetric parts, we obtain the cases in Figs. 4.12-3b and 4.12-3c. By superposing solutions of these two cases, we solve the original problem. Thus bending moments in Fig. 4.12-3a are $M_1 = M_4$, $M_2 = M_5 + M_7$, and $M_3 = M_4$. We have traded one solution of the entire structure for two solutions of half the structure. The possible advantage is that

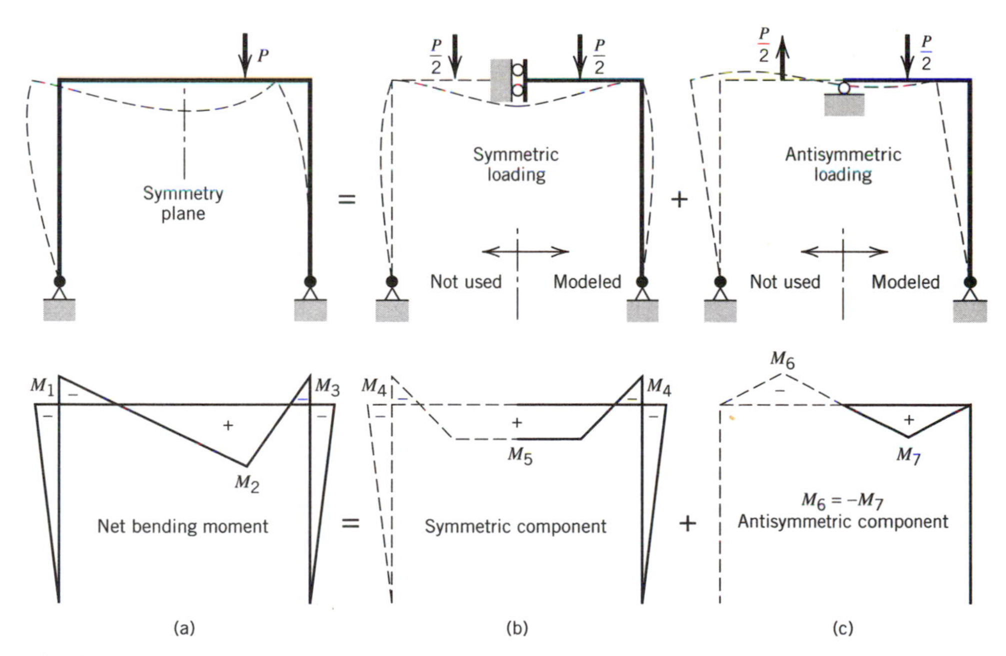

Fig. 4.12-3. Modeling a plane frame problem as the sum of symmetric and antisymmetric cases. (Reproduced from [5.4] by permission of the publisher.)

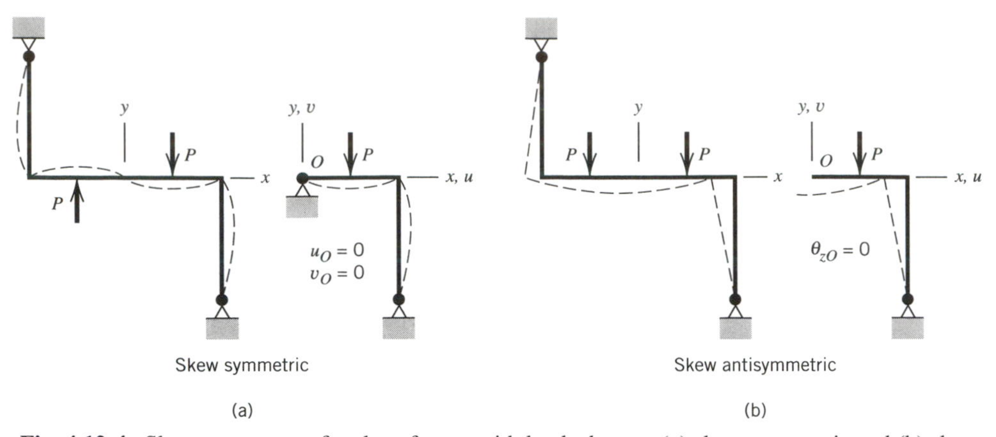

Fig. 4.12-4. Skew symmetry of a plane frame, with loads that are (a) skew symmetric and (b) skew antisymmetric.

the two solutions differ only in loads and support conditions. Such a trade may be advantageous if the structure is geometrically complicated and considerable effort is needed to prepare input data, or if the reduction in number of d.o.f. is important.

Skew or *inversion* symmetry is illustrated in Fig. 4.12-4. In Fig. 4.12-4a, a half-revolution of structure and loads about the z axis (normal to the paper) results in self-coincidence. In Fig. 4.12-4b, a half-revolution followed by reversal of loads results in self-coincidence. In both cases only half the structure need be analyzed, but support conditions at point O are not so readily stated as are support conditions for cases of reflective symmetry [4.7].

Axial symmetry prevails when a solid is generated by rotation of a plane shape about an axis in the plane. Although the structure is three-dimensional, the FE model need be only two-dimensional. Axially symmetric bodies are common and their analysis is discussed separately (Chapter 6).

A structure that is not axially symmetric may yet exhibit a rotational repetition of geometry, material properties, supports, and loads. This circumstance is called *cyclic* symmetry (or sectorial symmetry, or rotational periodicity). An example appears in Fig. 4.12-5a. A complete solution is obtainable by analysis of one repetitive portion, such as that in Fig. 4.12-5b. Other choices of representative repetitive portion are possible. Although only one such portion is needed, it is convenient to speak of "attachment" d.o.f. along AB and CD. Attachment d.o.f. along AB and CD must match exactly—in number, placement, type, and orientation—for the reason that d.o.f. along AB and CD must be constrained to have identical displacements. Specifically, nodes A and C must have the same displacement components in the respective n directions and the same displacement components in the respective s directions. If attachment d.o.f. carry externally applied loads, these loads must be applied on either AB or CD, but not both, as this would apply twice the load intended. In order to exploit cyclic symmetry, it is not necessary that the body be plane or that attachment d.o.f. lie on straight lines. In general, attachment d.o.f. lie on congruent curved surfaces in space, match exactly in position, and use d.o.f. that match in their orientations with respect to these surfaces. Concepts of cyclic symmetry need not be restricted to problems in which repetitions of form and loading appear with rotation about an axis. Similar repetitions may appear in a long slender structure. With appropriate loading, this would be possible in Fig. 4.11-1b, for

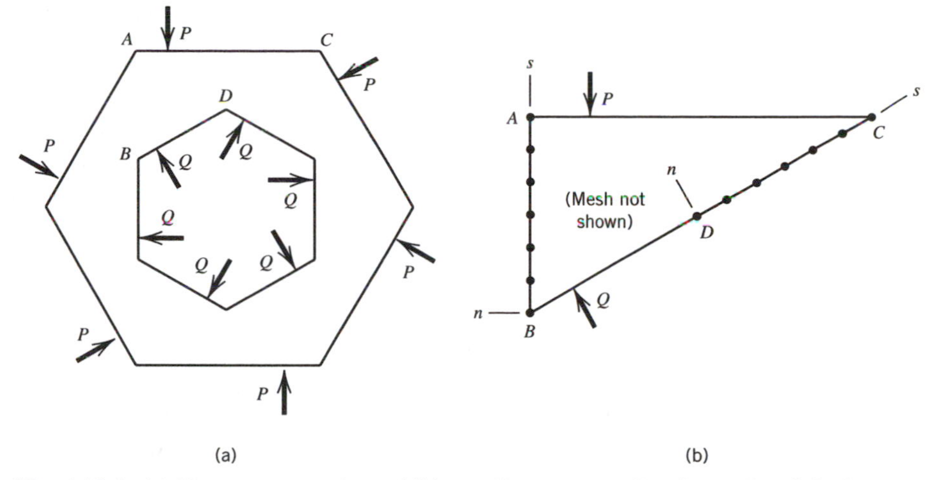

Fig. 4.12-5. (a) Plane structure that exhibits cyclic symmetry. Loads are P and Q. Supports (not shown) exert no force. (b) Typical repeating portion. Nodes on AB and CD are shown but the FE mesh is not shown.

example. This circumstance, less common than cyclic symmetry, may be called "repetitive" symmetry.

Caution. Symmetry concepts should be used sparingly and carefully in problems of vibration and buckling. For example, a uniform, simply supported beam has symmetry about its center but has *anti*symmetric vibration modes as well as symmetric vibration modes. If half the beam were analyzed, the support conditions of Fig. 4.12-1b would permit only symmetric vibration modes, while the support conditions of Fig. 4.12-1c would permit only antisymmetric vibration modes. Similarly, an axisymmetric solid or shell will have many vibration modes that are not axisymmetric. Caution is also needed in static problems that involve nonlinearity because symmetries present when loading begins may subsequently disappear.

4.13 CONSTRAINTS

A *constraint* may merely prescribe the numerical value of a d.o.f. and may then be called a "single-point constraint." The most common example is setting a d.o.f. to zero as a support condition. In the following discussion, "constraint" is used to mean a prescribed relation among d.o.f. (sometimes called a "multipoint constraint"). The problem of Fig. 4.3-3 is an example. In that problem, d.o.f. at nodes 1 and 2 are constrained to follow d.o.f. at nodes 3 and 4 and are replaced by d.o.f. at nodes 3 and 4 prior to assembly of elements. A constraint is roughly the opposite of a release (Section 2.3); however, d.o.f. in a constraint relation need not be physically adjacent.

One way to impose constraints is to use transformation, much as described below Eq. 4.3-2, to eliminate constrained d.o.f. prior to assembly of elements. For each equation of constraint, one d.o.f. can be eliminated. In what follows we describe how constraints may be applied to global equations $\mathbf{KD} = \mathbf{R}$, *after* assembly of elements, to override the elastic relation among d.o.f. to be constrained. We will describe two methods that are used in commercial software: the Lagrange multiplier method, which imposes constraints exactly, and the penalty method, which imposes constraints approximately. First, we illustrate constraint equations per se, as follows.

Constraint Equations. A constraint may be used in the plane problem of Fig. 4.13-1. The beam element has rotational d.o.f. but the plane elements do not. We wish to introduce moment communication at the left end of the beam element so that node 1 is not just a hinge connection. We may elect to make node 1 and edge 1-2 have the same rotation, that is, $\theta_{z1} = (u_2 - u_1)/b$. The equation of constraint is then

$$\begin{bmatrix} \frac{1}{b} & 0 & 1 & -\frac{1}{b} & 0 & 0 & 0 & \cdots \end{bmatrix} \mathbf{D} = 0 \tag{4.13-1}$$

where $\mathbf{D} = [u_1 \quad v_1 \quad \theta_{z1} \quad u_2 \quad v_2 \quad u_3 \quad v_3 \cdots]^T$ contains all d.o.f. active at the global level (here we save space by listing only the d.o.f. needed in the present example). As an alternative, we may elect to make node 1 and edge 3-1 have the same rotation, that is, $\theta_{z1} = (v_1 - v_3)/a$, for which the equation of constraint is

$$\begin{bmatrix} 0 & -\frac{1}{a} & 1 & 0 & 0 & 0 & \frac{1}{a} & \cdots \end{bmatrix} \mathbf{D} = 0 \tag{4.13-2}$$

Clearly, there are many plausible alternatives, such as using the u_i and v_i at nodes 2 and 3 instead, or enforcing the additional constraint that edges 1-2 and 1-3 remain perpendicular, and so on. One should not expect that any alternative will provide accurate stresses near node 1 in the plane body. (See Fig. 3.9-6 for a different treatment of this problem.)

An equation of constraint has the general form

$$\mathbf{CD} - \mathbf{Q} = \mathbf{0} \tag{4.13-3}$$

where $\mathbf{C}$ is an m by n matrix, m is the number of constraint equations ($m = 1$ in Eqs. 4.13-1 and 4.13-2), and n is the number of d.o.f. in the global vector $\mathbf{D}$. $\mathbf{Q}$ is a vector of constants. Often $\mathbf{Q} = \mathbf{0}$, as is the case in Eqs. 4.13-1 and 4.13-2. We will describe two ways to impose Eq. 4.13-3 on the global equations $\mathbf{KD} = \mathbf{R}$.

Lagrange Multiplier Method. We introduce as additional variables the Lagrange multipliers $\boldsymbol{\lambda} = [\lambda_1 \quad \lambda_2 \cdots \lambda_m]^T$. Each equation of constraint is written in homogeneous form and multiplied by the corresponding λ_i, which yields $\boldsymbol{\lambda}^T\{\mathbf{CD} - \mathbf{Q}\} = 0$. Next, the left-hand side of this equation is added to the usual energy terms, which produces the modified total energy expression

$$\Pi_p = \tfrac{1}{2}\mathbf{D}^T\mathbf{KD} - \mathbf{D}^T\mathbf{R} + \boldsymbol{\lambda}^T\{\mathbf{CD} - \mathbf{Q}\} \tag{4.13-4}$$

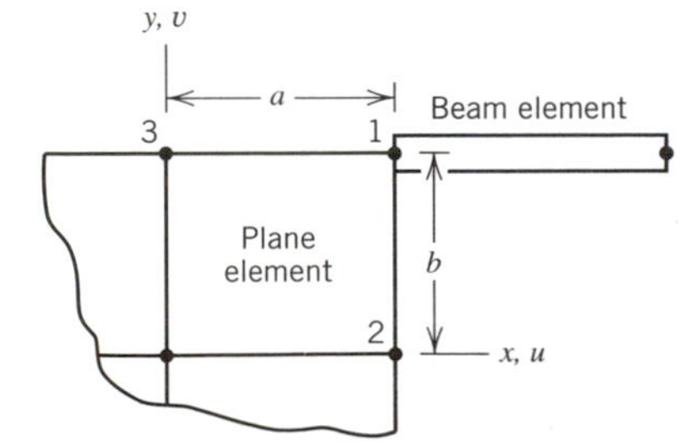

Fig. 4.13-1. A plane beam element joined to a plane quadrilateral element.

Derivatives of Π_p with respect to the D_i and the λ_i are set to zero, which yields

$$\begin{bmatrix} \mathbf{K} & \mathbf{C}^T \\ \mathbf{C} & \mathbf{0} \end{bmatrix} \begin{Bmatrix} \mathbf{D} \\ \boldsymbol{\lambda} \end{Bmatrix} = \begin{Bmatrix} \mathbf{R} \\ \mathbf{Q} \end{Bmatrix} \tag{4.13-5}$$

The lower partition states the m constraint conditions, Eq. 4.13-3. If $m = 0$ we obtain the usual result $\mathbf{KD} = \mathbf{R}$. Equation 4.13-5 is solved for both $\mathbf{D}$ and $\boldsymbol{\lambda}$. Despite the null submatrix, a Gauss elimination solution will not fail if eliminations are properly sequenced because eliminations introduce nonzero diagonal coefficients.

As an example, we impose the constraint $u_1 = u_2$ in Fig. 4.13-2a. After all support conditions have been imposed, but not the constraint condition, global equations are as shown in Fig. 4.13-2b. The equation of constraint is

$$\mathbf{C} \begin{Bmatrix} u_1 \\ u_2 \end{Bmatrix} = 0 \qquad \text{where} \qquad \mathbf{C} = [1 \quad -1] \tag{4.13-6}$$

Equation 4.13-5 becomes

$$\begin{bmatrix} k & -k & 1 \\ -k & 2k & -1 \\ 1 & -1 & 0 \end{bmatrix} \begin{Bmatrix} u_1 \\ u_2 \\ \lambda \end{Bmatrix} = \begin{Bmatrix} P \\ 0 \\ 0 \end{Bmatrix} \tag{4.13-7}$$

Solving, we obtain $u_1 = u_2 = P/k$, $\lambda = P$. The sign of λ is not significant, but its magnitude can be regarded as the force of constraint.

Penalty Method. Equation 4.13-3 is modified to read $\mathbf{t} = \mathbf{CD} - \mathbf{Q}$, so that $\mathbf{t} = \mathbf{0}$ implies satisfaction of the constraints. An energy expression analogous to Eq. 4.13-4 is

$$\Pi_p = \tfrac{1}{2}\mathbf{D}^T\mathbf{K}\mathbf{D} - \mathbf{D}^T\mathbf{R} + \tfrac{1}{2}\mathbf{t}^T\boldsymbol{\alpha}\mathbf{t} \tag{4.13-8}$$

where $\boldsymbol{\alpha} = \lceil \alpha_1 \quad \alpha_2 \cdots \alpha_m \rfloor$ is a *diagonal* matrix of "penalty numbers," chosen by the analyst and preferably dimensionless. Derivatives of Π_p with respect to the D_i are set to zero, which yields

$$[\mathbf{K} + \mathbf{C}^T\boldsymbol{\alpha}\mathbf{C}]\mathbf{D} = \mathbf{R} + \mathbf{C}^T\boldsymbol{\alpha}\mathbf{Q} \tag{4.13-9}$$

where $\mathbf{C}^T\boldsymbol{\alpha}\mathbf{C}$ is called a "penalty matrix." If $\boldsymbol{\alpha} = \mathbf{0}$, the constraints are ignored. As $\boldsymbol{\alpha}$ becomes large, the penalty of violating constraints becomes large, so that constraints are very nearly satisfied. Penalty numbers that are too large produce numerical ill-conditioning, which may make computed results unreliable and may even "lock" the mesh (e.g., if the material is incompressible; see below).

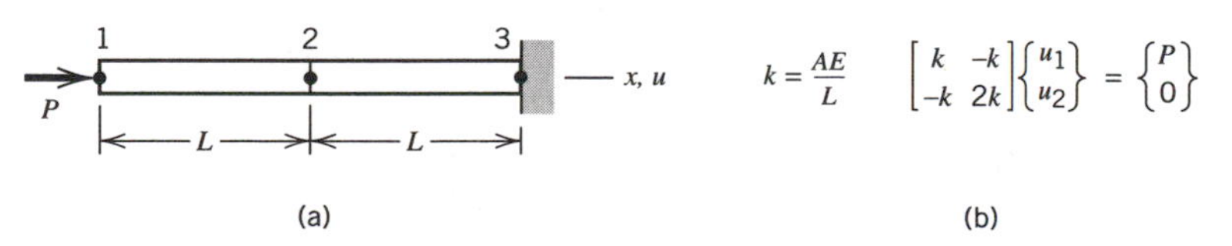

Fig. 4.13-2. (a) Two-element uniform bar. A = cross-sectional area; E = elastic modulus. (b) Global equations $\mathbf{KD} = \mathbf{R}$, with u_1 and u_2 the only nonzero d.o.f.

As an example, consider again the constraint $u_1 = u_2$ in Fig 4.13-2. There is only one constraint and therefore only one penalty number, which is dimensionless if we elect to write the constraint matrix as $\mathbf{C} = [\sqrt{k} \; -\sqrt{k}]$. Equation 4.13-9 becomes

$$\left(\begin{bmatrix} k & -k \\ -k & 2k \end{bmatrix} + \alpha \begin{bmatrix} k & -k \\ -k & k \end{bmatrix} \right) \begin{Bmatrix} u_1 \\ u_2 \end{Bmatrix} = \begin{Bmatrix} P \\ 0 \end{Bmatrix} \tag{4.13-10}$$

which has the solution

$$u_1 = \frac{2+\alpha}{1+\alpha}\frac{P}{k} \qquad u_2 = \frac{P}{k} \tag{4.13-11}$$

If $\alpha = 0$, then $u_1 = 2P/k$ and $u_2 = P/k$, which is the unconstrained elastic solution. If α becomes large, we approach the constrained solution $u_1 = u_2 = P/k$. Note that if α were infinite, the coefficient matrix in Eq. 4.3-10 would be singular. Thus we see that penalty numbers must be large enough to be effective but no so large as to cause numerical difficulties.

Very Stiff Elements. A stiff region or element in a comparatively flexible structure contributes a penalty stiffness matrix to **K**, as in the example of Eq. 4.13-10. Further examples appear in Section 5.10. A very stiff region may provoke serious ill-conditioning. Rather than attempting to manage the difficulty it is better to avoid it altogether by using (say) the Lagrange multiplier method to make the stiff region perfectly rigid. This is a common practical application of multipoint constraints.

Incompressible Materials. As Poisson's ratio approaches 0.5, a material approaches incompressibility. If an element is incompressible, its normal strains are constrained to sum to zero, which is the condition of no volume change. Accordingly, the number of constraint conditions in a FE model of an incompressible material is equal to the number of elements times the number of Gauss points used to integrate each element. These constraints are not imposed after **K** is formed; rather they arise naturally, are incorporated in **K**, and can be shown to have the form of a penalty matrix. If the penalty matrix is nonsingular the mesh "locks"; that is, computed d.o.f. may be orders of magnitude too small. A useful solution can be obtained if the penalty matrix is singular. Arguments too lengthy to repeat here [2.2] indicate that solutions are reliable if (a) penalty terms are integrated using fewer Gauss points than used for other terms in element stiffness matrices, (b) the ratio of the number of d.o.f. to the number of penalty Gauss points is approximately 2:1 for plane problems and 3:1 for solid problems, and (c) v is such that the reciprocal of $3(1 - 2v)$ is between $10^{p/3}$ and $10^{p/2}$, where p is the number of digits used in computer words.

The foregoing remarks about incompressibility do not apply to problems of plane stress, plate bending, and shells, for which thickness changes are unrestrained and the incompressibility condition is therefore not enforced.

ANALYTICAL PROBLEMS

4.1 Let the structures shown have two-node elements and one d.o.f. per node. Number the nodes so that there are as few coefficients as possible between the skyline and the diagonal of **K**. Also, write **K** to the extent of showing the locations of its nonzero coefficients for the numbering you choose.

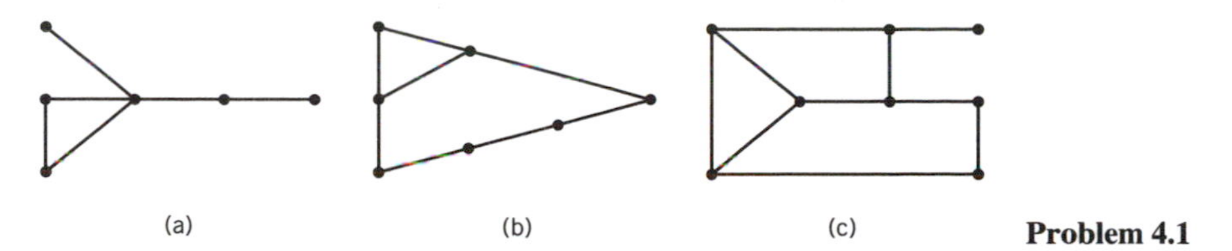

Problem 4.1

4.2 Imagine that the bar element in Fig. 4.3-1a is arbitrarily oriented in space, with its orientation defined by direction cosines l, m, and n of angles between the element axis and global axes XYZ. Write the appropriate transformation matrix **T** and obtain the resulting element stiffness matrix **k** that operates on nodal translation d.o.f. parallel to X, Y, and Z axes.

4.3 Imagine that the element in Fig. 4.3-1a is a *beam* element, having as nodal d.o.f. translations w_1 and w_2 normal to the xy plane and rotations θ_{y1} and θ_{y2} whose vectors are parallel to the y axis. Write the transformation matrix **T** that will convert the matrix $\mathbf{k}'$ that operates on these d.o.f. to a matrix **k** that operates on d.o.f. w_1, w_2, and rotations θ_{Xi} and θ_{Yi} $(i = 1, 2)$ about X and Y axes.

4.4 Generalize Fig. 4.3-3a so that there are the usual six d.o.f. per node and offsets a_i, b_i, and c_i $(i = 1, 2)$. Write the appropriate transformation matrix **T**.

4.5 Obtain numerical values of the four coefficients in matrix **J**, Eq. 4.4-4, for each of the elements shown. Also, compute $|\mathbf{J}|$ and explain its significance.

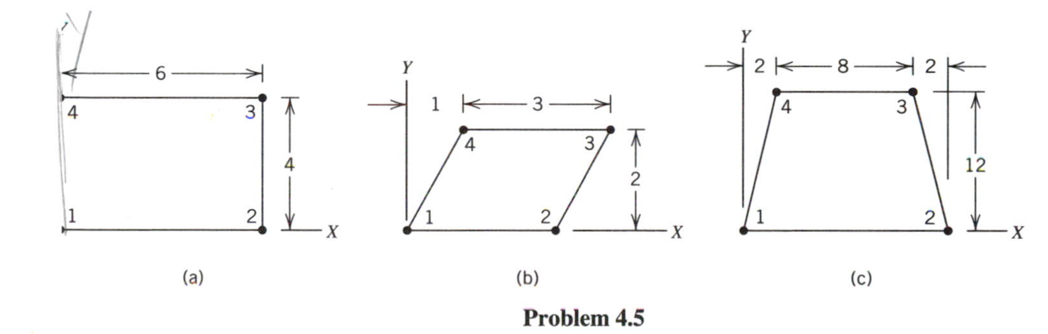

Problem 4.5

4.6 For the four-node plane element discussed in Section 4.4, write coefficients in the **B** matrix in terms of coefficients J^*_{ij} in $\mathbf{J}^{-1}$ and derivatives of the N_i with respect to ξ and η. Assume that nodal d.o.f. have the order $\{u_1 \quad v_1 \quad u_2 \cdots v_4\}$.

4.7 The "natural" coordinate system $\xi\eta$ in Fig. 4.4-1 is not unique. Another possible choice is the rs system shown in the sketch. Restate the N_i of Eq. 4.4-2 in terms of r and s.

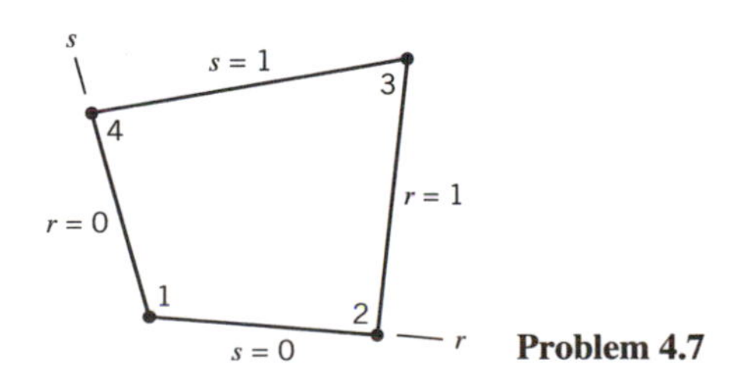

Problem 4.7

4.8 Gauss points and weights are symmetric with respect to the center of the integration interval, and a two-point rule integrates the polynomial $\phi = c_1 + c_2\xi + c_3\xi^2 + c_4\xi^3$ exactly. Use this information to derive the location and weight values of the two-point Gauss rule for the integration interval -1 to $+1$.

4.9 If we sum over all points of a Gauss quadrature rule, we obtain $\Sigma W_i = 2$ in one dimension and $\Sigma\Sigma W_i W_j = 4$ in two dimensions, for any order of rule. Why?

√**4.10** Use one-, two-, and then three-point Gauss quadrature rules to integrate the following functions over the interval $\xi = -1$ to $\xi = +1$. Compute the percentage error of each result.

(a) $\phi = \xi^2 + \xi^3$
(b) $\phi = \cos 1.5\xi$
(c) $\phi = (1 - \xi)/(2 + \xi)$

4.11 Use Gauss quadrature to evaluate the integral

$$I = \int_{-1}^{1}\int_{-1}^{1} \frac{3+\xi^2}{2+\eta^2}\, d\xi\, d\eta$$

Use (a) one point, (b) four points, and (c) nine points. Compare each of the results with the exact result obtained by use of a table of integrals.

4.12 (a) Let the element in Fig. 4.6-2b be a 2 by 2 square. The displacement mode shown is then $u = x(1 - 3y^2)$, $v = y(3x^2 - 1)$. Show that this mode produces zero strains at the four Gauss points.

(b) Sketch an adjacent eight-node element whose nodal displacements are of opposite sign to those in Fig. 4.6-2b. In what way are the two elements incompatible?

4.13 Imagine the stiffness matrix of a simple plane beam element (four d.o.f.) is to be integrated by Gauss quadrature.

(a) Sketch the deflected shape of the beam for the instability that is possible if a single Gauss point is used.
(b) How many Gauss points are needed to integrate **k** exactly? Why?
(c) Evaluate **k** using one-point quadrature. Show that it contains the correct bending stiffness, and explain its defects. *Note*: $|\mathbf{J}| = L/2$.

4.14 Consider a 24-d.o.f. solid element in the form of a cube. There are eight vertex nodes. Each has translational d.o.f. in x, y, and z directions. If integrated by a single Gauss point at the element center there are 12 instability modes. Using x-direction displacements only, sketch deformed elements for four of these modes.

4.15 If a bar element is formulated in isoparametric fashion, it extends from $\xi = -1$ to $\xi = +1$. Imagine that axial stresses are σ_{xA} and σ_{xB} at the respective Gauss points of an order 2 rule. Based on σ_{xA} and σ_{xB}, what is σ_x as a function of ξ, and what is the extrapolated value of σ_x at each end of the element?

4.16 For the uniform cantilever beam shown, assume that the lateral displacement field is given by $v = \Sigma\beta_i x^i$, where $i = 2, 3, \ldots, n$. Compute strain energy U by integration of $0.5EI(d^2v/dx^2)^2$ over length L [2.1] and the total energy as $\Pi_p = U - Pv_L$, where v_L is v evaluated at $x = L$. From the equations $\partial\Pi_p/\partial\beta_i = 0$, $i = 2, 3, \ldots, n$, compute the β_i and v_L for (a) $n = 2$ and (b) $n = 3$. Compare the computed v_L and bending moment at $x = 0$ with predictions of beam theory.

4.17 For the simply supported and uniformly loaded beam shown, assume that the lateral displacement field is $v = \beta x(L - x)$. Compute strain energy U as described in Problem 4.16 and the total energy as U minus the integral of $v(q\,dx)$ over length L. Compute β from $d\Pi_p/d\beta = 0$. At the ends, middle, and quarter-points of the beam, compare computed values of deflection, slope, and bending moment with values predicted by beam theory.

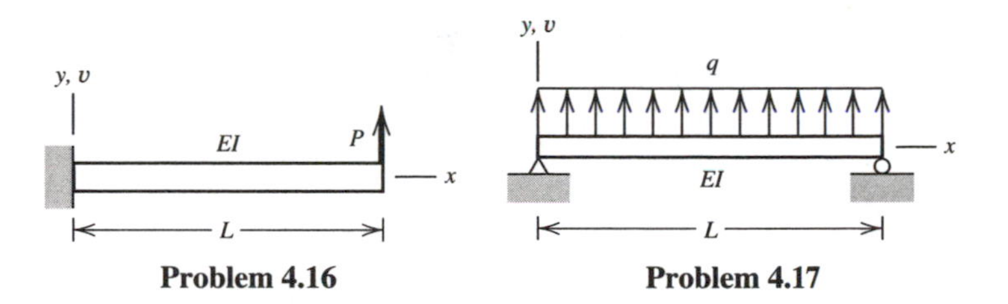

Problem 4.16 **Problem 4.17**

4.18 Repeat Problem 4.16 with load P replaced by a uniformly distributed load q. See Problem 4.17 for advice on calculating work associated with the load.

4.19 Proceeding as in Problem 4.17, compute the center deflections v_{0e} and v_{0a} in Problem 2.8. (Contrary to the simple assumption made in Problem 2.8, the present analysis will not yield $v_{0e} = v_{0a}$.)

4.20 Consider patch tests for constant σ_y and constant τ_{xy} in the FE model of Fig. 4.9-1a. Let the x-direction span be L. For convenience, place all side nodes of the patch at midsides. What are appropriate nodal loads in each case?

4.21 As an alternative form of the patch test, one could impose all d.o.f. at all nodes, using values consistent with a constant strain state, then calculate nodal loads $\mathbf{R} = \mathbf{KD}$. If the patch test is passed, what should be the calculated loads at the node internal to the mesh?

4.22 Let the equations that represent a substructure be

$$\begin{bmatrix} \mathbf{K}_{aa} & \mathbf{K}_{as} \\ \mathbf{K}_{as}^T & \mathbf{K}_{ss} \end{bmatrix} \begin{Bmatrix} \mathbf{D}_a \\ \mathbf{D}_{ss} \end{Bmatrix} = \begin{Bmatrix} \mathbf{R}_{aa} \\ \mathbf{R}_{ss} \end{Bmatrix}$$

By solving for $\mathbf{D}_{ss}$ from the lower partition and substituting into the upper partition, determine expressions for $\mathbf{K}_a$ and $\mathbf{R}_a$ in the reduced equations $\mathbf{K}_a\mathbf{D}_a = \mathbf{R}_a$.

4.23 (a) The beam shown is uniform and simply supported. Use elementary beam theory and the type of superposition method suggested in Fig. 4.12-3 to determine the deflection at the center point of the beam and at load P. Express answers in terms of P, a, E, and I.

(b) The problem depicted in Fig. 4.12-5a can be solved by the method of Fig. 4.12-3 rather than by the method of cyclic symmetry. Describe how.

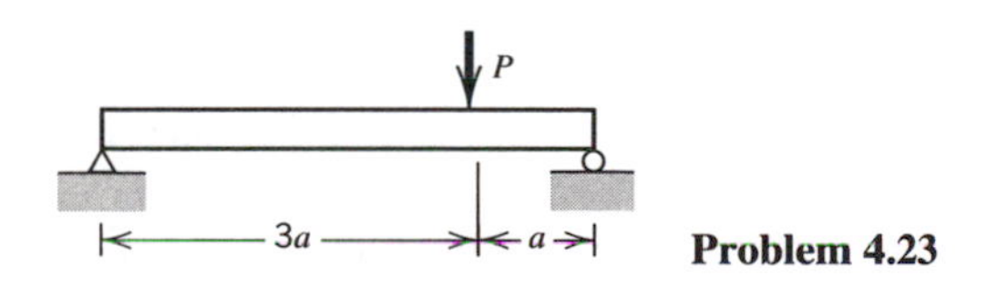

Problem 4.23

4.24 The sketch shows a plan view of a *grillage*, which is a planar arrangement of interconnected beams. Assume that all elements are identical and that nodal d.o.f. are w_i

(in the z direction), θ_{xi}, and θ_{yi}. The grillage is square and supports impose $w_i = 0$ at all nodes i on the boundary of the square. A z-parallel force of magnitude P acts on nodes as described below. Describe what portion of the grillage constitutes the smallest acceptable model, and what its boundary conditions are, if:

(a) All loads P act in the same direction.

(b) Loads P act upward for $y > 0$, downward for $y < 0$, and are omitted on $y = 0$.

(c) Loads P act upward in the first and third quadrants, downward in the second and fourth quadrants, and are omitted on x and y axes.

(d) Loads P alternate in direction by octants — for example, upward between $y = 0$ and $x = y$, and downward between $x = y$ and $x = 0$ — and are omitted on x and y axes and on the lines $x = y$ and $x = -y$.

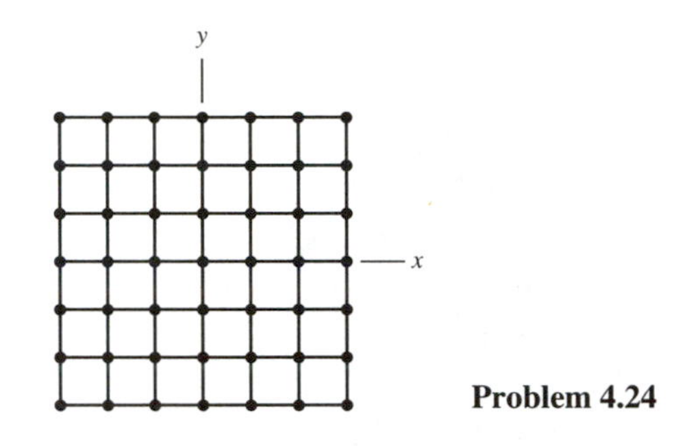

Problem 4.24

4.25 A uniform cantilever beam of constant thickness is tip-loaded by moment and/or transverse force. It is to be analyzed using plane elements. One need model only half the beam, using the portion on either side of the longitudinal axis through centroids of beam cross sections. Describe appropriate FE boundary conditions for such a model.

4.26 The sketch represents a uniform rectangular plate with force P applied normal to the plate at one corner. Imagine that the plate is supported by a uniform elastic foundation and that the FE mesh (not shown) is uniform. The d.o.f. at a typical node i are w_i, θ_{xi}, and θ_{yi}. Describe how the entire plate can be analyzed for lateral deflection $w = w(x, y)$ by analyzing a single quadrant four times, each time with appropriate loading and boundary conditions, then superposing results.

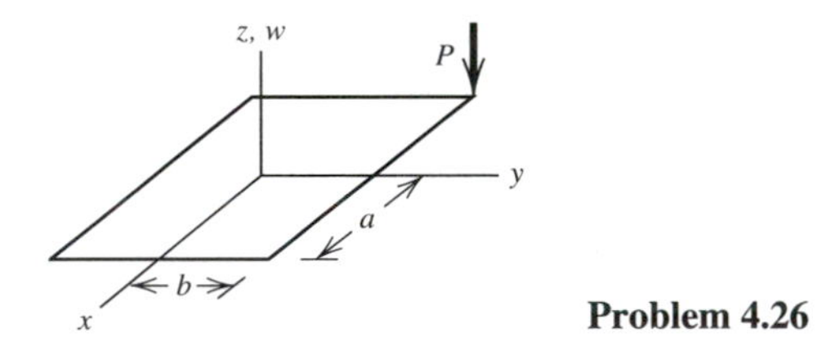

Problem 4.26

4.27 Write an equation in the form of Eq. 4.13-3 that states each of the following constraints.

(a) In Fig. 4.13-1, diagonal 2-3 has the rotation θ_{z1}. The d.o.f. of the FE model are stated below Eq. 4.13-1.

(b) Combine the constraint of part (a) with Eq. 4.13-2, so that $\mathbf{CD} = \mathbf{0}$ has two rows. Does $\mathbf{CD} = \mathbf{0}$ imply Eq. 4.13-1?

(c) In Fig. 4.13-1, line 2-1 rotates an amount e more than line 3-1, where e is a small angle.
(d) The right-hand edge of the mesh in Fig. 4.9-1a displaces as a straight line.

4.28 The uniform beam shown is simply supported and is loaded by moment M_o at its left end.
(a) Write an equation that constrains the two end rotations to be of equal magnitude but opposite sign.
(b) Solve for the end rotation by means of a Lagrange multiplier.
(c) Solve for the end rotation by means of the penalty method.

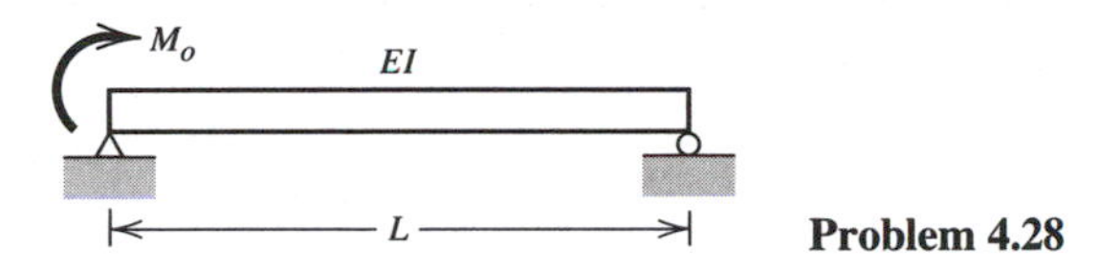

Problem 4.28

COMPUTATIONAL PROBLEMS

4.29 Model the T-section cantilever beam shown by using separate sets of beam elements, one set for the cross of the T and the other set for the stem. Use one, then two, then four elements along the length in each set. Make nodes of the stem slave to nodes of the cross. Let Poisson's ratio be zero. Compare computed values of tip displacement and rotation with values predicted by beam theory.

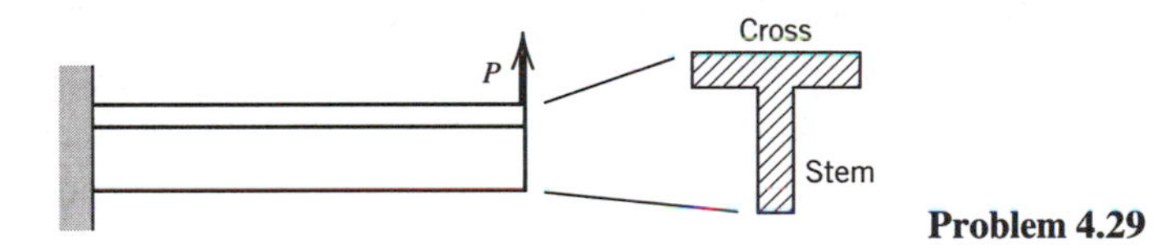

Problem 4.29

4.30 If software permits, reanalyze the beam problems of Fig. 3.11-1, using different orders of Gauss quadrature to form element stiffness matrices. Do changes in quadrature order have the expected effects?

4.31 If software permits, solve the problem depicted in Fig. 4.6-2a. For what value of L/H is the displacement of P greater in magnitude than L? Are stresses accurately computed anywhere in the model?

4.32 The structure shown is modeled by eight-node plane elements and is loaded by a force normal to one element side at B. Use four Gauss points to evaluate $\mathbf{k}$ of each element. Let elements 1 and 2 have elastic modulus E_a and element 3 have elastic modulus E_b. For what value of E_b/E_a is the relative displacement of nodes A and B greater in magnitude than dimension a? When this happens, are stresses accurately computed anywhere in the model? Use a small value for load P.

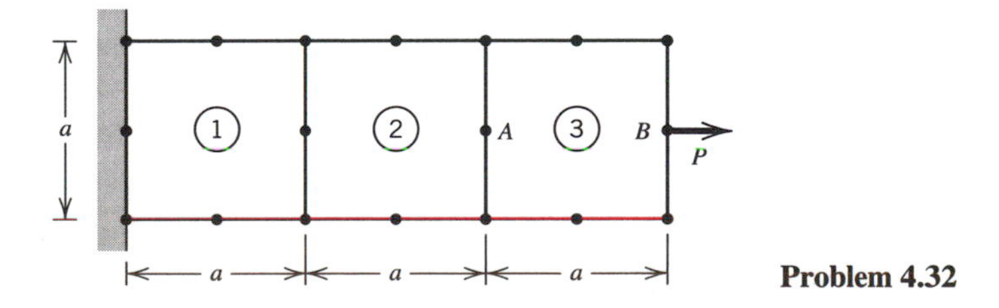

Problem 4.32

4.33 Assume that plane stress conditions prevail in Fig. 4.10-1b. Let *AB* be rigid and made to translate vertically downward a small amount. Compute stresses in the elastic medium *ABCD*. Investigate the effect of placing a rigid boundary *CD* close to *AB* and then further away. If the software includes infinite elements or boundary elements, use them to repeat the analysis that has the "close" placement of boundary *CD*.

4.34 The sketch shows a central crack of length $2a$ in a flat strip of material whose width is $2c$. If side nodes of isoparametric elements are moved to quarter points in the manner shown, stresses vary as $r^{-0.5}$ along certain radial lines [4.8]. The $r^{-0.5}$ variation accords with the theory of linear fracture mechanics. The mode I stress intensity factor K_I can be computed as

$$K_I = \frac{2G}{\kappa+1}\left(\frac{\pi}{2\ell}\right)^{0.5}[4\Delta_C - \Delta_D]$$

where G is the shear modulus, $\kappa = (3 - \nu)/(1 + \nu)$ for plane stress conditions or $\kappa = 3 - 4\nu$ for plane strain conditions, and Δ_C and Δ_D are the amounts of crack opening at C and D [4.9]. A handbook gives a formula for K_I:

$$K_I = \sigma_x \sqrt{\pi a}\,\frac{1-0.5\,(a/c)+0.326(a/c)^2}{[1-(a\ c)]^{0.5}}$$

Assign convenient dimensions, complete the FE model, do computations, and compare the computed and formula values of K_I. Also, use alternative methods for calculation of K_I if the software provides them.

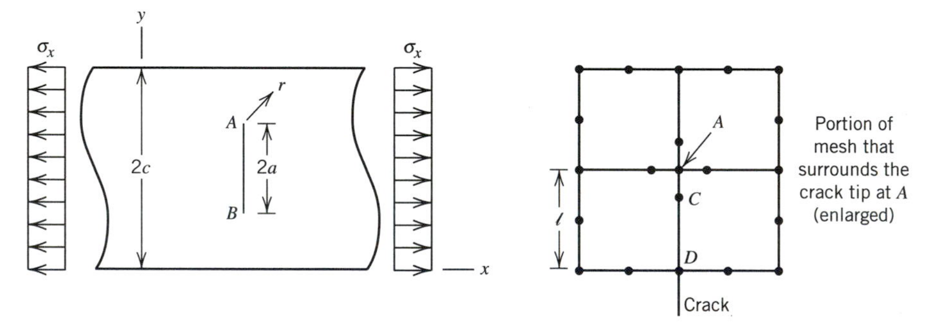

Problem 4.34

CHAPTER 5

Modeling, Errors, and Accuracy in Linear Analysis

This chapter is concerned with matching element behavior to anticipated structure behavior; treatment of loads, supports, and connections; planning the FE model; debugging; checking results; and convergence of successive solutions. Some relevant advice is discussed more fully in other chapters, especially for thermal analysis, dynamics, and nonlinear problems. The present chapter deals mostly with linear static problems of structural mechanics. Matters discussed are under the direct discretion and control of the analyst and require considerable thought, in contrast to topics in Chapter 4, which are largely internal to software and need only be properly invoked (such as a decision to use substructuring or to exploit symmetry).

Advice that follows does not fall neatly into categories. Accordingly, the division into sections is somewhat arbitrary. Neither the sections nor their contents should be regarded as checklists of procedural steps to be followed in every problem. Instead, the advice should be learned well enough that appropriate parts of it come automatically to mind when needed to deal with particular situations.

5.1 MODELING IN GENERAL

FE modeling is the simulation of physical behavior by a numerical process based on piecewise polynomial interpolation. In order to obtain a reliable FE solution, the analyst must first have a grasp of the problem area, be it stress analysis, thermal analysis, or whatever. Only then can one address questions that must be answered: What physical actions are important? Is the problem time independent? Are there nonlinearities? What are the boundary conditions? How will results be checked? And so on. If a FE analysis goes astray it is usually because the analyst's understanding of physical behavior, boundary conditions, limitations of theory, FE behavior, or options in the program is insufficient to prepare a satisfactory model. Clearly, *FE modeling is more than preparing a mesh and preprocessing.*

Skill in FE modeling is based on an ability to visualize physical behavior and relate it to element behavior. Skill is developed by practice and by critical evaluation of computed results. The necessary knowledge base includes statics, structural theory, and FE theory. Here the word "theory" does not imply something rarified and impractical; it means a system of knowledge and assumptions, with rules of procedure and having predictive value. Knowing the assumptions and limitations of a structural theory may keep us from using it inappropriately. As an example, in elementary beam theory only axial normal stress is taken into account, so elementary beam theory should not be used for very wide beams (use plate bending theory instead) or for short deep beams (use two-dimensional analysis instead). *One must understand the classic analysis tools*, specifically including the widely used (and widely misused) methods and formulas of elementary mechanics of materials. Assumptions *and restrictions* that underlie analysis tools are also incorporated in finite elements. For the practitioner, the main reason to study stress analysis theory is that assumptions and restrictions are revealed and one then knows when *not* to use a theory or a procedure. FE modeling is closely related to theory because FE is largely a way of implementing theory. The piecewise-polynomial way in which FE modeling implements theory requires that the analyst have a sense of how various elements respond to various loadings. A given element type might behave differently in different programs because of special restrictions, special features, and even input data defaults. One must also know how other aspects of the software behave. An assumption about how the software *should* behave may lead to great confusion and frustration. Therefore it is necessary to *study the documentation*, even though it is likely to be inscrutable in places. Increasingly, documentation is part of the software and can immediately be called to the screen.

A difficult problem or a large model should not be treated all at once. It is better to start with special cases and coarse meshes, then revise the model as necessary. If we resolve to *start simply and expect to revise*, we will have more confidence in the final results and may also reduce the total time spent on the project. Each FE model discloses information that improves the next one; for example, we can learn where stress gradients are large and refine the next model in that area.

Output cannot be accepted at face value. Six digits in numbers and pretty stress contours do not imply accuracy. It is necessary to *critically examine the computed results.* If checking is begun after FE computations are complete, there is a tendency to rationalize FE results already obtained, often at considerable effort. This tendency is reduced by having an approximate analysis in hand before FE analysis is begun. If this is done habitually, there will be an added benefit: skills in analysis and modeling will improve, because of thinking more deeply about problems at the outset and subsequently seeing how well approximate analyses agree with FE analyses.

5.2 STRUCTURE BEHAVIOR AND ELEMENT BEHAVIOR

What type of elements should I use—beam, shell, solid, or what? Triangular or quadrilateral? With or without side nodes? How many? How should the mesh be graded? Are there nonlinearities? Such questions inevitably arise. Answers may not come easily, especially for the initial FE model, but will not come at all without some understanding of how the structure is *likely* to behave and how elements are *able* to behave. In general, one remembers that *the essence of the FE method is piecewise polynomial interpolation* and tries to select elements of such a type and size that deformation of the structure over the region spanned by an element is closely approximated by deformation modes that the element can represent. Alternatively, but with similar intent, one could speak of trying to match the strain or stress field capabilities of an element to the strain or stress field in the region of the structure spanned by the element.

As a simple example, consider cases of pure bending, Fig. 5.2-1. We know that axial strain ε_x varies linearly with y across a straight member. Consider FE models in which dimension h is spanned by a single layer of plane elements (as in Fig. 3.11-1). From the discussion in Chapter 3 we know that this behavior would be modeled badly by CSTs but

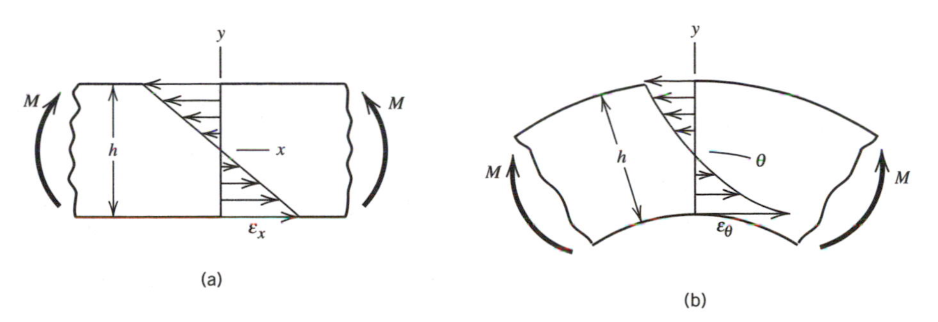

Fig. 5.2-1. Bending moment M applied to (a) straight member and (b) curved member.

would be modeled exactly by Q6 and Q8 rectangles. If the member is curved (Fig. 5.2-1b), circumferential strain ε_θ does not vary linearly with y [2.1]. Q6 elements would not be exact and would be inferior to Q8 elements because ε_θ can vary quadratically with y in a Q8 element but only linearly with y in a Q6 element. We could not have reached all these conclusions without understanding both structural behavior and element behavior.

In the remainder of this section we discuss aspects of structural behavior not usually discussed in a first course in mechanics of materials, and note how structural behavior is related to considerations in FE modeling. Similar behaviors and considerations may be presented by objects less conveniently shaped than those in the following examples.

A standard rolled section such as an I beam, Fig, 5.2-2a, usually carries transverse loads that cause bending. A beam is usually slender rather than short and deep. A proper model of a slender beam is built of standard two-node beam elements, which are discussed in Section 2.3. However, if the flanges are quite wide (Fig. 5.2-2b), the flexure formula $\sigma_x = My/I$ becomes inaccurate. It predicts that σ_x is independent of z. Actually, because of "shear lag," σ_x varies appreciably across a wide flange [1.5]. The physical action can be understood as follows. Each flange is loaded along its centerline by shear flow q applied by the web (Fig. 5.2-2c). The resulting axial deformation is not uniform across the flange, so neither is axial stress. A FE model that captures this behavior is built of two-dimensional elements, Fig. 5.2-2d. These elements could be membrane elements if the variation of σ_x through the thickness of a flange is negligible—but membrane elements have no y-direction stiffness, so plate elements would be more appropriate. Finally, compressive stresses in the lower flange may reduce its stiffness, even without reaching a local buckling condition. The effect is called "stress stiffening," even when it decreases stiffness rather than increasing it. Significant stress stiffening should be taken into account, but this will not happen automatically; the software must be *told* to do so.

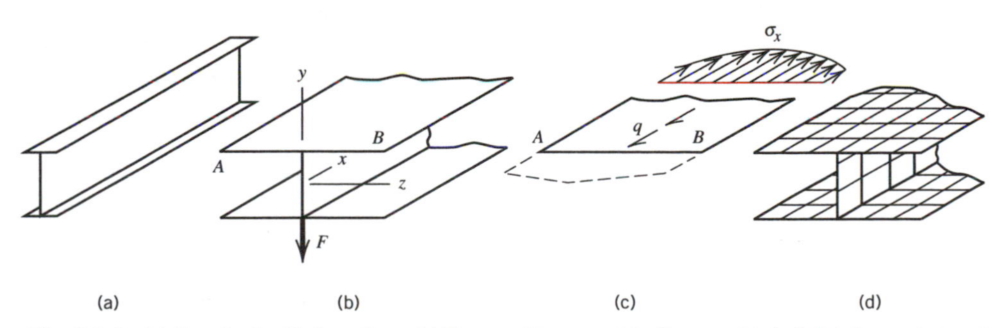

Fig. 5.2-2. (a) Standard rolled section. (b) Beam with very wide flanges. (c) Axial deformation and stress in upper flange. (d) FE model.

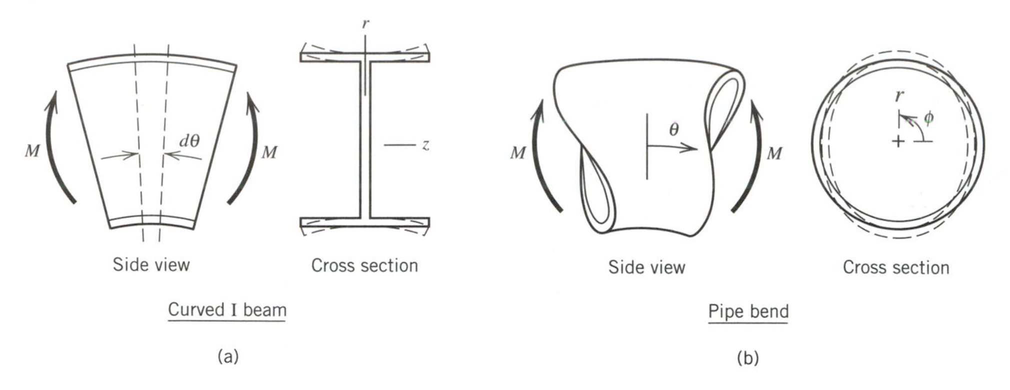

Fig. 5.2-3. Curved beams of thin-walled section under bending load. Dashed lines show deformation in the plane of a cross section.

If a curved beam has an I section, its flanges deflect radially when bending moment is applied (Fig. 5.2-3a). The physical action can be understood by considering a slice spanned by arc $d\theta$: circumferential flexural stress σ_θ has a radial component that pushes the outer flange outward and pulls the inner flange inward. This action is reversed if moment M is reversed. For either direction of M, flanges develop flexural stress σ_z directed normal to the web. Stress σ_z may be larger than the circumferential stress σ_θ. Also, radial motion of the flanges reduces the stiffness of the beam as seen by moments M. Standard beam elements do not account for these effects unless provided with correction factors [2.1, 5.1]. A FE model similar to that in Fig. 5.2-2d will be satisfactory if built of shell elements, that is, if each element has both membrane stiffness and bending stiffness.

Thin-walled pipe bends are subject to the same physical action as curved I beams. Their cross sections "ovalize" in response to bending moment (Fig. 5.2-3b). The resulting flexural stress in the ϕ direction may exceed stress in the θ direction. A pipe bend could be modeled by beam elements with correction factors [5.2]. Also, a FE model built of shell elements would be satisfactory, although tedious to prepare and having many d.o.f. However, piping systems are so often analyzed that special pipe-bend elements have been devised [5.3] and are often available in FE software.

A cross section of a prismatic beam has a "shear center," which is the point through which a transverse load must pass if the beam is to bend without twisting [2.1]. In Fig. 5.2-4a, load P on the thin-walled channel does not pass through the shear center of the cross section. In consequence, the channel twists as well as bends. Twisting produces shear stress. It also produces warping of cross sections; that is, it produces axial displacements such that initially plane cross sections do not remain plane. At end $x = 0$, warping is restrained by the support, which applies axial stresses proportional to the tendency to warp (Fig. 5.2-4c). If a beam formulation is to account for these effects, the usual six d.o.f. per node must be supplemented by a "warping" d.o.f. The standard two-node beam elements in most software cannot model warping effects. Instead, we must use a FE model similar to that in Fig. 5.2-2d.

Figure 5.2-5 shows a thin-walled cylindrical water tank with a fixed base. Support reactions M_0 and V_0 on the tank are uniformly distributed around its base. The tank can be modeled satisfactorily by cylindrical shell elements. We see that the axial flexural stress σ_z is large, quite localized near the base, and has steep gradients. A coarse mesh is satisfactory near the top of the tank, but the same coarseness near the bottom may portray the state of stress so poorly as to give little indication that a finer mesh is needed. In general, large flexural stresses and steep gradients are to be expected near "discontinuities" of

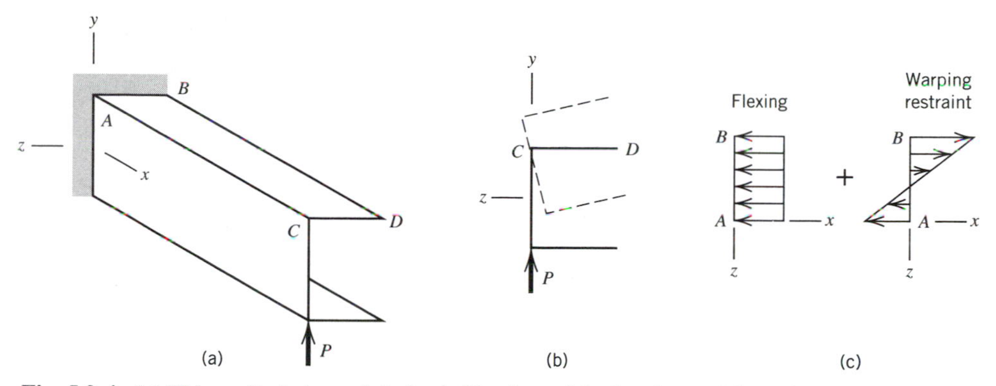

Fig. 5.2-4. (a) Thin-walled channel tip loaded by force P in the plane of the web. (b) Deflection of the tip cross section is shown by dashed lines. (c) Qualitative contributions to axial stress σ_x in the upper flange at the fixed end, viewed normal to the upper flange.

stress in shells. Discontinuities are associated with line loads, supports, reinforcements, and changes in curvature (e.g., where an ellipsoidal end cap is joined to a cylindrical pressure vessel). The reader is urged to learn enough about shell behavior to be able to anticipate where flexural stresses may be large [1.5, 2.1, 7.3].

Nonlinearity may appear in a variety of problems. In Fig. 5.2-6a, pressure p is carried mostly by bending action in the plate and there is an almost linear relation between lateral deflection and lateral pressure p until the center deflection is roughly half the thickness t [7.3]. This much deflection is reached quickly if the plate is thin. With greater deflection, membrane stresses support an increasing portion of the load. Thus the stiffness of the plate appears to increase. The plate need not have fixed supports: the same physical action arises whenever the deflected surface is *nondevelopable*, that is, cannot be unrolled into a flat sheet without producing membrane strains (cylindrical and conical surfaces are developable, spherical surfaces are not). To model large deflections of a plate we need shell elements, which account for both membrane and bending strains, and a nonlinear analysis procedure. (On the other hand, if the plate is *thick*, deflections will not be large enough for the foregoing nonlinearity to arise, but transverse shear deformation may become significant. One must then ask if plate elements in the software take transverse shear deformation into account). Figure 5.2-6b is a simple buckling problem. But if there is any imperfection, for example, an off-center load or a slight initial curvature of the col-

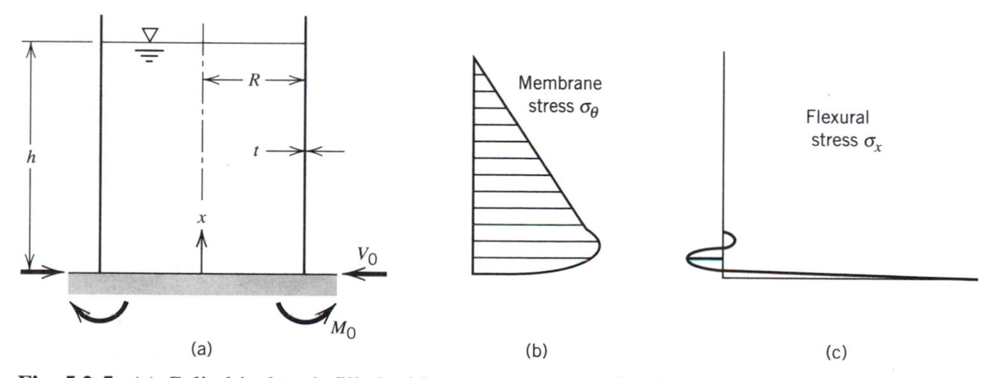

Fig. 5.2-5. (a) Cylindrical tank filled with water to depth h. (b) Circumferential membrane stress. (c) Longitudinal flexural stress.

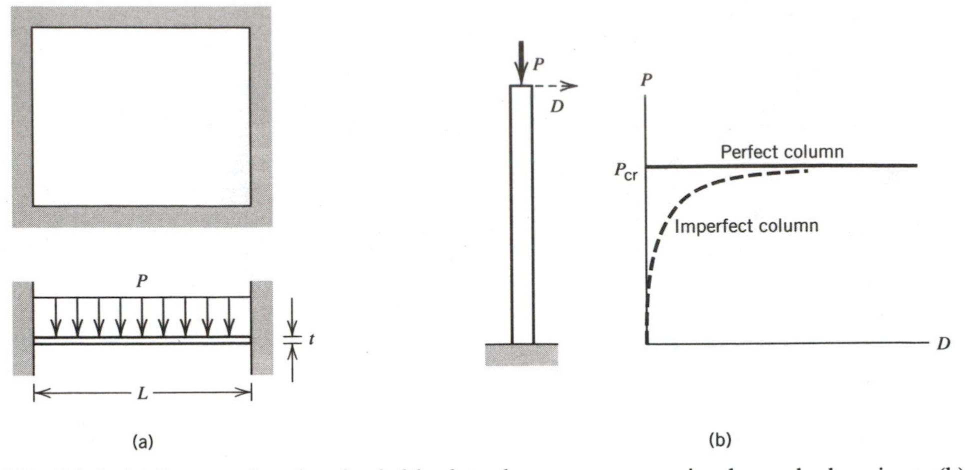

Fig. 5.2-6. (a) Rectangular plate loaded by lateral pressure p, seen in plan and edge views. (b) Column loaded by axial force P.

umn, it does not buckle but instead exhibits a nonlinear load versus deflection response. Most practical thin-walled structures are sufficiently complicated as to be "imperfect" in the sense that a classical linear buckling analysis may give a poor estimate of the actual collapse load. A nonlinear analysis is required instead.

In summary, one must both anticipate structure behavior and understand element behavior in order to make a suitable choice of elements, mesh, and analysis procedure. This cannot be done without a physical grasp of how various structural forms respond to various loads and support conditions, and a grasp of how elements behave *and how they are unable to behave* because of limitations in their displacement fields, restrictions of the structural theory on which they are based, or restrictions of analysis procedure (to linearity and time-independence, perhaps). Because nature is three dimensional, a decision to use bar, beam, plane, plate, or shell elements—in short *any* elements other than solid elements—constitutes an idealization, for which good judgment is needed.

5.3 ELEMENT TESTS AND ELEMENT SHAPES

How do elements of various shapes behave under various loads? A good way to find out is by computational testing, choosing problems for which the solution is already known. By doing computational tests we may incidentally learn how to use the software more effectively, and also resolve uncertainties about input conventions, defaults, output capabilities, coordinate systems used for stress output, symbols and abbreviations, and explanations in the documentation.

Two types of test have more to do with checking the validity of an element than showing how well it works. One is the *patch test*, described in Section 4.9. The other is the *eigenvalue test*, which proceeds as follows [2.2]. One computes eigenvalues of the stiffness matrix of a single unsupported element. This can be done in a standard FE program by assigning a unit mass to each d.o.f. in a vibration analysis (see Section 9.4). Squares of the computed vibration frequencies are the desired eigenvalues. Each eigenvalue is twice the strain energy of the element in the displacement mode corresponding to the eigenvalue. Accordingly, there should be exactly as many zero eigenvalues as there are possi-

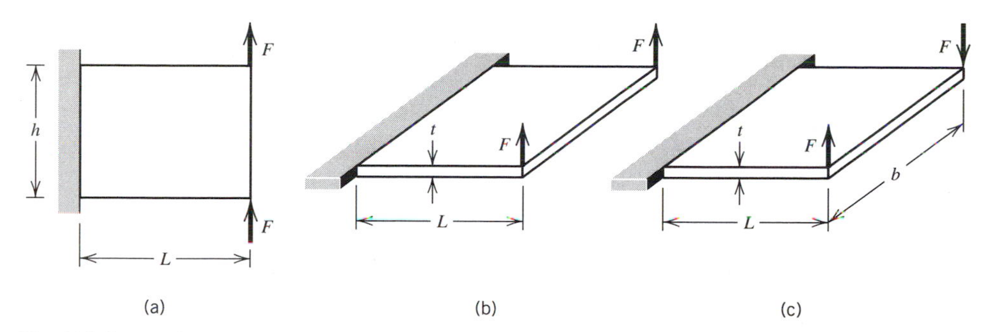

Fig. 5.3-1. Possible single-element tests. Node patterns are not shown. (a) Plane element. (b) Plate element in bending. (c) Plate element in twisting.

ble rigid-body motions (e.g., 3 and 6 zero eigenvalues in plane and three-dimensional problems, respectively). Too few indicates that one or more motions that should be rigid-body motions actually cause strain. Too many indicates that the element has one or more instabilities (Section 4.6). Ideally, an element is free of locking and instability in all situations.

A *single-element test* is a FE analysis like any other, except that the model consists of a single element. By varying the aspect ratio L/h in Fig. 5.3-1a, one can determine the sensitivity of an element to elongation. In Fig. 5.3-1b, one can determine if the plate element tends to "lock" as L/t becomes large and if transverse shear is taken into account as L/t becomes small. In Fig. 5.3-1c the effects of varying L/t and L/b can both be studied. If the element stiffness matrix is numerically integrated, the effect of changing the quadrature rule can be studied (if the software permits a choice in the matter).

The effects of element distortions other than aspect ratio can be studied with a FE model having two or more elements. Figure 3.11-1 contains the beginnings of such a study for each of several element types. Few such studies have been published [5.5], so the analyst must learn by means of trial computations.

Element shapes that are compact and regular usually give greatest accuracy. Accordingly, the ideal triangle is equilateral, the ideal quadrilateral is square, and so on. Of course other shapes must also be used in order to represent the structure geometry and grade a mesh from coarse to fine. But usually one should try to avoid shapes like those in Fig. 5.3-2. The elements shown are plane, but similar distortions of plate, solid, and shell elements are similarly detrimental. Such distortions usually reduce accuracy by making the element stiffer than it would be otherwise. More specifically, an element that has quadratic terms in its displacement field can be reduced to behaving like an element that has only linear terms if its shape is too greatly distorted [5.5]. The amount of degradation caused by a given distortion varies with element type, mesh arrangement, and physical problem. Distortion usually degrades stresses more than displacements, natural frequencies, mode shapes, and temperature fields [9.3]. Distorted elements can still display states of constant strain, but their ability to represent gradients declines. Therefore, distortion is likely to be especially detrimental in regions of stress concentration. Elements having side nodes in addition to corner nodes are usually less sensitive to shape distortion than elements having only corner nodes. Elements having corner nodes, side nodes, and internal d.o.f. are still better.

Deliberate distortion can be harmless, or even beneficial if used appropriately and with care: as examples, aspect ratio matters little if strain gradients are small, and a side node can be moved to the quarter-point location to model stresses near a crack tip [4.8]. Also, side curvature can be used to fit a curved boundary. As an example, consider Fig. 5.3-3.

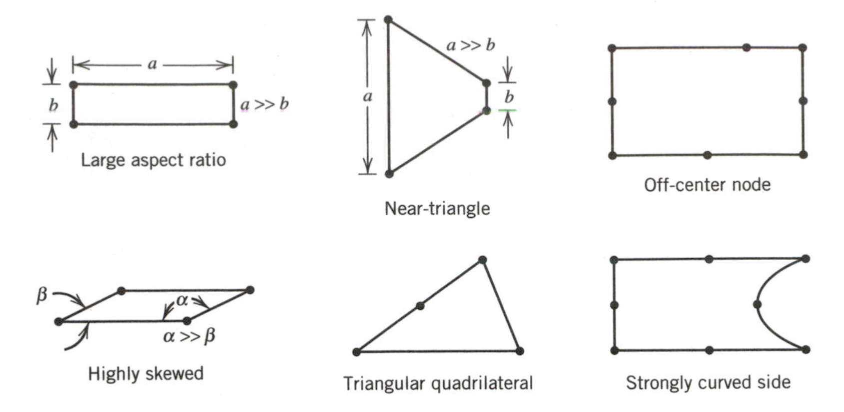

Fig. 5.3-2. Plane elements having shape distortions that usually reduce accuracy. (Reprinted from [2.2] by permission of John Wiley & Sons, Inc.)

Here the curved sides span 90°, a much larger angle than would ordinarily be considered acceptable, yet computed results are surprisingly good. The reason is that the element shape causes the isoparametric transformation to become singular at center of curvature O, which is exactly the point where curved-beam theory predicts infinite stress [5.5]. A similar match of transformation and stress field singularities occurs when quarter-point elements are used next to a crack tip (see Problem 4.34). If the element of Fig. 5.3-3 were used *within* a mesh where the stress field has no singularity at the center of curvature, or quarter-point elements were used where there is no crack tip, accuracy would be *decreased* by element distortions.

In three dimensions a "warping" distortion is possible. The four nodes of a quadrilateral shell element are usually not coplanar; that is, a typical quadrilateral shell element is warped. Similarly, warped quadrilateral membrane elements may appear in three-dimensional FE models of structures built of sheet metal and stiffeners. For shell and membrane elements alike, accuracy declines as the amount of warping increases. Warping may not be obvious during mesh generation, but software will probably check for warping and warn the user if it is excessive.

In addition to being careful with element shapes, one should not use abrupt changes of element size (Fig. 5.3-4). Even if element aspect ratios in Fig. 5.3-4 are satisfactory, the "poor" arrangement will produce a local disturbance in the stress field. Changes in element type (e.g., triangular to quadrilateral), abrupt changes in element size, poorly shaped elements, and poor interelement connections (Fig. 3.10-2) produce "artificial" disturbances in the stress field that may mistakenly be accepted as physically realistic. Particular care should be taken to avoid such changes, transitions, and distortions in regions where stress gradients are large and where accuracy is important.

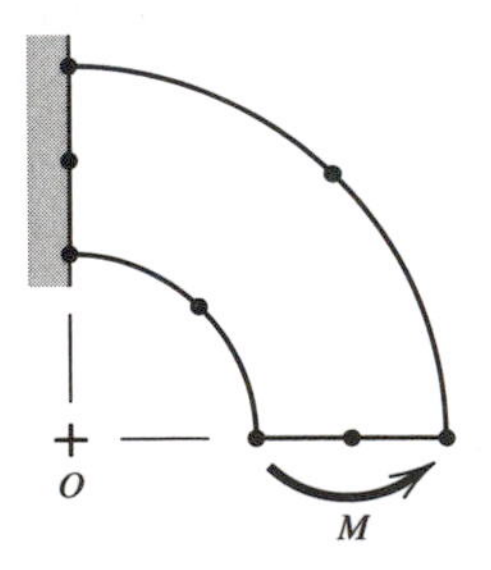

Fig. 5.3-3. Plane curved beam modeled by a single eight-node isoparametric element.

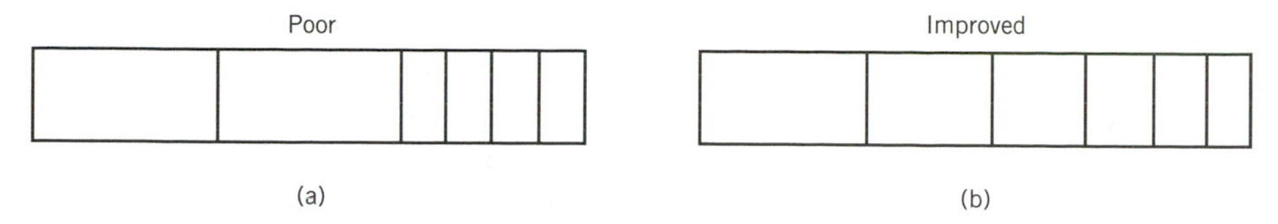

Fig. 5.3-4. Changes in element size are (a) too abrupt and (b) gradual (and much improved).

5.4 TEST CASES AND PILOT STUDIES

Test cases for which the answer is already known can be used to study convergence rate as well as sensitivities to element shape distortion and mesh arrangement. For example, consider a square plate with all four edges clamped, loaded in bending by uniform lateral pressure. Analytical results for stress and deflection are well known [1.5, 7.3]. Figure 5.4-1 shows some coarse FE meshes for this problem. By comparing results from the first two meshes and at least one additional refinement, one can determine the approximate rate of convergence toward correct results of a certain element type—for *one* particular test case. The rate may differ for other loadings and other support conditions. Results from two meshes can be extrapolated to provide an improved result if the convergence rate is known (see Section 5.15). Quite probably, stresses will converge more slowly than displacements. Moving a node (Fig. 5.4-1c) provides a test of sensitivity to element shape distortion. In the coarse meshes shown, changes in mesh arrangement can markedly change computed results: in Fig. 5.4-1d, two symmetries of the actual problem are lost; in Fig. 5.4-1e, the four corner elements do not deform because all of their d.o.f. are restrained by supports.

Different element types have different sensitivities to element distortions and mesh arrangements. Even a single element type, such as a four-node quadrilateral, may behave differently in different software packages because of differences in basic formulation or different choices of "add-on" refinements [3.2]. Knowledge of element behavior, gained from software documentation and computational testing, may enable us to use element types and shapes most appropriate to the problem at hand.

Test cases are often used in FE research papers. Authors want to compare their new formulations with existing methods. Accordingly, they calculate FE results for test cases already used by other authors. In this way a set of test cases has arisen by default rather than by design. This set of test cases has been criticized as reporting few if any bad results, perhaps because authors correct only bugs they happen to find, and because not enough different conditions are tested [3.7]. A good set of test cases exercises all behaviors that an element purports to model. In a set of standard test cases proposed in 1985 [3.7], one finds that a typical test case gives results ranging from poor to good, depending on the type of element used to solve it.

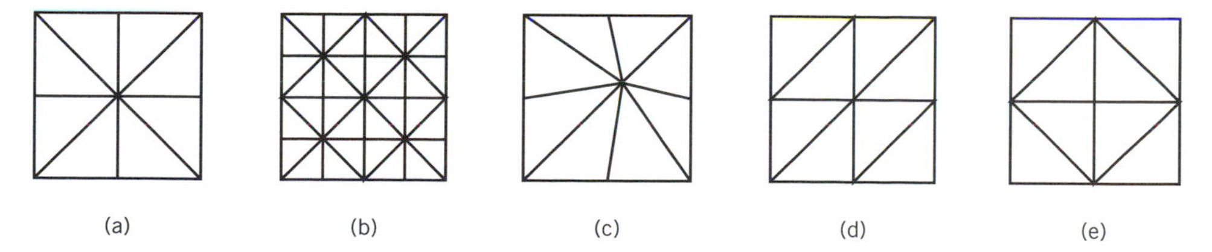

Fig. 5.4-1. Coarse meshes for a square plate that could be used in studies of convergence rate, mesh distortion, or mesh arrangement.

Software providers maintain extensive sets of test cases. They are used to "verify" a new version of the software by making sure it can solve all the test cases, and solve them at least as well as the previous version. Users can often purchase a manual that contains a software provider's test cases, perhaps without even purchasing the software, and use it to learn how to use the software. Some test cases may be simple, even including patch tests, but usually they involve more complicated geometry, loading, and support conditions.

An effort to establish a rational set of test cases has been undertaken in the United Kingdom by NAFEMS (National Agency for Finite Element Methods and Standards). Aims of the organization include setting FE standards and testing procedures and coordinating evaluation of FE software. The numerous NAFEMS test cases are called "benchmarks" and have the following characteristics: each uses a single element type; data preparation is straightforward; geometry, loading, and boundary conditions are unambiguous; and each has a single well-defined result (e.g., deflection or stress at a single point, or a set of natural vibration frequencies), known from theory or perhaps from *soundly justified* computation [5.7]. Commercial software packages can be compared by applying them to a given benchmark. Such comparisons occasionally appear in NAFEMS publications.

A *pilot study* is a simplified study of a larger problem, performed with a simplified model and perhaps also with limited analysis goals. Software capabilities likely to be used in the "real" problem can be tested. One might even insert intentional errors, to see if software error traps will detect them [5.4]. Benefits of a pilot study are a reduction in input and output data, a preview of structural behavior, a comparatively easy way to test modeling idealizations such as joints and supports, detecting blunders such as incorrect units for data, insight into what computational methods may be appropriate, and an indication of the type and amount of output to request. Pilot studies are particularly appropriate with dynamic or nonlinear problems, where structural behavior may be especially hard to foresee and the variety of computational options is large.

Simple test cases and pilot studies are highly recommended as a way to answer "what if" questions about modeling, or to test the software and discover how it *really* behaves. Software efficiency, accuracy, and ease of use for a certain type of problem can be tested. One is likely to see some behavior at odds with expectations, and perhaps even at odds with descriptions in the documentation.

5.5 MATERIAL PROPERTIES

Elastic constants for isotropic materials are usually easy to obtain and easy to convey properly to the software. Constants for anisotropic materials are more difficult on both counts. If x, y, and z are principal material directions, the stress–strain–temperature relation of an orthotropic material can be written

$$\begin{aligned}
\varepsilon_x &= +\frac{1}{E_x}\sigma_x - \frac{\nu_{yx}}{E_y}\sigma_y - \frac{\nu_{zx}}{E_z}\sigma_z + \alpha_x\,\Delta T & \gamma_{xy} &= \frac{\tau_{xy}}{G_{xy}} \\
\varepsilon_y &= -\frac{\nu_{xy}}{E_x}\sigma_x + \frac{1}{E_y}\sigma_y - \frac{\nu_{zy}}{E_z}\sigma_z + \alpha_y\,\Delta T & \gamma_{yz} &= \frac{\tau_{yz}}{G_{yz}} \\
\varepsilon_z &= -\frac{\nu_{xz}}{E_x}\sigma_x - \frac{\nu_{yz}}{E_y}\sigma_y + \frac{1}{E_z}\sigma_z + \alpha_z\,\Delta T & \gamma_{zx} &= \frac{\tau_{zx}}{G_{zx}}
\end{aligned} \tag{5.5-1}$$

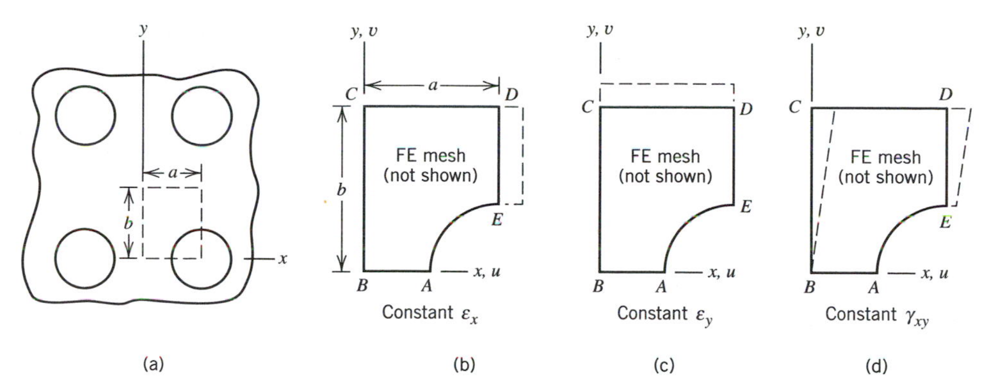

Fig. 5.5-1. (a) Portion of a plate with a regular pattern of holes. (*b*, *c*, *d*) A typical repeating geometry, showing deformations used in computing effective elastic properties. The FE mesh is not shown.

Not all constants in Eq. 5.5-1 are independent. Theory shows that the relations

$$E_x \nu_{yx} = E_y \nu_{xy} \qquad E_y \nu_{zy} = E_z \nu_{yz} \qquad E_z \nu_{xz} = E_x \nu_{zx} \tag{5.5-2}$$

must be satisfied by any real material. Nevertheless, there are still nine independent elastic constants and three independent coefficients of thermal expansion. A general anisotropic material has 21 independent elastic constants. It may not be easy to obtain all necessary constants and state them properly as input, not getting them mixed up and with due regard to principal material directions that may be differently oriented in different parts of the structure.

As a minor point, incompatible and underintegrated elements may display a dependence on Poisson's ratio in problems that should be independent of Poisson's ratio, such as plane beam problems.

Corrugations, indentations, or perforations can have an appreciable effect upon the stiffness of a plate or a shell. If geometric disturbances are numerous and have a regular pattern, they can be "smeared" to produce a substitute plate or shell without geometric disturbances but having modified elastic constants. The constants are readily available in some cases [5.8]. If not, they can be calculated by the procedure now described by means of an example [5.9].

Consider in-plane behavior of a perforated plate, Fig. 5.5-1a. Isolate a typical repeating portion such as *ABCDE*, and model it by an FE mesh (not shown in Fig. 5.5-1). Consider three displacement states of the FE model, i = 1, 2, and 3, with each state i described by a set of d.o.f. $\mathbf{D}_i$ associated with a particular state of constant strain in the *substitute* plate. In what follows, we elect to use constant strain states of unity. Accordingly, on boundary *ABCDE* of the FE model of the actual plate, some d.o.f. in each $\mathbf{D}_i$ are prescribed as follows (see Section 4.12 for a discussion of symmetry and antisymmetry conditions):

Strain	D.o.f	On *AB*	On *BC*	On *CD*	On *DE*
$\varepsilon_x = 1$	$\mathbf{D}_1$	$v = 0$	$u = 0$	$v = 0$	$u = a$
$\varepsilon_y = 1$	$\mathbf{D}_2$	$v = 0$	$u = 0$	$v = b$	$u = 0$
$\gamma_{xy} = 1$	$\mathbf{D}_3$	$u = 0$	$v = 0$	$u = b$	$v = 0$

All d.o.f. not prescribed, whether internal to the mesh or on its boundary, are unrestrained. Next solve three FE problems $\mathbf{KD}_i = \mathbf{R}_i$, $i = 1,2,3$, for d.o.f. in each $\mathbf{D}_i$ that are not prescribed and for nodal forces in each $\mathbf{R}_i$ associated with prescribed d.o.f. For internal d.o.f. and for boundary d.o.f. that are not prescribed, nodal forces in $\mathbf{R}_i$ are zero. Therefore the nonzero entries in $\mathbf{R}_i$ are the externally applied boundary forces associated with prescribed boundary d.o.f. For $i = 1$,

$$\mathbf{R}_1 = \mathbf{KD}_1 \tag{5.5-3}$$

When subjected to the same boundary displacements, nodal forces in the substitute (unperforated) plate are

$$\mathbf{R}_1^* = \mathbf{K}^*\mathbf{D}_1; \qquad \text{that is,} \quad \mathbf{R}_1^* = \left[\sum \int \mathbf{B}^T\mathbf{E}\mathbf{B}\,dV\right]\mathbf{D}_1 \tag{5.5-4}$$

where $\mathbf{K}^*$ is the stiffness matrix of the substitute plate, here formed as the sum of element stiffness matrices (see Eq. 3.1-10). FE discretization of the substitute plate is used here only as a conceptual convenience; no such FE model need actually be constructed to obtain material properties of the substitute plate. Also, when writing $\mathbf{K}^*\mathbf{D}_1$ we imagine that $\mathbf{D}_1$ contains d.o.f. on the sides of an a by b rectangle of the substitute plate. We require that work done by $\mathbf{D}_1$ be the same in actual and substitute plates. Thus $\mathbf{D}_1^T\mathbf{R}_1/2 = \mathbf{D}_1^T\mathbf{R}_1^*/2$, and since $\mathbf{BD}_1 = \boldsymbol{\varepsilon}_x = [1 \quad 0 \quad 0]^T$, Eq. 5.5-4 yields

$$\mathbf{D}_1^T\,\mathbf{R}_1 = \sum \int \boldsymbol{\varepsilon}_x^T\mathbf{E}\boldsymbol{\varepsilon}_x dV = E_{11}V \tag{5.5-5}$$

from which $E_{11} = \mathbf{D}_1^T\mathbf{R}_1/V$, where V is the volume of the substitute plate; ab times thickness t in the present example. In similar fashion we write $\mathbf{D}_2^T\mathbf{R}_1/2 = \mathbf{D}_2^T\mathbf{R}_1^*/2$, and since $\mathbf{D}_2^T\mathbf{B}^T = \boldsymbol{\varepsilon}_y^T = [0 \quad 1 \quad 0]$, Eq. 5.5-4 yields

$$\mathbf{D}_2^T\,\mathbf{R}_1 = \sum \int \boldsymbol{\varepsilon}_y^T\mathbf{E}\boldsymbol{\varepsilon}_x dV = E_{21}V \tag{5.5-6}$$

from which $E_{21} = \mathbf{D}_2^T\mathbf{R}_1/V$. Proceeding similarly for the remaining terms, we obtain

$$\mathbf{E} = \frac{1}{V}\begin{Bmatrix}\mathbf{D}_1^T \\ \mathbf{D}_2^T \\ \mathbf{D}_3^T\end{Bmatrix}[\mathbf{R}_1 \quad \mathbf{R}_2 \quad \mathbf{R}_3] \tag{5.5-7}$$

as the material property matrix of the substitute plate that has no geometric disturbances. Analogous arguments can be made if properties are referred to polar or cylindrical coordinates [5.9]. Flexural stiffness coefficients of a substitute plate can be determined in similar fashion, by applying unit curvature states and calculating associated nodal moments.

5.6 LOADS

A concentrated load must be applied at a node. This is required by practice, not theory. With the possible exception of beam elements, standard software is not structured to accept non-nodal concentrated loads as input data. In practice, one merely arranges the mesh so that there is a node at each location where a concentrated load must be applied.

According to classical linear theories of beams, plates, and solids, at a point loaded by concentrated normal force there is

- finite displacement and finite stress in a beam,
- finite displacement and infinite stress in a plate, and
- infinite displacement and infinite stress in a two- or three-dimensional solid.

These confusing assertions are consequences of differing premises about the nature of stress fields in standard linear theories of beams, plates, and elasticity. Also, a truly concentrated force would cause material beneath it to yield, and a linear theory must rule out yielding. Physically, a concentrated force does not exist; it is a mathematical convenience that represents a distributed load of high intensity that acts on a small area. Moreover, when a concentrated force is applied to a node of a FE model, infinite displacement or stress will never be computed. Indeed, a concentrated force on a plane FE model has a nonunique distributed equivalent (Fig. 5.6-1a), which one would certainly not expect to produce infinite displacements or infinite stresses. Infinite values can only be *approached* as the mesh is repeatedly refined.

If axisymmetric conditions prevail, what appears to be a concentrated load on a solid or shell of revolution shown in cross section is really a line load on a nodal circle. Software may require such a load to be input for a 1 radian slice or perhaps for the entire circumference of the nodal circle. In the latter case a radially directed line load q, whose dimensions are [force/length], is described as a force $2\pi rq$ on a circle of radius r, even though the net force is statically equivalent to zero.

A concentrated moment cannot be applied to a node that has only translational d.o.f. The moment can be applied as couple forces (Fig. 5.6-1b) or, alternatively, can be distributed to a group of nodes by use of constraint techniques (Section 4.13).

Distributed loads are applied to nodes as concentrated nodal loads that are statically equivalent or perhaps work-equivalent. This matter is discussed in Section 2.5 for bars and beams and in Section 3.9 for plane elements. Usually software can generate equiva-

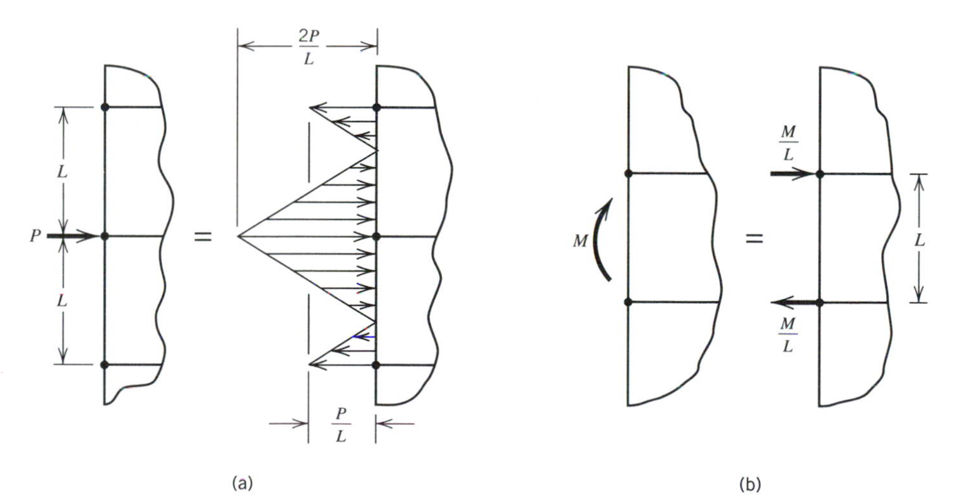

Fig. 5.6-1. (a) A concentrated force and a statically equivalent line load on a linear edge of a plane FE model (see also Fig. 3.9-1). (b) Application of a couple when nodes have only translational d.o.f.

lent nodal loads from input data that describe distributed loading. If rotational d.o.f. are present, as in beam and plate elements, software may or may not include nodal moments in its vector of nodal equivalent loads. One can discover what the software does by running suitable one-element test problems, and comparing computed results with theoretical results.

In linear problems, loads maintain their original orientations in space, regardless of the magnitudes of the computed displacements. Sometimes a problem is nonlinear because of loads called *follower forces*, whose directions change as the structure deforms. An example is pressure on a membrane. Pressure always acts normal to the membrane. A nonlinear analysis is required if deflection is appreciable.

In describing temperatures for a thermal-stress analysis, one may be concerned with whether the software uses nodal temperatures or element temperatures. The distinction becomes important if one wishes to describe a step change in temperature across an interelement boundary, as might be done to simulate a shrink fit. Temperatures interpolated from nodal values do not describe a step change.

5.7 CONNECTIONS

Connections between parts are made by bolting, welding, gluing, and so on. Realistic modeling of connections is usually difficult, with FE or any other analysis method, because of geometric complexity and the possibilities of slippage, gap closure, and partial loss of contacts. A fine-mesh FE model of a connection may capture its behavior accurately, but such a model is not practical unless the connection itself is the object of study. More often, connections are modeled only to the extent needed to represent their effect on the rest of the structure. If data about connection stiffness are known, perhaps from experiment, one might approximate a connection by using standard elements that have modified elastic properties. For example, a bolted connection between two beams might be modeled by a short beam element of reduced flexural stiffness EI. The following remarks describe other connection problems and ways of modeling them.

Section 4.3 describes how rigid offsets may be used to attach a reinforcing beam to a plate. Constraint equations can be used for the same purpose (Section 4.13). Either technique can be applied to connections depicted in Fig. 5.7-1. The approximate models cannot represent bending action that develops because of axial loads. The improved models can. They use rigid links AB to model offsets and transmit moment as well as force. The rigid-link model is still approximate: in reality AB is an elastic path, although a comparatively stiff one [5.4].

If done properly, plane and solid elements having only translational d.o.f. can be connected. In Fig. 5.7-2a, externally applied loads on the plane portion must act in the yz plane because this portion has no bending stiffness. To support loads having an x component, bending stiffness and rotational d.o.f. must be added to the plane portion, which can then be attached to the solid portion by methods described in connection with Fig. 3.9-6 or 4.13-1.

With rare exceptions, plane and *axisymmetric* solid elements *cannot* be connected. In Fig. 5.7-2b the axisymmetric portion is seen in cross section. Thus plane and axisymmetric meshes look similar, and both typically use two d.o.f. per node. But axisymmetric elements are rings, not plane quadrilaterals, and what appear to be node points are actually nodal circles. In a formulation restricted to axial symmetry, displacements in the axisymmetric portion are independent of the circumferential coordinate. Accordingly, if some-

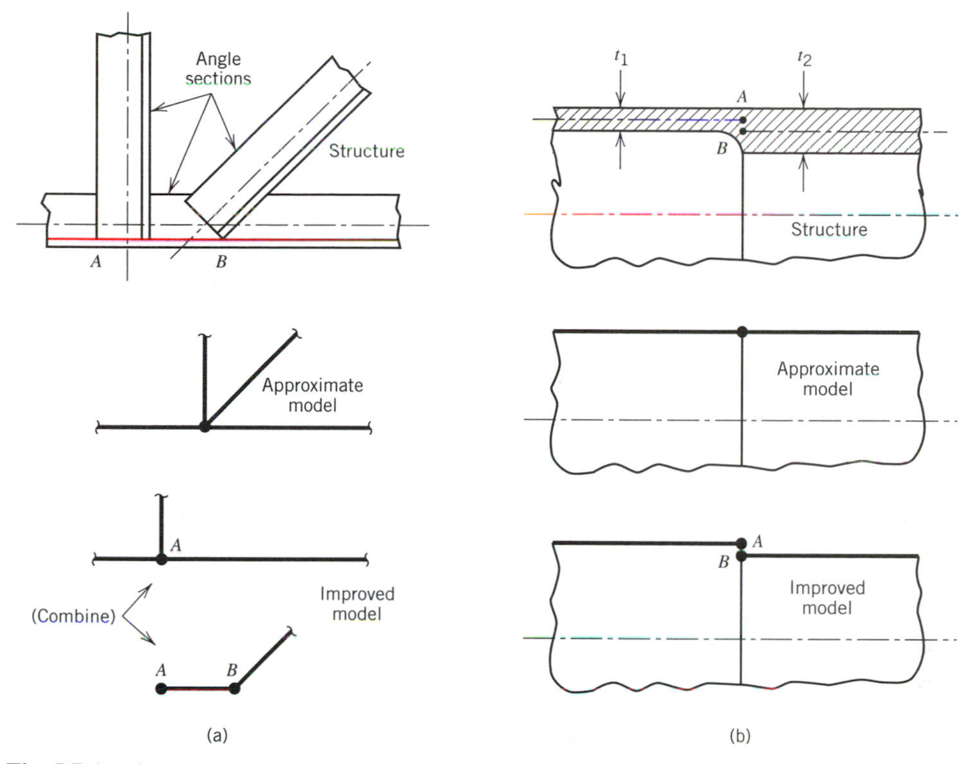

Fig. 5.7-1. Offset connections and possible models. (a) Joint in a plane frame. (b) Cylindrical tubes of different thicknesses, $t_1 \neq t_2$. (Reprinted from [2.2] by permission of John Wiley & Sons, Inc.)

thing is to be attached, it must also be axisymmetric in both geometry and loading. The mismatch in Fig. 5.7-2b is roughly similar to the "poor connections" in Fig. 3.10-2.

However, an axisymmetric model can carry nonaxisymmetric loads if Fourier series are used (see Chapter 6). By extending this procedure, axisymmetric and generally shaped solids can be connected. The calculations are not simple and are not part of standard FE software, but they provide the only way of making a connection like that in Fig. 5.7-2b even approximately correct.

A bolted joint in a pipe, Fig. 5.7-3, is axisymmetric in geometry except for the bolts. As an approximate representation of the bolts, one can "smear" them around the bolt circle. The trick is to replace the bolts by an axisymmetric solid of radius r and length L that has the same stiffness in the axial direction as bolts it replaces but *zero* stiffness in the circumferential direction θ (because bolts provide no circumferential stiffness). Consider axial loads on the pipe in Fig. 5.7-3. Let there be n bolts around the bolt circle, each of elastic modulus E_b. Their combined axial stiffness is $k_b = A_b E_b / L$, where $A_b = n(\pi d^2/4)$. The replacement solid has axial stiffness $k_s = A_s E_s / L$. The condition $k_b = k_s$ yields $A_s E_s = n(\pi d^2/4)E_b$. The replacement solid is connected only to nodes on flange surfaces (nodal *circles* A and B in Fig. 5.7-3c). Accordingly, the replacement solid can be a single element. Indeed, input data can describe it as a two-node bar element between nodes A and B, having cross-sectional area A_s and elastic modulus E_s. This does not violate the rule that bodies with and without axial symmetry cannot be connected because here we are only using the *description* of a nonaxisymmetric element as a device to obtain the desired type of axisymmetric element. Axial stress in a bolt is computed by multiplying its com-

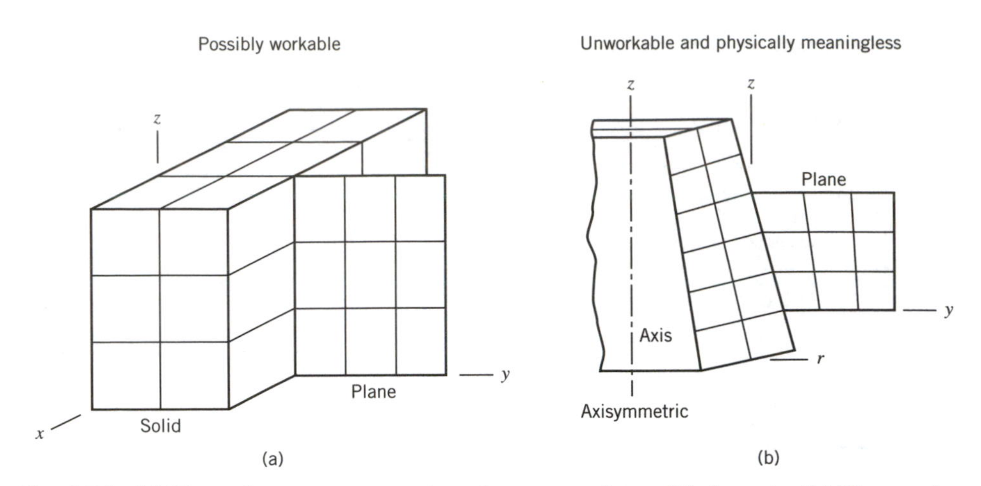

Fig. 5.7-2. (a) Plane elements can sometimes be connected to solid elements. (b) Plane and axisymmetric FE models *cannot* be connected.

puted axial strain by its *actual* elastic modulus. In similar fashion one might smear the bending stiffness of individual bolts and arrange to recover their flexural stresses from computed displacements.

Bolt pretension can be simulated by reducing the temperature of the bolts an amount ΔT. To compute ΔT we argue as follows, using notation of the preceding paragraph and α as the coefficient of thermal expansion of the bolt material. Let P be the known tensile force required in *all* bolt material, so that P/n is the pretension in one bolt. The temperature drop ΔT in the bolts produces force P in the bolt material and also a clamping force P in the pipe flange spanned by bolts. Both parts must contract the same amount, that is, with contraction positive

$$\alpha L \, \Delta T - \frac{PL}{A_b E_b} = \frac{PL}{A_f E_f} \tag{5.7-1}$$

where $A_f E_f / L$ is the stiffness of the flange as seen by the bolts. To calculate this stiffness one can remove the bolts, apply an axisymmetric clamping force F to the flange alone,

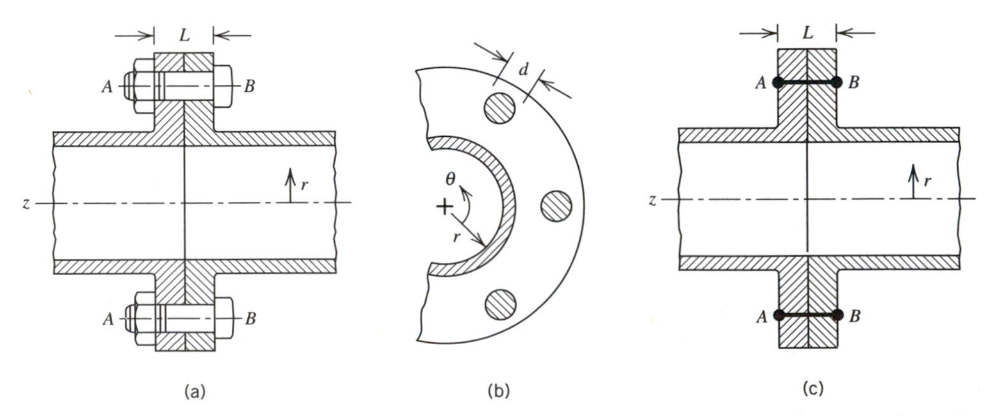

Fig. 5.7-3. (a) Bolted pipe connection. (b) Axial view. (c) Axisymmetric connector AB replaces the bolts.

and compute the resulting displacement δ_{AB}. Then

$$\delta_{AB} = \frac{FL}{A_f E_f} \quad \text{from which} \quad \frac{A_f E_f}{L} = \frac{F}{\delta_{AB}} \tag{5.7-2}$$

Now that $A_f E_f / L$ is known, Eq. 5.7-1 yields the required ΔT. Some rearrangements of these calculations are possible [5.10].

The foregoing bolt representation can be useful in calculating the change in bolt stress produced by loading. The elastic stiffness of the joint is produced more by flanges than bolts because flanges are more massive. However, if there is a gasket between flanges, tightening the bolts tends to rotate the flanges relative to the gasket, perhaps resulting in partial loss of contact even before external load is applied. A load-dependent contact area renders the problem nonlinear. If flanges are connected without the gasket, even with a large clamping force the joint will not be as stiff as continuous material would be, owing to the near impossibility of achieving a perfect fit and preventing slippage.

The behavior of connections can be quite important in dynamic analysis. Connections (e.g., in a frame) are usually stiffest in deformation modes activated by the static load that must be carried. Displacements in dynamic analysis can load connections in their more flexible modes. If little is known about these lesser stiffnesses, so that they are carelessly modeled, there may be appreciable disagreement between computed behavior and actual behavior [5.9].

Structures may contain parts that can make or break contact. This situation renders the problem nonlinear: stiffness is displacement dependent, loads are not directly proportional to displacements, and an iterative solution is required. A nonlinear spring or "gap element," Fig. 5.7-4a, is one of many tools available for analysis of such problems. A nonlinear spring can be used to connect nodes on adjacent parts that may come in contact. A spring occupies the gap between parts. The gap may be zero, but this presents no obstacle to modeling. A spring may span the gap anyway: if the gap is zero the spring has stiff-

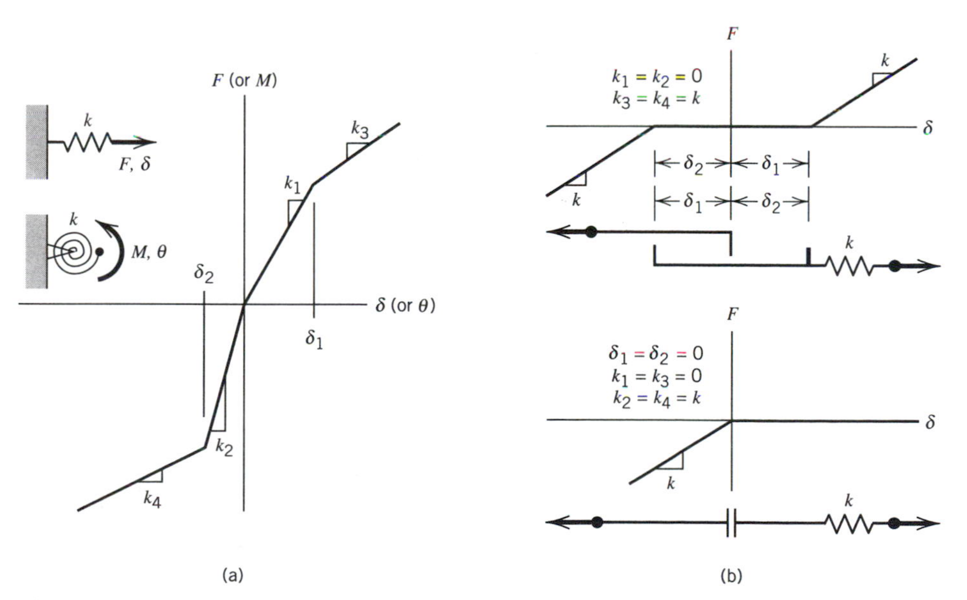

Fig. 5.7-4. (a) Possible way to describe the stiffness of a nonlinear spring. In practice, a spring may resist force F or moment M. (b) Two examples that model gaps and contact.

ness but no length, unless one wishes to imagine a length as a conceptual aid. The respective examples in Fig. 5.7-4b depict a gap element for elastic contact with initial slack and a gap element for compression only with initial contact. Take care: a large spring stiffness that appears when a gap closes constitutes a *penalty method*, which may provoke numerical difficulties (Section 4.13). Gap conditions are discussed more fully in Section 10.6. Commercial software may allow different combinations of springs, gaps, sliders, and dampers to produce a bewildering variety of nonlinear two-node elements.

5.8 BOUNDARY CONDITIONS

Boundary conditions are also called support conditions in structural mechanics. They are often misrepresented or not described properly as input data. Care is needed because changes in support conditions that appear minor can have a major effect on computed results. For example, in Fig. 5.8-1 the change from a roller to a hinge at B allows the supports to apply horizontal forces to the beam. The beam in Fig. 5.8-1a could be modeled by standard two-node beam elements that lie on the line between A and B. The beam in Fig. 5.8-1b must be modeled by plane elements, or by beam elements along the centerline of the actual beam with vertical links to connect end nodes of the FE model to supports A and B below them.

Some support conditions are dictated by FE technology rather than by physical considerations. A restraint, such as a prescription of zero displacement or zero rotation, must appear *at* a node rather than between nodes. The d.o.f. not active in the FE model must be suppressed, whether or not they are on the boundary of the FE model. For example, typical plane elements resist two in-plane translational d.o.f. per node, but software makes six global d.o.f. available at every node. To prevent singularity of the structure stiffness matrix, rotational d.o.f. and out-of-plane translational d.o.f. must be suppressed, whether or not loads are applied to these d.o.f. This matter is discussed more fully in Section 2.4.

Boundary conditions are often misrepresented because of carelessness or because the physical situation does not present a clear choice. Input data as understood by the software can be checked easily: graphic capabilities of preprocessors can depict boundary conditions at each supported node, using symbols that show the direction of restraint and its type (displacement or rotation). These plots should be inspected carefully for data input blunders. When boundary conditions of the physical problem are unclear, it may be possible to bound the correct solution by two analyses, each based on a different set of boundary conditions. As a simple example, rotations at ends of a uniformly loaded beam may be elastically restrained to an uncertain degree. Two analyses, one with simple supports and the other with fixed supports, will respectively overestimate and underestimate the actual magnitude of bending moment at midspan.

On occasion a FE model is connected to another FE model or to a support by springs

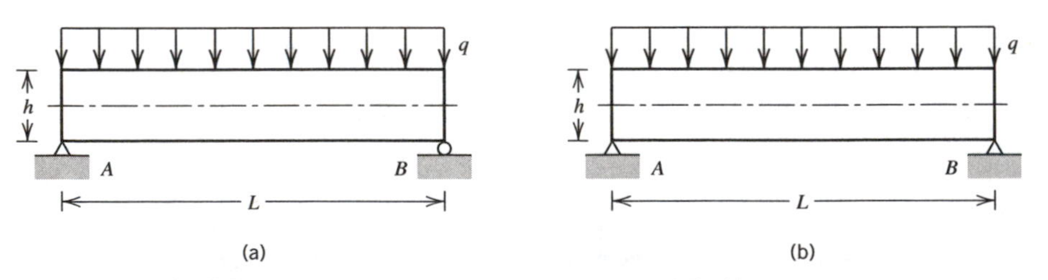

Fig. 5.8-1. A beam with (a) simple supports and (b) hinge supports.

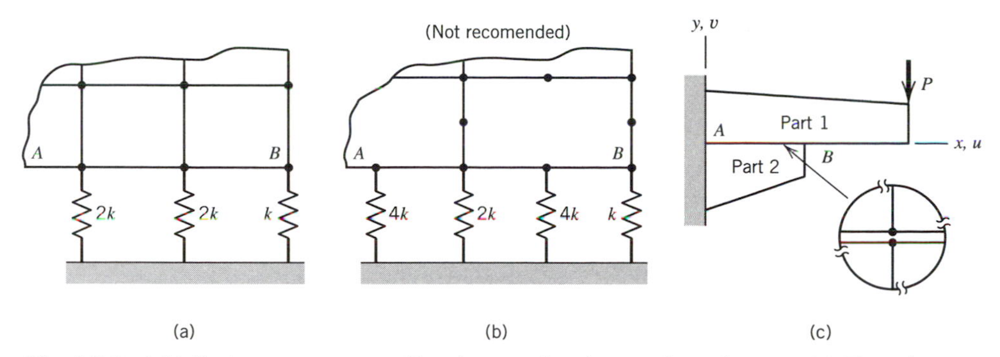

Fig. 5.8-2. (a,b) Spring supports at uniformly spaced nodes on plane elements. (c) Two plane regions with sliding contact along *AB*. FE meshes are not shown, but inset shows typical adjacent nodes in parts 1 and 2.

(Fig. 5.8-2a,b). It is reasonable to require that springs exert nodal loads consistent with uniform stress in the FE model when edge *AB* is translated vertically downward. These loads, discussed in Section 3.9, demand that stiffnesses of *uniformly spaced* springs have the relative values shown in Fig. 5.8-2a,b. If springs are used to connect the face of a *solid* FE model to a parallel plane support, the proper model may be counterintuitive. Solid elements having side nodes require oppositely directed nodal loads to represent a uniform pressure, as suggested by Fig. 3.9-5d. This means that springs attached to corner nodes should have negative stiffness! Or, if all springs were assigned the same stiffness, it means that computed tensile force in a spring would not necessarily imply tensile stress in the solid at the spring location. Clearly, elements having side nodes present ample opportunity for confusion, so it is recommended that they not be used in association with connecting springs.

The nature of a support or a contact between parts can sometimes be determined by numerical trial. Imagine that parts 1 and 2 in Fig. 5.8-2c are modeled by plane elements and that sliding is possible along *AB*. Should adjacent nodes in the two parts have independent u but the same v along *AB*? Or does a gap appear, so that there is contact only *at* *A* and *B*? One way to find out is to connect adjacent nodes in the two parts along *AB* by springs or two-node bar elements, oriented vertically. The bar element that connects adjacent nodes at *B* should be very stiff so the two parts will not interpenetrate (but take note of the cautionary remarks in Section 5.10 and remarks associated with Eq. 4.13-9). The remaining vertical bar elements between *A* and *B* can be very soft. Thus, in effect, we have placed a roller between the two parts at *B* and left the rest of *AB* unconnected. If computed results show that the weak vertical bars between *A* and *B* carry (small) tensile force, one concludes that a gap opens along *AB* and that the model is correct. If instead the weak vertical bars carry compressive force, the model is incorrect; either these bars must be made very stiff to prevent interpretation or, as a better alternative, pairs of adjacent nodes across *AB* should be constrained to have the same vertical displacement.

It is sometimes necessary to impose a *nonzero* displacement or rotation. In the event that software accepts only zero values, the trick shown in Fig. 5.8-3 can be used *with caution*. In this example, point *A* is required to have the known displacement component δ_p in the s direction. The displacement component of *A* in the n direction is to be computed. A spring of very large stiffness k_p, say 10^6 times the largest diagonal coefficient in the structure stiffness matrix, is added to the FE model. A large force, $F_p = k_p\delta_p$, is applied as a load in the s direction. Force F_p is resisted by the stiff spring and the comparatively flimsy FE model. Point *A* will have a computed deflection component in the s direction

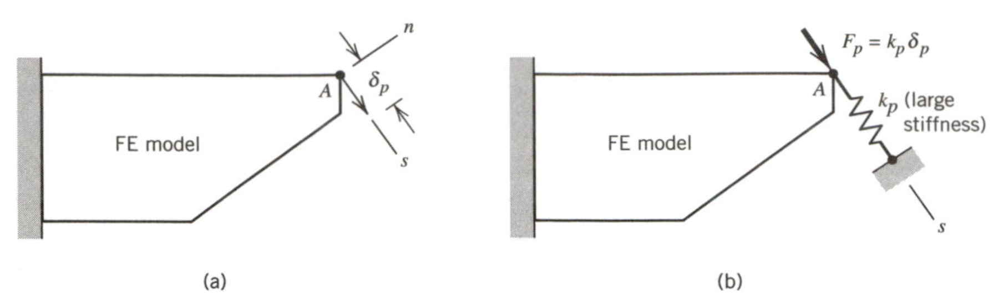

Fig. 5.8-3. (a) Displacement component δ_p is to be imposed at A. (b) "Penalty method" approximation. *Caution*: may introduce numerical errors.

very slightly less than δ_p. This trick is a penalty method (discussed in Section 4.13). It invites numerical errors (see also Section 5.10).

Boundary conditions are VERY OFTEN misrepresented. Be careful with them.

5.9 PLANNING THE ANALYSIS

In order to plan an analysis project, the analyst must have a good grasp of the physical problem, the behavior and limitations of finite elements, and the options and limitations of the software. Otherwise the problem may not be well defined, the FE model may not capture its important features, loading and support conditions may be inappropriate, and too much faith may be placed in computed results. Preprocessors can easily generate a fine mesh where it is not needed, and postprocessors can use attractive display to hide a need for greater refinement. Many who do FE analysis overestimate their competence and underestimate the care required. It is possible that most FE analyses are so flawed that an approximation based on statics and elementary mechanics of materials would be more accurate.

An analyst competent in the problem area should plan the work to be done. One probably knows the purpose of the analysis project, which may range from analysis of a trial design to analysis of an existing product that has failed. Accordingly, one probably has some knowledge of the accuracy required. This knowledge may affect the number and kind of assumptions and simplifications that will be incorporated in FE models. These decisions must be made by an analyst who understands the behavior of the actual structure. It is better to start with such considerations and a plan, which may be revised as more is learned, than to start without a clear direction. Planning an analysis will involve the following considerations.

Understanding the Problem. The broad goal of any analysis is to obtain adequate answers at reasonable cost in time and effort. The first step is to define the problem, with attention to what is known and what is desired, what physical actions are involved, and what simplifying assumptions are appropriate. This may sound tedious but it may save more time than it takes, especially when a problem is complicated—and it probably *is* complicated if a FE analysis is being undertaken.

Major questions about behavior include the following. Is static analysis appropriate? Or does the problem involve vibration or shock loading? If dynamic, can damping be ignored? If not, how should it be represented? Might there be local or global buckling? If material properties are temperature dependent or anisotropic, are material data available? Are there nonlinearities, due to material yielding, gaps that may open or close, or displacements large enough to change the way loads are transmitted or applied? To pose and answer such questions one must have an ability to visualize physical behavior and structural interactions. The answers will decide the general nature of the analysis project.

More detailed questions follow, which influence the specifics of FE models. What are the load cases? Do they involve concentrated or distributed loads, or body forces from self-weight or spinning? Are loads fixed in direction or do their directions change as load increases? Do "loads" include prescribed nonzero displacements? Can symmetry be exploited? Are there elastic supports or connections of uncertain stiffness? Are there cutouts that act as stress raisers, perhaps on a scale below mesh size? How reliable are data about geometry, loads, supports, and material properties?

Preliminary Analysis. Prior to FE analysis, some results should be anticipated, qualitatively or quantitatively and preferably both. An approximate preliminary analysis may be based on statics, mechanics of materials, formulas from handbooks, or experimental results. In almost all situations there is some way to obtain approximate results that can be compared with FE results subsequently obtained. Even a crude analysis should be adequate to detect a strange displacement pattern or stress field, or a numerical result in error by orders of magnitude because of a blunder in data preparation. There is a tendency to trust the computed results because FE analysis requires considerable time and effort to accomplish. There is also a tendency to regard existing results as correct until proved otherwise. Therefore predictions should be in hand *before* doing FE analysis, to promote the viewpoint that FE results are the results on trial. A preliminary solution serves the purpose, even if it must be crude, and may have the added benefit of providing insight that improves the FE model. Other benefits of preliminary analysis include a sharpening of analytical skills and perhaps even discovering that a FE analysis is not needed after all.

Start with Simple FE Models and Improve Them. What types of elements should I use and how many of them? This may be the question suggested by a decision to use FE analysis. A conclusive answer cannot be given at the outset, but based on anticipated structure behavior and the known behavior and limitations of various finite elements, the analyst can prepare a trial FE model. An adequate FE model develops from a sequence of FE models, each of which guides development of the next, so that the last has enough elements of the proper type. The term "sequence of models" may suggest a great amount of effort. However, the sequence may not be long and some models may differ little from one another. The sequential approach builds confidence in the final result. It also takes less time overall than an attempt to construct a very detailed FE model at the outset, only to find that it is inappropriate or inadequate because some aspects of behavior were not foreseen. As FE software becomes more widely available, pre- and postprocessors improve, and computing costs decline, there is a tendency to use more and more elements in FE models. This is unwise if done as a substitute for understanding.

There are exceptions to most rules for FE modeling. This said, the following rules are usually helpful. Include all of the structure in the model; do not omit part of it on the assumption that it is lightly stressed or does not influence the remainder of the structure. Use a finer mesh to obtain stresses (or mode shapes of vibration) than to obtain displacements (or natural frequencies of vibration). If the problem involves nonlinearity or anisotropy, analyze a linear or isotropic version of the problem first. If there are dynamic effects, do a static analysis first, using loads that approximate the major dynamic load. A linear static analysis is easier to perform and interpret and may disclose flaws in the FE model. A linear analysis may also disclose that local buckling is possible, or that stresses are so large that plastic action will develop. Software will not *automatically* proceed to analyze for a buckling load or do an elastic–plastic analysis. The *user* must decide what type of analysis is required, select appropriate options in the program, and launch the analysis.

The foregoing suggestions lead naturally to a sequence of FE models. If a structure is an assemblage of distinct parts, it is sometimes possible to begin with a "stick model," which is a model built of a few bar and beam elements. This does not mean that the initial model should be as crude as possible, only that it should be comparatively simple. Very likely, the analyst can foresee where stresses may be largest and use more elements in these areas, even in the initial model. Each successive model serves to improve the next, by showing more clearly where stresses and stress gradients are large. The sequence also may show that there are appreciable changes from one FE model to the next, for example, changes of stresses in a certain region or changes in statically indeterminate support reactions. Such changes suggest that convergence is not yet adequate and that mesh refinement is needed. The sequence of FE models may include two-dimensional models as steps toward three-dimensional models. If possible, three-dimensional models should be the latter models in the sequence because they are the most tedious and time consuming to prepare and the most demanding of computer resources.

Check the Model and the Results. Modeling defects that prevent execution will probably be identified by error messages from the software. Defects that produce unreliable results must be detected by the user. Computed results must be critically examined. These important matters are discussed in sections that follow.

Numerical Experiments for Design Purposes. In an effort to improve a structural design, one may wish to understand the effects of changes in certain design variables. For example, one might ask how the largest stress in a trial design is related to changes in a certain thickness t, a certain hole radius r, and a certain length L. Analyses using various choices for t, r, and L can be undertaken after an acceptable FE model has been generated, provided that changes in the design variables are not so large as to invalidate the FE model. Imagine that it has been decided to examine the peak stress using thicknesses t_1 and t_2, radii r_1 and r_2, and lengths L_1 and L_2. There are eight possible combinations of these design variables. An analysis (a "numerical experiment") should be performed for each of the eight, in order that the combined results may be used to predict the values of t, r, and L most likely to reduce the peak stress. The procedures are part of the study called *design of experiments*, which has been used for years in planning a productive set of physical experiments. It can also be used for planning numerical experiments and interpreting the results. See [5.17] for an introduction.

Such numerical experiments are facilitated by software having the ability to revise the FE mesh automatically when one of the design variables is changed. Automated mesh revision facilitates automated optimization, in which software seeks the values of design variables that minimize a function such as structural weight, subject to limits on stress, deflection, or other quantities. Software with optimization capability is becoming commonplace.

5.10 NUMERICAL ERROR: SOURCES AND DETECTION

We distinguish between errors inherent in the FE process and outright mistakes. The "mistakes" category includes choosing the wrong data, forgetting loads or supports, and having bugs in the software. The "errors" category includes errors always present to some degree: modeling error (because reality is replaced by mathematical theory), discretization error (because mathematical theory is implemented in piecewise fashion by FE methods), and numerical error (because the computer does not use an infinite number of bits to represent each number). The present section is concerned with numerical error and how it is related to modeling choices.

Ill-Conditioning. A set of equations is ill-conditioned if small changes in the coefficient matrix or the vector of constants produce large changes in the solution vector. Consider the two-d.o.f. structure in Fig. 5.10-1. It is described by the equations

$$\mathbf{KD} = \mathbf{R} \quad \text{is} \quad \begin{bmatrix} k_1 & -k_1 \\ -k_1 & k_1 + k_2 \end{bmatrix} \begin{Bmatrix} u_1 \\ u_2 \end{Bmatrix} = \begin{Bmatrix} P \\ 0 \end{Bmatrix} \tag{5.10-1}$$

Each equation plots as a straight line in a u_1u_2 coordinate system. These are the solid lines in Fig. 5.10-1. Shaded bands along the lines suggest inexactness associated with use of a finite number of bits to represent each number in computer memory. The exact solution of Eqs. 5.10-1 is represented by the intersection of the solid lines. The numerical solution is represented by a point somewhere in the region where the shaded bands overlap. This region is large when $k_1 >> k_2$ but small when $k_2 >> k_1$.

We can again conclude that the case $k_1 >> k_2$ may be troublesome by considering an elimination solution of Eqs. 5.10-1. Addition of the first equation to the second eliminates and converts the second equation to

$$[(k_1 + k_2) - k_1]u_2 = P \tag{5.10-2}$$

which would be exactly $k_2u_2 = P$ if the computer used an infinite number of bits to represent k_1 and k_2. But if $k_1 = 1.000000$ and $k_2 = 4.444444(10)^{-6}$ and the computer were to carry (say) seven digits per word, the subtraction in Eq. 5.10-2 would yield $1.000004 - 1.000000 = 0.4(10)^{-6}$; that is, only one significant digit would remain. If the computer were to carry six digits per word, the result would be $1.00000 - 1.00000 = 0.00000$, and

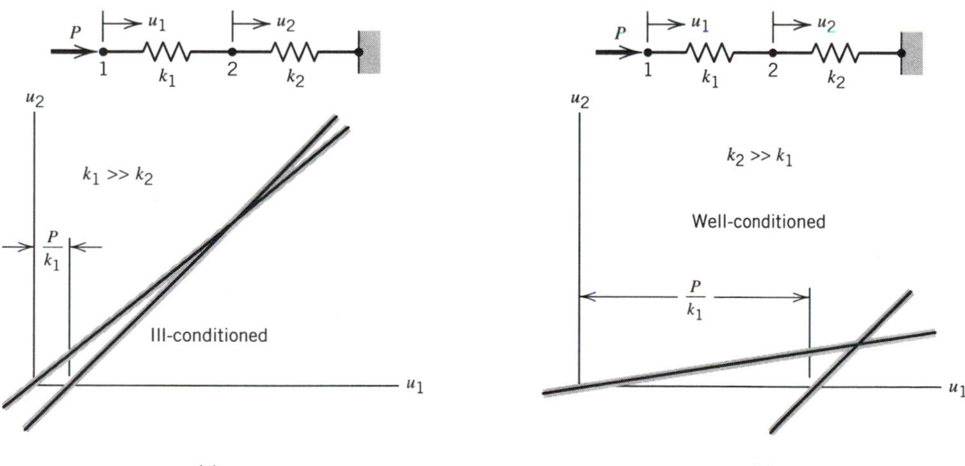

Fig. 5.10-1. A two-d.o.f. structure. (a) Stiff part supported by flexible part. (b) Flexible part supported by stiff part.

the software would probably complain that the stiffness matrix is singular. Note that this trouble does not arise when $k_2 >> k_1$.

The foregoing example shows that the case $k_1 >> k_2$ in Fig. 5.10-1 produces ill-conditioned equations, and that the solution may be inaccurate if computer words have too few bits. This conclusion is true for FE models in a general way: numerical error becomes more likely when elements or regions in a FE model have large differences in stiffness, with the stiffer part supported by the more flexible part. In the present context, "support" means elastic resistance. The conclusion is also plausible on physical grounds. If stiffness differences were exaggerated without limit, the stiffer part would become unsupported, so that a static equilibrium solution would not be possible. Analogous difficulties may occur in nonstructural problems. The main danger of ill-conditioning is not that equation-solving may fail, but that it may *succeed* yet produce a solution whose errors are serious but not large enough to make it obvious that something is wrong.

Problems Susceptible to Ill-Conditioning. Some modeling practices avoid numerical trouble while others invite it. In Fig. 5.10-2a, node A is supported by a roller that allows motion in only the y' direction. This type of support can be treated without numerical trouble by constraint transformations (Section 4.3) or Lagrange multipliers (Section 4.13). An approximate but physically reasonable alternative model is that of Fig. 5.10-2b: by adding a very stiff spring along the x' axis, we allow motion in the y' direction but almost prevent motion in the x' direction. Thus we create a penalty constraint (Section 4.13), which creates ill-conditioning if spring stiffness k is large because there is then a stiff part (the spring) supported by a flexible part (the three-bar truss). Figure 5.10-2c has

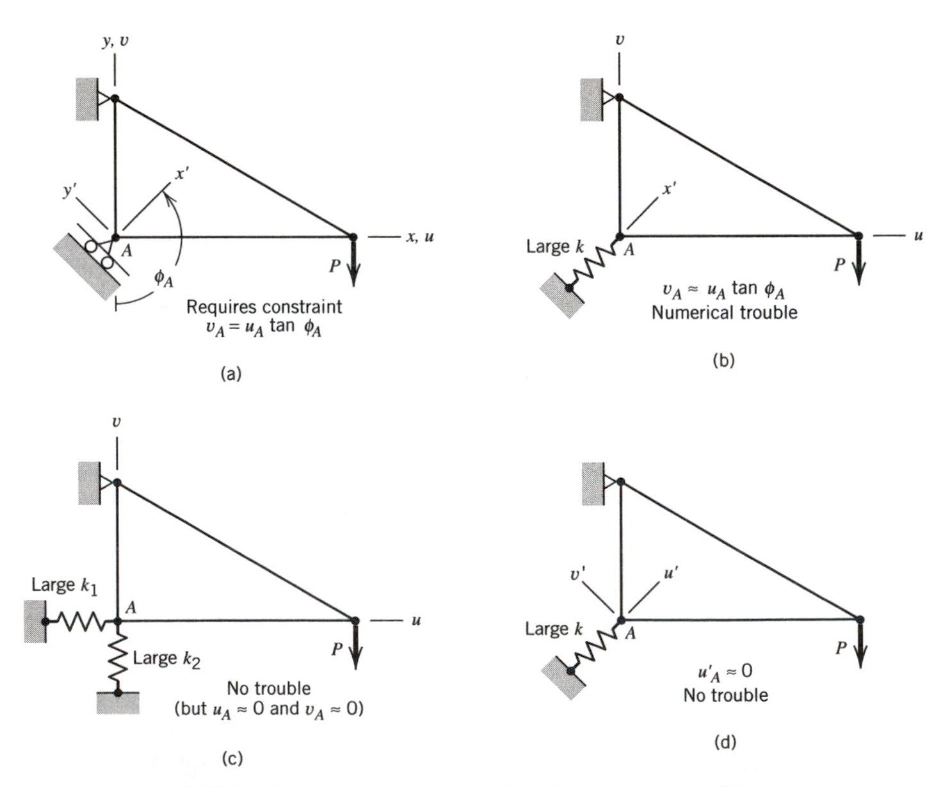

Fig. 5.10-2. Three-bar plane truss with various support conditions.

a different support condition at A, but this arrangement does *not* create numerical trouble even if k_1 and/or k_2 is large. The difference between Fig. 5.10-2b and Fig. 5.10-2c is that an inclined stiff spring to ground contributes large diagonal *and* off-diagonal coefficients to $\mathbf{K}$, while x- and y-parallel stiff springs to ground contribute only large diagonal coefficients (d.o.f. u_A and v_A are x-parallel and y-parallel in both cases). Large diagonal coefficients alone cause no trouble; this is the case for $k_2 >> k_1$ in Eq. 5.10-1. One can make the inclined spring contribute to only the diagonal of $\mathbf{K}$, and thus avoid numerical trouble, by adopting x'-parallel and y'-parallel d.o.f. at node A (u'_A and v'_A in Fig. 5.10-2d). In other words, d.o.f. at node A are reoriented in going from Fig. 5.10-2b to Fig. 5.10-2d. If very stiff elements appear *within* a model rather than only as boundary support elements, large off-diagonal coefficients in $\mathbf{K}$ are inevitable and numerical trouble is likely. This is why offsets in Fig. 4.3-3 are made *rigid* and are treated by constraint transformations.

Thin-walled structures tend to produce ill-conditioned equations because their membrane stiffness is much larger than their bending stiffness. The structure in Fig. 5.10-3 is an example. It is an arch-like structure of uniform thickness t that resembles a folded strip of paper. It is modeled by four-node elements that resist both membrane and bending deformations. Forces of magnitudes 1 and 2 were applied to nodes of element A, as shown. The x-parallel membrane stress in element A was computed for several values of thickness t, and the following results were obtained: for $t = 0.6(10)^{-6}$, stress error was noticeable (roughly 5%); for $t = 0.2(10)^{-6}$, stress error was severe; for $t = 0.1(10)^{-6}$, the stiffness matrix was declared singular by the software. The value of t for which trouble appears depends on the software and computer used. Conventionally, computed stresses are more seriously degraded by ill-conditioning than are displacements because stresses $\boldsymbol{\sigma} = \mathbf{EBd}$ are computed from *differences* in displacements, and element A has small strains but large rigid-body motion. In other words, with regard to horizontal displacements, element A is almost rigid but is lightly supported by the rest of the structure. One can argue that the foregoing thicknesses t are ridiculously small. Nevertheless, one should be aware of the nature of the problem, because numerical difficulty becomes more likely as the number of d.o.f. increases. This means that if a large FE model tends to produce ill-conditioned equations, mesh refinement may make results worse rather than better.

Another possible cause of ill-conditioning is a Poisson ratio near 0.5 (Section 4.13).

Testing for Trouble. Before solving equations, one can look for the largest and smallest diagonal coefficients of $\mathbf{K}$ and suspect ill-conditioning if the ratio of largest to smallest

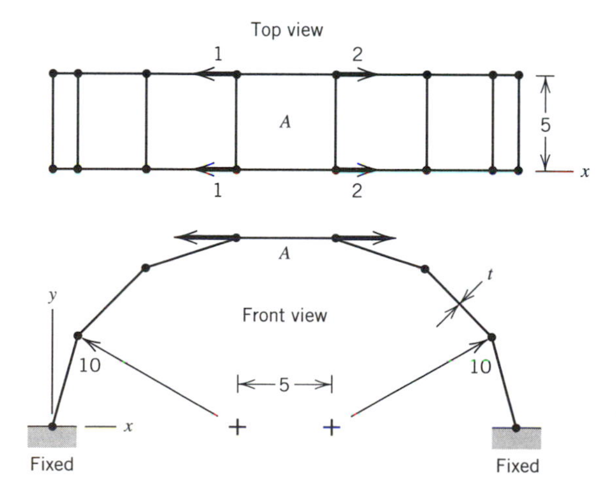

Fig. 5.10-3. Arch-like structure modeled by flat shell elements. Loads are applied to element A.

exceeds some large number. Some software packages do this automatically. However, such a test may be pessimistic (e.g., $k_2 >> k_1$ in Fig. 5.10-1 and Fig. 5.10-2c,d). The following test does not have this defect but is not an *a priori* test.

As coefficients of **K** are being processed by a direct (noniterative) equation solver, the software may apply a *diagonal decay test* [5.11]. Processing an equation involves subtractions that reduce the magnitudes of diagonal coefficients K_{ii} in equations not yet processed. This is seen in Eqs. 5.10-1 and 5.10-2, where elimination of d.o.f. u_1 reduces K_{22} from $k_1 + k_2$ to k_2. The program can store the original value of each K_{ii} and divide it by its reduced value just before the reduced value acts as a pivot in processing the ith equation. If this ratio is 10^n, then about n digits of accuracy have been lost from K_{ii}, leaving $p - n$ accurate digits if computer words store p digits each. A small value of $p - n$ can either trigger a warning message or terminate execution. The test is simple and cheap.

Users occasionally examine computed support reactions to see if they are in static equilibrium with applied loads. If they are, we have some evidence that results are not contaminated by numerical error. However, if the structure is statically indeterminate, incorrect reactions may still satisfy equilibrium. Satisfactory reactions may even give a false sense of security by suggesting that *everything* is correct, while in fact the mesh may be utterly inadequate (e.g., Fig. 3.2-2).

There appears to be no single test for numerical error that is always reliable. Even the diagonal decay test can fail to detect a significant loss of accuracy [2.2]. To *avoid* numerical error, an analyst must understand the modeling practices that promote it and choose alternatives where possible. In particular, it is usually best to change a very stiff region in a model to a *perfectly* stiff region by exactly imposing constraint relations that make the stiff region move as a rigid body.

5.11 COMMON MISTAKES

Mistakes include errors of judgment in FE modeling and blunders in data preparation. Both are noted in the present section. Section 5.12 discusses more systematic searches for errors in the model and the data. Corrections for errors are either obvious or are discussed in other sections.

We remark (yet again) that a major mistake, which generates errors of omission and commission, is insufficient familiarity with the physical problem, element behavior, and analysis limitations. It is also a major mistake to ignore warning messages produced by the software. Vigilance is needed in every step of an analysis.

Division by zero will occur as element **k** matrices are being generated if Poisson's ratio is 0.5 in a plane strain, axisymmetric, or three-dimensional solid problem, and if the thickness of a plate or shell element is zero. If unspecified, the thickness of a plane element may default to unity, depending on the software.

A singular global stiffness matrix **K** may be caused by any of the following:

- Material properties such as elastic moduli are zero.
- One or more nodes are not connected to any element.
- There are no supports, or supports are insufficient to prevent all rigid-body motions.
- A mechanism is created because part of the model is inadequately restrained (e.g., the beam in Fig. 4.13-1 with no rotational connection at node 1, or no restraint of lateral translation d.o.f. in a flat part of a structure modeled by plane elements having only translational d.o.f.).
- A mechanism is created because too many releases are prescribed at a joint.

- There are large stiffness differences, as discussed in Section 5.10.
- Part of the structure has buckled. (This is possible if the "stress stiffening" effect is included and negative stiffening has reduced the net stiffness to zero or less.)
- In nonlinear analysis, supports or connections have reached zero stiffness, so that all or part of the structure is inadequately supported.

A singular **K** usually triggers an error message and stops execution of the analysis. If execution stops, or if execution continues but the results are bizarre, it is clear that something is wrong and a search for the cause is needed. It is much more dangerous if there are errors that lead to plausible but quite inaccurate results. Some errors in this category are as follows.

- Elements are of the wrong type, for example, shell elements are used where solid elements are needed, or axisymmetric elements are used where plane elements are needed.
- Supports are wrong in location, type, or direction. (Supports can be too many as well as too few; for example, complete fixity instead of a hinge, or too many d.o.f. constrained in an attempt to impose symmetry conditions.)
- Loads are wrong in location, type, direction, or magnitude. If symmetry is exploited, a load in a plane of symmetry may not have been divided by 2. Similarly, the stiffness of a beam that straddles a plane of symmetry may not have been divided by 2.
- Other data may be incorrect. It is easy to be off by a power of 10 or use inconsistent units. For example, feet may not have been converted to inches, or angular velocity may have been stated as revolutions per second instead of radians per second (or vice versa).
- An element may have been defined twice. The duplication is hard to detect because it cannot be seen when elements are plotted. The result is a "hard spot" in the model.
- A connection may be physically meaningless (e.g., Fig. 5.7-2b).

A record of the status and progress of the analysis project should be maintained. Records become important if work must be resumed after an interruption. Confusion is the result of poor record keeping: Where did data used to prepare the model come from? Did I remember to make changes X, Y, and Z or not? Which data files correspond to which model? Does the title line of the analysis refer to the current model or the previous model?

5.12 CHECKING THE MODEL

A model should be checked prior to computation, both to make success more likely and to avoid making the checking task more distasteful by postponing it. Indeed, the model should be checked *as it is being prepared*, using graphical features of preprocessors. It is easier to correct mistakes as soon as they appear than to locate and correct them later. Mistakes can be made anywhere, even with simple data, so everything should be checked. Undetected mistakes can prevent execution, or lead to bizarre results, or lead to results that are plausible but wrong. Some checking is done by the analyst and some is done automatically by software.

Checking Done By the Analyst. Nodes and elements may be generated simultaneously. However, if nodes are entered or generated separately, they usually appear on the screen, with or without numbers at the user's option. Only when the node pattern appears satisfactory should elements be entered; thus a misplaced node is less likely to be obscured by elements. When the element layout appears satisfactory the boundary conditions can be applied, and so on.

Many graphical devices are available, with of course some differences between software packages. A shrink plot (Fig. 5.12-1), in which individual elements are reduced in size about 20%, shows immediately if an element is missing. A missing element may not be apparent in a standard mesh plot such as Fig. 3.12-3, especially for three-dimensional models. Other graphical devices, no less useful, include slices (sectioning), hidden lines present or removed, views from various directions, windowing, zoom, and perspective. These options become particularly useful with three-dimensional models. One may also be able to scale selectively, for example, to exaggerate the smaller dimension of a slender model. Support conditions are usually identified by special symbols that convey location, type (displacement or rotation), and direction of the support. Loads may be plotted in an analogous way. One may be able to plot the boundary of the model. If part of it looks like a crack in the material, then some nodes are adjacent or coincident but unconnected, perhaps intentionally but perhaps not.

Some checking must be done by examining a list rather than graphically. For example, it may be important to verify the location of some nodes more accurately than a plot will show. Material properties must be listed and also cross-sectional properties used for beams.

Checking Done By the Software. Commercial software does some checking automatically. These checks involve computing a numerical value from the input data and comparing it with one or more stored values that define limits of acceptability. Data items may be graded "pass" or "fail" or in some cases "pass with warning." Anything but "pass" produces a warning message. "Fail" also prevents execution. These "error traps" complement but do not replace checking done by the analyst. Software often allows a "check run" that stops short of solving global equations or even generating them. The check run applies automatic data checks and may also estimate storage requirements and solution time that will be required by actual analysis. Errors that may be detected by automatic data checks include the following.

- A node is not connected to any element.
- Nodes are close together or coincident but not connected. The analyst must decide if this is intentional or not.
- Elements share a node but do not use the same set of d.o.f. at the node.
- Poisson's ratio is not in the range $0 < \nu < 0.5$. An analogous test may detect impossible properties of an orthotropic material.
- Elements have too large an aspect ratio or corner angles that differ too greatly.
- A side node (of an element that has them) may curve the side too greatly or be too far from midside.
- A four-node element in space is too greatly warped; that is, its nodes are too far above and below the mean plane.
- The dihedral angle between three- or four-node elements in space is too far from 180°.
- A curved shell element spans too great an arc.

Tests for excessive element shape distortion are arbitrary. There are no universally applicable criteria. What is acceptable in one situation may be unacceptable in another [9.8].

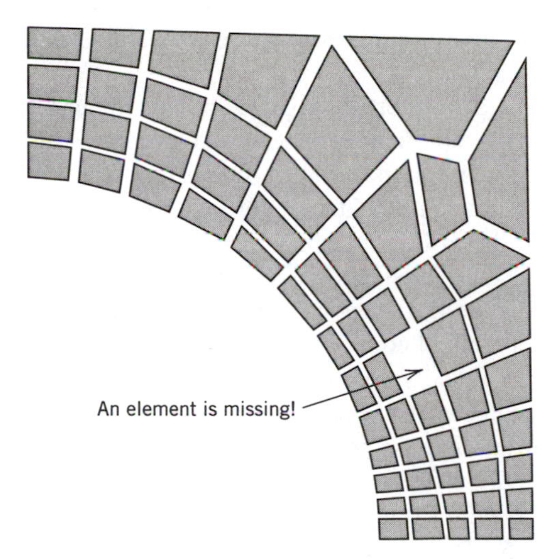

Fig. 5.12-1. A "shrink plot" of elements is but one of many graphical tools that can be used for checking. The mesh shown (without the missing element) is used in Section 3.12.

Accordingly, the absence of a warning does not guarantee that shapes are satisfactory, nor does the issuance of a warning necessarily mean that shapes are unacceptable. Nevertheless, all warning messages should be read, and action taken where needed.

Automatic checking cannot disclose whether elements are of the appropriate type and size, whether units are consistent, whether loads and supports are properly located, and so on. The analyst is responsible for these matters and for the quality of the work.

5.13 CRITIQUE OF FE RESULTS

With the help of graphical tools in the postprocessor, one first examines results qualitatively to see if they "look strange." For this initial examination, displacement results may be the most informative, as described in the following paragraph. If no flaws are obvious, results are examined in more detail and quantitatively compared with expectations. Several expectations should already be available from preanalysis planning (Section 5.9). In comparing FE results with results obtained otherwise—from approximate solution, handbook formulas, alternative software, existing similar structures, or experiment—one must be sure that the physical situations that produce the results are substantially the same. In comparisons between FE and experiment, for example, it is unfortunately common for there to be differences in supports, loading, and even structure geometry, especially if analysts and experimentalists do not communicate well. If FE results pass such comparisons and also pass a critique as suggested in what follows, one must decide if further analysis is required, and, if so, how it should be influenced by the current analysis.

Displacements should be examined first, plotted and scaled so as to be easily visible. Typically, software plots only straight lines between nodes, so that curved shapes assumed by deformed beam elements and deformed edges of plate and shell elements are not visible. Animation of the plot makes the directions of nodal displacements apparent. One should see that actual displacements agree with intended supports, for example, that displacements are tangent to roller supports, have only rotation at a hinge, and are zero at fixed boundaries. If symmetry is expected, it should be visible in the displacement field. Usually it is obvious on physical grounds that some points will displace more than others; this also should be visible. A gap should not "overclose" so that adjacent parts interpenetrate. Sectioning and views from different directions should be used as necessary. One may also plot contours of displacements and rotations.

Deformations produced by temperature change are plotted by software as if the undeformed configuration exists at zero temperature. Accordingly, temperature changes should be stated relative to a reference temperature at which the model is considered undeformed. The reference temperature becomes the zero temperature as far as deformation plots are concerned.

Support reactions can be examined to see if they satisfy statics, for example, to see if the sum of computed x-direction reactions balances the intended x-direction load. If not, it is more likely that the intended load was not applied correctly than that reactions have been incorrectly computed. Software may automatically compute the sum of support reactions in each coordinate direction and the moment of the reactions about each coordinate axis. Note that all reactions must be referred to the same coordinate system, and that constraint equations (if used) may introduce fictitious forces.

It is well to recall that a linear solution is based on equilibrium equations written with respect to the *undeformed* geometry. In Fig. 5.13-1, physically possible displacements may be large enough that the spring must carry tension if static equilibrium is to prevail. A standard linear analysis takes no notice of this possibility; it calculates a compressive force $F = Ph/b$ in the spring, even if computed displacements are very large, so that the displaced configuration resembles Fig. 5.13-1b. If displacements are *actually* this large, a correct solution corresponding to this configuration can only be obtained by doing a *nonlinear* analysis. However, if a linear (small displacement) analysis has been performed and Fig. 5.13-1b is the result of great exaggeration of displacements solely for plotting purposes, the linear solution may be quite accurate even though a plot such as Fig. 5.13-1b may confuse and mislead the analyst. Another example of how plots can mislead appears in Fig. 5.13-2. The deformed shape suggests that the beam has gotten longer and that depth h has increased toward the right end. This is only an impression that results from great exaggeration of computed displacements. In the linear solution displacements of the beam axis are entirely *vertical*; horizontal displacement components of the beam axis are *zero*. In Fig. 5.13-2b the linear solution has merely been scaled up. In any cross section, points on top and bottom surfaces (such as A and B) still have a vertical separation h. Figure 5.13-2b does *not* represent the deformed shape of a real beam whose deflections are truly large. A large-deformation shape can only be computed by an analysis that uses the deformed shape in constructing the equations to be solved. This is *nonlinear* analysis, in which displacements and stresses are not directly proportional to load. The deformed shape produced by scaling up the usual *linear* solution $\mathbf{D} = \mathbf{K}^{-1}\mathbf{R}$ is *not correct* if deflections are indeed large.

In standard software, stresses are computed from displacement differences. Accordingly, stresses are usually less accurate than displacements, although sometimes they are *as* ac-

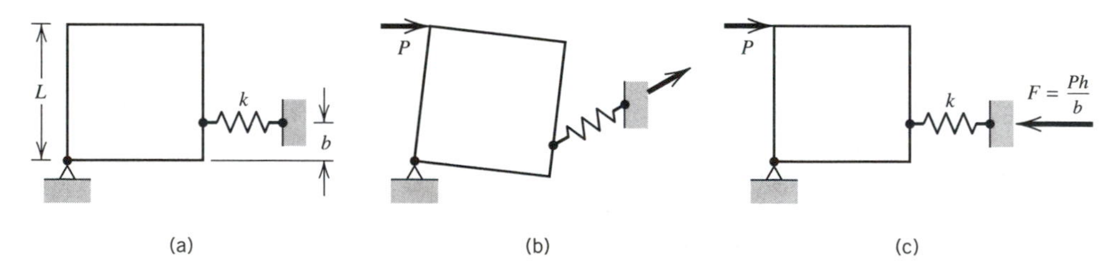

Fig. 5.13-1. (a) Block supported by a hinge and a soft spring. (b) Possible displaced shape produced by load P. (c) Reaction F computed by linear analysis.

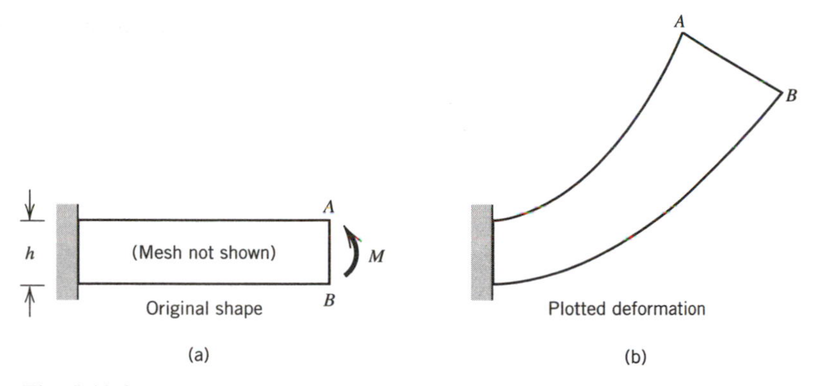

Fig. 5.13-2. (a) Cantilever beam, loaded by tip moment. (b) Deformation from linear analysis, greatly exaggerated, as may be plotted by FE software.

curate as displacements. In either case, stresses cannot be trusted if displacements are suspect. In vibration analysis, mode shapes cannot be trusted if natural frequencies are suspect.

An array of graphical tools is available for viewing computed stresses. Tabulated results can also be useful. Before examining stresses one must ask how they are presented by the software. Are they referred to global or to local axes? If the latter, how are these axes oriented? Are stress resultants (e.g., bending moments) reported? They must not be mistaken for stresses. In beam, plate, and shell elements, stresses may be available at upper, middle, and lower element surfaces. Which is desired, and which surface of an element is called its upper surface by the software? Are stresses averaged at nodes? This is incorrect if coordinate systems do not match or if there are discontinuities of thickness or material properties (Fig. 3.10-1). One can plot contours or shades of any individual normal or shear stress, a principal stress, the "stress intensity," or the von Mises stress (Eqs. 3.10-1 and 3.10-2). Principal stress trajectories (lines tangent to a principal stress direction) can be plotted as dashed lines, with the length of each dash proportional to the magnitude of the principal stress at that location. Trajectories show the "flow" of stress and can be used to identify the primary load-carrying path in a FE model. (Note that a stress trajectory is not a stress contour. A stress contour is the locus of points that have the same stress. It says nothing about the stress direction.) Stresses can be viewed on user-defined cross sections of solid models. Usually one can scale, window, and choose different viewpoints.

Advice that bears repeating, even belaboring, is that stress plots should be based on *un*averaged nodal stresses, so as to retain interelement discontinuities in the plotted contours or shades. Discontinuities are an obvious qualitative measure of discretization error. Contours plotted from nodal average stresses are interelement-continuous. Their appearance is pleasing, but they convey discretization error only by how much they change in direction across interelement boundaries. Examples of averaged and unaveraged contours appear in Figs. 3.12-2 and 3.12-3. Note that contours are interelement-continuous across the line of symmetry $x = y$ in Fig. 3.12-2c. Accordingly, a mesh *may* be too coarse even when an unaveraged stress plot shows continuity. See also the last paragraph of Section 7.6.

Some characteristics of an accurate stress field are as follows. Stress contours should be normal to a plane of reflective symmetry (of loads as well as geometry). At a free boundary, one of the principal stresses should be zero. At a boundary loaded only by pressure p, one of the principal stresses should be $-p$. Principal stress trajectories should be normal or tangent to free boundaries, boundaries loaded only by pressure p, and planes of reflective symmetry. In an axially symmetric problem, radial and circumferential normal stresses should be equal on the axis of revolution. None of these conditions is likely to be met perfectly. The amount of imperfection is a measure of discretization error, or is perhaps a warning of an error in the FE model. One would also suspect a error in modeling if there are unexpected stress gradients or stress concentrations in unreasonable places.

The foregoing inspection is largely qualitative. Quantitative inspection involves comparing computed displacements and stresses with preliminary analytical results (prepared in advance!), experimental results, predictions from formulas in textbooks and handbooks, and whatever else may be appropriate and reasonably available. Such checks are likely to be most useful for the earliest models in a sequence, when blunders are most likely and the appropriateness of some assumptions may still be in question. Inevitably, there will be disagreements between FE results and other results used for comparison. Reasons for any substantial disagreement must be sought. FE results are not necessarily at fault when there is disagreement, but experience shows that most users are entirely too willing to accept computed results at face value [1.7].

Close inspection of results shows how the FE model can be improved. A need for mesh refinement is indicated in regions where stress contours display considerable interelement discontinuity. The *closeness* of stress contours is another guide: if plotted stress contours have equal increments between them, elements that span several stress contours should be refined more than elements that span few contours. In addition to revising the FE model, it may be necessary to alter the scope of the analysis. This may happen if initial assumptions such as no buckling, no gap closure, or no plastic action are inconsistent with the computed magnitudes of displacements and stresses.

5.14 STRESS CONCENTRATIONS. SUBMODELING

Stress Concentrations. The FE method is not very good at calculating peak stresses at holes, fillets, and so on. Often a stress raiser is small, being roughly the size of an element that would be used if the stress raiser were absent. Surrounding the stress raiser by a greatly refined mesh would be a considerable chore. Sometimes a tabulated stress concentration factor (SCF) can be used instead, as follows. The stress raiser (e.g., a small hole) is not modeled, but nominal stresses at its location are calculated by FE analysis. Then, if a tabulated SCF for the local geometry and stress field is available, one need only multiply the nominal stress by the SCF to obtain the peak stress.

If the needed SCF is not tabulated, the following alternative may be available [5.12]. The discontinuity is modeled by a coarse "local" mesh and the peak stress is computed. To compensate for the coarseness of the mesh, the peak stress must be scaled by a factor. The factor is computed by using the same local mesh to solve a "secondary" problem for which results are known. The factor is equal to the ratio of exact stress to computed stress in the secondary problem. The success of the method depends on the availability of a secondary case that is "close" to the primary case, and the ability of the analyst to recognize it.

As an example of this method, consider stress at point E in Fig. 5.14-1a. The mesh used is *very* coarse. For the load P used, the computed stress at point E is 221 (the units do not matter here). The same local mesh is embedded in a tensile strip, Fig. 5.14-1b, as a suitable secondary case for which the SCF is known. At point E, for the load applied, the secondary case yields an exact stress of 130 by using the SCF and a computed stress of 92.8 from FE analysis. Hence the correction factor is 130/92.8 = 1.40. The final estimate

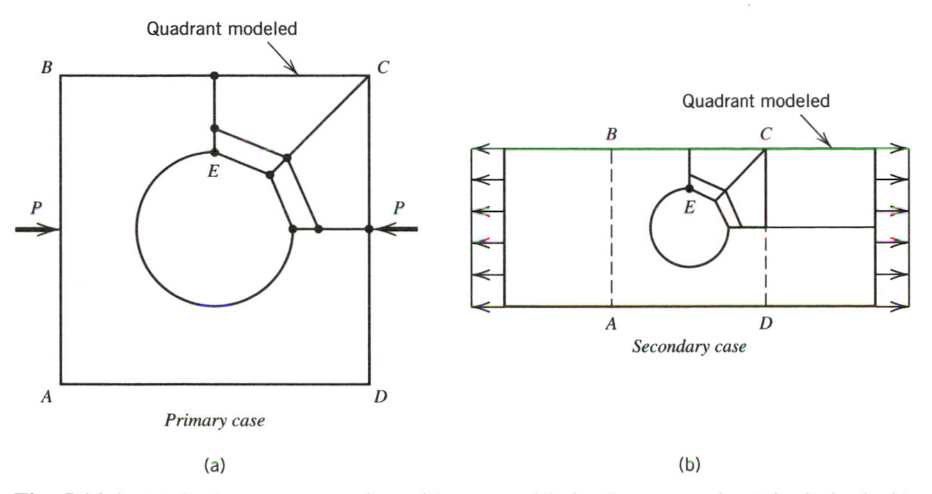

Fig. 5.14-1. (a) A plane square region with a central hole. Stress at point E is desired. (b) "Secondary" case of locally similar geometry for which exact results are known.

of stress at point E in Fig. 5.14-1a is therefore 221(1.40) = 310. The exact stress in Fig. 5.14-1a is very nearly 337 (as computed from a highly refined mesh). To obtain the final estimate of 310 *without* the correction factor, the 2 by 2 mesh in Fig. 5.14-1a must be replaced by an 8 by 8 mesh, with still greater refinement needed for greater accuracy.

Submodeling. A SCF is not tabulated for every kind of stress raiser. A hole or other discontinuity may be oddly shaped or so close to boundaries or loads that it does not lie in the simple kind of stress field for which a SCF is tabulated. Then a refined-mesh study is needed in order to determine the peak stresses. However, it is not necessary to revise and reanalyze the entire FE model. Mesh refinement can be strictly local. This technique is called *submodeling*.

As an example, consider the structure shown in Fig. 5.14-2a. Only a portion of the entire FE mesh is shown. We assume that this mesh is too coarse to give accurate stresses

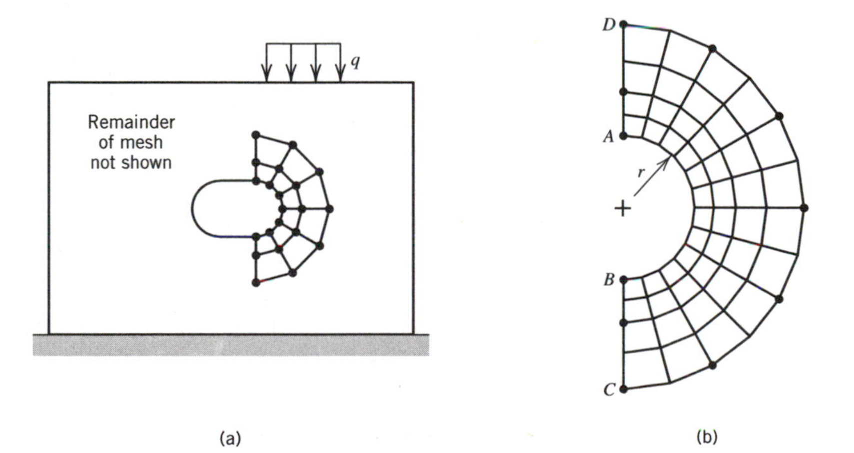

Fig. 5.14-2. (a) A portion of the FE mesh in a coarse-mesh plane FE model. (b) A submodel. Dots show nodes on the cut boundary that also appear in the coarse mesh.

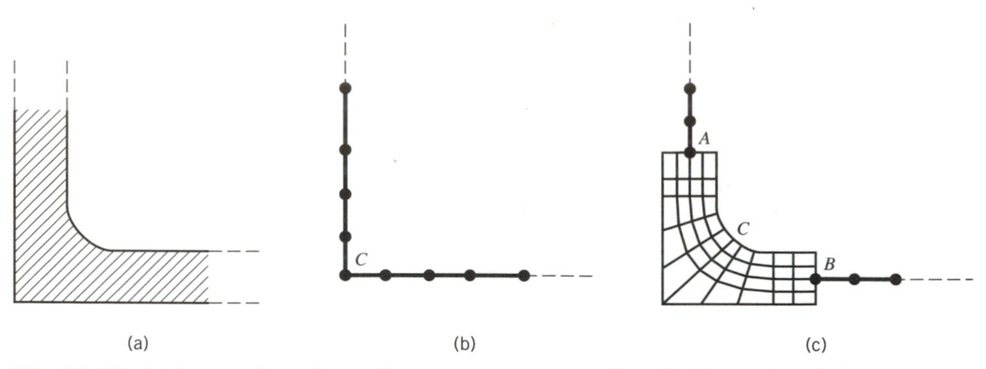

Fig. 5.14-3. (a) Intersection of two plates, seen in cross section. (b) FE model of plate elements, seen edge-on. (c) FE model that combines plate and solid elements.

on the boundary of the cutout, but fine enough to give reasonably accurate displacements *near* the cutout. A possible submodel is shown in Fig. 5.14-2b. It is loaded by prescribed *displacements* at all nodes along *BC*, *CD*, and *DA*. The required displacements are obtained from the coarser-mesh solution. Some of these nodal displacements must be interpolated, because not all nodes along *BC*, *CD*, and *DA* also appear in the coarser mesh. The necessary interpolation capability is included in some commercial software. Displacements are imposed only at nodes along the "cut boundary" *BCDA*. Thus d.o.f. are *not* imposed at any internal nodes of the submodel or at any nodes on the arc of radius r except nodes A and B. The submodel may be refined repeatedly without ever changing the coarser model.

Submodeling is reminiscent of substructuring, but submodeling does not require that the coarse mesh and the submodel have identical node patterns along the cut boundary. Submodeling also makes no provision for updating the "attachment" d.o.f. by connecting the coarse mesh and the submodel together and solving the entire system.

A form of submodeling can be used for intersections. The intersecting plates of Fig. 5.14-3a can be regarded as part of a larger structure (not shown). A plate element model, Fig. 5.14-3b, does not do well at resolving detail near corner C because plate elements lie on midsurfaces of the actual plates. If bending dominates, one might apply a stress concentration factor to computed bending moment in the plate elements at C. Alternatively, the submodel approach in Fig. 5.14-3c might be used. Plate and solid elements meet at A and B and can be connected by methods discussed in Sections 3.9 and 4.13.

Clearly, submodeling requires skill in order to construct a coarse mesh that is not *too* coarse and to place boundaries of the submodel far enough from the stress raiser. In Fig. 5.14-2, "far enough" means that d.o.f. along arc *CD* would be almost unaffected by a decrease in radius r. As a partial check, stresses before and after submodeling can be compared: if submodeling does little to change stresses on the cut boundary of the submodel, we have some assurance that the cut boundary placement is acceptable. Even when skillfully done, stresses may be underestimated because the coarser-mesh model is likely to err by being too stiff, which means that displacements imposed on the submodel will be too small.

5.15 CONVERGENCE WITH MESH REFINEMENT

FE results should converge toward exact results as a mesh is repeatedly refined. This will indeed happen if there are no blunders in FE modeling and if elements pass patch tests.

Convergence is toward results of the corresponding mathematical model, for example, toward results of beam theory if beam elements are used. Convergence will be "from below"—meaning that each FE model errs by being too stiff—*provided that* conditions noted in Section 4.8 are met. If, in addition, each refinement is by subdivision of existing elements, with existing nodes retained and not repositioned, convergence will be *monotonic* from below. The mesh in Fig. 5.15-1b is such a refinement of the mesh in Fig. 5.15-1a. Mathematically, one says that this kind of subdivision retains the old trial field as a subset of the new one. None of this says anything about the *rate* of convergence. There is little advantage to monotonic convergence with one type of element if a different type provides nonmonotonic convergence but considerably greater accuracy for the same number of d.o.f.

In order to say more about convergence some terms must be defined. Let h be an approximate linear size measure of an element; that is, the actual length of a bar or beam element, or $A^{1/2}$ where A is the area of a plane or plate element, or $V^{1/3}$ where V is the volume of a 3D solid element. Let p be the degree of the highest *complete* polynomial in the element displacement field. Thus $p = 1$ for the CST (Eq. 3.2-1), $p = 1$ for the basic four-node quadrilateral (Eq. 3.4-1), and $p = 2$ for the eight-node quadrilateral (Eq. 3.5-1).

Common parlance refers to "h-refinement" and "p-refinement," in which h or p is changed in going from the old mesh to the new. An h-refinement changes element sizes without changing element types (so p remains constant). A p-refinement changes element types without changing element sizes (so h remains constant). In p-refinement, nodes may be added to existing elements and/or d.o.f. may be added to existing nodes. Examples appear in Fig. 5.15-1. These examples are *uniform* refinements, in which the positions of existing nodes are not changed. Another possibility is " r-refinement," Fig. 5.15-1d, in which r means "rearrange"; that is, existing nodes are moved without changing the number of elements or the number of d.o.f. Because the number of d.o.f. is not increased, r-refinement can provide only limited improvement in accuracy. Of course none of these refinement methods need be used in isolation. Commonly, nodes are rearranged when doing h- or p-refinement, or when doing h- and p-refinement in combination. It appears that for problems containing singularities, such as reentrant corners or cracks, p-refinement converges much faster than h-refinement, especially if combined with r-refinement so that mesh density is most greatly increased around singularities [5.13].

Some commercial programs are *self-adaptive*, which means that they are able to estimate the error of a FE solution, revise the mesh, reanalyze, and repeat this cycle *automatically* until a prescribed convergence tolerance is met. More is said about this in Section 5.16. A self-adaptive program may be based on h-refinement or on p-refinement.

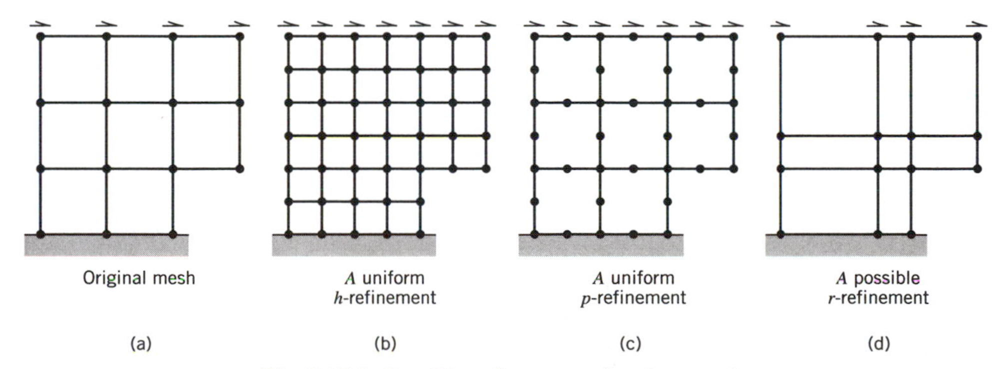

Fig. 5.15-1. Possible refinements of a plane mesh.

Adaptive h-refinement capability has been obtained by additions to existing codes: error-calculation and mesh-revision modules were added, and a driver was written to repeat the entire analysis, error-checking, mesh-revision sequence until the convergence tolerance is met. Refinement can continue until limits of computer capacity or numerical noise are reached. Adaptive p-refinement requires more sophisticated programming because element types are changed in successive mesh revisions and nodes may have more than the usual number of d.o.f. Refinement can continue until the highest-order elements coded in the program have been used.

Sometimes computed results from two analyses can be extrapolated to yield an improved result. The argument is as follows, with reference to the h method. We assume that convergence is *monotonic* and that the convergence rate is known; for example, if error in a certain quantity is quartered when size of elements is halved, then error is proportional to h^2. Figure 5.15-2a illustrates the general case of error proportional to h^q, which plots as a straight line when the abscissa is h^q. The ordinate, ϕ, represents the quantity of interest, such as displacement or stress at a certain point. By simple linear extrapolation of ϕ versus h^q, we obtain

$$\phi_\infty = \frac{\phi_1 h_2^q - \phi_2 h_1^q}{h_2^q - h_1^q} \tag{5.15-1}$$

as the value of ϕ expected at infinite mesh refinement, when $h = 0$. Figure 5.15-2b shows that for nonmonotonic convergence, extrapolation based on values of ϕ at points such as C and D may produce a *worse* result rather than an improved result. Note also that at least three analyses are required in order to determine q, and that all of them may have to be based on at least moderate mesh refinement if q is to be determined with certainty. If successive mesh refinements are nonuniform, it is not clear how h_1, h_2, and so on are to be measured. Perhaps the h_i should then pertain to the element in each mesh nearest the point of interest. Results from three or more analyses may fail to plot as a straight line for any value of q. Then one might opt for linear extrapolation, by using a least-squares of a straight line to the results of three or more analyses, but with no guarantee of an improved result.

One sometimes examines results from two different meshes, notes that they are in substantial agreement, and concludes that convergence is almost complete. The conclusion is

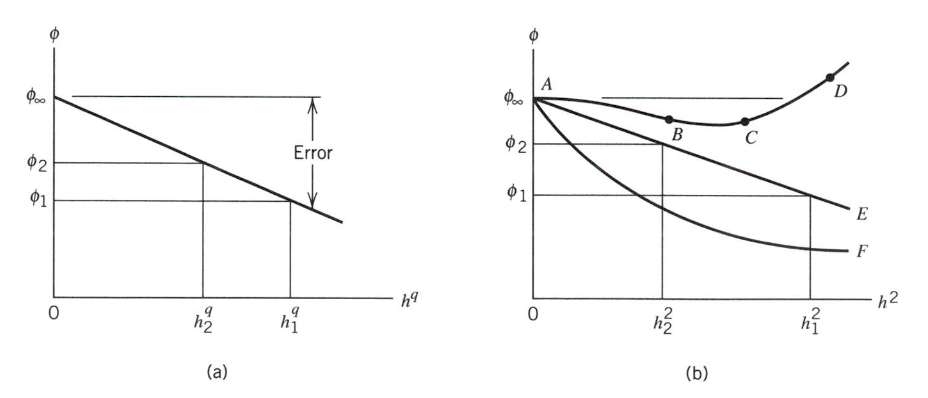

Fig. 5.15-2. (a) Extrapolation with error proportional to h^q. (b) Error is nonmonotonic in *ABCD* and proportional to h^2 in *AE*. *AF* would be a straight line if the abscissa were h.

plausible but may be false. It is possible that results from two meshes plot as points B and C in Fig. 5.15-2b, which have the same error.

5.16 ERROR MEASURES AND ADAPTIVITY

A FE solution contains enough information to estimate its own discretization error. An error estimate can be calculated by the postprocessor and used to guide mesh revision, so that the next analysis will be more accurate. Cycles of analysis and mesh revision can be repeated until a convergence test is satisfied. These ideas can be implemented in different ways. One method uses the Z^2 error estimator, named after its authors [5.14]. It pertains to the error in computed stresses and is summarized as follows.

First, it is necessary to discuss stress fields. Consider the simple example of a uniformly loaded bar, Fig. 5.16-1. The element-by-element stress field is discontinuous between elements, but the stress field constructed from nodal average stresses is continuous. Indeed, in this particular example (but not in general) the continuous field is exact, except in elements at either end of the bar. There are other ways of constructing a continuous stress field [5.15], but what matters for the error estimate is that the continuous field, however constructed from discontinuous element stresses, is regarded as the most accurate portrayal of the exact stress field that the current discretization can provide. Accordingly, the *difference* between the element-by-element field and the continuous field can be regarded as an approximate error field. This error field is indicated by shading in Fig. 5.16-1. Whatever the element type, when discontinuities of stress (or bending moment) appear between elements, the amount of discontinuity is regarded as a measure of error. We identify the various stresses as follows.

σ = the element-by-element stress field (discontinuous)

σ^* = the averaged or smoothed stress field (continuous)

$\sigma_E = \sigma - \sigma^*$, the "error" stress field

A strain energy can be associated with each stress.

$$U = \sum_{i=1}^{n} U_i \qquad \text{where} \qquad U_i = \int_0^L \frac{E^{-1}}{2} \sigma^2 A\, dx \tag{5.16-1a}$$

$$U^* = \sum_{i=1}^{n} U_i^* \qquad \text{where} \qquad U_i^* = \int_0^L \frac{E^{-1}}{2} \left(\sigma^*\right)^2 A\, dx \tag{5.16-1b}$$

$$U_E = \sum_{i=1}^{n} U_{Ei} \qquad \text{where} \qquad U_{Ei} = \int_0^L \frac{E^{-1}}{2} \sigma_E^2 A\, dx \tag{5.16-1c}$$

where A is the cross-sectional area, L is the element length, and summation signs indicate that energy contributions of all n elements of the mesh are added. With elements of arbitrary type, σ, σ^*, and σ_E become stress *vectors*, elastic modulus E becomes the matrix $\mathbf{E}$ of elastic constants, and integration is over element volumes. Thus a typical integrand, in Eq. 5.16-1a for example, becomes $\frac{1}{2}\boldsymbol{\sigma}^T\mathbf{E}^{-1}\boldsymbol{\sigma}\, dV$. Element energy errors U_{Ei} do not indicate

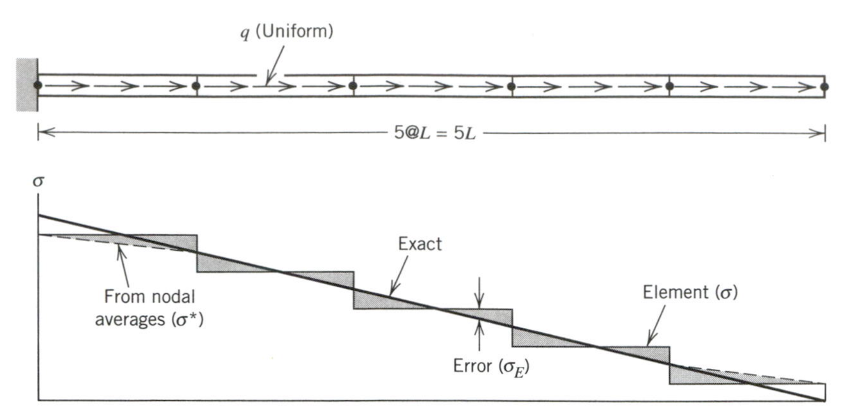

Fig. 5.16-1. A uniform bar modeled by two-node bar elements of equal length. Stresses associated with uniform axial load q are shown.

the accuracy of stresses in individual elements; instead, they are used to guide mesh revision. A global quantity η is used to test for convergence, where η is the relative energy error

$$\eta = \left(\frac{U_E}{U+U_E}\right)^{1/2} \qquad \text{where} \qquad 0 \leq \eta \leq 1 \tag{5.16-2}$$

The denominator uses $U + U_E$ as an approximation of the exact strain energy, in recognition of the probable overstiffness of the FE model, which makes U smaller than the exact strain energy. Apparently, U^* could be used instead of $U + U_E$. The square root serves to associate η with the stress field, as strain energy is proportional to squares and products of stresses. Note that η is a *global* quantity and does not measure error at any particular point. Indeed, if an analogous quantity η_i were defined for each element it too would be unreliable as an element error measure (e.g., it could be large just because U_i is small). As examples of relative energy error, (global) values of η are 0.373 in Fig. 3.12-2 and 0.183 in Fig. 3.12-3. Neither value of η is small enough to indicate that the solution is satisfactory.

In practice, an adaptive solution proceeds as follows. An initial analysis is followed by postprocessing that yields U_{Ei} for every element and the global quantity η. If η is less than a prescribed value, say, 0.05, the procedure terminates. Otherwise the mesh is revised, by h and/or p methods and possible repositioning of nodes, so that more elements or more d.o.f. are placed in existing elements where U_{Ei} is comparatively large. Another cycle of the procedure is begun by analysis of the revised FE model. With suitable coding all of this can proceed *automatically*, beginning with a user-supplied FE model and terminating after perhaps four cycles with a refined FE model and presumably accurate results—which the analyst must check, as usual. Figure 5.16-2 is an example of this process. Note that there appear to be stress concentrations at A and B. It may be wise to exclude such singularities from the region to be treated by self-adaptive analysis, as there is little purpose in seeking improved accuracy at points where stress is known to be infinite. Again, neither U_{Ei} nor η works as an indicator of percentage error in a local stress value.

The procedure by which element energy errors U_{Ei} are used to revise the mesh [5.14]

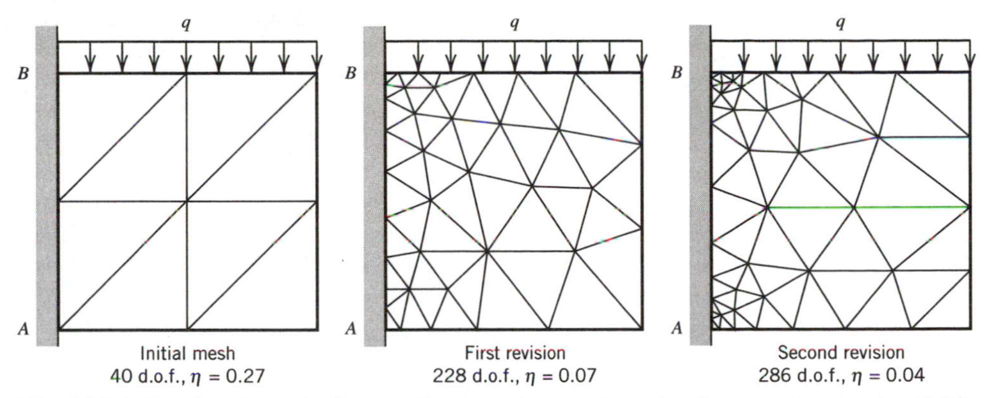

Fig. 5.16-2. Results of an adaptive solution in a plane region using linear strain triangles [5.16]. Poisson's ratio is 0.3. All d.o.f. along AB are set to zero. Each revision aimed at $\eta = 0.05$.

tends to produce a mesh in which U_{Ei} is the same in all elements. The first mesh revision does not make all U_{Ei} values exactly the same because the computed stress field is somewhat changed by mesh revision. The analyst can monitor error estimates in successive mesh revisions by looking at plots of U_{Ei} values, using different colors for different magnitudes of U_{Ei}. The postprocessor will color each element according to its U_{Ei} value. Accordingly, the plot will be multicolored after the first analysis, when U_{Ei} values are quite different. At full convergence all U_{Ei} values would be equal and the plot would be monochrome. Another way to monitor the iterative process is to inspect unaveraged stress contours, as has been repeatedly advocated.

The foregoing arguments are built on the assumption that discontinuities in element-by-element stress field are indicative of error. This is not necessarily so: stresses *should* be discontinuous in some situations, such as at an abrupt change in thickness or modulus or at a shrink fit interface. *In such cases the region over which the stress error estimate is computed should exclude the known discontinuities.*

What if there are multiple load cases? The foregoing mesh revision procedure will produce a different mesh for each load case. It would be more convenient to produce a single mesh that constitutes an improvement for all load cases. A possible strategy is to predict improved element sizes from each load case separately, using U_{Ei} values appropriate to each load case, then generate a mesh in which element size at every location is the smallest of the several sizes predicted.

Software having automatic adaptive capability does much to free the analyst from the labor of preparing meshes and altering them to make the next analysis more accurate. The analyst must still understand the physical problem and the FE method well enough to create a correct and adequate initial FE model. Mistakes in loads, support conditions, and so on will propagate through adaptive cycles and produce an improved solution to the wrong problem. Also, poor choices of element types or an initial mesh that is too coarse may not disclose enough detail to permit the revised mesh to be an improvement. Automatic adaptivity *seems* to guarantee that final results will be adequate, but of course there can be no such guarantee. It remains the duty of the analyst to do the work properly and to critically examine computed results.

Concluding Remarks. The analyst is not to blame for everything that may go wrong. Software contains errors, despite the best efforts of software vendors to sweep them out.

Errors exist even in software developed by vendors that meet the demanding quality control and verification procedures of the Nuclear Regulatory Commission (NRC). And for every vendor that meets NRC requirements there are many that do not. One author [5.6] puts it strongly: "Beware of computers. And, especially beware of developers of engineering software." Regardless of the source of trouble, the engineer who uses the software is held responsible for the results.

CHAPTER 6

3D Solids and Solids of Revolution

This chapter considers solid elements, first for the general 3D case, then for the special (but very common) case of axial symmetry. Each of these two cases is followed by an example application. Axisymmetric geometry with nonaxisymmetric loading is described last.

6.1 INTRODUCTION

The term "3D solid" is used to mean a three-dimensional solid that is unrestricted as to shape, loading, material properties, and boundary conditions. A consequence of this generality is that all six possible stresses (three normal and three shear) must be taken into account (Fig. 6.1-1). Also, the displacement field involves all three possible components, u, v, and w. Typical finite elements for 3D solids are tetrahedra and hexahedra, with three translational d.o.f. per node. Figure 6.1-1b shows a hexahedral element, about which more will be said in Section 6.2.

Problems of beam bending, plane stress, plates, and so on, can all be regarded as special cases of a 3D solid. Why then not simplify FE analysis by using 3D elements to model everything? In fact, this would not be a simplification. 3D models are the hardest to prepare, the most tedious to check for errors, and the most demanding of computer resources. Also, some 3D elements would become quite elongated in modeling beams, plates, and shells; this invites locking behavior and ill-conditioning (Sections 3.6 and 5.10).

A solid of revolution, also called an axisymmetric solid, is generated by revolving a plane figure about an axis in the plane. Common examples include a hose nozzle and a light bulb, although the light bulb has a very thin wall and would be properly classed as a *shell* of revolution for stress analysis purposes. Loads and supports may or may not have axial symmetry. Initially, we will consider the case where geometry, elastic properties, loads, and supports are all axisymmetric. Consequently, nothing varies with the circumferential coordinate θ, material points displace only radially and axially, and shear stresses $\tau_{r\theta}$ and $\tau_{\theta z}$ are both zero. Thus the analysis problem is *mathematically* two-dimensional. Axisymmetric finite elements are often pictured as plane triangles or quadrilaterals, but these plane shapes are actually *cross sections* of annular elements, and what appear to be nodal points are actually nodal circles (Fig. 6.1-2).

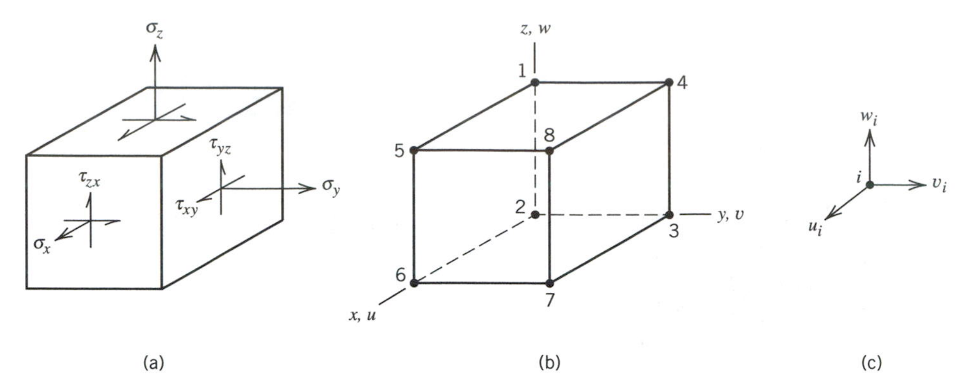

Fig. 6.1-1. (a) 3D state of stress. (b) An eight-node hexahedron FE. (c) The d.o.f. at a typical node ($i = 1, 2, \ldots, 8$).

Stress–Strain–Temperature Relations. As usual, the constitutive relation of a linearly elastic material is written as

$$\boldsymbol{\sigma} = \mathbf{E}\boldsymbol{\varepsilon} + \boldsymbol{\sigma}_0 \tag{6.1-1}$$

For an isotropic material in three dimensions, with initial stress $\boldsymbol{\sigma}_0$ produced by temperature change, Eq. 6.1-1 symbolizes the relation

$$\begin{Bmatrix} \sigma_x \\ \sigma_y \\ \sigma_z \\ \tau_{xy} \\ \tau_{yz} \\ \tau_{zx} \end{Bmatrix} = \begin{bmatrix} (1-\nu)c & \nu c & \nu c & 0 & 0 & 0 \\ & (1-\nu)c & \nu c & 0 & 0 & 0 \\ & & (1-\nu)c & 0 & 0 & 0 \\ & & & G & 0 & 0 \\ & \text{symmetric} & & & G & 0 \\ & & & & & G \end{bmatrix} \begin{Bmatrix} \varepsilon_x \\ \varepsilon_y \\ \varepsilon_z \\ \gamma_{xy} \\ \gamma_{yz} \\ \gamma_{zx} \end{Bmatrix} - \frac{E\alpha\,\Delta T}{1-2\nu} \begin{Bmatrix} 1 \\ 1 \\ 1 \\ 0 \\ 0 \\ 0 \end{Bmatrix} \tag{6.1-2}$$

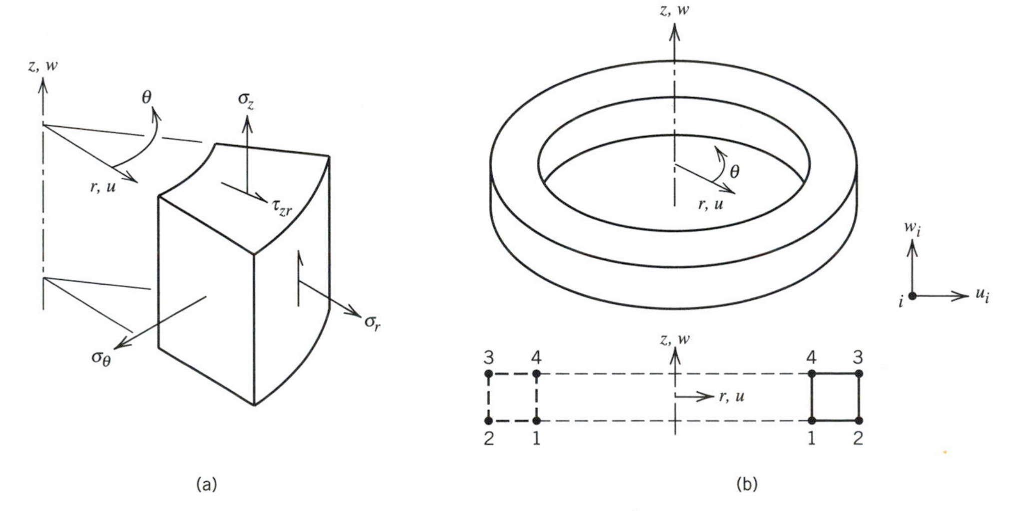

Fig. 6.1-2. (a) Axisymmetric state of stress. (b) A four-node axisymmetric element and d.o.f. at a typical node ($i = 1, 2, 3, 4$).

where E is the elastic modulus, ν is Poisson's ratio, α is the coefficient of thermal expansion, ΔT is temperature change, and

$$c = \frac{E}{(1+\nu)(1-2\nu)} \quad \text{and} \quad G = \frac{E}{2(1+\nu)} \tag{6.1-3}$$

In the same notation, for axial symmetry and an isotropic material, Eq. 6.1-1 symbolizes the relation

$$\begin{Bmatrix} \sigma_r \\ \sigma_\theta \\ \sigma_z \\ \tau_{zr} \end{Bmatrix} = \begin{bmatrix} (1-\nu)c & \nu c & \nu c & 0 \\ & (1-\nu)c & \nu c & 0 \\ & & (1-\nu)c & 0 \\ \text{symmetric} & & & G \end{bmatrix} \begin{Bmatrix} \varepsilon_r \\ \varepsilon_\theta \\ \varepsilon_z \\ \gamma_{zr} \end{Bmatrix} - \frac{E\alpha\,\Delta T}{1-2\nu} \begin{Bmatrix} 1 \\ 1 \\ 1 \\ 0 \end{Bmatrix} \tag{6.1-4}$$

The resemblance between Eqs. 3.1-4, 6.1-2, and 6.1-4 is obvious. In particular, note that if ν approaches 0.5 a division by zero impends, inviting the troubles of locking and ill-conditioning.

The FE method is not restricted to isotropic materials, but we will not discuss anisotropic materials here. For a solid of revolution, software may require that material properties not depend on θ and that θ be a principal material direction of an orthotropic material.

Strain–Displacement Relations. Let $u = u(x, y, z)$, $v = v(x, y, z)$, and $w = w(x, y, z)$ be displacement components of an arbitrary material point in the x, y, and z directions, respectively. If strains and rotations are small, strains and displacement gradients in Cartesian coordinates are related by the equations

$$\begin{aligned} \varepsilon_x &= \frac{\partial u}{\partial x} & \gamma_{xy} &= \frac{\partial u}{\partial y} + \frac{\partial v}{\partial x} \\ \varepsilon_y &= \frac{\partial v}{\partial y} & \gamma_{yz} &= \frac{\partial v}{\partial z} + \frac{\partial w}{\partial y} \\ \varepsilon_z &= \frac{\partial w}{\partial z} & \gamma_{zx} &= \frac{\partial w}{\partial x} + \frac{\partial u}{\partial z} \end{aligned} \tag{6.1-5}$$

For a solid of revolution we switch from Cartesian coordinates to cylindrical coordinates. If deformations are axially symmetric, the circumferential displacement component v is zero, the radial displacement component is $u = u(r, z)$, and the axial displacement component is $w = w(r, z)$. Shear strains $\gamma_{r\theta}$ and $\gamma_{\theta z}$ are zero. Nonzero strains in the case of *axial symmetry* are

$$\begin{aligned} \varepsilon_r &= \frac{\partial u}{\partial r} & \varepsilon_\theta &= \frac{u}{r} \\ \varepsilon_z &= \frac{\partial w}{\partial z} & \gamma_{yz} &= \frac{\partial w}{\partial r} + \frac{\partial u}{\partial z} \end{aligned} \tag{6.1-6}$$

The expression $\varepsilon_\theta = u/r$ is derived in Fig. 6.1-3. Note that zero circumferential displacement does not imply zero circumferential strain. If desired, Eqs. 6.1-5 and 6.1-6 can be stated in matrix format, like Eq. 3.1-6.

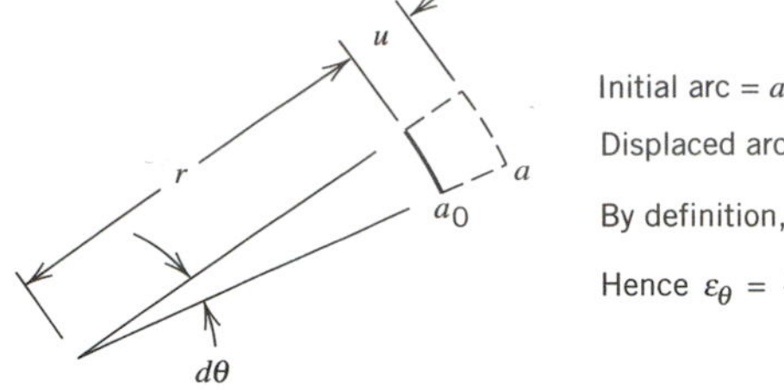

Fig. 6.1-3. Circumferential strain ε_θ when a differential arc displaces radially an amount u.

Displacements $\mathbf{u}$ within an element are interpolated from nodal d.o.f. $\mathbf{d}$ in the usual way; that is, $\mathbf{u} = \mathbf{Nd}$, where $\mathbf{N}$ is the shape function matrix. If nodes have only translational d.o.f. and n is the number of nodes per element, $\mathbf{N}$ has $3n$ columns for a 3D element and $2n$ columns for an axisymmetric element. Thus, for 3D solids, $\mathbf{u} = \mathbf{Nd}$ is

$$\begin{Bmatrix} u \\ v \\ w \end{Bmatrix} = \begin{bmatrix} N_1 & 0 & 0 & N_2 & 0 & 0 & \cdots \\ 0 & N_1 & 0 & 0 & N_2 & 0 & \cdots \\ 0 & 0 & N_1 & 0 & 0 & N_2 & \cdots \end{bmatrix} \begin{Bmatrix} u_1 \\ v_1 \\ w_1 \\ u_2 \\ v_2 \\ w_2 \\ \vdots \end{Bmatrix} \tag{6.1-7}$$

Similarly, for axisymmetric displacements in a solid of revolution, $\mathbf{u} = \mathbf{Nd}$ is

$$\begin{Bmatrix} u \\ w \end{Bmatrix} = \begin{bmatrix} N_1 & 0 & N_2 & 0 & \cdots \\ 0 & N_1 & 0 & N_2 & \cdots \end{bmatrix} \begin{Bmatrix} u_1 \\ w_1 \\ u_2 \\ w_2 \\ \vdots \end{Bmatrix} \tag{6.1-8}$$

Formulas for k. Substitution of $\mathbf{u} = \mathbf{Nd}$ into the strain–displacement relation yields the strain–displacement matrix $\mathbf{B}$, which in turn enters the integrand of the formula for element stiffness matrix $\mathbf{k}$, as explained in connection with Eqs. 3.1-8 and 3.1-10. With n the number of nodes per element, and translational d.o.f. only, these relations for 3D solids and solids of revolution are as follows:

General solids:	Solids of revolution with axisymmetric stress field:	
$\underset{6\times1}{\boldsymbol{\varepsilon}} = \underset{6\times3n}{\mathbf{B}}\ \underset{3n\times1}{\mathbf{d}}$	$\underset{4\times1}{\boldsymbol{\varepsilon}} = \underset{4\times2n}{\mathbf{B}}\ \underset{2n\times1}{\mathbf{d}}$	(6.1-9a)
$\underset{3n\times3n}{\mathbf{k}} = \int\int\int \mathbf{B}^T \underset{6\times6}{\mathbf{E}}\, \mathbf{B}\, dx\, dy\, dz$	$\underset{2n\times2n}{\mathbf{k}} = \int\int\int \mathbf{B}^T \underset{4\times4}{\mathbf{E}}\, \mathbf{B}\, r\, dr\, d\theta\, dz$	(6.1-9b)

If nodal rotation d.o.f. are also present, additional columns appear in $\mathbf{N}$ and in $\mathbf{B}$, and $\mathbf{k}$ is of larger order.

Integration with respect to θ in an axisymmetric problem produces a factor 2π, which is a common multiplier of both $\mathbf{K}$ and $\mathbf{R}$ in the global equation $\mathbf{KD} = \mathbf{R}$. In some software the 2π multiplier is discarded. Then loads in $\mathbf{R}$ pertain to a 1-radian segment.

Remarks. To prevent singularity of **K**, boundary conditions on a 3D solid must suppress six rigid-body motions: translation along, and rotation about, each of the three coordinate axes. In a solid of revolution with axisymmetric deformations, translation w along the z axis is the only possible rigid-body motion. Accordingly, **K** will be nonsingular if w is prescribed at only one node (or, stated more properly, around one nodal circle).

An axisymmetric radial component of load is statically equivalent to zero, but this does not mean that it can be discarded from the load vector. It still produces deformation and stress. Over the circumference, a radial line load of q units of force per unit of (circumferential) length is regarded as contributing a radial force $2\pi rq$ of units to the load vector, where r is the radius at which q acts. Likewise, a moment of M N·m per unit of (circumferential) length is statically equivalent to zero but is regarded as applying a moment about the θ direction of $2\pi rM$ N·m. Similar remarks can be made for the radial body force load associated with spinning about the z axis.

An unrestrained body that is homogeneous and either isotropic or rectilinearly orthotropic is unstressed by temperature change if the temperature field is either constant or linear in Cartesian coordinates xyz. An unrestrained solid of revolution that is either isotropic or cylindrically orthotropic is *not* unstressed by a temperature field that is linear in radius r of cylindrical coordinates. The solid of revolution would remain stress-free if the temperature field is either constant or a linear function of axial coordinate z only.

Although a plane FE model and the cross section of an axisymmetic FE model look alike, and each uses the same pattern of nodal d.o.f., it is physically meaningless to couple them together (Fig. 5.7-2b). Physically, such a connection would not produce axisymmetric deformations in the solid of revolution. If this kind of connection is actually intended, it will usually be necessary to model the solid of revolution by 3D elements.

Some software allows the analysis of a solid of revolution under loading without axial symmetry. The technique is summarized in Section 6.6. This is usually a "stand-alone" analysis; attachment to a 3D solid or a plane structure is not allowed.

Caution. In problems of buckling or vibration, axial symmetry of geometry, material properties, loading, and support conditions does not guarantee axial symmetry of displacement.

6.2 ELEMENTS FOR 3D SOLIDS

Most solid elements are direct extensions of plane elements discussed in Chapter 3. The extensions consist of adding another coordinate and another displacement component. The behavior and the limitations of specific 3D elements largely parallel those of their 2D counterparts.

Constant Strain Tetrahedron. This element (Fig. 6.2-1a) has three translational d.o.f. at each of its four nodes, for a total of 12 d.o.f. In terms of generalized coordinates β_i, its displacement field is

$$\begin{aligned} u &= \beta_1 + \beta_2 x + \beta_3 y + \beta_4 z \\ v &= \beta_5 + \beta_6 x + \beta_7 y + \beta_8 z \\ w &= \beta_9 + \beta_{10} x + \beta_{11} y + \beta_{12} z \end{aligned} \qquad (6.2\text{-}1)$$

Like the constant strain triangle (Eq. 3.2-1), the constant strain tetrahedron is accurate only when strains are almost constant over the span of an element. The element is poor at

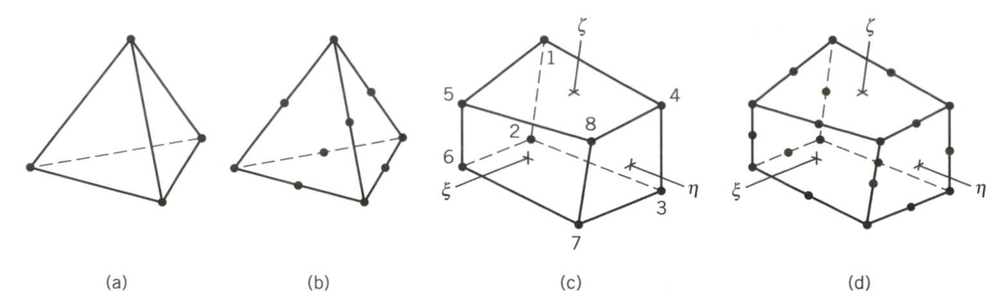

Fig. 6.2-1. Common 3D elements. (a) Constant strain (four-node) tetrahedron. (b) Linear strain (ten-node) tetrahedron. (c) Trilinear (eight-node) hexahedron. (d) Quadratic (20-node) hexahedron.

representing fields of bending or twisting if the axis of bending or twisting either intersects the element or is close to it.

Linear Strain Tetrahedron. This element (Fig. 6.2-1b) has ten nodes, each with three translational d.o.f., for a total of 30 d.o.f. Its displacement field in terms of generalized coordinates can be obtained by adding the six quadratic modes x^2, y^2, z^2, xy, yz, and zx to each of the expressions for u, v, and w in Eqs. 6.2-1. Like the six-node triangle (Eq. 3.3-1), the ten-node tetrahedron has a strain field that is linear in the coordinates. The element can therefore represent fields of pure bending exactly. Depending on the coordinates assigned to edge nodes, edges of undeformed elements can be straight or curved.

Trilinear Hexahedron. This element is also called an eight-node brick. Its rectangular form, shown in Fig. 6.1-1b, has the displacement field

$$\begin{aligned} u &= \beta_1 + \beta_2 x + \beta_3 y + \beta_4 z + \beta_5 xy + \beta_6 yz + \beta_7 zx + \beta_8 xyz \\ v &= \beta_9 + \beta_{10} x + \beta_{11} y + \beta_{12} z + \beta_{13} xy + \beta_{14} yz + \beta_{15} zx + \beta_{16} xyz \\ w &= \beta_{17} + \beta_{18} x + \beta_{19} y + \beta_{20} z + \beta_{21} xy + \beta_{22} yz + \beta_{23} zx + \beta_{24} xyz \end{aligned} \tag{6.2-2}$$

Each of the three displacement expressions contains all modes in the expression $(c_1 + c_2 x)(c_3 + c_4 y)(c_5 + c_6 z)$, which is the product of three linear polynomials in which the c_i are constants. Each of Eqs. 6.2-2 contains all linear modes, some of the quadratic modes (x^2, y^2, and z^2 are missing), and one of the cubic modes (xyz). The resemblance of Eqs. 3.4-1 and 6.2-2 is obvious.

The hexahedral element can be of arbitrary shape if it is formulated as an isoparametric element (Section 4.4). The coordinates used are shown in Fig. 6.2-1c. The six faces of the element are defined by $\xi = \pm 1$, $\eta = \pm 1$, and $\zeta = \pm 1$. Displacement expressions can be written as

$$u = \sum N_i u_i \qquad v = \sum N_i v_i \qquad w = \sum N_i w_i \tag{6.2-3}$$

Index i runs from 1 to 8 in each summation. Shorthand for the shape functions is

$$N_i = \tfrac{1}{8}(1 \pm \xi)(1 \pm \eta)(1 \pm \zeta) \tag{6.2-4}$$

in which all signs are negative for N_2, all signs are positive for N_8, and so on. The formula for the element stiffness matrix in isoparametric coordinates is

$$\mathbf{k} = \int_{-1}^{1}\int_{-1}^{1}\int_{-1}^{1} \mathbf{B}^T \mathbf{E} \, \mathbf{B} |\mathbf{J}| \, d\xi \, d\eta \, d\zeta \tag{6.2-5}$$

As in Eq. 4.4-8, $|\mathbf{J}|$ can be regarded as a scale factor. Here it expresses the volume ratio of the differential element $dX \, dY \, dZ$ in global Cartesian coordinates to its representation $d\xi \, d\eta \, d\zeta$ in isoparametric coordinates. Equation 6.2-5 is integrated numerically, usually by a 2 by 2 by 2 Gauss quadrature rule.

Like the bilinear quadrilateral, the trilinear hexahedron cannot model beam action well because its sides remain straight as the element deforms. If elongated, it suffers from shear locking when bent. A remedy for locking, described for plane elements in Section 3.6, is also applicable in three dimensions. By extension of Eqs. 3.6-2, we add to each of the three displacement fields in Eqs. 6.2-3 the incompatible modes $(1 - \xi^2)$, $(1 - \eta^2)$, and $(1 - \zeta^2)$, each multiplied by a generalized coordinate g_i. Thus a total of nine internal d.o.f. are introduced. Thus augmented, the element is incompatible, but it is valid in the same way the element described in Section 3.6 is valid.

Quadratic Hexahedron. This element, shown in Fig. 6.2-1d, is a direct extension of the quadratic quadrilateral described in Section 3.5. Like the linear strain tetrahedron, edges of undeformed elements can be straight or curved. If the element is rectangular it can model linear strain fields exactly. Equation 6.2-5 is the formula for its stiffness matrix in isoparametric coordinates, where $\mathbf{B}$ is now a 6 by 60 rectangular matrix. If $\mathbf{k}$ is integrated by a 2 by 2 by 2 Gauss quadrature rule, three "hourglass" instabilities of the type shown in Fig. 4.6-2b are possible, one involving u, another v, and the third w displacements. Three additional hourglass instabilities are also possible, in each of which displacements on opposite faces of the element have opposite sign. In plane problems an hourglass instability of a quadratic quadrilateral is noncommunicable and causes no difficulty. But in 3D problems, it is conceivable that elements will be strung end to end as in Fig. 6.2-2. Even if one end of the model is restrained as shown, there may be a near-instability analogous to that in Fig. 4.6-2a, as shown in Fig. 6.2-2. This possibility is avoided in commercial software by using a stabilization device, a special 14-point rule, or even a 3 by 3 by 3 rule (27 points) to integrate $\mathbf{k}$.

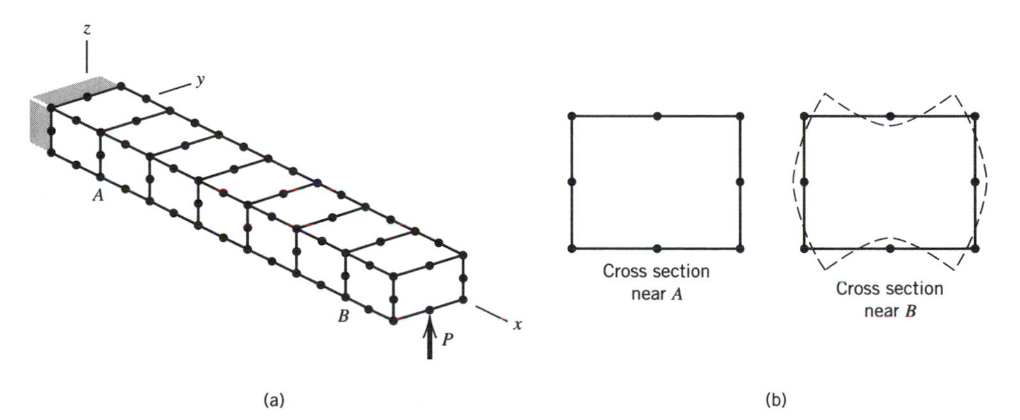

Fig. 6.2-2. (a) FE model composed of quadratic solid elements. (b) Near-instability is possible far from the fixed end if elements are integrated by a 2 by 2 by 2 Gauss rule.

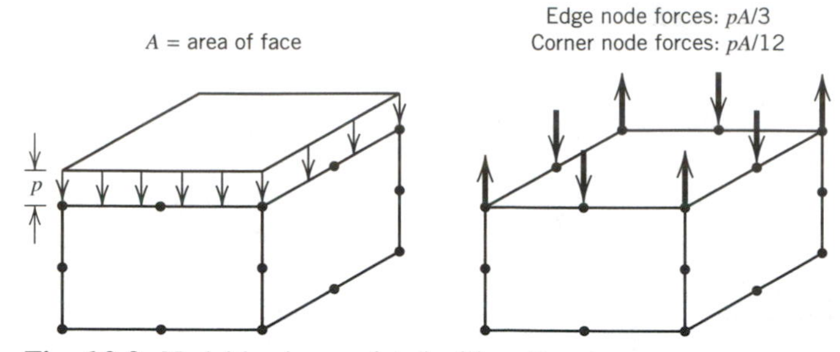

Fig. 6.2-3. Nodal loads associated with uniform pressure p on a rectangular face of a hexahedron with midedge nodes.

Remarks. Additional 3D elements are of course possible and are described in the documentation of commercial software. Some elements may have nodes at the middle of each face, in addition to midedge nodes. The analyst should consult the documentation, in case even the basic elements described above have special features or "add-ons."

Software can be expected to compute automatically the nodal forces that represent body force loading, or pressure loading on a face of a 3D element. This is fortunate, as appropriate nodal forces are often not obvious. As an example, uniform pressure p on a face of a linear strain tetrahedron with midedge nodes should be applied as forces $pA/3$ at each midedge node on the face, where A is the area of the face. This distribution is work-equivalent and is shown in Fig. 3.9-5c for an analogous plane problem. Another example appears in Fig. 6.2-3 (note the resemblance to Fig. 3.9-5d).

Typical 3D elements do not use rotational d.o.f. Accordingly, rotational d.o.f. must be suppressed in the global equations. Software may or may not do so automatically.

Patch tests for 3D solid elements are entirely analogous to patch tests for plane elements. However, the requirement that at least one node be within the mesh means that at least one node must be within the volume, not just along an edge or on a surface of the solid.

Surface views do not reveal the internal structure of a 3D mesh. The user should therefore make full use of preprocessor graphics to check the FE model prior to analysis. To reduce storage requirements and execution time, bandwidth or wavefront reducers should be applied if software does not do so automatically. In examining computed displacements and stresses it is helpful to view results from different directions and on different cross sections. Most often, peak stresses appear on the surface of a solid; therefore a plot of surface stresses should be examined.

6.3 A 3D APPLICATION

A curved beam is bent in its own plane, as in Fig. 5.2-1b. More precisely, the structure is a portion of a ring, symmetric about an axis of revolution, spanning an arbitrary number of degrees about the axis, and loaded by a bending moment whose vector is parallel to the axis. Stresses of greatest magnitude are sought. The analysis problem is *not* axisymmetric because radial cross sections rotate with respect to one another and radial displacements are not independent of θ. Accordingly, we use 3D solid elements but also use cylindrical coordinates because they conveniently fit the geometry.

The particular shape of the cross section and the FE mesh chosen are shown in Fig. 6.3-1. Stresses do not vary with θ, so only a typical slice between radial planes need be analyzed (Fig. 6.3-1b). Bending moment M must be applied "indirectly" because we do

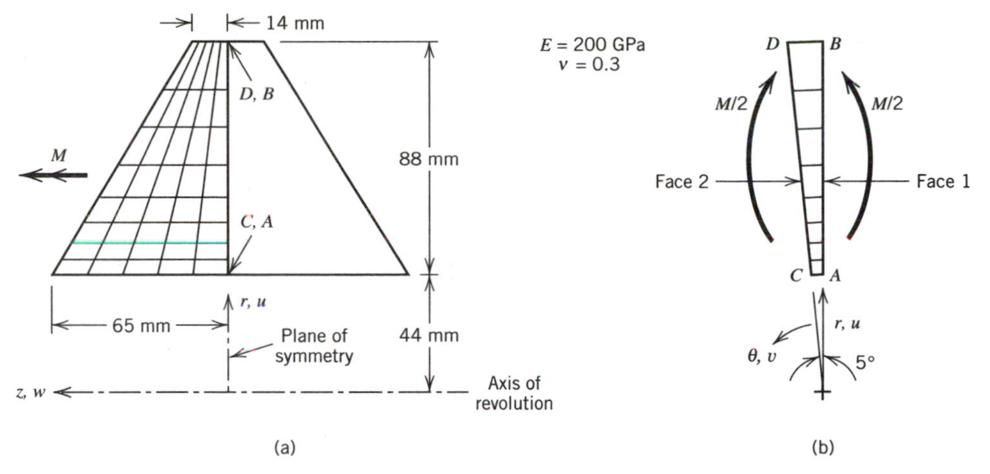

Fig. 6.3-1. (a) Cross section of a curved beam, showing FE mesh in the left half. (b) The FE model viewed parallel to the z axis.

not know what stresses it produces and therefore cannot impose appropriate nodal loads. Instead, we will prescribe *displacements*, such that radial plane sections remain plane and moment load is applied without a net force, then compute M as the moment produced by computed circumferential stresses.

Preliminary Analysis. The straight beam flexure formula Mc/I yields the stress $8.94(10)^{-6}M$ at the inside edge, in units MPa if M is in N·mm. A formula for circumferential stress in a curved beam is readily available [1.5, 2.1]. According to the formula, stress at the inside edge is

$$\text{curved beam theory, along } r = 44\text{mm:} \qquad \sigma_\theta = 13.46(10)^{-6}M \tag{6.3-1}$$

in MPa if M is in N·mm. For comparison with FE results we must return to this formula after M is known. Might Eq. 6.3-1 be adequate? It may be, but possibly a FE analysis will show otherwise. In deriving the formula it is assumed that a cross section does not distort in its own plane, so that stresses do not vary in the z direction. A FE analysis contains no such restriction and may therefore yield different results.

FE Model and Analysis. There is symmetry about a z-constant plane that contains points $ABCD$, so only half the cross section need be meshed. Curved beam theory tells us that stress gradients will be highest on the edge nearest the center of curvature. Accordingly, the mesh is graded so that elements near the inner edge span a smaller radial distance. In Fig. 6.3-1b, face 1 and its nodes are merely rotated 5° to generate face 2. The wedge between the two faces contains a single layer of eight-node 3D elements. Each element contains nine internal d.o.f. associated with incompatible displacement modes. Prescribed nodal displacements in the radial, circumferential, and axial directions are as follows:

Face 1	Face 2
$u = 0$ at node A	
$v = 0$ at all nodes	$v = 0.0001(r_c - r)$ at all nodes
$w = 0$ at nodes along AB	$w = 0$ at nodes along CD

All remaining nodal d.o.f. are unrestrained. Setting $u = 0$ at A prevents rigid-body translation in the r direction, and setting $w = 0$ on $ABCD$ imposes symmetry about the $r\theta$ plane. The expression $v = 0.0001(r_c - r)$ causes face 2 to remain plane as it rotates about a z-parallel axis at $r = r_c$. The number 0.0001 is arbitrary and r_c is a number such that node A exerts no radial force on the FE model. At the outset the appropriate value of r_c is unknown. Therefore *two* preliminary FE analyses are performed, respectively using the arbitrarily chosen values $r_c = 60$ mm and $r_c = 70$ mm. The respective radial reactions at A are computed by the software as 2001 N and 357 N. By linear extrapolation, the radial reaction at A should be zero when $r_c = 72.2$ mm. The value $r_c = 72.2$ mm is used in a third and final FE analysis, which provides a radial reaction of essentially zero at A, as expected. Circumferential support reactions on face 1 produce a moment about a z-parallel axis, which is automatically calculated by the software. This moment is doubled to yield moment M on the *entire* cross section. If stresses for a prescribed moment M_p are required, one need only multiply computed stresses by the ratio M_p/M.

Critique of FE Results. The deformed shape of the cross section is shown in Fig. 6.3-2a. Animation shows that the intended boundary conditions have indeed been enforced. On physical grounds we argue that the deformed shape is reasonable, as follows. Radial stress σ_r is known to be tensile, so it is proper that the 88-mm dimension becomes larger. Circumferential stress σ_θ is, respectively, tensile and compressive on inner and outer portions of the cross section, while axial stress σ_z is small, so that the Poisson effect should cause inner and outer portions, respectively, to contract and expand in the z direction, as is indeed observed. Circumferential tensile stress on the inner portion pulls material toward the center of curvature. Outer corners of a cross section are more flexible than the central part, so it is proper that corner E moves inward relative to central point A. This effect is discussed in Section 5.2 with reference to thin-walled cross sections, whose radial deflections are of course much more significant.

Material that moves radially inward while bounded by faces 1 and 2 in Fig. 6.3-1 must shorten circumferentially. Thus a compressive strain is superposed on the tensile strain due to flexing. Radial deflection provides greatest "stress relief" to material that deflects farthest. Accordingly, it is reasonable that Fig. 6.3-2a shows lower circumferential stress at E than at A. At A and E, respectively, FE analysis yields circumferential stresses of 146

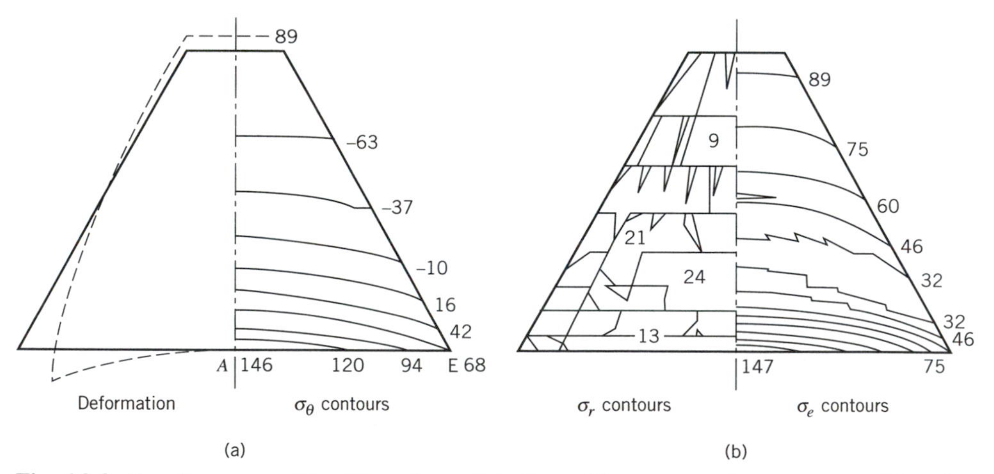

Fig. 6.3-2. Results computed by FE analysis: distortion of the cross section, exaggerated for plotting, and unaveraged stress contours. Stress units are MPa. (Stresses shown in the right half were actually computed by FE analysis in the left half.)

MPa and 65 MPa. The software uses σ_θ stresses to compute the moment $M/2$ on face 1 of the FE model in Fig. 6.3-1. Thus we obtain $M = 8.804(10)^6$ N·mm. Hence Eq. 6.3-1 yields $\sigma_\theta = 119$ MPa, in approximate agreement with the average of FE stresses at A and E, which is $(146 + 65)/2 = 106$ MPa. As expected, σ_θ contours are more closely spaced at smaller values of r. The neutral axis is the locus of points where $\sigma_\theta = 0$. We see from Fig. 6.3-2a that the neutral axis is curved, in contradiction of the assumption made in mechanics of materials theory.

Figure 6.3-2a shows little or no interelement discontinuity of σ_θ contours. This is not surprising, because circumferential strain is $\varepsilon_\theta = (u/r) + (\partial v/\partial\theta)/r$ [6.1]. This means that ε_θ is essentially a plot of the displacements of face 2, which are of course interelement-continuous. Figure 6.3-2b shows that σ_r contours are badly discontinuous. But the largest σ_r is about 25 MPa, much less than the largest σ_θ. Accordingly, the plot of von Mises stress σ_e, Fig. 6.3-2b, shows small to moderate discontinuities. The stress field yields the relative energy error $\eta = 0.050$, an acceptably small value. We conclude that results are reliable, at least for stress σ_θ.

6.4 AXISYMMETRIC SOLID ELEMENTS

Except for having to account for circumferential strain ε_θ, axisymmetric elements are very similar to plane elements. Available element shapes (in cross section) and nodal patterns are as described in Chapter 3. Capabilities and shortcomings of a specific element type are much the same as for the corresponding plane element. However, it is necessary to discuss the effects of the additional strain term $\varepsilon_\theta = u/r$. Consider the very simplest axisymmetric solid element, a three-node triangle. Its displacement field for axisymmetric deformation is

$$\begin{aligned} u &= \beta_1 + \beta_2 r + \beta_3 z \\ w &= \beta_4 + \beta_5 r + \beta_6 z \end{aligned} \tag{6.4-1}$$

which is identical to Eq. 3.2-1 except for r in place of x, z in place of y, and w in place of v. From Eqs. 6.1-6 and 6.4-1, the element strain field is

$$\begin{aligned} \varepsilon_r &= \beta_2 \qquad & \varepsilon_\theta &= \beta_1 \frac{1}{r} + \beta_2 + \beta_3 \frac{z}{r} \\ \varepsilon_z &= \beta_6 \qquad & \gamma_{zr} &= \beta_5 + \beta_3 \end{aligned} \tag{6.4-2}$$

Unlike its plane relative, this element is not a constant strain element because the ε_θ expression contains r and z. The only possible rigid-body motion is axial translation, $w = \beta_4$. Strain is present if any other β_i is nonzero. Rotation of the element cross section in the rz plane is resisted by the $\beta_3(z/r)$ term in the ε_θ expression. Such a "ring rolling" deformation is produced by moment M in Fig. 6.4-1a, where M is a moment uniformly distributed around the circumference of the element. The dimensions of M are [force·length/length] or simply [force]. Due to M, the cross section shown rotates counterclockwise a small amount; circumferential strains appear that are tensile in the lower part of the element and compressive in the upper part. A *plane* triangular element would not resist M because the rotation would be a rigid-body motion.

An element of arbitrary quadrilateral shape, such as that in Fig. 6.4-1b, would be for-

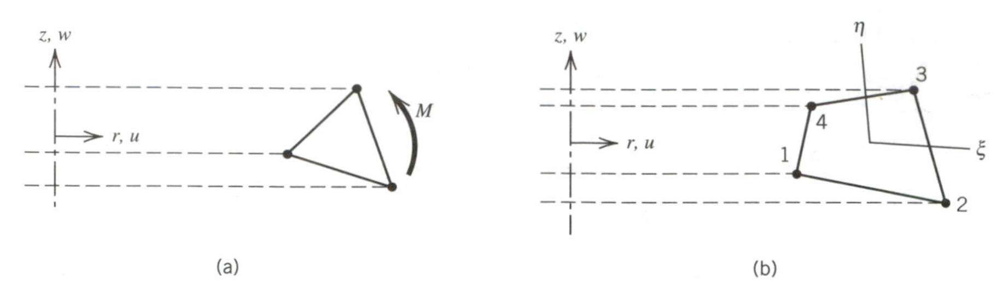

Fig. 6.4-1. (a) Three-node axisymmetric element loaded by circumferentially distributed moment M. (b) Four-node axisymmetric element.

mulated as an isoparametric element. The formula for element stiffness matrix **k** is written

$$\mathbf{k} = \int_0^{2\pi}\int_{-1}^{1}\int_{-1}^{1} \mathbf{B}^T\mathbf{E}\mathbf{B}r|\mathbf{J}|\,d\xi\,d\eta\,d\theta \tag{6.4-3}$$

which is like Eq. 4.4-8 except for the presence of r in place of t, integration with respect to θ, use of a 4 by 4 array **E**, and the addition of a row to **B** that states the relation $\varepsilon_\theta = u/r$. For an element with translational d.o.f. only, this relation is

$$\varepsilon_\theta = \frac{1}{r}\left(N_1u_1 + N_2u_2 + \cdots + N_nu_n\right) \tag{6.4-4}$$

where the N_i are shape functions and n is the number of nodes per element. The multiplication $\mathbf{B}^T\mathbf{E}\mathbf{B}r$ produces terms that contain $1/r$. This poses no difficulty for numerical integration provided that no integration points are placed on the z axis.

As for boundary conditions, prescription of w on a single nodal circle is sufficient to prevent rigid-body motion of the FE structure and hence prevent singularity of **K**. However, one should also set $u = 0$ at all nodes on the axis of revolution. Nonzero u at $r = 0$ would mean that either a small hole appears or the material overlaps itself, both of which are physically unreasonable. A "pinhole" would also provide a stress concentration factor of 2.0 for circumferential stress σ_θ.

Points on the axis of revolution have zero radial coordinate and zero radial displacement. If stresses are computed *at* $r = 0$, instead of being extrapolated *to* $r = 0$ from Gauss point values, we obtain the strain calculation $\varepsilon_\theta = 0/0$. This is an awkward situation that can be avoided by calculating $\varepsilon_r = \partial u/\partial r$ instead, then equating ε_r to ε_θ. This trick exploits the theoretical requirement that $\varepsilon_r = \varepsilon_\theta$ for points at $r = 0$. (The requirement is met by Eqs. 6.4-2: with $\beta_1 = \beta_3 = 0$ so that $u = 0$ for points on $r = 0$, we obtain $\varepsilon_r = \partial u/\partial r = \beta_2$ and $\varepsilon_\theta = u/r = \beta_2$.) Stresses follow from Eq. 6.1-4. In commercial software these considerations are hidden from the user. Nevertheless, it is of interest to run a simple test case to discover if computed stresses σ_r and σ_θ are indeed equal at $r = 0$.

Axisymmetric four-node quadrilateral elements may contain incompatible modes, which are described in Section 3.6 as a way to improve bending response. In an axisymmetric element these modes may be activated even when there is no bending. Thus there is a spurious radial bulge of each element, whose effect is to produce a spurious shear strain γ_{zr} everywhere in the element except at $\xi = \eta = 0$. The bulge tends to disappear if the element has a small cross section far from the axis of revolution. If the effect remains troublesome, the analyst can tell the software to omit the incompatible modes.

Ill-conditioning of the global equations is promoted by very slender elements; that is, elements whose cross section is very small and very distant from the axis of revolution. The stiffness of a slender cross section in resisting strains ε_r, ε_z, and γ_{zr} is much greater than the stiffness that resists circumferential strain ε_θ. An even larger stiffness difference is possible when slender elements have the ability to bend circumferentially, as described in Section 6.6. Thus very slender elements may present the error-prone case of large stiffnesses embedded in small stiffness.

From Eq. 6.4-2 we see that strains are constant in the three-node element only if $\beta_1 = \beta_3 = 0$, in which case $\varepsilon_r = \varepsilon_\theta$. Constant but independent strain states are therefore not possible. Similar tendencies appear in many other element types, so that commonly used axisymmetric elements fail patch tests. This does not invalidate the elements because constant strain conditions in an element are approached as the mesh is refined, that is, as elements become slender.

6.5 AN AXISYMMETRIC APPLICATION

Figure 6.5-1 shows a cross section of an axisymmetric structure, already meshed with the elements we propose to use. The structure consists of an outer disk *BEFC* of constant thickness, attached to a tapered inner disk *DABE* of the same material by means of a shrink fit. Physically, the shrink fit is accomplished by heating the outer disk, slipping it over the inner disk, and allowing both disks to return to a uniform temperature. Dimensions are such that when the outer disk is 100°C hotter than the inner disk, the inner radius of the outer disk and the outer radius of the inner disk are both precisely 400 mm. We ask:

1. What contact stresses along *BE* are produced by the shrink fit?
2. If the assemblage is set spinning about the z axis, at what angular velocity will the shrink fit loosen?

In seeking answers we will find that question 1 has aspects that are not anticipated unless we think ahead very carefully, and that question 2 is not well posed and requires some analytical thinking.

Preliminary Analysis. We first ask for the contact pressure p caused by the shrink fit (Fig. 6.5-2). For a simple approximation we assume that the inner disk has no taper and that the outer disk can be treated by formulas applicable to a ring that is thin in the radial

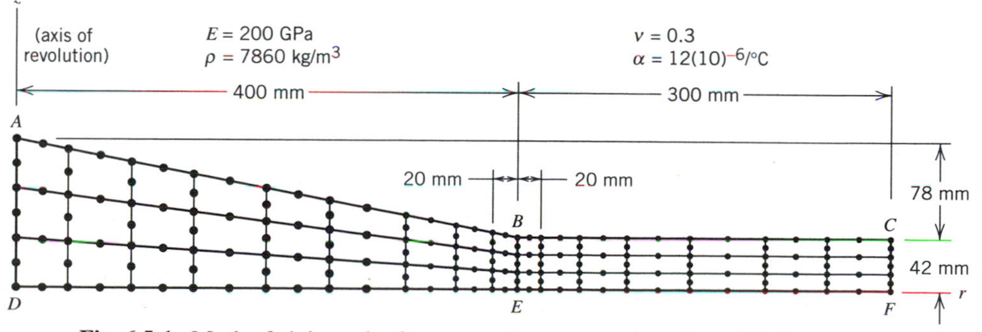

Fig. 6.5-1. Mesh of eight-node elements on the cross section of a solid of revolution.

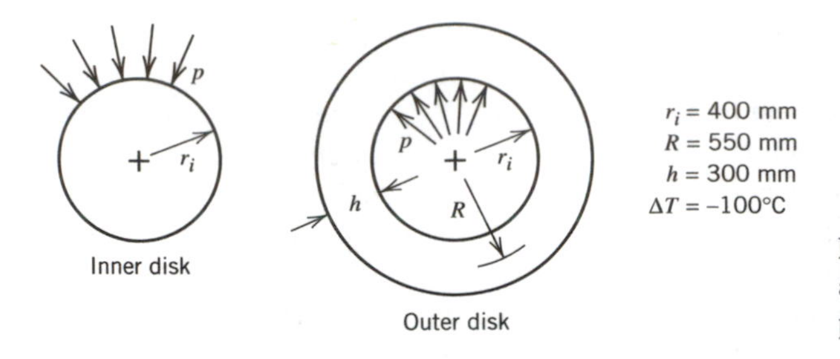

Fig. 6.5-2. Analytical model for approximate analysis of shrink fit pressure p.

direction. Thus, in the inner disk, $\sigma_r = \sigma_\theta = -p$, circumferential strain is $\varepsilon_\theta = (\sigma_\theta - \nu\sigma_r)/E = -p(1-\nu)/E$, and radial displacement of the outer edge is $u_i = r_i\varepsilon_\theta = -pr_i(1-\nu)/E$. In the outer disk (treated as a slender ring), $\sigma_\theta = pR/h$, $\sigma_r = 0$, $\varepsilon_\theta = (\sigma_\theta - \nu\sigma_r)/E = pR/Eh$, and $u_i = r_i\varepsilon_\theta = pRr_i/Eh$, where the mean radius R has been used in calculating σ_θ. In addition, the shrink fit produces radial displacement $u_i = \alpha r_i \Delta T$ at $r = r_i$ in the outer disk, where $\Delta T = -100°\text{C}$ in our case. Equating u_i values at $r = r_i$ in inner and outer disks, we obtain

$$-\frac{pr_i}{E}(1-\nu) = \frac{pRr_i}{Eh} + \alpha r_i \Delta T \tag{6.5-1}$$

Solving for p and using the data of Figs. 6.5-1 and 6.5-2, we obtain

$$p = -\frac{E\alpha\,\Delta T}{(1-\nu) + \dfrac{R}{h}} \qquad \text{from which} \qquad p = 95\text{ MPa} \tag{6.5-2}$$

We elect to analyze spinning by using a ready-made formula rather than working from first principles. The radial stress in a solid disk of *constant* thickness, outer radius a, mass density ρ, and spinning at angular velocity ω is [2.1]

$$\sigma_r = \frac{3+\nu}{8}\rho\omega^2(a^2 - r^2) \tag{6.5-3}$$

Using $a = 0.700$ m and setting $\sigma_r = -95$ MPa at $r = 0.400$ m, we obtain $\omega = 298$ rad/s as the approximate angular velocity at which the shrink fit should loosen.

Of course the inner disk is thicker and therefore stiffer than we have assumed, which means that the actual p can be expected to be larger than stated in Eq. 6.5-2. Also, the structure is not symmetric about a z = constant plane. Axial components of deflection will respond accordingly: the outer part should bend *downward* due to the shrink fit and *upward* due to spinning. These flexing actions will introduce bending stresses, which near point E will increase the magnitude of (compressive) shrink fit contact stress and increase the magnitude of (tensile) radial stress due to spinning.

FE Model and Analysis. The cross section is represented by eight-node elements, as shown in Fig. 6.5-1. This kind of element is known to work well, so that the mesh may appear overly refined for an initial model, but the geometry is so simple that the mesh is easy to generate. There are two displacement d.o.f. per node. Both are set to zero at node D. Only the radial displacement is set to zero at other nodes along AD. All nodes not on the z axis are unrestrained. Shrink fit loading is produced by stating that portion $BEFC$ is uniformly decreased in temperature by 100°C. Spinning is treated as a separate load case, using ω = 298 rad/s as the prescribed angular velocity. Spinning creates inertia (body force) loading, for which appropriate forces on individual nodes are computed automatically by the software.

Critique of FE Results. By inspection, deflected shapes (not shown) are found to be reasonable. Animation shows that nodes along *AD* move only axially, as intended. The cross section deflects like a beam cantilevered from *AD*, downward for the shrink fit and upward for spinning. In addition, tip *CF* moves radially a small amount, inward for the shrink fit and outward for spinning.

Radial stresses σ_r produced by the shrink fit are shown in Fig. 6.5-3a. Near *BE* their average magnitude is in approximate agreement with Eq. 6.5-2, but contours show severe interelement discontinuities. What is wrong? It is that Eqs. 6.5-1 and 6.5-2 have considered only radial stresses; they have ignored other stresses that interact with radial stresses. Temperature decrease in the outer disk causes it to contract axially as well as radially. Additionally, circumferential stresses in inner and outer disks are, respectively, compressive and tensile. The associated Poisson-effect axial strains are, respectively, tensile and compressive and therefore add to the axial thermal strains. To maintain identical z-direction lengths along *BE*, as the foregoing FE analysis requires, the disks must apply z-direction friction forces to one another along *BE*. These forces perturb the stress field and require more elements in the neighborhood of *BE* if details are to be resolved. It should be noted that our analysis assumes that a bond is created as soon as the disks make contact, and that the bond is thereafter unbroken.

Circumferential stresses σ_θ produced by the shrink fit, Fig. 6.5-3b, appear much more reliable. Interelement continuity of stress contours is fair to good near *BE*. Continuity across *BE* is neither seen nor expected. However, the relative energy error is $\eta = 0.28$, an alarmingly high value. But this is for the *entire* FE model, which represents a misapplication of the error estimate. It should not be applied across discontinuities such as the discontinuity of σ_θ across *BE*. The error estimate can legitimately be applied to inner and outer disks separately, for which the respective values are $\eta = 0.17$ and $\eta = 0.09$. If we exclude from each disk the three elements nearest *BE* (six elements altogether), we obtain

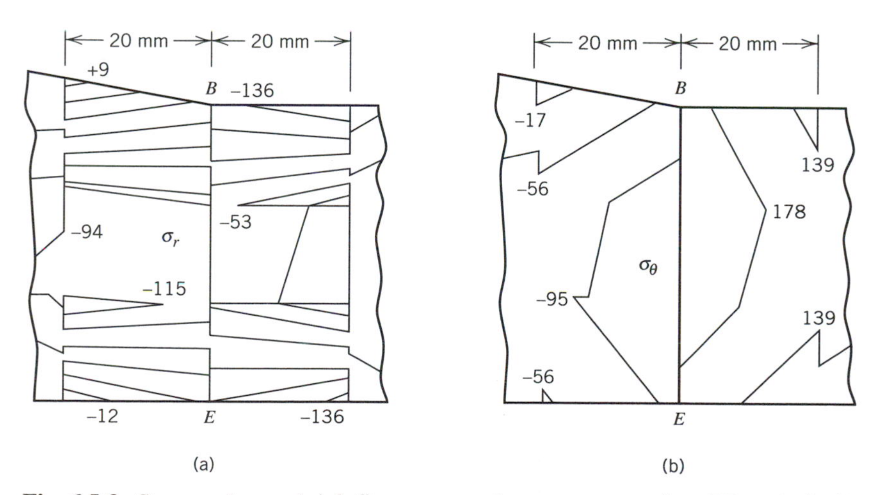

Fig. 6.5-3. Stresses due to shrink fit: unaveraged stress contours from FE analysis, in units of MPa. (a) Radial stress σ_r. (b) Circumferential stress σ_θ.

the respective values $\eta = 0.04$ and $\eta = 0.02$. Clearly, the mesh is too coarse for accurate stress calculation near *BE*. For the case of spinning, discussed later, an error estimate for the entire FE model is acceptable, as there is no discontinuity of stresses in this loading case (the value calculated is $\eta = 0.004$).

Contours of stresses associated with spinning at $\omega = 298$ rad/s are shown in Fig. 6.5-4. Interelement continuity is good and these plots give no reason to doubt the validity of the results. Radial stress near *BE* is about 100 MPa on the average, about what is needed to cancel the compressive radial stress that FE analysis gives for the shrink fit. Does this mean that 298 rad/s is indeed the ω at which the shrink fit will loosen? Since σ_r is not uniform along *BE*, we now see that we have not been clear about what is meant by "loosen." Is it when contact along *BE* is completely broken, or when the superposed results of shrink fit and spinning give zero σ_r at some point along *BE*? The speed for complete separation along *BE* can be obtained by writing an equation analogous to Eq. 6.5-1 and solving for ω:

$$\omega^2\delta_1 = \omega^2\delta_2 + \alpha r_i \Delta T \tag{6.5-4}$$

where δ_1 is the largest radial deflection along *BE*, to be computed by a FE analysis of the inner disk *alone* spinning at $\omega = 1$ rad/s, and δ_2 is the radial deflection along *BE* for $\omega = 1$ rad/s in the outer disk, modeled *alone* by FE or preferably by use of a simple and readily available analytical formula for a spinning disk of constant thickness with a central hole [1.5, 2.1]. We see ω^2, not ω, in Eq. 6.5-4 because stresses and displacements are proportional to the square of angular velocity. As in Eq. 6.5-1, $\Delta T = -100°C$.

The other question, about the value of ω for which loosening begins, requires that accurate values of σ_r along *BE* be known for the shrink fit loading. At a node *j* along *BE*, let us symbolize these radial stress values by $[(\sigma_r)_{\Delta T}]_j$. Also, let $[(\sigma_r)_\omega]_j$ represent radial stress at node *j* along *BE* due to spinning (of the *entire* disk) at angular velocity $\omega = 1$ rad/s. This value is easily computed by dividing the σ_r stresses depicted in Fig. 6.5-4a by the ω^2 used to compute them. The ω at which loosening would appear at node *j* is obtained by solving for ω in the equation

$$\omega^2[(\sigma_r)_\omega]_j + [(\sigma_r)_{\Delta T}]_j = 0 \tag{6.5-5}$$

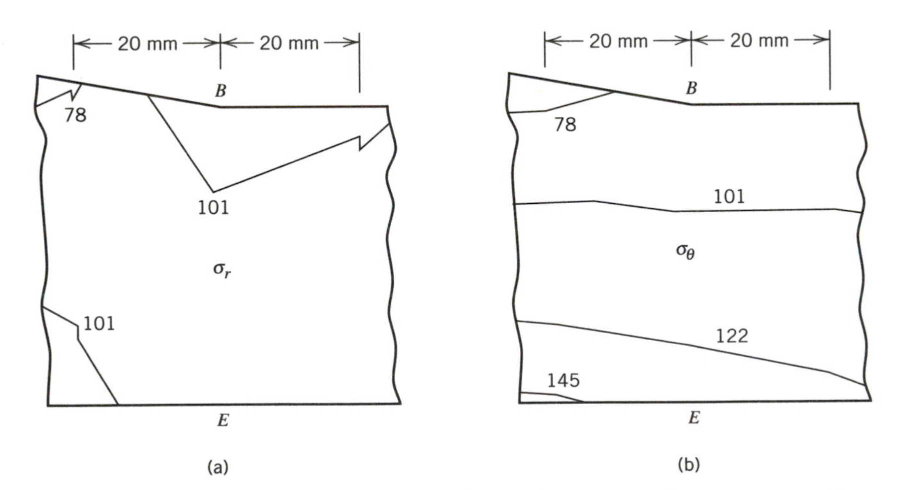

Fig. 6.5-4. Stresses due to spinning about the z axis: unaveraged stress contours from FE analysis, in units of MPa. (a) Radial stress σ_r. (b) Circumferential stress σ_θ.

The calculation must be done for all nodes j along BE and the lowest of the several ω values chosen as the desired result. Note that Eq. 6.5-5 is applicable only to the node of first loosening. As soon as contact is broken, the continuity presumed by the FE model is lost.

6.6 NONAXISYMMETRIC LOADS

If the body is axisymmetric but loads and/or boundary conditions are not, the problem is three-dimensional in the sense that all three displacement components are in general nonzero. Thus displacement components u (radial), v (circumferential), and w (axial) are each functions of r, θ, and z. Similarly, all six possible stress components are in general nonzero and are functions of r, θ, and z (Fig. 6.6-1). It what follows we summarize how the loading can be broken into components, an analysis made for each component separately, and results combined to produce the solution for the original loading [2.2]. The advantage of this approach over a fully three dimensional FE analysis is that no discretization in the θ direction is required; elements remain annular, as in Fig. 6.4-1. Thus the problem is easier for the analyst and is less demanding of computer resources.

Loads. Any loading on a solid or shell of revolution can be described as a sum of its Fourier series components. Figure 6.6-2 shows a particular example, in which a radial load of intensity q_o acts outward for $0 < \theta < \pi$ and inward for $\pi < \theta < 2\pi$. For stress analysis purposes, it may be possible to represent this load with sufficient accuracy by only the first three terms of the infinite series. In general, an arbitrary load is represented by the Fourier series

$$q = \sum_{n=0}^{\infty} q_{cn} \cos n\theta + \sum_{n=1}^{\infty} q_{sn} \sin n\theta \tag{6.6-1}$$

where q_{cn} and q_{sn} are amplitudes that depend on n (but not on θ). Here n represents the harmonic number, *not* the number of nodes per element. In Fig. 6.6-2, $q_{cn} = 0$ and $q_{sn} = 4q_o/n\pi$ is evaluated for n odd. Figure 6.6-3 depicts the first three terms of this series.

Additional examples appear in Fig. 6.6-4. Figure 6.6-4a is the axisymmetric radial loading produced by $n = 0$ in the cosine series. Figure 6.6-4b is the $n = 1$ term of the cosine series, which might be used without any other series terms to approximate wind

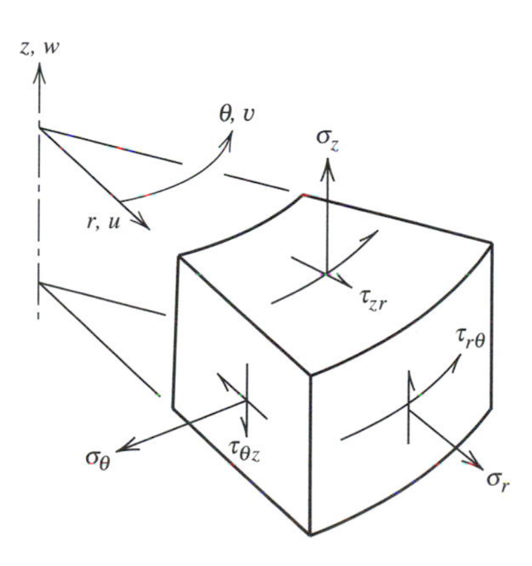

Fig. 6.6-1. Stresses and displacements in a solid of revolution with nonaxisymmetric loads.

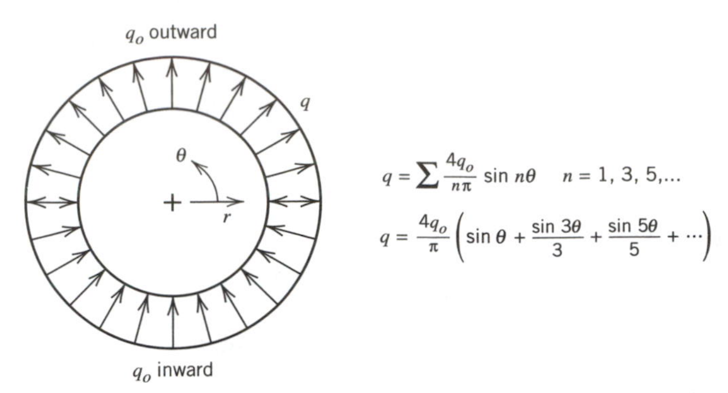

Fig. 6.6-2. A uniform load, alternately outward and inward, and its representation by Fourier series.

loading. Figure 6.6-4c is again produced by $n = 0$ in the cosine series, but now associated with the tangential direction rather than the radial direction.

In general, q in Eq. 6.6-1 represents load in *any* of the directions r, θ, or z and can be used as a line load on a nodal circle. A distributed surface loading on the body is represented by Eq. 6.6-1 on each of several nodal circles; a concentrated load is represented by Eq. 6.6-1 on a single nodal circle. The software user can either provide series terms themselves as input data or can ask the software to calculate series terms from a description of the loading. Boundary conditions can also be described by Fourier series. Thus, for example, one might load a solid of revolution by prescribing axial displacement $w = \sum w_{cn} \cos n\theta$ on one or more nodal circles. Usually the loading is such that one or the other series in Eq. 6.6-1 will suffice; both will not be needed in a single problem.

Displacements. Like loads, displacement fields can be expanded in Fourier series. For radial, circumferential, and axial displacement field components, respectively,

$$u = \sum_{n=0}^{\infty} u_{cn} \cos n\theta + \sum_{n=1}^{\infty} u_{sn} \sin n\theta$$

$$v = \sum_{n=1}^{\infty} v_{cn} \sin n\theta + \sum_{n=0}^{\infty} v_{sn} \cos n\theta \tag{6.6-2}$$

$$w = \sum_{n=0}^{\infty} w_{cn} \cos n\theta + \sum_{n=1}^{\infty} w_{sn} \sin n\theta$$

where u_{cn}, u_{sn}, v_{cn}, v_{sn}, w_{cn}, and w_{sn} are displacement field *amplitudes* that depend on r, z, and n but are independent of θ. Displacements (and loads) that vary as $\cos n\theta$ or as $\sin n\theta$ are called "symmetric" or "antisymmetric," respectively, with reference to the plane $\theta = 0$. In an axisymmetric problem, $n = 0$ and all terms with subscript s are zero, so that $v = 0$ and the displacement field is described by $u(r, z) = u_{c0}$ and $w(r, z) = w_{c0}$. In pure torsion, $n = 0$ and all terms with subscript c are zero, so that $u = w = 0$ and the displacement field is described by $v(r, z) = v_{s0}$.

An element displacement field is obtained by interpolation from nodal values of displacement field amplitudes. For example, taking only the first series terms on the right-

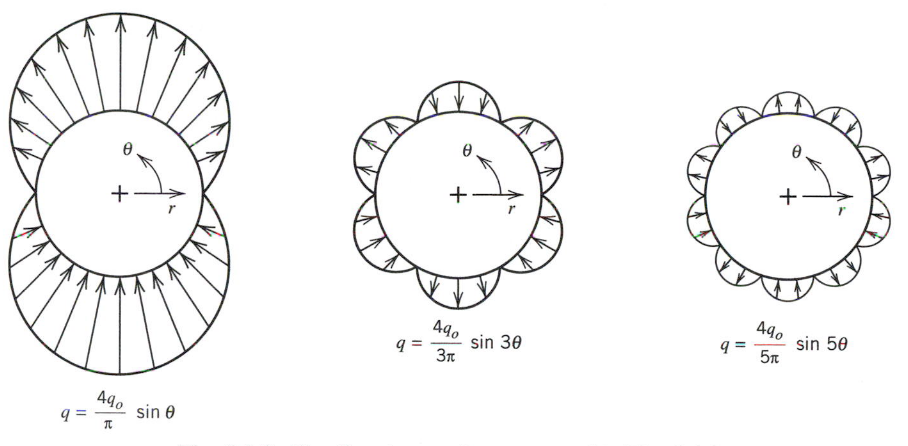

Fig. 6.6-3. The first three series terms used in Fig. 6.6-2.

hand sides of Eqs. 6.6-2, and using them as amplitudes at an element node i, displacement field amplitudes within an element are

$$u_{cn} = \sum N_i u_{cni} \qquad v_{cn} = \sum N_i v_{cni} \qquad w_{cn} = \sum N_i w_{cni} \tag{6.6-3}$$

where the N_i are shape functions, for example, Eqs. 4.4-2 for a four-node element. Equations 6.6-3 resemble Eqs. 6.1-7, but to obtain displacements u, v, and w themselves we must substitute Eqs. 6.6-3 into Eqs. 6.6-2. The next step in formulating an element is to substitute u, v, and w into the 3D strain-displacement relations in cylindrical coordinates (which we do not bother to write out). The result, for a typical harmonic n, is a strain–displacement matrix $\mathbf{B}$ that contains six rows and whose coefficients B_{ij} are functions of r, z, n, and θ. Finally, integration of $\mathbf{B}^T\mathbf{E}\mathbf{B}$ over the element volume yields an element stiffness matrix $\mathbf{k}$ that contains either n or n^2 in several of its terms. Assembly of elements yields the global equations for the nth harmonic

$$\mathbf{K}_n\mathbf{D}_n = \mathbf{R}_n \tag{6.6-4}$$

in which the arrays are of order $3N$ for a FE model that has N nodes and translational d.o.f. only. As for boundary conditions, a prescribed displacement of zero on a nodal circle dictates that the nodal displacement amplitudes be zero in every harmonic. Also,

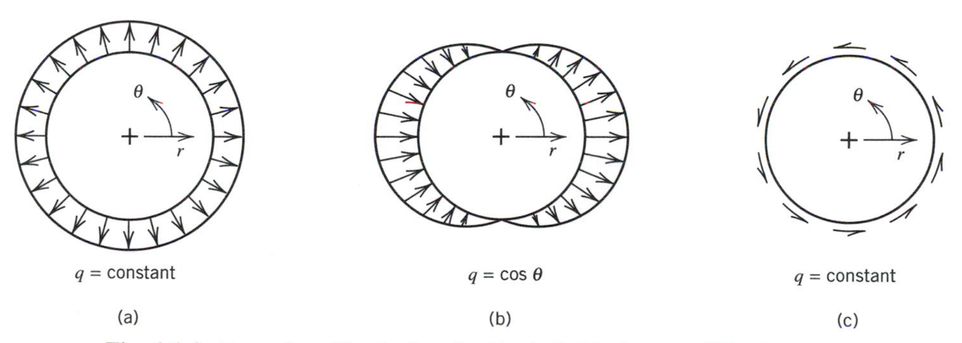

Fig. 6.6-4. Examples of loads described by individual terms of Fourier series.

when $n = 0$ the nodal d.o.f. v_{cni}, u_{sni}, and w_{sni} have no stiffness associated with them and must therefore be suppressed.

The important feature of representing loads and displacements by their Fourier series components is this: *a single harmonic of load produces only the corresponding harmonic of displacement.* For example, the load harmonic $q = q_{c3} \cos 3\theta$ in Eq. 6.6-1 produces the displacement amplitudes u_{c3}, v_{c3}, and w_{c3} *but no other displacement amplitudes in Eqs. 6.6-2.* Thus the response to nonaxisymmetric load "decouples" into as many separate analyses as there are terms in the load series. Each separate analysis uses the same size $\mathbf{K}_n$ and three d.o.f. per node, namely, the nodal displacement amplitudes u_{cni}, v_{cni}, and w_{cni} in Eqs. 6.6-3.

Remarks. In summary, an axisymmetric solid under nonaxisymmetric loads can be analyzed by breaking the load into terms of an infinite series, calculating the response to each term, and superposing results during postprocessing. Typically, only the first few terms of a series are sufficient for the required accuracy. Computational demands are greatly reduced as compared with the requirements of a 3D discretization.

The Fourier series approach can also be applied to other problems, most notably to plate bending, where it is known as the *finite strip method.* In effect, division into finite elements in one direction is replaced by separate analyses corresponding to terms of a Fourier series. The method seems best suited to analysis of box beams and folded plates.

ANALYTICAL PROBLEMS

6.1 State Eqs. 6.1-5 and 6.1-6 in the matrix product format of Eq. 3.1-6.

6.2 Imagine that an isotropic cylinder of solid circular cross section is loaded by torque about its axis and is modeled by linear strain tetrahedra. Is the correct strain field represented exactly? Explain.

6.3 Write the eight shape functions N_i separately for the element of Fig. 6.2-1c. Thus assign the proper algebraic signs to each N_i in Eq. 6.2-4.

6.4 A four-node axisymmetric element has the rectangular cross section shown. No incompatible displacement modes are used. Write the $\mathbf{B}$ matrix. Express the B_{ij} in terms of a, b, R, r, and z.

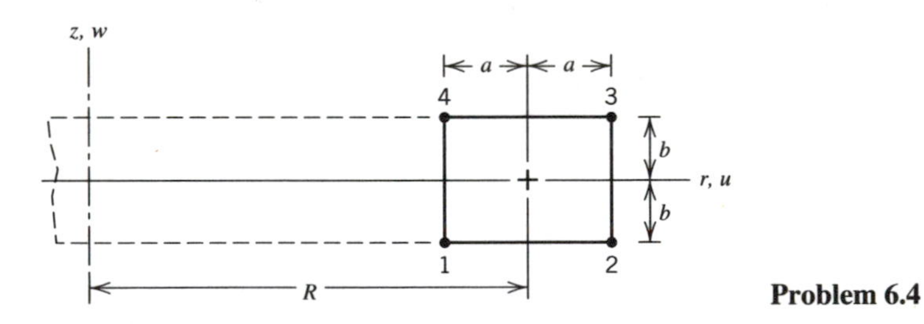

Problem 6.4

6.5 Dashed lines in the sketch show independent displacement modes of a four-node rectangular element having two displacement d.o.f. per node. Which of these modes are associated with strain energy in the element and which are not? Answer for each of the following situations. (Sketch reprinted from [2.2] by permission of John Wiley & Sons, Inc.)

(a) The element is plane and $\mathbf{k}$ is integrated by one Gauss point.

(b) The element is plane and $\mathbf{k}$ is integrated by four Gauss points.

(c) The element is axisymmetric and $\mathbf{k}$ is integrated by one Gauss point.
(d) The element is axisymmetric and $\mathbf{k}$ is integrated by four Gauss points.

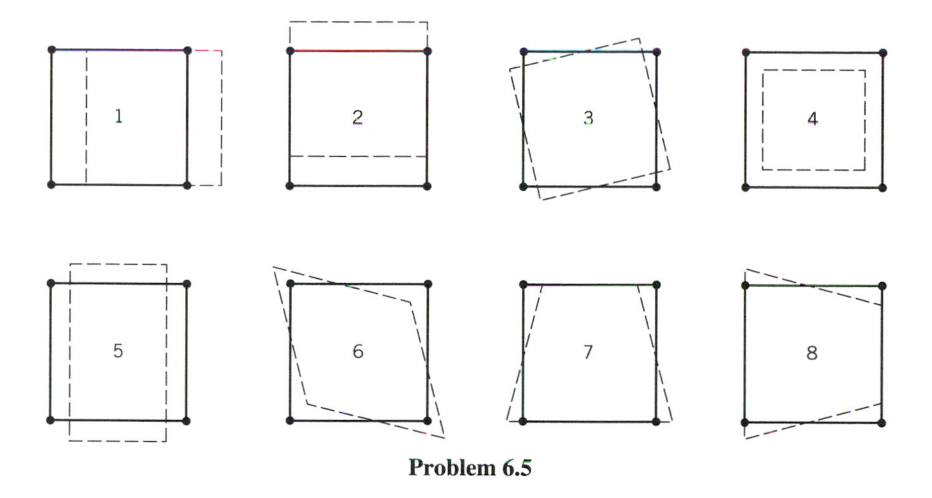

Problem 6.5

6.6 At $r = 0$ in an axisymmetric problem, strains ε_r and ε_θ must be equal. Why?

6.7 Devise an argument or an example that explains the large difference in stiffness noted for slender elements near the end of Section 6.4.

6.8 (a) In Fig. 6.6-2, rotate the r axis 90° counterclockwise, so that the $\theta = 0$ plane is vertical on the page. Do not reposition the load. Write the Fourier series for load that applies to this arrangement.
(b) Hence write a load series that describes the following radial load: $q = q_o$ outward over $0 < \theta < \pi/2$, $q = q_o$ inward over $\pi < \theta < 3\pi/2$, and $q = 0$ over the remaining two quadrants, where q_o is a constant.

6.9 It can be shown that a sinusoidal loading $q_n = q_{sn}\sin(n\pi x/L)$ on the uniform simply supported beam shown produces a lateral deflection v that is also sinusoidal and is given by

$$v_n = \frac{q_{sn}L^4}{EIn^4\pi^4}\sin\frac{n\pi x}{L}$$

A Fourier series for a load q is $q = \Sigma q_n$ and the associated beam deflection is $v = \Sigma v_n$. For each of the following loadings, evaluate the lateral deflection and bending moment at $x = L/2$. Use one, then two, then three series terms. Compute the percentage error of each result.
(a) Uniformly distributed load q_o, for which $q_{sn} = 4q_o/n\pi$ and $n = 1, 3, 5, \ldots$.
(b) Concentrated center force P, for which $q_{sn} = (2P/L)\sin(n\pi/2)$ and $n = 1, 2, 3, \ldots$.

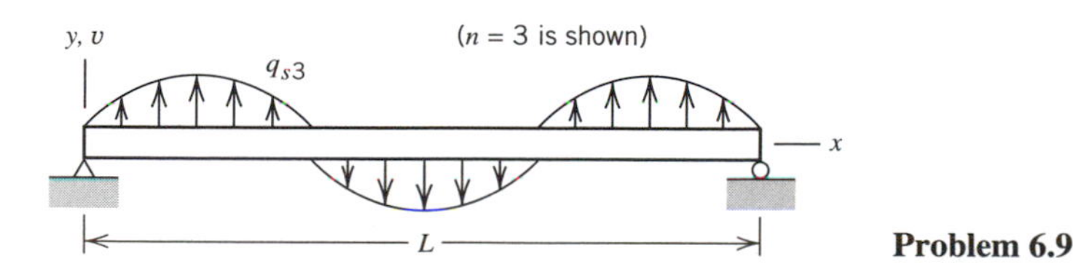

Problem 6.9

6.10 Write the special form of Eq. 6.6-2 that describes each of the following rigid-body motions:

(a) Translation in the z direction.

(b) Translation in a radial direction and parallel to the plane $\theta = 0$.

(c) Translation in a radial direction and parallel to the plane $\theta = \pi/2$.

(d) Small rotation about an r axis in the plane $z = 0$. (The r axis is at $\theta = 0$.)

6.11 (a) A flat plate of uniform thickness t contains a small circular hole of radius R. The plate is loaded by uniaxial tensile stress σ_o, as shown. Imagine that stresses in the neighborhood of the hole are to be analyzed using an axisymmetric FE model of outer radius c, with nonaxisymmetric loads applied at $r = c$. For the full thickness t, what should these loads be, in terms of t, σ_o and θ? *Suggestion*: Consider stress transformation equations or Mohr's circle.

(b) Repeat part (a), but replace σ_o by the linearly varying normal stress associated with pure bending in which moment vectors are directed normal to the $r\theta$ plane.

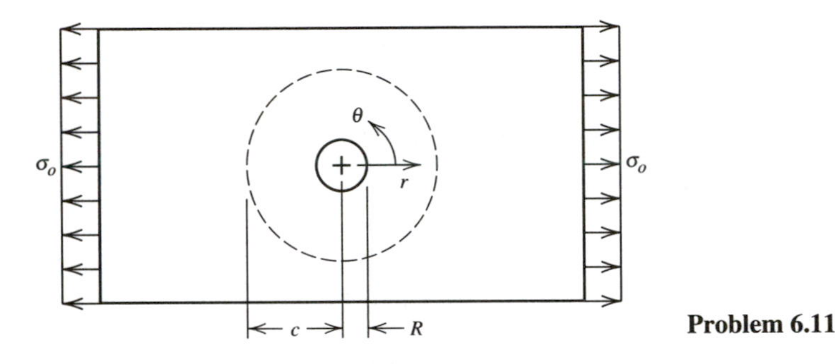

Problem 6.11

COMPUTATIONAL PROBLEMS

In the following problems compute significant values of stress and/or displacement, as appropriate. Exploit symmetry if possible. Choose convenient numbers and consistent units for material properties, dimensions, and loads. When mesh refinement is used, estimate the maximum percentage error of FE results in the finest mesh. Unless directed otherwise, assume that the material is isotropic.

A FE analysis should be preceded by an alternative analysis, probably based on statics and mechanics of materials, and oversimplified if necessary. If these results and FE results have substantial disagreement, we are warned of trouble somewhere.

6.12 The cantilevered rectangular block of material shown is loaded by a uniformly distributed z-direction load q along line AA. As support conditions in the plane $y = 0$, apply q in the negative z direction along the line $z = h/2$, set $u = v = w = 0$ at $x = y = z = 0$, $w = 0$ at $x = b$ on the x axis, and $v = 0$ at all nodes. Include the case $b >> h$ as one of the configurations analyzed, and be sure to examine results near $x = 0$ or $x = b$.

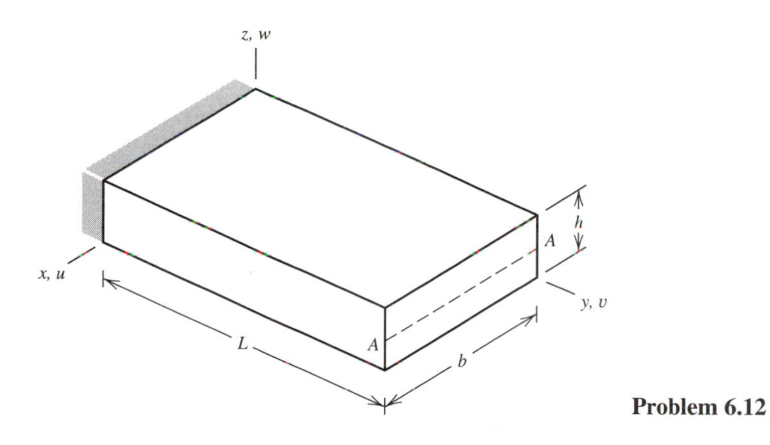

Problem 6.12

6.13 Consider the application of a torque to the rectangular block sketched for Problem 6.12. Let the torque act about the y-parallel centroidal axis. As support conditions in the plane $y = 0$, set $v = 0$ at the midpoint of each edge, and set $u = w = 0$ at all nodes. Devise a simple (if approximate) way to apply the torque to the end $y = L$.

6.14 Consider a small circular hole in a plate, such that radius R may be considerably less than the plate thickness t (see sketch). Well away from the hole, the plate is in a state of pure shear. Investigate the stress concentration factor. Take note of the effect of Poisson's ratio.

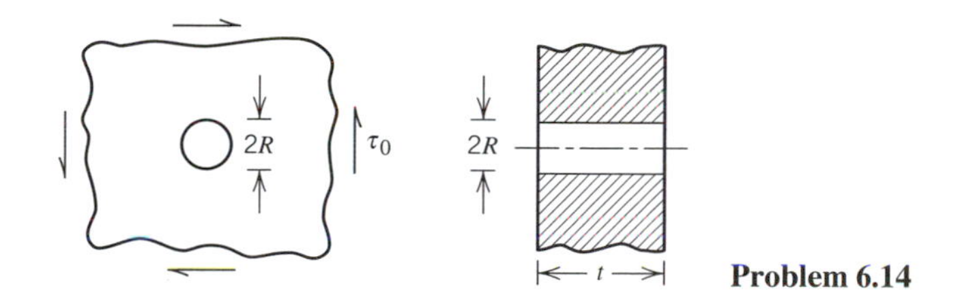

Problem 6.14

6.15 The sketch shows a prismatic beam whose cross section is a right triangle. The beam is cantilevered from end $z = 0$. Apply x-direction force to the plane $z = L$. Place the force at $y = 0$, then at $y = b$. In addition to stress analysis, determine the y location of the force for which cross sections have no rotation about the z axis.

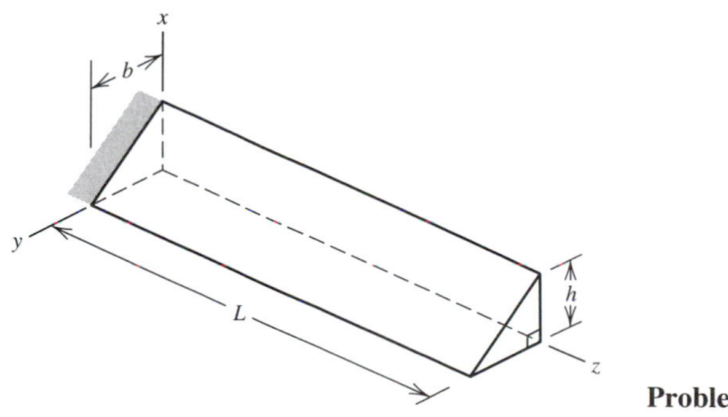

Problem 6.15

6.16 Replace the trapezoidal cross section of Fig. 6.3-1 by a solid circular cross section.

6.17 The sketch shows a massive coil spring, formed by bending a bar of rectangular cross section into a helix. Loading is by axial forces F at the ends (ends are not shown). Assume that the spring is closely coiled, which means that angle α is small. Then one may isolate a wedge of one coil spanning a small angle $\Delta\theta$ and assume that $u = 0$ on the cross sections thus exposed [6.1]. Also, $w = 0$ on $\theta = 0$ and $w = c\,\Delta\theta$ on $\theta = \Delta\theta$, where c is a constant. Circumferential displacement is $v = v(r, z)$. Impose u and w displacement boundary conditions, solve for stresses, and compute F from computed reactions at nodes.

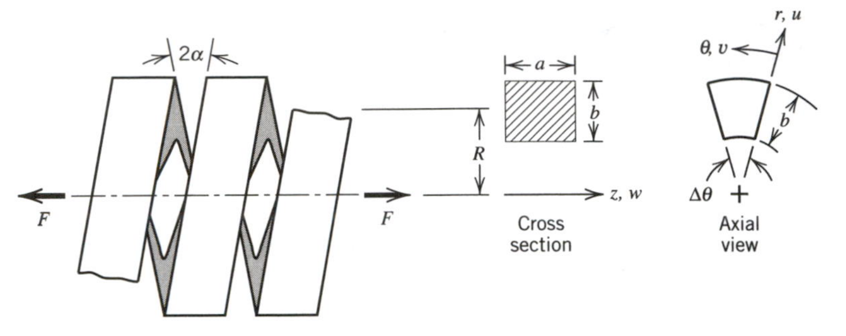

Problem 6.17

6.18 (a) The sketch shows an axisymmetric FE model of a circular disk. The outer edge is prevented from rotating in the rz plane and is loaded by total axial force $2P$, uniformly distributed around the circumference. The mesh is poor in that the element size change is abrupt. Compare results given by the mesh shown with results given by a more gradual size change, as in Fig. 5.3-4b.

(b) Are stresses σ_r and σ_θ equal at $r = 0$? They should be.

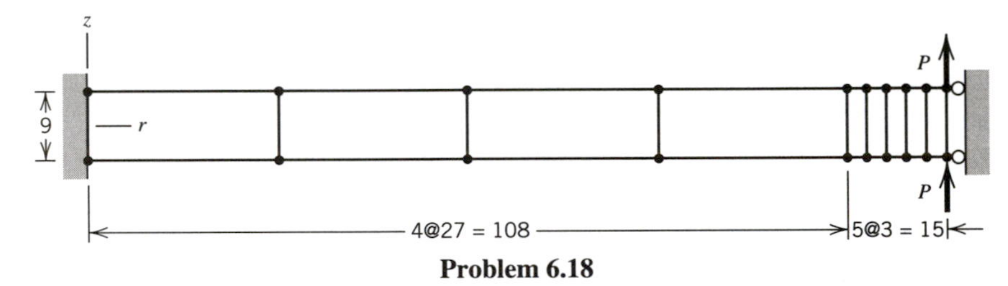

Problem 6.18

6.19 The sketch shows the cross section of a truncated cone built of two different materials. The cone becomes a cylinder if $R_1 = R_2$, and a disk if $R_1 - R_2 = L_1 + L_2$, with a central hole of radius R_2. Loading is by temperature change.

(a) Let the temperature change be uniform.

(b) Let the temperature change vary linearly with s.

(c) Let the temperature change vary linearly in the direction of thickness t.

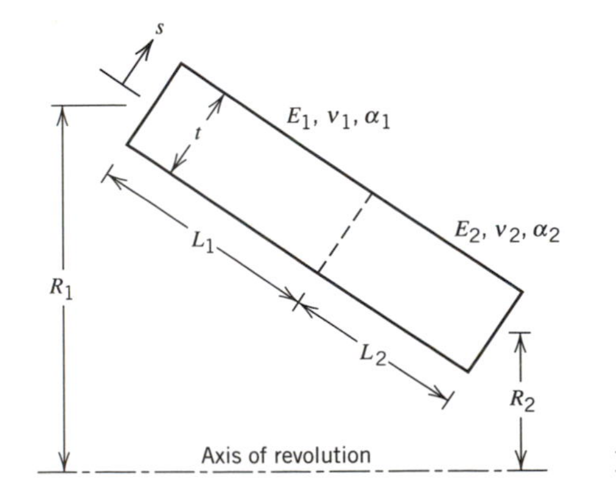

Problem 6.19

6.20 In Problem 6.19 let there be a single material. Also, omit the thermal load. Instead, apply the mechanical load shown in the sketch ($F_1R_1 = F_2R_2$, where each F_i is a uniform circumferential line load with dimensions [force/length]).

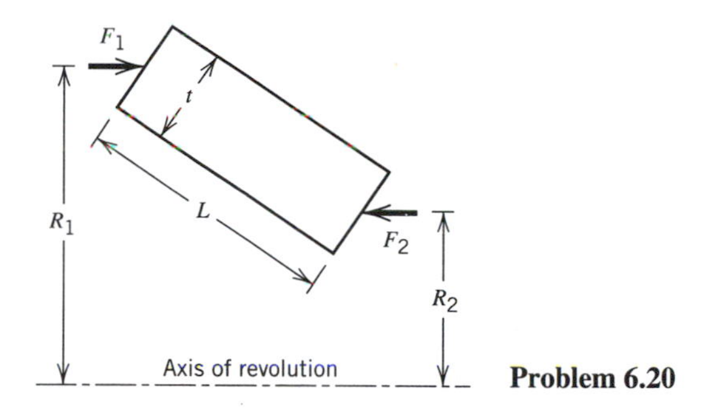

Problem 6.20

6.21 The sketch shows the cross section of a long circular cylinder capped by a hemisphere. The juncture at A contains quarter-circle fillets of radius r. Loading is by internal pressure. To approximate peak stresses at the juncture, use stress concentration factors rather than great mesh refinement.

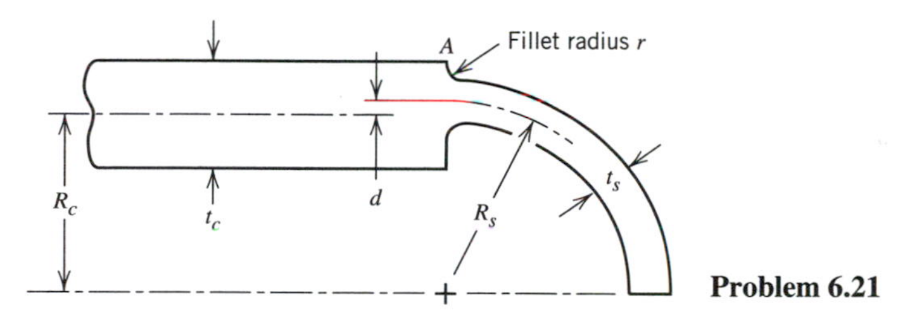

Problem 6.21

6.22 (a) A thick-walled cylinder is loaded by internal pressure p. The sketch shows a cross section, viewed axially. Calculate the largest stresses. Also, examine the axial stress and strain, and from them draw a conclusion about the necessary length of the FE model in the axial direction.

(b) Consider a compound cylinder, built by shrink-fitting a jacket of inner radius c on an inner cylinder of outer radius $c + \Delta c$, where $\Delta c << c$. Internal pressure p is subsequently applied. For given values of p, a, b, and c, what Δc is needed to produce the same maximum von Mises stress in the inner cylinder and in the jacket? [2.1] (Model Δc by a uniform temperature change in the inner cylinder or in the jacket.)

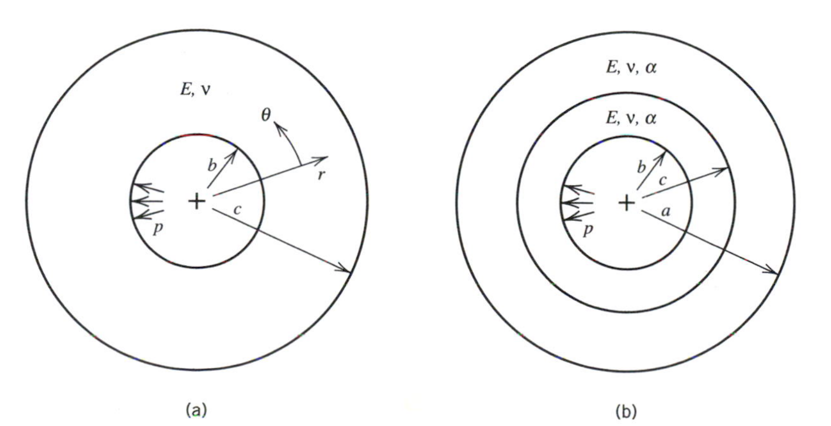

Problem 6.22

6.23 Let the sketch for Problem 6.22a represent the cross section of a thick-walled *sphere* under internal pressure. Investigate the state of stress.

6.24 A spherical vessel has a radially directed cylindrical outlet, as shown in cross section. The inside and/or outside may be reinforced by the enlargements shown: on the inside by AB perpendicular to the axis of the cylindrical part; on the outside by CD tangent to the outer surface of the spherical part. Let $t_c = 2t_s r_c / r_s$. Apply internal pressure p. Compute the largest tensile stress as a percentage of p. Consider the following configurations.

(a) $r_s = 400$ mm, $t_s = 100$ mm, $r_c = 200$ mm, $h_o = 80$ mm, $h_i = 0$.
(b) $r_s = 400$ mm, $t_s = 100$ mm, $r_c = 200$ mm, $h_o = 0$, $h_i = 80$ mm.
(c) $r_s = 400$ mm, $t_s = 100$ mm, $r_c = 200$ mm, $h_o = 80$ mm, $h_i = 80$ mm.
(d) $r_s = 400$ mm, $t_s = 40$ mm, $r_c = 100$ mm, $h_o = 40$ mm, $h_i = 0$.
(e) $r_s = 400$ mm, $t_s = 40$ mm, $r_c = 100$ mm, $h_o = 0$, $h_i = 40$ mm.
(f) $r_s = 400$ mm, $t_s = 40$ mm, $r_c = 100$ mm, $h_o = 40$ mm, $h_i = 40$ mm.

6.25 (a) A flywheel of the cross section shown spins with constant angular velocity ω about the z axis. Investigate the state of stress.
(b) Similarly, investigate stresses due to spinning the cone of Problem 6.19 about its axis of revolution.

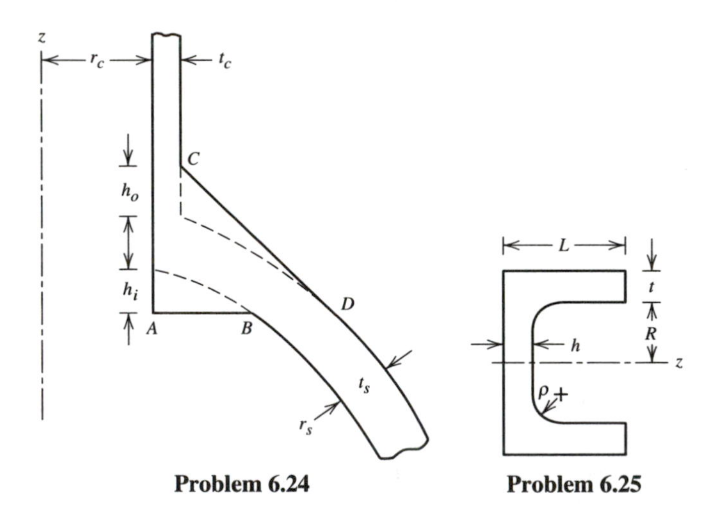

Problem 6.24 **Problem 6.25**

6.26 Consider the problem depicted in Fig. 5.7-3a and the approximate axisymmetric idealization suggested in Fig. 5.7-3c. Devise reasonable dimensions, number of bolts, and bolt pretension forces.
(a) Solve for stresses due to axial load on the pipe.
(b) If the software you use can treat nonaxisymmetric loads, solve for stresses due to bending load on the pipe.

6.27 Numerically solve for the maximum stress in Problem 6.11, using the Fourier series approach.

CHAPTER 7

Plates and Shells

Elementary concepts of plate behavior are discussed, followed by remarks about plate elements, FE modeling for plates, and an example application. This is followed by a similarly organized discussion of shells, with example applications for axisymmetric geometry and for more general geometry.

7.1 PLATE DISPLACEMENTS, STRAINS, AND STRESSES

A plate can be regarded as the two-dimensional analogue of a beam. Beams and plates both carry transverse loads by bending action, but they have significant differences. A beam can be straight or curved; a plate is flat (a curved geometry would make it a shell). A beam typically has a single bending moment; a plate has two bending moments and a twisting moment. Deflection of a beam need not stretch its axis; with few exceptions, deflection of a plate will strain its midsurface, and this may have important consequences. A more detailed summary of plate behavior is presented in what follows. We exclude anisotropic, composite, and layered plates from our discussion, although FE formulations for them are available.

Thin-Plate Theory. Consider a plate of thickness t that straddles the xy plane. Plate surfaces are at $z = \pm t/2$ and the plate "midsurface" is at $z = 0$. A differential slice cut from the plate by planes perpendicular to the x axis is shown in Fig. 7.1-1. Loading causes the plate to have lateral displacement $w = w(x, y)$ in the z direction. The differential slice moves to the position shown in Fig. 7.1-1b, with right angles preserved in cross sections because transverse shear deformation is neglected. Thus $\gamma_{yz} = 0$ and $\gamma_{zx} = 0$. In general $\gamma_{xy} \neq 0$; right angles in the plane of the plate are *not* preserved. An arbitrary point P has displacement $u = -z(\partial w/\partial x)$ in the x direction. An analogous argument with a differential slice cut from the plate by parallel planes normal to the y axis yields $v = -z(\partial w/\partial y)$ as the y-direction displacement of point P. Hence Eqs. 3.1-5 yield the strains

$$\varepsilon_x = -z\frac{\partial^2 w}{\partial x^2} \qquad \varepsilon_y = -z\frac{\partial^2 w}{\partial y^2} \qquad \gamma_{xy} = -2z\frac{\partial^2 w}{\partial x\,\partial y} \tag{7.1-1}$$

The first of these equations is used in beam theory, where it is usually written in the form $\varepsilon_x = -y/\rho$ for a beam lying along the x axis and bent to radius of curvature ρ in the xy plane. The second and third of Eqs. 7.1-1 are not used in beam theory.

Normal stress σ_z is considered negligible in a thin plate. Accordingly, we may substi-

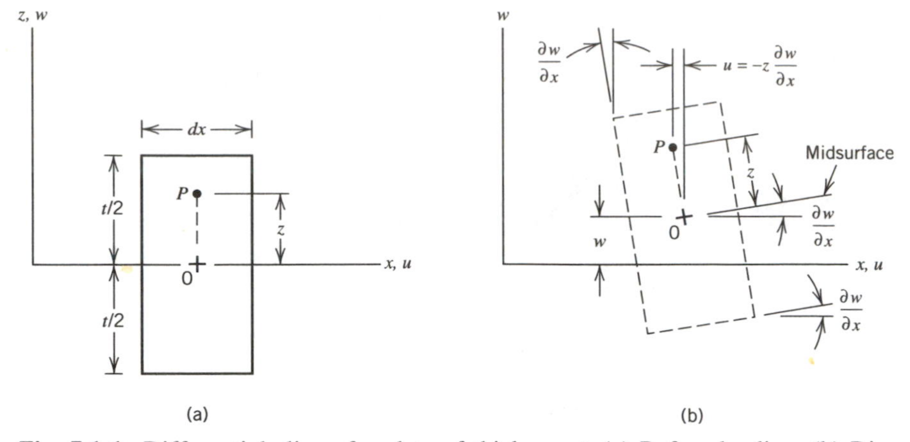

Fig. 7.1-1. Differential slice of a plate of thickness t. (a) Before loading. (b) Displacements after loading, according to Kirchhoff theory. Transverse shear deformation is neglected, so right angles in the cross section are preserved. Displacements in a yz-parallel plane are similar. (Reprinted from [2.2] by permission of John Wiley & Sons, Inc.)

tute Eqs. 7.1-1 into the plane stress equations, which are Eqs. 3.1-2 and 3.1-3 for an isotropic material. The resulting stresses are

$$\begin{Bmatrix} \sigma_x \\ \sigma_y \end{Bmatrix} = -z\frac{E}{1-\nu^2}\begin{bmatrix} 1 & \nu \\ \nu & 1 \end{bmatrix}\begin{Bmatrix} \partial^2 w/\partial x^2 \\ \partial^2 w/\partial y^2 \end{Bmatrix} \qquad \tau_{xy} = -2zG\frac{\partial^2 w}{\partial x\, \partial y} \tag{7.1-2}$$

These stresses are depicted in Fig. 7.1-2. Like flexural stress in a beam, they vary linearly with distance from the midsurface. Transverse shear stresses τ_{yz} and τ_{zx} are also present, even though transverse shear *deformation* is neglected. Transverse shear stresses vary quadratically through the thickness. The stresses of Eq. 7.1-2 give rise to bending moments M_x and M_y and twisting moment M_{xy}, as shown in Fig. 7.1-2b. Moments are functions of x and y and are computed *per unit length* in the plane of the plate. For example, an increment of M_x is $dM_x = z(\sigma_x\, dA)$, where dA is $dA = (1)\, dz$. Accordingly,

$$M_x = \int_{-t/2}^{t/2} \sigma_x z\, dz \qquad M_y = \int_{-t/2}^{t/2} \sigma_y z\, dz \qquad M_{xy} = \int_{-t/2}^{t/2} \tau_{xy} z\, dz \tag{7.1-3}$$

Stresses in Eq. 7.1-3 vary linearly with z. Thus, for example, if $\bar{\sigma}_x$ is the magnitude of σ_x at $z = t/2$, then $\sigma_x = 2\bar{\sigma}_x z/t$, and Eq. 7.1-3 yields $\bar{\sigma}_x = 6M_x/t^2$. This formula can be regarded as the flexure formula $\bar{\sigma}_x = M_x c/I$, applied to a unit width of the plate and with $c = t/2$. Similarly, maximum magnitudes of σ_y and τ_{xy} are $6M_y/t^2$ and $6M_{xy}/t^2$ respectively.

We can now see how moments and stresses in a plate differ from their counterparts in a beam. One difference is the presence of M_{xy} in plate theory. M_{xy} is the only moment present if the plate has deflection $w = cxy$, where c is a constant. This is called a state of pure twist. It can be realized by applying upward forces to two diagonally opposite corners of a rectangular plate and downward forces to the two remaining corners.

Another difference between beam and plate stresses appears if we apply a moment such as M_x to a beam and also to a plate. Let the beam lie along the x axis and be bent so that it deforms in the xz plane (e.g., Fig. 6.2-2). The beam has a narrow cross section, so that stress σ_y is zero on its sides and almost zero in between. Due to the Poisson effect, top and bottom edges of a cross section become curved, that is, they become arcs in the yz

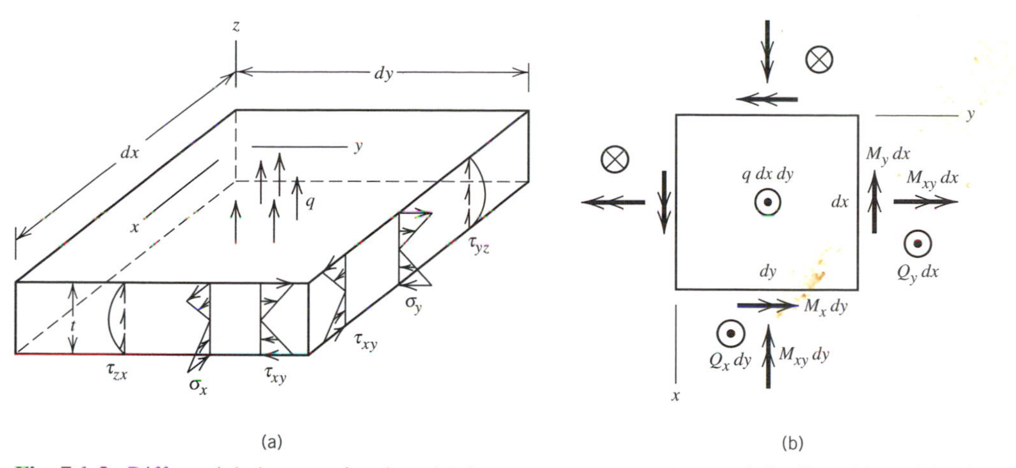

Fig. 7.1-2. Differential element of a plate. (a) Stresses on cross sections and distributed lateral load $q = q(x, y)$. (b) Differential forces and moments. Arrows that represent forces normal to the plate midsurface are viewed end-on. (Reprinted from [2.2] by permission of John Wiley & Sons, Inc.)

plane. In contrast, a plate has a wide cross section, and top and bottom edges of a cross section remain straight y-parallel lines when M_x is applied. This difference between beam and plate behavior is easily seen by bending a rubber eraser, which behaves like a beam, and then a sheet of paper, which behaves like a plate. We conclude that $\partial^2 w/\partial y^2 = 0$ when a plate is bent to a cylindrical surface $\partial^2 w/\partial x^2 \neq 0$ by M_x alone. Hence Eqs. 7.1-2 show that flexural stress σ_x is accompanied by stress $\sigma_y = \nu\sigma_x$. Stress σ_y constrains the plate against the deformation $\partial^2 w/\partial y^2$ the beam would have, thereby stiffening the plate. The amount of stiffening is proportional to $1/(1 - \nu^2)$, so that a unit width of plate has "flexural rigidity" $D = Et^3/[12(1 - \nu^2)]$. A *beam* of unit width has flexural stiffness $EI = Et^3/12$. Thus $D = EI/(1 - \nu^2)$ for a unit width.

The theory outlined above is classical thin-plate theory. It is also called "Kirchhoff" plate theory after its principal developer. It neglects transverse shear deformation, which can be significant if the plate is thick, that is, if t is more than roughly one-tenth the span of the plate. An alternative to thin-plate theory, called "Mindlin" plate theory, not only accounts for transverse shear deformation but produces finite elements for plates more easily.

Mindlin Plate Theory. To account for transverse shear deformation, the assumption that right angles in a cross section are preserved must be abandoned. This means that planes initially normal to the midsurface may experience rotations different from rotations of the midsurface itself. Thus the differential element in Fig. 7.1-1a has the deformations depicted in Fig. 7.1-3, where θ_x and θ_y are rotation components of a line initially normal to the midsurface. Combining these displacements with Eqs. 3.1-5, we obtain

$$u = z\theta_y \qquad \varepsilon_x = z\frac{\partial \theta_y}{\partial x} \qquad \gamma_{xy} = z\left(\frac{\partial \theta_y}{\partial y} - \frac{\partial \theta_x}{\partial x}\right)$$

$$v = -z\theta_x \qquad \varepsilon_y = -z\frac{\partial \theta_x}{\partial y} \qquad \gamma_{yz} = \frac{\partial w}{\partial y} - \theta_x \qquad \gamma_{zx} = \frac{\partial w}{\partial x} + \theta_y \tag{7.1-4}$$

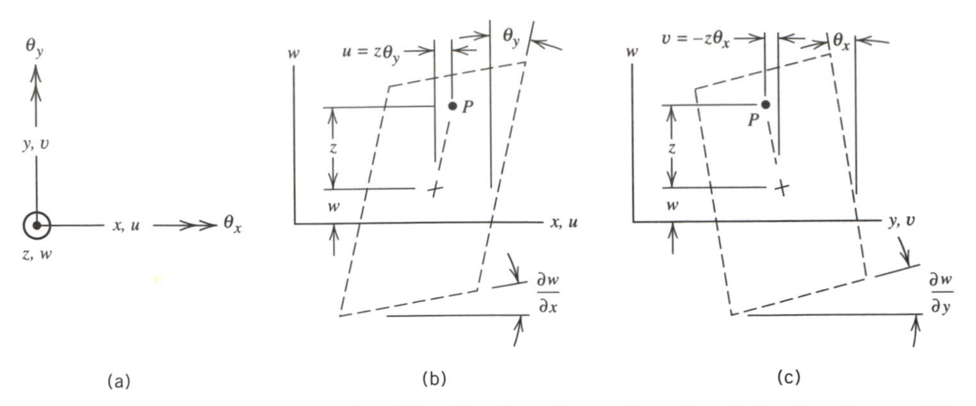

Fig. 7.1-3. (a) Displacements and positive directions for rotations θ_x and θ_y, viewed normal to the xy plane. (b) Mindlin theory displacements in an xz-parallel cross section. (c) Mindlin theory displacements in a yz-parallel cross section.

instead of Eqs. 7.1-1. Equations 7.1-4 are the main equations of Mindlin plate theory. If $\theta_x = \partial w/\partial y$ and $\theta_y = -\partial w/\partial x$, transverse shear deformation vanishes and Eqs. 7.1-4 reduce to Eqs. 7.1-1.

Loads and Supports. Loads in the z direction, either distributed or concentrated, may be applied to lateral surfaces $z = \pm t/2$ or to edges of a plate. Such loads are called "lateral" loads. Distributed load has dimensions [force/length^2] on a lateral surface or [force/length] on an edge. A plate edge may also be loaded by a bending moment whose vector is tangent to the edge. The same kinds of edge loads may be applied to the plate by supports. Support conditions for FE plate models are discussed in Section 7.2.

At the point where a concentrated lateral (z direction) force is applied, Kirchhoff theory predicts infinite bending moments. Mindlin theory predicts infinite bending moments *and* infinite displacement. In reality no force can be truly concentrated, and in plate theory the infinities disappear if the "concentrated" load is applied over a small area instead. In FE work no infinite quantities will be computed, nor will they even be approached by any practical amount of mesh refinement.

Large Displacements and Membrane Forces. Internal force resultants in the plane of the plate have been omitted from the foregoing discussion, as is customary in standard plate theories. These "membrane forces" can develop as a consequence of the deflection and can significantly influence the response of the plate to load. By using the following beam example we can explain the effect of membrane forces in a simple way.

The beam in Fig. 7.1-4 is assumed to have hinge supports that remain exactly a distance L apart, regardless of how much load is applied. When the beam is loaded it develops the usual flexural stresses, and because the supports are immovable it also develops membrane force N that supports part of the applied load by "string action." This action is seen in Fig. 7.1-4b: forces N are not collinear and therefore develop a z-direction component. Let us assume that the deflected shape is a parabola with center deflection w_c, where $w_c << L$. Then it can be shown that the uniformly distributed load supported by string action is

$$q_s = \frac{64}{3}\frac{Ebt^4}{L^4}\left(\frac{w_c}{t}\right)^3 \tag{7.1-5}$$

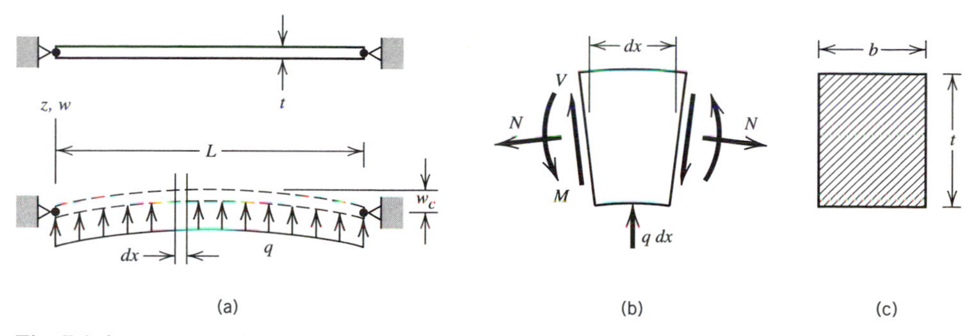

Fig. 7.1-4. (a) Beam of length L with immovable hinge supports. (b) Differential element, showing "string action" axial force N. (c) Cross section.

From a table, the center deflection of a simply supported *beam* under uniformly distributed load q_b is $w_c = 5q_bL^4/384EI$, from which we can solve for q_b in terms of w_c. We now state that the total load q supported by string and beam actions occurring simultaneously is the sum of q_s and q_b.

$$q = q_s + q_b \approx \frac{Ebt^4}{L^4}\left[21.3\left(\frac{w_c}{t}\right)^3 + 6.40\left(\frac{w_c}{t}\right)\right] \tag{7.1-6}$$

This argument is not exact because the actual deflected shape is a single function of x, not the different shapes assumed above for separate string and beam actions. Nevertheless, Eq. 7.1-6 is a useful approximation. It shows that string and beam actions each support about half the total load when $w_c/t = 0.5$, which is not a large deflection if t is small in comparison with L.

The foregoing argument is of little value for beams because immovable supports are not found in practice. The value of the argument is its implication for problems of thin plates. The counterpart of string action in a beam is strain of the midsurface in a plate. Deflection $w = w(x, y)$ of a plate produces no strain of the midsurface *only* if w describes a "developable" surface, such as a cylinder or a cone. In general, loading produces a deflected shape $w = w(x, y)$ that is not developable. Accordingly, in general there *are* strains at the midsurface, and membrane forces appear that carry part of the load. The linear plate theories outlined earlier in this section are not valid if w/t is "large," that is if w/t exceeds a few tenths. The practical limit of w/t is case dependent. Large w/t makes the problem nonlinear, necessitating an iterative solution, and yielding deflections and stresses significantly different than those predicted by linear theory.

Membrane forces may arise because of deflection, as described earlier, or may be present at the outset because of load components tangent to the midsurface. Either way, membrane forces have a "stress stiffening" effect: if tensile they effectively increase the flexural stiffness; if compressive they decrease it. Compressive membrane forces may become large enough to produce buckling.

7.2 FINITE ELEMENTS FOR PLATES

A plate is a thin solid and might be modeled by 3D solid elements (Fig. 7.2-1a). But a solid element is wasteful of d.o.f., as it computes transverse normal stress and transverse shear stresses, all of which are considered negligible in a thin plate. Also, thin 3D ele-

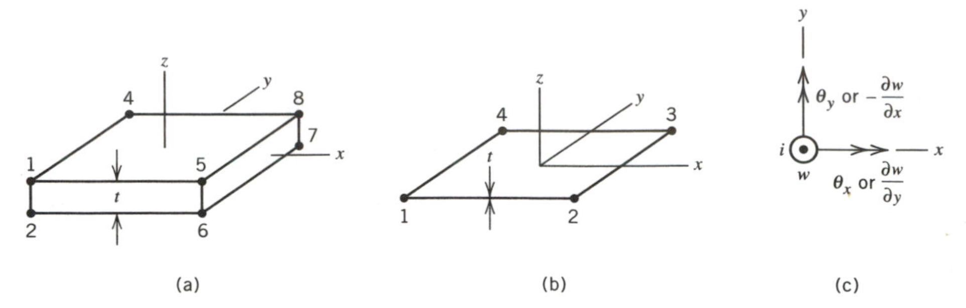

Fig. 7.2-1. (a) A 3D solid element. (b) The comparable plate element. (c) Plate d.o.f. at a typical node i, viewed normal to the xy plane.

ments invite troubles produced by ill-conditioning because stiffness associated with thickness-direction strain ε_z is very much larger than other stiffnesses. The plate element in Fig. 7.2-1b has half as many d.o.f. as the comparable solid element and omits ε_z from its formulation. In sketches, thickness t may appear to be zero, as in Fig. 7.2-1b, but the physically correct value is of course used in formulating element stiffness matrices.

A *plane* element must be able to display states of constant σ_x, σ_y, and τ_{xy} if it is to pass patch tests. A *plate* element must be able to display these states in each z = constant layer, which means that a valid plate element must pass patch tests for states of constant M_x, M_y, and M_{xy}. Figure 7.2-2 depicts a patch test for constant M_x, which requires constant $\partial^2 w/\partial x^2$ in Kirchhoff theory and constant $\partial\theta_y/\partial x$ in Mindlin theory if the test is to be passed. If $\nu \neq 0$, rotations about the x axis must be prevented at nodes 1, 3, 4, 6, and 7 in Fig. 7.2-2.

Although Cartesian coordinates are used in our discussion, this is not a limitation of plate theory or of FE theory. Classical plate theory uses polar coordinates for circular plates. In FE analysis, a circular plate can be modeled by shell of revolution elements, simply by making shell elements flat rather than (say) cylindrical or conical. Each such element is thus a flat annular ring, joined to adjacent annular elements at its inner and outer radii.

Kirchhoff Plate Elements. The stiffness matrix of a Kirchhoff plate element can be calculated from an expression analogous to Eq. 3.1-10, in which **E** is replaced by a matrix of flexural rigidities and **B** is contrived to produce curvatures when it operates on nodal d.o.f. that describe a lateral displacement field $w = w(x, y)$. Thus the behavior of a Kirchhoff element depends entirely on the assumed w field, which is a polynomial in x and y whose d.o.f. are nodal values of w, $\partial w/\partial x$, and $\partial w/\partial y$.

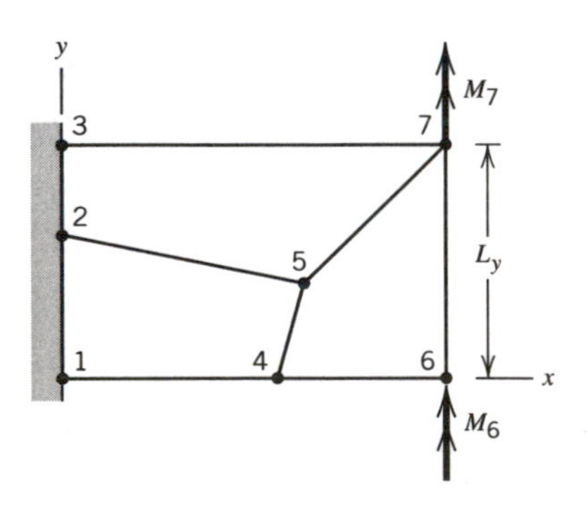

Fig. 7.2-2. A patch test for constant curvature (or for constant M_x). Nodal moment loads are $M_6 = M_7 = M_x L_y/2$.

A 12-d.o.f. rectangular Kirchhoff element (Fig. 7.2-1b) can be formulated rather easily. It is an incompatible element; that is, if n is a direction normal to an element edge, $\partial w/\partial n$ is not continuous between elements for some loading conditions. Accordingly, the element cannot guarantee a lower bound on computed displacements, so results may converge "from above" rather than "from below." A *compatible* rectangular element with corner nodes only requires that twist $\partial^2 w/\partial x \partial y$ also be used as a nodal d.o.f., which is undesirable.

A restriction to rectangular shapes is unacceptable, so many triangular elements have been devised. It is surprisingly difficult to obtain a triangular Kirchhoff plate element that can represent states of constant curvature and twist, has no preferred directions, and gives good results when applied to a variety of test cases [3.2]. Experience in formulating plate elements has shown that Eqs. 7.1-4 are more productive than Eqs. 7.1-1. Accordingly, Mindlin plate elements are in common use, as are "discrete Kirchhoff" elements, which also use Eqs. 7.1-4 as the starting point.

Mindlin Plate Elements. A Mindlin element is based on three fields: $w = w(x, y)$, $\theta_x = \theta_x(x, y)$, and $\theta_y = \theta_y(x, y)$. Each is interpolated from nodal values. If all interpolations use the same polynomial, then for an element of n nodes,

$$\begin{Bmatrix} w \\ \theta_x \\ \theta_y \end{Bmatrix} = \sum_{i=1}^{n} \begin{bmatrix} N_i & 0 & 0 \\ 0 & N_i & 0 \\ 0 & 0 & N_i \end{bmatrix} \begin{Bmatrix} w_i \\ \theta_{xi} \\ \theta_{yi} \end{Bmatrix} = \mathbf{Nd} \tag{7.2-1}$$

From Eqs. 7.1-4 and 7.2-1 we can form a strain–displacement matrix **B**. In the stiffness matrix formula, Eq. 3.1-10, **E** is a 5 by 5 matrix that includes the 3 by 3 **E** of plane stress and also shear moduli associated with the two transverse shear strains. Integration of $\mathbf{B}^T\mathbf{EB}$ with respect to z is done explicitly. Integration in the plane of the element is done numerically if the element is isoparametric.

The N_i are given by Eqs. 4.4-2 for a four-node quadrilateral element. An eight-node quadrilateral is also popular, based on the same N_i used for a plane eight-node element. In any z = constant layer, strains vary in the same way as in the corresponding plane element. Accordingly, the behavior of a Mindlin plate element can be deduced from the behavior of the corresponding plane element *provided that* all terms of the integrand are integrated by the same quadrature rule. But this is usually not done, for reasons now discussed.

Consider the bending mode shown in Fig. 7.2-3. Element strains ε_x are independent of x. Therefore *any* order of quadrature will report the same strain energy of pure bending. However, this element is like the element of Fig. 3.4-2c in that it displays spurious shear strain. If a/t is large, transverse shear strain γ_{zx} becomes large and the element is much too stiff in bending, *unless* γ_{zx} is evaluated at $x = 0$, where γ_{zx} vanishes. But one-point quadrature for all strains would introduce four instability modes. This observation suggests "selective" integration, in which one-point quadrature is applied to transverse shear terms and four-point quadrature is applied to bending terms. Then two instability modes remain. They may be controlled by "stabilization" matrices. Other Mindlin elements, such as the eight-node element, have analogous shortcomings and may also be treated by selective integration and other fix-ups. Calculated stresses are usually most accurate at Gauss points.

With so many options available in element formulation, element behavior cannot be deduced solely from the element displacement field. One expects that all elements in

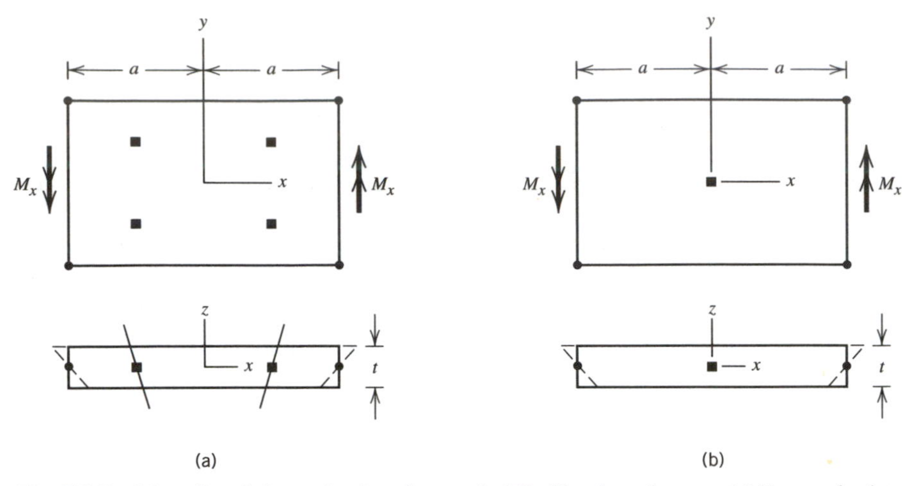

Fig. 7.2-3. A bending deformation in a four-node Mindlin plate element. (a) Four-point integration rule. (b) One-point integration rule.

commercial software will be free of instabilities and will be able to represent constant-moment states, as this is merely the minimum capability that a reliable element must have. The eight-node element can also represent linearly varying bending moments if its shape is rectangular. The analyst should learn how a plate element behaves by using it in test cases. In particular, one may wish to know if accuracy is lost when the element becomes very thin, and how accuracy declines as aspect ratio increases and as quadrilateral shapes become distorted from a rectangle.

Discrete Kirchhoff Elements. The essential feature of a discrete Kirchhoff element is that transverse shear strain is set to zero at a finite number of points in the element, rather than at every point as in the classical theory of thin plates. The result is a thin-plate element, usually triangular in shape, without side nodes, and incompatible in displacements, which usually converges rapidly with mesh refinement [7.1]. A reason for the coarse-mesh accuracy of discrete Kirchhoff elements is that lateral displacement w along each element edge is cubic in the edge-tangent coordinate. This is a more competent interpolation than prevails in Mindlin elements having corner nodes only, where w along an edge is only linear in the edge-tangent coordinate.

For a triangular element, one starts the formulation process by interpolating rotation components of a midsurface-normal line from nodal values at vertices and midsides:

$$\theta_x = \sum N_i \, \theta_{xi} \quad \text{and} \quad \theta_y = \sum N_i \, \theta_{yi} \tag{7.2-2}$$

where $i = 1, 2, \ldots, 6$. For a triangular element, the N_i can be shape functions of the six-node plane triangle discussed in Section 3.3. Thus far 12 nodal rotation d.o.f. have been introduced. Lateral displacement w along each edge can be interpolated in terms of nodal values w_i, $(\partial w/\partial x)_i$, and $(\partial w/\partial y)_i$ at vertices i, where $i = 1, 2, 3$. A total of 21 nodal d.o.f. have now been introduced. We seek a nine-d.o.f. element having vertex nodes only and three d.o.f. per node. Accordingly, 12 d.o.f. must be eliminated. For a particular triangular element known as the DKT element [7.1], this is done by using Eqs. 7.1-4 to set γ_{yz} and γ_{zx} to zero at the vertices, setting the edge-tangent transverse shear strain to zero at midsides, and constraining edge-normal rotations to vary linearly along each edge. The

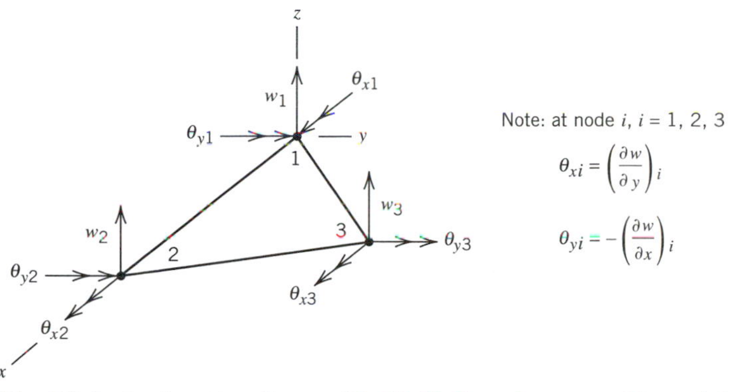

Fig. 7.2-4. A triangular discrete Kirchhoff plate element and its nodal d.o.f.

results of all this are interpolations for θ_x and θ_y over the element in terms of the nodal d.o.f. shown in Fig. 7.2-4. These interpolations and the expressions for ε_x, ε_y, and γ_{xy} in Eqs. 7.1-4 are sufficient to establish the strain–displacement matrix **B**. The element stiffness matrix then follows in standard fashion.

In some software, an element that the user perceives as quadrilateral is in fact built of triangles. The software may use two triangles that share the diagonal of the quadrilateral, then overlay this construction with two more triangles that share the *other* diagonal, and finally divide the resulting stiffness matrix by 2. This "overlapping triangle" arrangement avoids a small directional bias associated with using only two triangles and having to choose one diagonal or the other to divide the quadrilateral.

After so many manipulations it is not apparent how a discrete Kirchhoff plate element will behave. As with Mindlin plate elements, the analyst should use numerical experiments to learn about element behavior.

Loads and Supports. For lateral pressure applied to surfaces z = constant, and for z-direction line load along an edge, work-equivalent nodal loads consist of forces and moments for Kirchhoff elements but only forces for Mindlin elements. Ad hoc formulas for nodal moments might be devised. However, in practice nodal moments associated with surface pressure are usually omitted, for reasons discussed in Section 2.5. A distributed moment along an edge produces nodal moments, as shown in Fig. 7.2-2. These moments must be retained. Most software is capable of computing appropriate nodal loads from input data that define a distributed loading.

Support conditions are classed as clamped, simple, or free (no support), in direct analogy to the possible support conditions of a beam. Nodal d.o.f. that must be prescribed for these support conditions are as follows, with reference to the notation in Fig. 7.2-5:

Edge condition	Prescribed d.o.f.	Natural condition	
Clamped	$w = \theta_n = \theta_s = 0$	None	
Simply supported	$w = 0$	$M_n = 0$	(7.2-3)
Free	None	$Q = M_n = M_{ns} = 0$	

"Natural" boundary conditions are stress boundary conditions. The user does not prescribe them to FE software. They will not be computed exactly by FE but will be approached as a mesh is repeatedly refined.

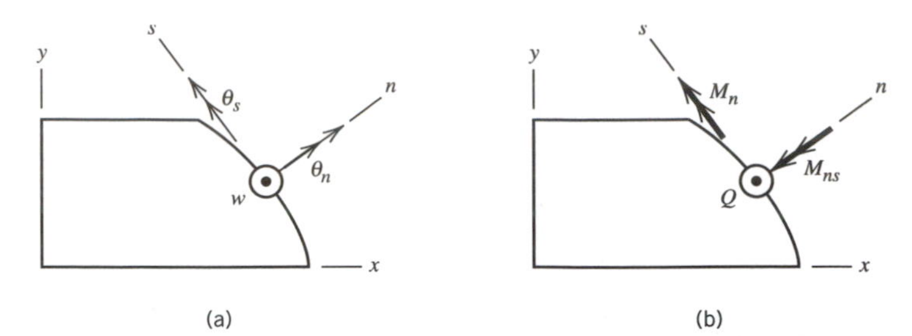

Fig. 7.2-5. (a) The d.o.f. along an arbitrarily oriented edge of a plate. Rotation vectors θ_s and θ_n describe rotational motion, in *nz* and *sz* planes, respectively, of a midsurface-normal line. (b) Transverse shear force Q, bending moment M_n, and twisting moment M_{ns}, each per unit length.

In classical plate theory, the simply supported condition requires $\theta_n = 0$ as well as $w = 0$. If used with Mindlin plate elements, this "hard" support condition has been found to be very detrimental to accuracy if a simply supported boundary includes a corner where edges meet at an angle other than 90° [7.2]. The "soft" simple support condition of Eqs. 7.2-3, in which θ_n is unrestrained, is much better for such cases and does no harm when edges intersect at right angles.

Test Cases. Some commonly used test cases for thin-plate bending are depicted in Fig. 7.2-6. The square plate in Fig. 7.2-6a has side length L and all edges are either simply supported or clamped. The plate is laterally loaded by either a uniformly distributed load q or a concentrated center force P. Center deflections can be stated as either $\alpha qL^4/D$ or $\alpha PL^2/D$, where α is a case-dependent constant [7.3]. With $D = Et^3/[12(1 - \nu^2)]$ and $\nu = 0.3$, α is 0.00406 (simply supported, uniform q), 0.0116 (simply supported, concentrated P), 0.00126 (clamped, uniform q), or 0.00560 (clamped, concentrated P). Bending moments for these cases are reported in [7.3]. For FE analysis, only one quadrant (or even only one octant) need be modeled. Typically, a uniform mesh is used and one examines the convergence of solutions as the mesh is refined.

In a rhombic plate with simply supported edges, Fig. 7.2-6b, theory predicts that bending moments M_x and M_y are infinite but of opposite sign at an obtuse corner such as the one at $x = y = 0$ [7.2]. Considerable mesh refinement may be needed to approximate this

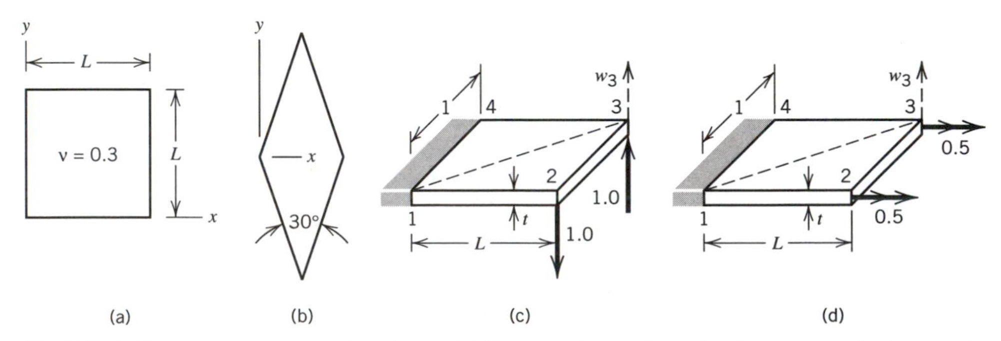

Fig. 7.2-6. Common test cases for plates. (a) Square, clamped or simply supported edges, uniformly distributed load or concentrated center lateral load. (b) Rhombic, simply supported edges, uniformly distributed lateral load. (c,d) Twisting of a flat strip of unit width, with nominally equivalent loadings. (c and d reprinted from [2.2] by permission of John Wiley & Sons, Inc.)

behavior. If "hard" simple supports are used with Mindlin elements (which is not recommended), even center deflections of the rhombic plate may display considerable error despite a fine mesh.

The twisting test case, Fig. 7.2-6c,d, has alternative loads that produce the same torque, as shown. With $E = 10^7$, $\nu = 0.25$, and $t = 0.05$, "benchmark" results give $w_3 = 0.0029L$ as the displacement of one corner [7.1]. Usually the structure is modeled by one rectangular or two triangular elements, to test the effect of element aspect ratio. Some of the many proposed plate elements do badly in this test.

7.3 A PLATE APPLICATION

A large flat plate of constant thickness is loaded by a concentrated force. The plate is supported entirely by a "Winkler" elastic foundation that lies between the plate and a flat base that is assumed to be perfectly rigid (Fig. 7.3-1a). A Winkler foundation exerts a pressure k_0 MPa on the plate for each millimeter of z-direction deflection. Data are as follows:

$$E = 200 \text{ GPa} \qquad L = 1650 \text{ mm}$$

$$\nu = 0.3 \qquad t = 30 \text{ mm}$$

$$k_0 = 0.2 \text{ MPa/mm} \qquad P = 1.0 \text{ N}$$

The deflection and state of stress in the plate are required. The unit load is convenient because results need only be multiplied by the actual load if it is not unity. The choice $L = 1650$ mm is somewhat arbitrary and is explained in what follows.

Preliminary Analysis. It is stated that the plate is "large," implying that its edges are far distant from the load at $x = y = 0$. Analysis can be confined to the neighborhood of the load. Thus there is no need for an extensive FE model whose outer portion would be essentially undeflected and unstressed. But how large should the model be? Textbooks and handbooks provide useful formulas for beams on an elastic foundation but not for plates on an elastic foundation. Accordingly, to plan the initial FE analysis, we assume that a cross section of the plate has the same extent of downward deflection as an infinitely long *beam* on elastic foundation loaded by a concentrated force. A beam whose cross section

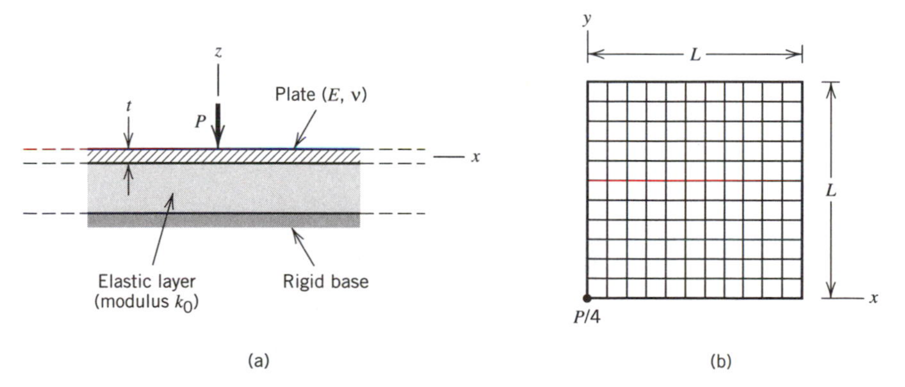

Fig. 7.3-1. (a) Cross section of a plate on an elastic foundation. (b) FE mesh on a quadrant, with z-direction force $P/4$ at $x = y = 0$.

is of width b and depth 30 mm has moment of inertia $I = 2250b$ mm^4 and a beam foundation modulus $k = bk_0$. Pursuing this analysis in its usual terminology [2.1], we obtain $\beta = 0.00325$/mm and $x = 3\pi/4\beta = 726$ mm as the distance from the load at which the deflection of the beam changes from downward to upward. For an initial FE analysis of the plate we guess that the mesh need span no more than roughly twice this distance. We choose $L = 1650$ mm as a convenient side length for the FE model.

To obtain an estimate of maximum deflection for comparison with subsequent FE results, we will approximate the deflected shape of the plate, then write an equation requiring that load P be equal to the force provided by foundation pressure on the plate. A simple assumption for displaced shape, in terms of radial distance r from load P, appears in Fig. 7.3-2a. It is here assumed that $w = 0$ for $r > R$. The foundation pressure is $p = -k_0 w$. The negative sign is needed so that negative (downward) w will produce positive (upward) p. For equilibrium of vertical forces,

$$P = \int_0^{2\pi} \int_0^R p\, r\, dr\, d\theta \qquad \text{hence} \qquad w_o = \frac{2\pi P}{(\pi^2 - 4) k_0 R^2} \tag{7.3-1}$$

With $R = 726$ mm, as calculated above, we obtain $w_o = 10.2(10)^{-6}$ mm as the approximate deflection of load P. It seems pointless to also estimate stresses because to do so the second derivative of $w = w(x)$ is required and $w = w(x)$ itself is already approximate (see remarks under "Stresses" in Section 2.5). For example, the assumed shape in Fig. 7.3-2a has the same magnitude of second derivative at $r = 0$ and at $r = R$, but in reality bending moments certainly will not be equal at these two locations.

FE Model and Analysis. The problem has axial symmetry, so it would be fitting to use polar coordinates and axisymmetric elements. For illustrative purposes we elect instead to use Cartesian coordinates and quadrilateral elements. Figure 7.3-1b shows an 11 by 11 mesh of four-node plate elements, each a 150-mm square built of four overlapping discrete Kirchhoff triangular elements. The software used allows an elastic foundation to be included as an option for this type of plate element. We use a uniform mesh because it is very easy to prepare and at this stage we are unsure about the appearance of an improved mesh, although we can safely predict that a graded mesh should have greatest refinement near $x = y = 0$. A load $P/4$ is applied at $x = y = 0$ on the quadrant modeled. Symmetry is imposed about x and y axes; that is, $\theta_x = 0$ along the x axis and $\theta_y = 0$ along the y axis. Displacement w is unrestrained at all nodes.

If available software does not include an elastic foundation option, the user might instead connect each node of the plate to the rigid base by a linear spring. This would be an

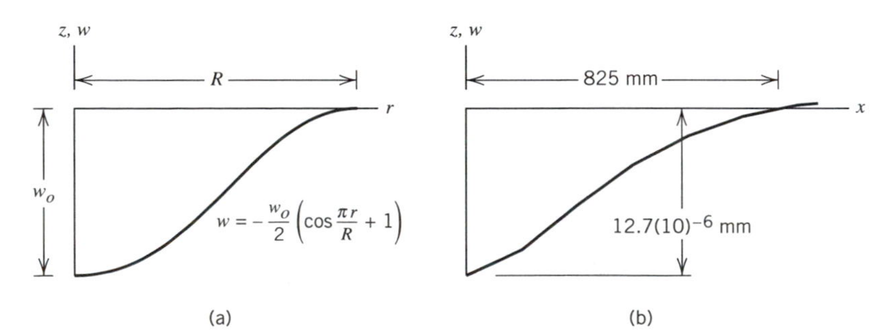

Fig. 7.3-2. (a) Deflected shape assumed in preliminary analysis. (b) Deflection along $y = 0$ computed by FE with the mesh of Fig. 7.3-1b.

especially tedious procedure with a graded mesh, for which the appropriate spring stiffnesses might all differ from one another. A better option would be to place the plate on a layer of 3D solid elements, whose material is anisotropic with only E_z nonzero if a Winkler elastic foundation is to be simulated.

Critique of FE Results. The deflected shape is shown in Fig. 7.3-2b as software plots it, with straight lines between nodes rather than the curved shapes that element edges actually have. The computed deflection at $x = y = 0$ is $w = 12.7(10)^{-6}$ mm downward, in remarkably good agreement with the deflection $10.2(10)^{-6}$ mm from preliminary analysis. The computed $w = w(x)$ passes through zero at about $x = 825$ mm, also in good agreement with the preliminary guess of 726 mm. The largest upward deflection is $0.2(10)^{-6}$ mm and occurs far from $x = y = 0$, but within the quadrant modeled rather than at its boundaries.

Stress contours appear in Fig. 7.3-3. As expected, contours of von Mises stress σ_e are symmetric about the line $x = y$, which implies only that we have not blundered in imposing boundary conditions. Contours of σ_e should be concentric circles, which is not the case near $x = y = 0$. Interelement continuity of the contours is fair. Were we to model the entire plate by a mesh symmetric with respect to both x and y axes, we would see *perfect* interelement continuity across the x and y axes, but this continuity would have nothing to do with whether results are accurate or not. Similar remarks can be made with reference to σ_x contours in Fig. 7.3-3b, although symmetry with respect to the line $x = y$ is neither seen nor expected. The relative energy error of the stress field is $\eta = 0.20$, too large for comfort. (Note that if we had chosen to use axisymmetric annular plate elements, visual inspection of computed stress fields would not provide such clear indications of an overly coarse mesh.)

We conclude that results are reasonable for the initial coarse-mesh analysis. Another analysis is required, with greater mesh refinement near $x = y = 0$. Span L could be reduced, say to 1200 mm or 1000 mm, with little loss of accuracy. Theory says that bending moments and flexural stresses are infinite at a concentrated load; accordingly, we should determine how P is actually applied. Perhaps it should be represented by a distributed load of large intensity acting on a small area.

Note that the foundation is assumed to pull down on the plate where it deflects upward. A plate that lifts off its foundation demands a nonlinear analysis because the zone

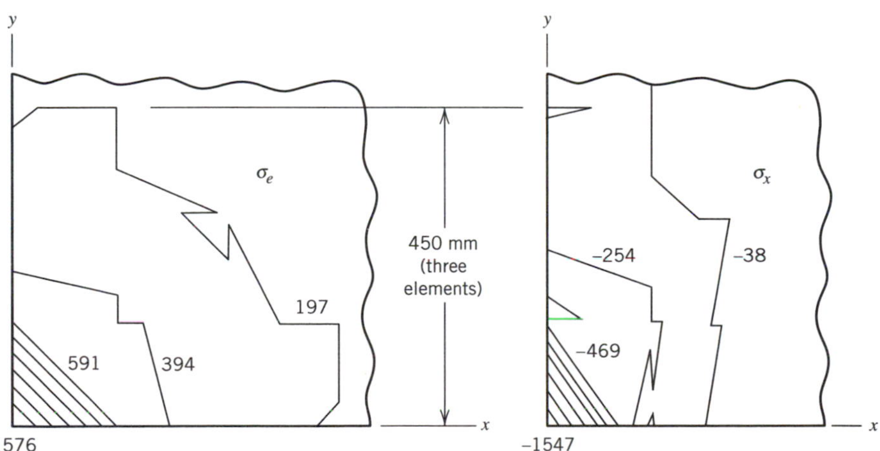

Fig. 7.3-3. Contours of stresses σ_e and σ_x on the top surface of the plate near $x = y = 0$ in Fig. 7.3-1b. Stress units are Pa.

of lift-off is unknown at the outset. In the foregoing problem, lift-off might make little difference, judging by the relatively small magnitude of upward deflection.

7.4 SHELLS AND SHELL THEORY

The geometry of a shell is defined by its thickness and its midsurface, which is a curved surface in space. Load is carried by a combination of membrane action and bending action. A thin shell can be very strong if membrane action dominates, in the same way that a wire can carry great load in tension but only small load in bending. A wire must have a different shape for every different distribution of lateral load if there is to be no bending; in contrast, a shell of a given shape can carry a variety of distributed loadings by membrane action alone.

However, no shell is completely free of bending stresses. They appear at or near point loads, line loads, reinforcements, junctures, changes of curvature, and supports. In short, *any* concentration of load or geometric discontinuity can be expected to produce bending stresses, often much larger than membrane stresses, but usually quite localized in a "boundary layer" near the load or discontinuity. Figure 7.4-1 shows examples of loads and geometries that produce bending. Axial force G must be transferred through the structure to the support at the left. Consequently the simple support around the base AA applies axially directed line load, which has a shell-normal component that causes bending. Similarly, around BB, the cylindrical and conical parts exert shell-normal load components on one another. Shell-normal load is also transferred across FF because the cylindrical and spherical shells try to expand different amounts under internal pressure (unless they are suitably matched, say, by appropriate choice of thicknesses). Line load EE is obviously shell-normal, as is the restraint provided by reinforcing ring DD. Internal forces and moments are shown in more detail in Fig. 7.4-2, where their action–reaction nature is displayed. The dimensions of V_0 are [force/length]; the dimensions of M_0 are [force·length/length]. *All* lettered regions in Fig. 7.4-1 are regions where bending is important. Another example appears in Fig. 5.2-5. The foregoing examples are axisymmetric, but similar remarks apply also to shells of general shape.

Flexural stress and bending moment in a shell are related in the same way as for a plate (see Eq. 7.1-3 and below). Thus, for the cylindrical shell in Fig. 7.4-2b, meridional flexural stress σ_x has magnitude $6M_0/t^2$ on the shell surfaces at $x = 0$, compressive on the outside of the shell for the direction of M_0 shown. Circumferential stress $\sigma_\theta = \nu\sigma_x$ is also present. If M_0 were zero, meridional flexural stress $6M/t^2$ would be present because $M =$

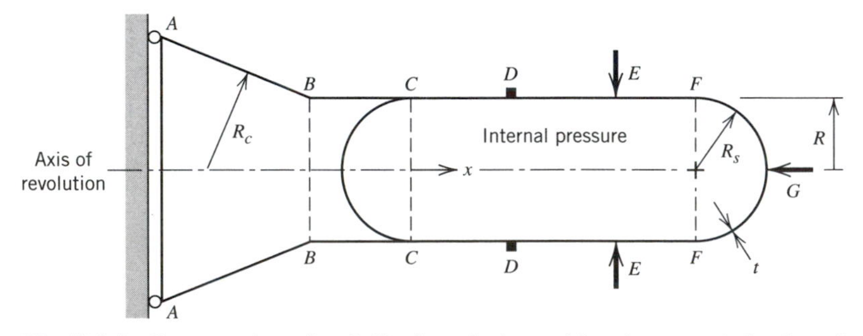

Fig. 7.4-1. Cross section of a shell of revolution, with axisymmetric loads and supports. The construction consists of a cylindrical vessel with hemispherical end caps, supported by a cylindrical shell $BBCC$ and a conical shell $AABB$.

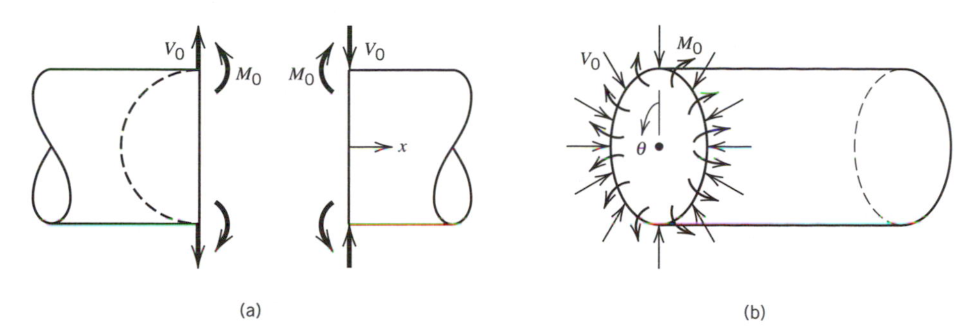

Fig. 7.4-2. (a) Forces and moments associated with bending at a discontinuity such as *CC* in Fig. 7.4-1. (b) Isometric view of the cylinder at *CC*.

$M(x)$ is created by V_0. This M would be zero at $x = 0$ and would reach its greatest magnitude for $x > 0$ but within the boundary layer. Again, circumferential stress $\sigma_\theta = \nu\sigma_x$ would be present. Membrane stresses, constant through the thickness, would be superposed on the flexural stresses. In Fig. 7.4-2b, both M_0 and V_0 contribute to circumferential membrane stress σ_θ at $x = 0$. The contribution is tensile due to M_0 but compressive due to V_0, because M_0 tends to enlarge the end while V_0 tends to shrink it (for the directions of M_0 and V_0 shown in Fig. 7.4-2b).

How large is the boundary layer in which bending may be important? A simple approximation can be obtained from the theory of a shell of revolution. If a radial line load or a bending moment is uniformly distributed around one end of a circular cylindrical shell (as in Fig. 5.2-5 or 7.4-2), analytical solutions for radial displacement and bending moment as functions of axial distance x from the end contain terms of the forms $e^{-\lambda x}\cos\lambda x$ and $e^{-\lambda x}\sin\lambda x$, where

$$\lambda = \left[\frac{3(1-\nu^2)}{R^2t^2}\right]^{1/4} \tag{7.4-1}$$

and R and t are, respectively, the radius and thickness of the shell [7.3]. For $\lambda x = 3^{0.25}$, $e^{-\lambda x}\cos\lambda x = 0.07$. That is, if we regard $e^{-\lambda x}\cos\lambda x$ as the dominant term in the solution, we conclude that the end displacement or bending moment declines to 7% of its peak value when $\lambda x = 3^{0.25}$. For this λx and with $\nu = 0$, Eq. 7.4-1 yields $x = \sqrt{Rt}$. Analysis of a spherical shell yields a similar result. As a convenient *approximate* guideline for *any* shell, useful for mesh layout in an initial FE analysis, one may estimate that $\sqrt{Rt}$ is the span of the boundary layer. As an example, approximate dimensions of an aluminum beverage can are $R = 33$ mm and $t = 0.10$ mm, hence $\sqrt{Rt} = 1.8$ mm, measured axially along the cylinder. In FE analysis, a coarse initial mesh might use *at least* two shell elements to span the boundary layer. Much larger elements can probably be used outside the boundary layer, where displacements and stresses have smaller gradients.

A shell has *two* principal radii of curvature at every point. Each is measured *normal to the shell* and each is the radius of a small arc drawn in the shell midsurface. The small arcs intersect at right angles. One radius is the largest, the other the smallest, of the radii of all possible arcs in the midsurface at the point where arcs intersect. In Fig. 7.4-1, the cylinder has principal radii R and infinity. Both principal radii are R_s in the sphere. In the cone, principal radius R_c is a function of x; the other principal radius is infinite. In the approximation that $\sqrt{Rt}$ is the span of the boundary layer, R should be taken as the radius of curvature, measured normal to the shell, of a small arc that lies in the shell midsurface

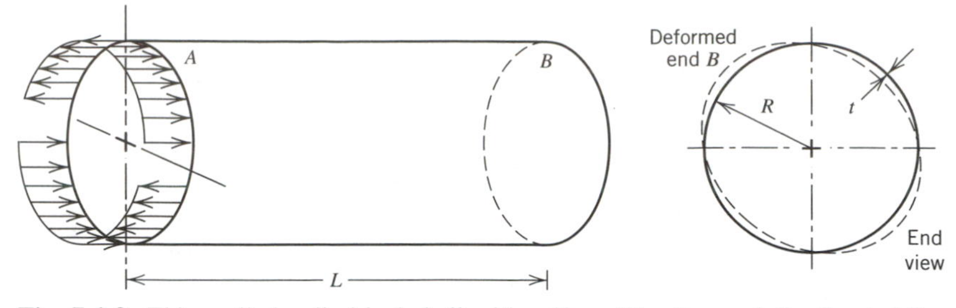

Fig. 7.4-3. Thin-walled cylindrical shell with self-equilibrating, axially directed line loads applied to one end.

and is tangent to a line load or a geometric discontinuity. In a shell of revolution this radius may be larger than the distance from the discontinuity to the axis of revolution.

Shell-*tangent* edge loads produce actions that are not confined to a boundary layer. In Fig. 7.4-3, axial loads act on end A of the unsupported cylindrical shell, but the largest displacements appear at end B, in apparent contradiction of Saint-Venant's principle. Another case in point is an I beam under torsional load with all displacements prohibited at one end: at a distance of several times the beam depth from the fixed end, the effects of end restraint are still important. One should remember that Saint-Venant's principle is applicable to relatively massive isotropic bodies. Thin-walled structures and highly anisotropic structures may behave quite differently.

Equations that describe the behavior of a circular cylindrical shell under axisymmetric loading have the same form as equations that describe the behavior of a beam on an elastic foundation. Accordingly, an understanding of either problem is almost immediately transferable to the other. Equations that describe the behavior of shells of other shapes are considerably more complicated, so that shell solutions for engineering purposes must usually be obtained by FE analysis.

7.5 FINITE ELEMENTS FOR SHELLS

Shell Elements. The most direct way to obtain a shell element is to combine a membrane element and a bending element. Thus a simple triangular shell element can be obtained by combining the plane stress triangle of Fig. 3.2-1 or Fig. 3.7-1c with the plate bending triangle of Fig. 7.2-4. The resulting element is flat and has five or six d.o.f. per node, depending on whether or not the shell-normal rotation θ_{zi} at node i is present in the plane stress element. Regardless of how a shell element is formulated, if θ_{zi} is present as a global d.o.f. but omitted from the element formulation, the global stiffness matrix will be singular if all elements that share node i happen to be flat and coplanar at node i. This difficulty can be avoided by including a simple "stabilization matrix," and one expects that commercial software will do so.

A quadrilateral shell element can be produced in similar fashion, by combining quadrilateral plane and plate elements. Remarks of the preceding paragraph again apply. However, a four-node "flat" quadrilateral is in general a warped element because its nodes are not all coplanar. A modest amount of warping can seriously degrade the performance of an element. Commercial software may allow only a very small amount of warping. Much greater warping is allowable if a simple "fix" is included in the element formulation [3.2, 7.4], but not all software does so.

Advantages of a flat element include simplicity of element formulation, simplicity in the description of element geometry, and the element's ability to represent rigid-body motion without strain. Disadvantages include the representation of a smoothly curved shell surface by flat or slightly warped facets, so that there are fold lines where elements meet. There is discretization error associated with the lack of coupling between membrane and bending actions within individual elements. Membrane-bending coupling arises *globally* because adjacent elements are not coplanar: membrane force in one element is transferred to a neighboring element with an element-normal component, which produces bending. Discretization error can of course be reduced by using smaller elements. Common advice is that a flat shell element should span no more than roughly 10° of the arc of the actual shell.

Curved elements based on shell theory avoid some shortcomings of flat elements but introduce other difficulties. More data are needed to describe the geometry of a curved element. Formulation is complicated, as it invokes a shell theory (of which there are many). Membrane and bending actions are coupled within the element, so it is harder to avoid membrane locking, that is, harder to avoid great overstiffness in bending because details of the element formulation cause membrane strains to appear in association with bending action, and membrane stiffness is far greater than bending stiffness if the shell is thin.

Isoparametric shell elements occupy a middle ground between flat elements and curved elements based on shell theory. One begins with a 3D *solid* element such as that shown in Fig. 7.5-1a. The element can model a shell if thickness t is small in comparison with other dimensions. However, such an element has the defects noted at the outset of Section 7.2 in connection with plate bending. Accordingly, we transform the element, reducing the number of nodes from 20 to 8, by expressing translational d.o.f. of the 20-node element in terms of translational and rotational d.o.f. at the midsurface of an 8-node element (Fig. 7.5-1b). Figure 7.5-1c shows these d.o.f. in local coordinates at a typical midsurface node b. Nodes a and c appear in the solid element but not in the shell element. A local z axis is directed through a, b, and c. Displacements at a and c in terms of d.o.f. at b are

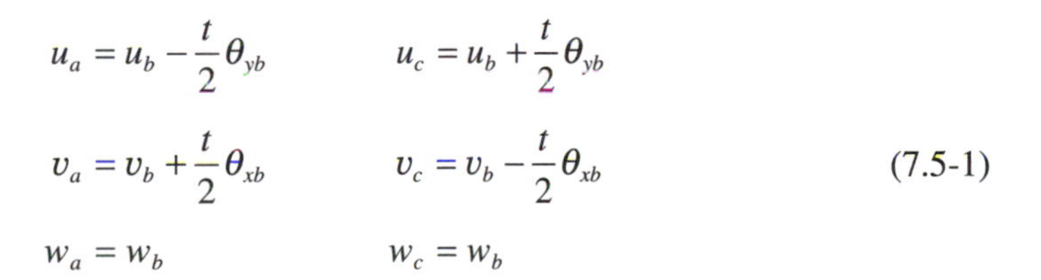

$$u_a = u_b - \frac{t}{2}\theta_{yb} \qquad u_c = u_b + \frac{t}{2}\theta_{yb}$$

$$v_a = v_b + \frac{t}{2}\theta_{xb} \qquad v_c = v_b - \frac{t}{2}\theta_{xb} \tag{7.5-1}$$

$$w_a = w_b \qquad w_c = w_b$$

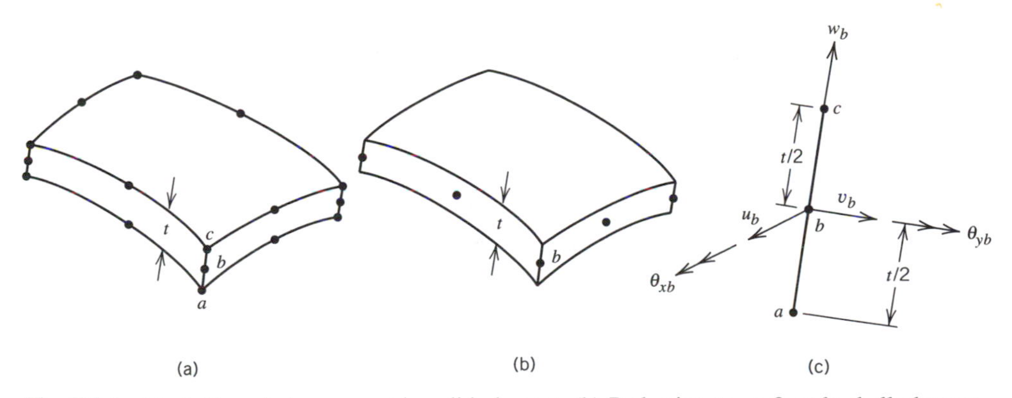

Fig. 7.5-1. (a) A 20-node isoparametric solid element. (b) Reduction to an 8-node shell element. (c) The d.o.f. at a typical node b.

With relations like these for all thickness-direction lines of nodes, shape functions of the 20-node solid are transformed so as to operate on the three translations and two rotations at each node of the 8-node element. Thus we obtain a Mindlin shell element. Thickness-direction normal stress is taken as zero, and the stress–strain relation **E** is written in a way that reflects this condition. The element stiffness matrix is integrated numerically. A reduced or selective integration scheme may be used to avoid transverse shear locking and membrane locking. A less palatable remedy for locking is to arbitrarily reduce the transverse shear and/or membrane stiffness, on the grounds that shell behavior will still be dominated by the comparatively small bending stiffness.

Test Cases. Shell elements are perhaps the most difficult elements to formulate, and shell behavior is often difficult to anticipate. One may wish to test shell elements in the software to be sure of their validity, sensitivity to shape distortion, and behavior in example problems for which results are known in advance. If the geometry is flat, a shell element should be able to model both plane stress and plate bending and should pass tests used to evaluate plane and plate elements. Next, one might use shell elements to solve plane arch problems. Many arch solutions are tabulated [1.5].

Commonly used test cases for shell elements are shown in Fig. 7.5-2. Details appear in [7.5, 7.6]. The shell roof is loaded by its own weight $q = 90$ per unit area. Straight edges are free and curved edges have "diaphragm" support, meaning that translational d.o.f. parallel to the plane containing the curve are prohibited but translational d.o.f. normal to this plane and all rotational d.o.f. are unrestrained. Ends of the pinched cylinder have diaphragm support. The load is $F = 1.0$. The hemisphere has a free edge and is restrained only against rigid-body motion. Loads are $F = 2.0$. The twisted strip is cantilevered and there are two load cases, $F_1 = 10^{-6}$ and $F_2 = 10^{-6}$. Commonly used numerical data, and accepted displacements of point A in the direction of the load (Δ_A), are as follows:

Problem	R or b	L	t	E	ν	Δ_A
Roof	25	50	0.25	$432(10)^6$	0.0	0.3024
Cylinder	300	600	3.00	$3(10)^6$	0.30	$0.1825(10)^{-4}$
Hemisphere	10	—	0.04	$68.25(10)^6$	0.30	0.0924
Strip	1.1	12	0.0032	$29(10)^6$	0.22	$5256(10)^{-6}$
						$1294(10)^{-6}$

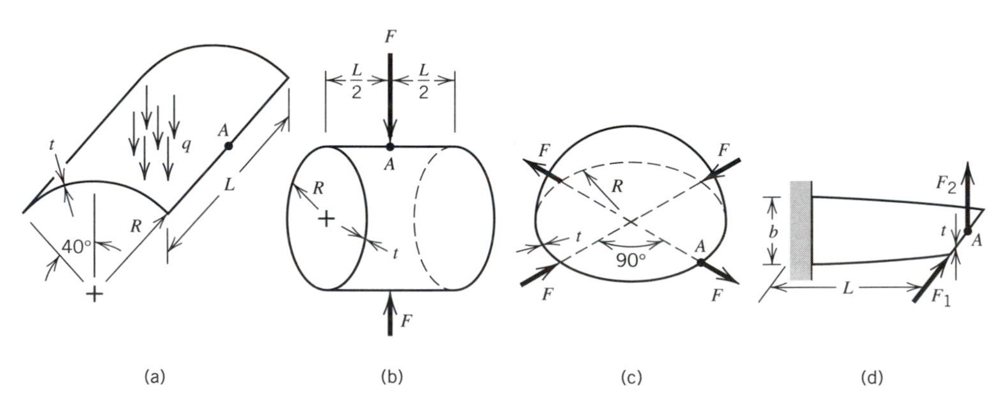

Fig. 7.5-2. (a) Cylindrical shell roof loaded by its own weight. (b) Pinched cylinder. (c) Hemisphere with four uniformly spaced radial forces in the equatorial plane. (d) Strip with a 90° twist from end to end.

Another useful test case is a thin-walled cylindrical tube, slit lengthwise along one generator, and twisted about its axis [7.5]. Displacements and stresses may be calculated by mechanics of materials theory.

The foregoing cases test different aspects of element behavior. Membrane action dominates in the shell roof, bending action dominates in the hemisphere, and both actions are active in the pinched cylinder. Sensitivity to warping is tested if quadrilateral elements are used to model the twisted strip. The slit cylindrical tube tests ability to model pure twisting moment. Experience has shown that it is possible for a particular element to do very well in most test cases but quite poorly in one or two others.

Shells of Revolution. In cross section, an element for a shell of revolution resembles a beam element. Like an element for a solid of revolution (Section 6.4), an element for a shell of revolution has nodal circles rather than nodal points. Typically there are two nodal circles per element. A shell of revolution element may be "flat" (conical) or curved. The simplest formulation resembles the 2D beam element discussed in Section 2.3, in that it uses a cubic lateral displacement field, a linear meridional displacement field, and each nodal circle has two translations (one radial, the other axial) and one rotation as d.o.f. However, the strain–displacement relation (Eq. 6.1-6 for axial symmetry) and integration (the second of Eqs. 6.1-9b) produce a stiffness matrix rather different from Eq. 2.3-9. In most software higher-order terms are added to the element displacement field by means of internal d.o.f.

Conical shell elements have advantages and disadvantages like those of other flat shell elements. Consider a special case (Fig. 7.5-3). The spherical shell is modeled by axisymmetric conical shell elements, which are "flat" in the sense that one principal radius of curvature is infinite. Internal pressure loading applies outwardly directed force to each nodal circle. Nodal moment loads are omitted for reasons discussed in Section 2.5. The FE response depends on the nature of the element formulation and stress recovery procedures, but a coarse mesh may display spurious bending moments, namely, the moment distribution seen in Fig. 7.5-3, which of course is not present in a pressurized sphere. If the structure were indeed a stack of conical frusta, bending action would be expected and would indeed be computed if each frustum in Fig. 7.5-3 were modeled by more than one conical shell element. Postprocessing devices that avoid spurious bending moments include evaluating moments only at Gauss point locations and calculating moments from expressions written for curved shell elements. Software can be expected to include one or more of these devices.

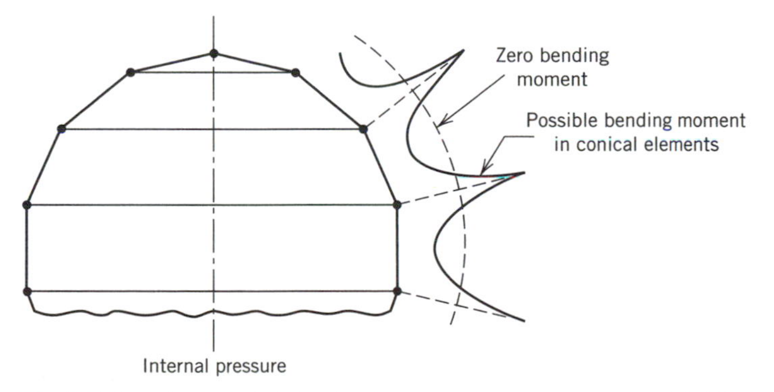

Fig. 7.5-3. Spherical shell loaded by internal pressure and modeled by a coarse mesh of conical frusta shell elements.

Like axisymmetric solids, axisymmetric shells under nonaxisymmetric loading can be analyzed by use of Fourier series. Thus the load is represented as the sum of its series components, an analysis is performed for each component, and results are superposed. This procedure is useful for wind loading on axisymmetric tanks and towers and could also be applied to the pinched cylinder and hemisphere problems in Fig. 7.5-2.

Remarks. A shell may be reinforced by stiffeners attached to one side of the shell. Then the shell midsurface and the stiffener axis are not coincident. Rigid offsets can be used to couple stiffener nodes to shell nodes, as described in Section 4.3.

Normal stress at a point on the surface of a shell is the sum of a membrane component σ_m and a bending component σ_b, for example,

$$\sigma = \sigma_m + \sigma_b \qquad \text{where} \qquad \sigma_m = \frac{N}{t} \qquad \text{and} \qquad \sigma_b = \pm\frac{6M}{t^2} \tag{7.5-2}$$

where N is a membrane force (dimensions [force/length]) and M is a bending moment (dimensions [force·length/length]). The expression $6M/t^2$ is the flexure formula Mc/I, applied to a unit width and thickness $2c = t$. Software may report σ, σ_m, σ_b, N, M, or any one of these at the user's option, most probably in local coordinates xy tangent to the shell midsurface. Software documentation must be consulted for details, such as the global orientation of local coordinates, and the algebraic sign in Eq. 7.5-2 (i.e., which surface of the shell is understood by the software as being the top surface).

7.6 AN AXISYMMETRIC SHELL APPLICATION

The structure we consider is a segment of a spherical shell, Fig. 7.6-1a. It is fixed to a rigid horizontal foundation at the equator and is loaded by a uniformly distributed vertical line load around the top, which is unrestrained. The state of stress is to be investigated.

Preliminary Analysis. Membrane forces exist throughout the structure. Bending moments also exist, but if this shell behaves like the water tank in Fig. 5.2-5, we expect that

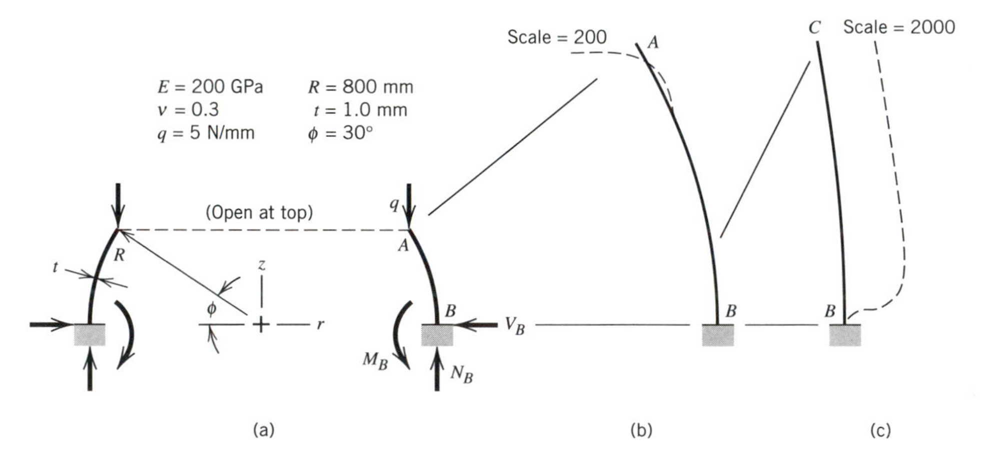

Fig. 7.6-1. (a) Geometry, loading, and properties of a shell of revolution. (b) Computed displacements along AB, multiplied by 200. (c) Computed displacements along lower quarter of AB, multiplied by 2000.

they will be localized near A and B. Meridional membrane force and membrane stress at B are easy to calculate. By summation of vertical forces,

$$2\pi RN_B = 2\pi(R \cos 30°)q \qquad \text{hence} \qquad N_B = 0.866q \qquad (7.6\text{-}1)$$

Hence N_B = 4.33 N/mm, and membrane stress N_B/t = 4.33 MPa (compressive) should be found on the shell midsurface at B. A simple equilibrium equation of membrane shell theory [2.1] tells us that if the support at B were a frictionless surface, the *circumferential* stress at B would also be 4.33 MPa but tensile. Accordingly, in our problem the equator tries to expand but is restrained by the support, which applies an inward shear force V_B distributed around the equator. But V_B also acts to rotate the bottom of the shell inward, and to prevent rotation the support applies distributed meridional moment M_B. Computed results should produce V_B and M_B in the directions shown in Fig. 7.6-1a. Around the top, load q has a midsurface-normal component that bends the shell inward. There is no support near A to oppose this action, so displacements of the shell should be large and inward at A. Also, bending moment at A should be zero, because the top is neither loaded by externally applied moment nor restrained against rotation.

FE Model and Analysis. The axisymmetric shell elements chosen are truncated cones with a nodal circle at each end. Each node has two translations and one rotation as d.o.f. Four additional d.o.f. are added internally for purposes of stiffness matrix formulation. According to the argument associated with Eq. 7.4-1, boundary layers near A and B each span roughly $\sqrt{800(1)}$ = 28 mm, or about 2.0° of arc. This distance should be spanned by two elements at least, even in an initial analysis. For our analysis, the software is instructed to mesh the arc from A to B automatically, using element lengths of 4 mm at A and B and 20 mm at the middle of the arc. The result is 65 nodes and 64 elements with graduated element lengths along the arc. All d.o.f. at node B are restrained. All other d.o.f. are unrestrained.

Critique of FE Results. Computed displacements are found to have the qualitative behavior expected (Fig. 7.6-1b,c), and reactions N_B, V_B, and M_B are found to have the directions expected. Computed meridional membrane stress at B is –4.33 MPa, exactly as predicted. Meridional bending moment M near the top and bottom is shown in Fig. 7.6-2, with the sign convention that positive M creates tensile flexural stress on the inside of the shell. As expected, M = 0 at the very top. The largest flexural stress is $\pm 6M/t^2$ = $\pm 6(17.7)/1^2 = \pm 106$ MPa. Around the top of the shell, the downward force $2\pi(R \cos 30°)q$ is resisted by the upward force $2\pi(R \cos 30°)(N_A \cos 30°)$ from the membrane force N_A at A. Hence $N_A = q/\cos 30°$ = 5.77 N/mm in compression, and the associated membrane stress is N_A/t = –5.77 MPa. Combining membrane and bending components, we conclude that the meridional stress of largest magnitude in the shell is about 106 + 6 = 112 MPa compressive, on the inside about 17 mm from the top. The meridional flexural stress at B, $\pm 6M_B/t^2$, is only ±10 MPa. At the top of the shell, the large radial displacement creates a large circumferential membrane strain ε_θ (see Eq. 6.1-6). The associated circumferential membrane stress is $\sigma_\theta = -175$ MPa. So large a compressive stress suggests that buckling is possible, for which a separate analysis would be required.

The software does not provide a relative energy error based on interelement stress discontinuities for this type of element. If it were available, it might not be able to alert us to trouble if we were to use elements spanning so large an arc that the localized behavior near A and B would not be captured. However, judging by Fig. 7.6-2, the mesh used is adequate. What we see is a variation of the form $e^{-\lambda x}\sin \lambda x$, as noted above Eq. 7.4-1.

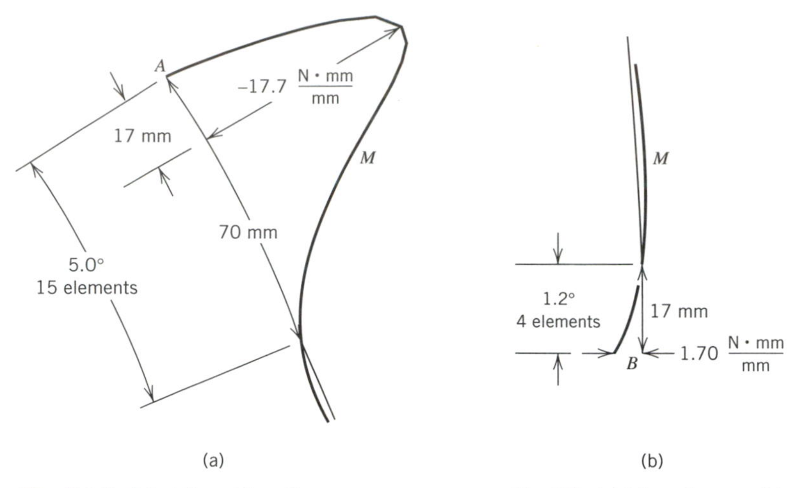

Fig. 7.6-2. Meridional bending moments computed by FE. (a) Near the top. (b) Near the base.

Displacements and stresses in a shell do not vary with steeper gradients than this. Since this variation is portrayed with reasonable smoothness in Fig. 7.6-2, it does not appear that mesh refinement is necessary.

Remarks. It happens that there are tabulated formulas applicable to this problem, although they are tedious to apply by hand [1.5, 2.1]. The formulas give values of V_B, M_B, and largest M less than 1% different from the values computed by FE.

Imagine that a small circular hole must be drilled in the shell midway between A and B. Will it be necessary to abandon axisymmetric analysis, use general shell elements instead, and greatly refine the mesh around the hole? No; none of these. At the location of the hole, only membrane stresses are present. Elementary membrane shell theory shows that meridional and circumferential membrane stresses are, respectively, –4.64 MPa and 4.64 MPa. Standard formulas for stress concentration [1.5] yield a peak stress of 3(4.64) – (–4.64) = 18.6 MPa at the hole. FE analysis is not needed.

Plots of bending moment, such as Fig. 7.6-2, show no interelement discontinuity regardless of the coarseness of the mesh. Thus, when using shell of revolution elements, we must do without this helpful warning of a need for mesh refinement. The same is occasionally true of plate bending: in Fig. 7.3-1b, if a uniform line load were applied along the left edge of the plate, contours of flexural stress on the upper surface would be lines parallel to the y axis, and interelement discontinuity in the x direction would not be apparent.

7.7 A GENERAL SHELL APPLICATION

A thin-walled tube of circular cross section is fixed at one end and is loaded by transverse force P at the other end (Fig. 7.7-1a). In order to distribute load around the end of the tube at $z = L$, a thin flat disk of radius R is attached to the end. The tube is slit open lengthwise along one side. Deflection at the loaded end and significant stresses are required, using $P = 1.0$ N for the load.

Preliminary Analysis. At first glance the structure appears to be nothing more than a cantilever beam. Thus for comparison with FE results, values of tip deflection and root

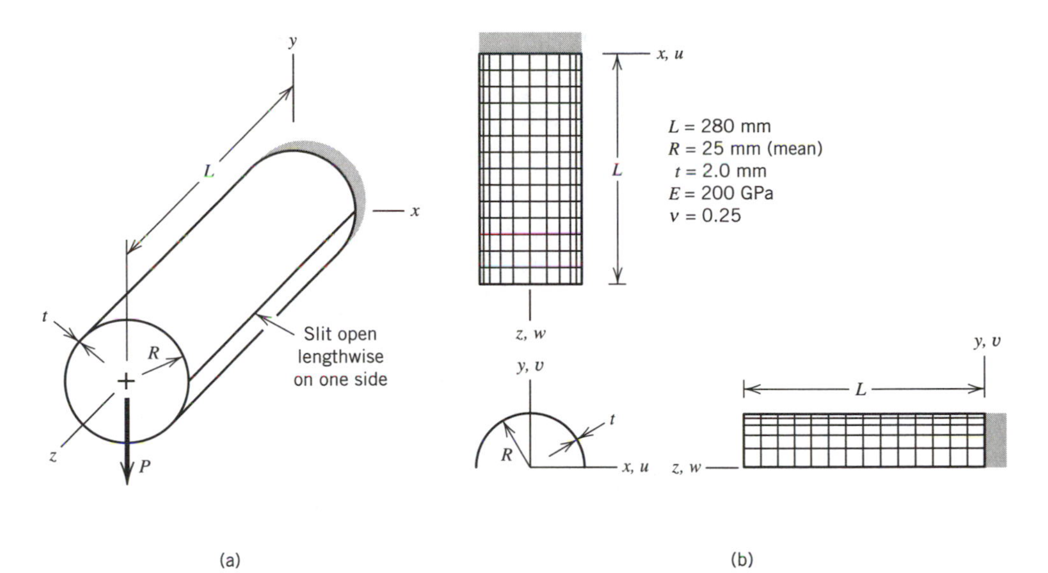

Fig. 7.7-1. (a) Cantilever beam in the form of a slit cylindrical tube. (b) FE mesh on the upper half of the tube (sketch not to scale).

flexural stress should suffice. With $I = \pi R^3 t$ the centroidal moment of inertia of the annular cross-sectional area, elementary beam deflection and flexure formulas yield

$$v = -\frac{PL^3}{3EI} = -373(10)^{-6}\,\text{mm} \qquad \text{and} \qquad \sigma_z = \frac{PLR}{I} = 0.0713\,\text{MPa} \qquad (7.7\text{-}1)$$

An analyst who next obtains FE results is in for a surprise. Equations 7.7-1 do not tell the whole story or even the major part of it because the slit strongly influences behavior.

Further Preliminary Analysis. Because of the slit, the tube will twist as well as bend. Cross sections will rotate because there is a moment about the axis of shear centers, which according to theory is the line defined by $x = -2R$, $y = 0$. Looking along the z axis toward the origin, the $z = L$ end of the tube will rotate clockwise. Material just above and just below the slit tends to move in $-z$ and $+z$ directions, respectively. This "warping" tendency is easily demonstrated by twisting a rolled-up sheet of paper. Warping is prevented at the support by axial stresses, compressive just above the slit and tensile just below, which are superposed on flexural stresses σ_z. This discussion is qualitative, but with Eq. 7.7-1 it may be sufficient for checking purposes. Readers who seek further pre-FE analysis may proceed as follows.

In the present problem an analytical solution is reasonably simple, although an explanation of it is beyond the scope of our treatment. Results from beam theory, Eqs. 7.7-1, are augmented by results from restrained-warping torsion theory of thin-walled cross sections [2.1]. The theory presumes that cross sections do not warp at $z = 0$ and warp freely at $z = L$. Results are as follows. Shear centers of each cross section lie on the line defined by $x = -2R$, $y = 0$. Therefore the torque of the load about the shear center is $T = -2RP$, represented by a vector in the negative z direction. Properties of the cross section are $J_R = 2\pi R t^3/3$, $J_\omega = 2\pi(\pi^2 - 6)R^5 t/3$, and $k^2 = GJ_R/EJ_\omega$, where G and E are, respectively, the shear and elastic moduli. In the notation of [1.5], J_R is called K, J_ω is called C_w (the "warping constant"), and k is called β. Rotation of the loaded end is

$$\theta = -\frac{TL}{GJ_R}\left(1 - \frac{\tanh kL}{kL}\right) = -418(10)^{-6} P(0.0268) = -11.2(10)^{-6} \text{ rad} \tag{7.7-2}$$

in which TL/GJ_R is what the rotation would be if warping were unrestrained by the support. Rotation θ is about the shear center, so that at $x = y = 0$ on the loaded end a displacement $v = 2R\theta = -560(10)^{-6}$ mm is algebraically added to the displacement v of Eq. 7.7-1, for a total of $v = -933(10)^{-6}$ mm. Axial stress due to restraint of warping at $x = R$, $y = z = 0$, is

$$\sigma_z = \pm E(\pi R^2)\frac{Tk}{GJ_R}\tanh kL = \pm 0.169 \text{ MPa} \tag{7.7-3}$$

negative above the slit and positive below. This stress is more than double the flexural stress on top of the tube predicted by Eq. 7.7-1.

FE Model and Analysis. For an initial FE analysis we choose a uniform mesh because of its simplicity. The mesh is coarse. The problem displays antisymmetry with respect to the plane $y = 0$, so only half the tube need be modeled. Figure 7.7-1b shows a 10 by 14 mesh of four-node shell elements, each formed by combining the quadrilateral membrane element of Section 3.6 with four overlapping discrete Kirchhoff plate-bending triangular elements. Combination is carried out automatically by the software. All d.o.f. at nodes on the plane $z = 0$ are restrained. Along the line of nodes at $x = -R$, $y = 0$ we set $u = w = \theta_y = 0$ and leave v, θ_x, and θ_z unrestrained (see Fig. 4.12-2). The load applied to the FE model is 0.5 N. The half-circular area at $z = L$ is filled with ten triangular elements that share a common vertex at the load point (Fig. 7.7-2a). These elements maintain the circular shape of the end but do very little to restrain warping of the tube cross section.

Critique of FE Results. The expected directions of axial warping displacement were visible in an animation of the displaced shape. Vertical tip displacements at $x = \pm R$ are shown in Fig. 7.7-2a. The average of these two values is $967(10)^{-6}$ mm, not far from the predicted magnitude of $933(10)^{-6}$ mm. The difference of the two values, divided by $2R$, yields the rotation $\theta = -11.4(10)^{-6}$ rad (clockwise), which agrees well with Eq. 7.7-2. Axial membrane stresses are shown in Fig. 7.7-2b. The largest magnitude of stress agrees

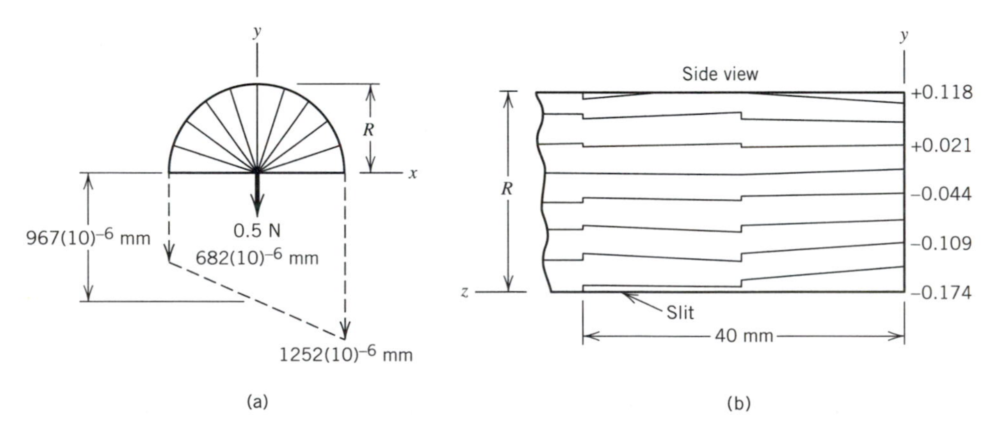

Fig. 7.7-2. (a) Computed vertical deflections at the loaded end. (b) Side view, showing contours of axial stress σ_z on the outside surface near the supported end. Stress units are MPa.

well with Eq. 7.7-3. The stress at $y = R$ is rather different from the value given by Eq. 7.7-1. Note that stresses produced by restraint of warping, Fig. 7.7-2b, are not localized; instead, they decay slowly with distance from the supported end, in seeming contradiction of Saint-Venant's principle. Stress contours show fair interelement continuity. The relative energy error of the stress field is $\eta = 0.11$. Another analysis with a finer mesh should be undertaken.

Remarks. The foregoing critique is not very demanding if the analyst can anticipate the results, as in the "further preliminary analysis" subsection. Otherwise FE results may be puzzling and even alarming. This example shows the value of a good understanding of how structures respond to load. Such knowledge suggests that a thorough critique should include examination of shear stresses, which may also have significant magnitudes.

We have not mentioned flexural stresses σ_b of Eq. 7.5-2. Their contribution may be small, but the mesh used is too coarse to model them properly. According to the argument associated with Eq. 7.4-1, the width of the boundary layer is roughly $\sqrt{Rt} \approx 7$ mm, about one-third the axial span of the elements used. Also, if stresses σ_b were to be almost independent of the circumferential coordinate, their contours would be circumferential rings whether the mesh is coarse or fine. In such a situation a plot of unaveraged stress contours would not convey information that the mesh is too coarse.

Through what point should load P act if the end is not to rotate? In other words, where is the shear center? Because the amount of twist θ is directly proportional to applied torque, θ is linearly related to P and to the x coordinate of the point through which load P passes. Thus, with x_P the x coordinate of P,

$$\theta = (a_1 + a_2 x_P)P \tag{7.7-4}$$

where a_1 and a_2 are constants to be determined. One condition that can be used to determine them is $\theta = -11.4(10)^{-6}$ rad when $x_P = 0$, from the foregoing FE analysis. Another condition is supplied by analyzing the same FE model under the alternative loading shown in Fig. 7.7-3a, which is statically equivalent to a single downward force of 0.5 N at the theoretical shear center, $x_P = -2R = -50$ mm. This loading yields $\theta = 1.1(10)^{-6}$ rad. Solving for a_1 and a_2, then setting $\theta = 0$ in Eq. 7.7-4, we obtain $x_P = -a_1/a_2 = -45.6$ mm as the x coordinate of the shear center of the FE model. Why does this value differ from

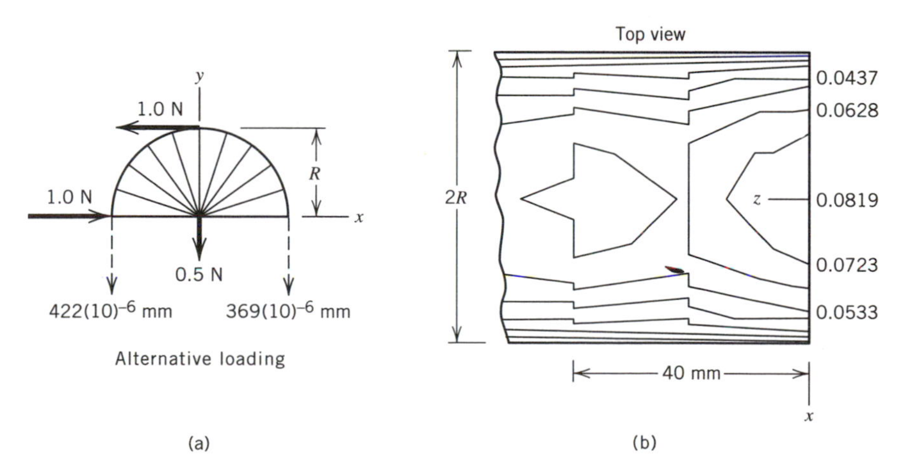

Fig. 7.7-3. (a) Computed vertical deflections at the loaded end, for the alternative loading shown. (b) Top view, showing contours of axial stress σ_z on the outside surface near the supported end, for the alternative loading. Stress units are MPa.

the theoretical value? The theoretical value is based on beam theory that omits transverse shear deformation; FE analysis includes it except at the fixed end, where it is prevented. Also, the theoretical value is unrelated to the support conditions. (Figure 7.7-3b shows axial stresses produced by the alternative loading. Because the twist is not quite zero, contours are not quite symmetric about the plane $x = 0$, but the largest stress is on top and is in reasonable agreement with Eq. 7.7-1.)

ANALYTICAL PROBLEMS

7.1 Determine coefficients in the matrix that operates on the curvature vector $[\partial^2 w/\partial x^2 \quad \partial^2 w/\partial y^2 \quad 2\partial^2 w/\partial x\,\partial y]^T$ to produce the vector of moments $[M_x \quad M_y \quad M_{xy}]^T$. Express these coefficients in terms of ν and the flexural rigidity $D = Et^3/[12(1-\nu^2)]$.

7.2 Sketch the deflected shape and describe the state of stress if the rectangular plate in Fig. 7.2-1b has each of the following lateral deflections, where the c_i are constants:
(a) $w = c_1(x^2 + y^2)$
(b) $w = c_2(x^2 - y^2)$
(c) $w = c_3 xy$

7.3 Derive Eq. 7.1-5. *Suggestions*: Let $w = 4w_c x(L - x)/L^2$, assume that tensile force is independent of x, and note that for $w_c << L$ the change in length is the integral of $0.5(dw/dx)^2 dx$ from $x = 0$ to $x = L$.

7.4 (a) For each of the following displacement modes, sketch the deformed shape of a rectangular four-node plate element. The c_i are constants. Use expressions for u and v shown in Fig. 7.1-3.
(1) $w = 0$, $\theta_x = c_1 x$, $\theta_y = c_1 y$
(2) $w = c_2 xy$, $\theta_x = 0$, $\theta_y = 0$
(3) $w = 0$, $\theta_x = 0$, $\theta_y = c_3 xy$
(4) $w = 0$, $\theta_x = c_4 xy$, $\theta_y = 0$
(b) Verify that these modes are zero-energy modes if all terms of **k** are integrated by one-point Gauss quadrature.
(c) Some other displacement modes of the element are *not* zero-energy modes under one-point Gauss quadrature. For each of these, write an expression like one of the four listed under part (a).
(d) Which modes cited in part (a) cease to be zero-energy modes if terms associated with transverse shear strain are integrated by one-point quadrature while the remaining terms are integrated by four-point quadrature?

7.5 Let an eight-node Mindlin plate element be square and have the deformation mode $w = c(3\xi^2\eta^2 - \xi^2 - \eta^2)$, $\theta_x = \theta_y = 0$, where c is a constant, $\xi = x/a$ and $\eta = y/b$ (see sketch). Sketch the deformed shape, and show that strains are zero at the Gauss points of a 2 by 2 integration rule.

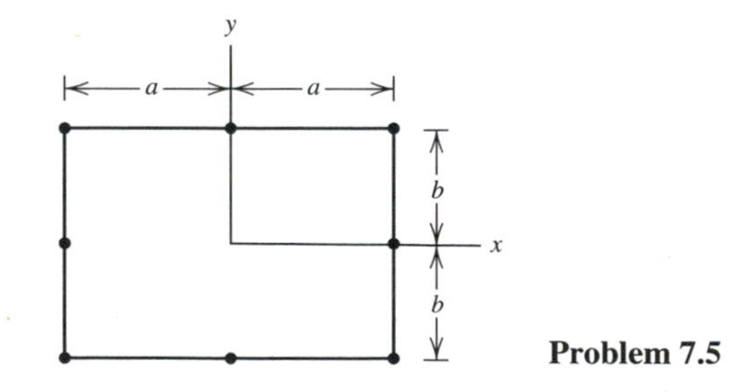

Problem 7.5

7.6 Consider one of the constraints used in formulating a discrete Kirchhoff plate element, as follows. In Fig. 7.2-4, let side 1-2 have length L. Assume that θ_y along this side depends on θ_{y1} and θ_{y2} and is linear in x. Also assume that w is cubic in x along this side and depends on nodal values of w and its first derivative at nodes 1 and 2. If $\gamma_{zx} = 0$ at the midpoint of the side, what equation relates the six nodal d.o.f. mentioned here?

7.7 For the strip under twisting load in Fig. 7.2-6, what corner deflection w_3 is predicted by mechanics of materials theory for the twisting of a strip having a narrow rectangular cross section? Use data provided at the end of Section 7.2.

7.8 It can be shown that the radial (outward) displacement produced by moment and shear at the base of the thin-walled cylindrical tank in Fig. 5.2-5 has the form $w = Ae^{-\lambda x}\cos \lambda x + Be^{-\lambda x}\sin \lambda x$, where A and B are constants and λ is given by Eq. 7.4-1. At $x = 0$, this "bending" w nullifies the "membrane" w produced by the fluid contained. Also $dw/dx = 0$ at $x = 0$ (neglecting a small membrane rotation term). Evaluate A and B. Hence, obtain the axially directed flexural stress $\sigma_b = \pm[Et^3/\{12(1 - \nu^2)\}](d^2w/dx^2)$, and determine an expression for the ratio σ_b/σ_m, where $\sigma_m = pR/t$ is the circumferential membrane stress that would exist at $x = 0$ in the absence of bending. Numerically evaluate σ_b/σ_m for the case $R = 16$ m, $t = 0.02$ m, $\nu = 0.3$, and water to a depth $h = 10$ m.

7.9 (a) Consider internal pressure loading on the shell shown in Fig. 7.4-1. (Omit the other loads shown.) If no bending stresses are to arise at or near juncture FF, what equation relates the elastic moduli, thicknesses, and Poisson ratios in the cylindrical and hemispherical portions?
(b) In Fig. 7.4-1, increase radius R_s of the spherical end cap FGF so that it becomes "flatter", that is, so that $R_s > R$ and the cap subtends an arc less than 180°. If E and t are the same for cylinder and cap, and only membrane stresses are considered, how should R and R_s be related so that displacements normal to the axis of revolution will be the same when internal pressure is applied? (Other loads shown in Fig. 7.4-1 are absent.)
(c) In part (b), will bending action nevertheless be present at or near FF? Explain.
(d) Express the shell-normal component of line load around AA in terms of the force at G, the radius of circle AA, and the inclination of AB with respect to the axis of revolution.

7.10 The sketch represents the cross section of a Dewar flask, that is, inner and outer axisymmetric shells connected at one end and with vacuum in the small space between.
(a) In terms of atmospheric pressure p and the dimensions shown, what are the membrane stresses in the cylindrical and hemispherical portions?
(b) Identify the locations where bending stresses will arise.

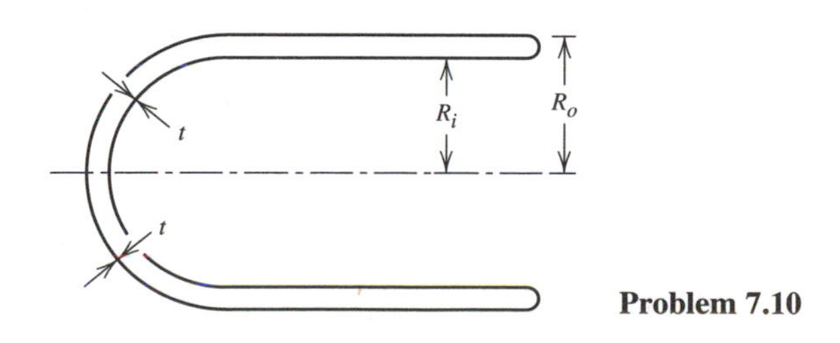

Problem 7.10

COMPUTATIONAL PROBLEMS

When specific instructions are not stated in the following problems, compute significant values of moment, stress, or displacement, as appropriate. Exploit symmetry if possible. Choose convenient numbers and consistent units for material properties, dimensions, and loads. When mesh refinement is used, estimate the maximum percentage error of FE results in the finest mesh. Unless directed otherwise, assume that thicknesses are uniform and the material is isotropic.

A FE analysis should be preceded by an alternative analysis, probably based on statics and mechanics of materials, and oversimplified if necessary. If these results and FE results have substantial disagreement we are warned of trouble somewhere.

7.11 (a) Patch-test the plate elements in the software you use. You may wish to adapt Fig. 7.2-2 to cases of pure bending and pure twist.

(b) Do plate elements in your software take transverse shear deformation into account? Find out by computational testing of a cantilever. Plan carefully the loading, support conditions, number of elements required, material properties, and how numerical results will be interpreted in order to answer the question.

7.12 The sketch is a plan view of a cantilever plate of thickness t. Apply equal forces in the z direction to nodes 2 and 3. Compare computed moments and tip deflection with cantilever beam (or plate) theory. Use one quadrilateral element, then two triangular elements. Also use elements having side nodes, if available. Does accuracy decline as thickness t becomes extremely small, or as L/b becomes large? Do the element formulations appear to take transverse shear deformation into account?

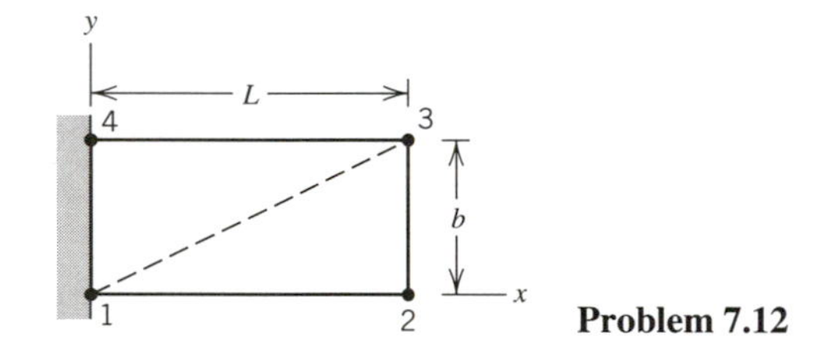

Problem 7.12

7.13 Analyze the twisting problems in Fig. 7.2-6c and Fig. 7.2-6d using each of the loadings shown. Otherwise, follow the instructions of Problem 7.12.

7.14 (a) Construct plots of w, M_x, and M_y along the x axis in Fig. 7.2-6b. Use "soft" simple supports.

(b) Repeat part (a), but use "hard" simple supports.

7.15 For the square plate of Fig. 7.2-6a, compute the center deflection, center bending moments, and bending moment of largest magnitude at the middle of one edge. Use four uniform meshes, starting with the coarsest possible mesh on one quadrant, then increasing the number of elements by a factor of 4 in each successive mesh refinement. Consider the following cases.

(a) Simply supported edges, uniformly distributed load.

(b) Simply supported edges, concentrated center force.

(c) Clamped edges, uniformly distributed load.

(d) Clamped edges, concentrated center force.

7.16 Explore the effect of thickness t in Problem 7.15 as follows.

(a) Decrease t in successive analyses by a factor of (say) 100 each time. How many of the computed results can be trusted?

(b) Increase t in successive analyses. Do results from the finer meshes continue to agree with theory? Why or why not? How large do you think t can be, as a fraction of side length L, if plate elements are used for analysis?

(c) In part (a), choose a case for which t is small but results are reliable. Replace the plate elements by 3D solid elements, and repeat the calculation.

7.17 Repeat Problem 7.15 with the addition of four reinforcing beams to the lower surface of the plate. Let the beams each have length $L/2$, meet at right angles at the center of the plate, and be oriented parallel to x and y axes. A possible choice for beam cross sections is width $b = 2t$ and depth $h = 5t$, where t is the plate thickness.

7.18 The plate shown has two supported edges, which are clamped and include angle 2ϕ, and two free edges. A uniform line load in the z direction is applied along span $2c$ of the longer free edge.

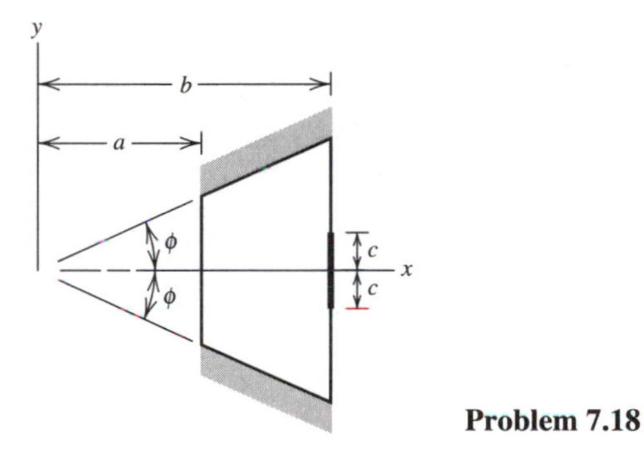

Problem 7.18

7.19 Problems 7.15 and 7.18 can be repeated, using thermal load instead of mechanical load. A simple thermal load is $\Delta T = (2z/t)T_o$, where T_o is the surface temperature relative to the initial stress-free temperature.

7.20 (a) Alter the plate problem discussed in Section 7.3 by letting the load act on an edge. For example, apply P at $x = y = 0$ and let the plate extend to infinity in only the positive y direction.

(b) Similarly, apply P to one corner. For example, apply P at $x = y = 0$ and let the plate extend to infinity in the first quadrant only.

7.21 Add reinforcing beams of cross section b by h to edges of the plates in Problem 7.20. Attach the beams to the lower surfaces of the plates. One might prescribe the ratio b/h, then seek values of b and h that make the largest stresses in the beams equal to the largest stresses in the plates.

7.22 Two circular plates, one having a central hole, are shown in cross section, with some possible cases of axisymmetric mechanical load. The temperature field $\Delta T = (2z/t)T_o$, where T_o is the surface temperature relative to the initial stress-free temperature, is another possible load case. The outer edge $r = a$ may be simply supported or clamped. Additional problems result if an elastic foundation is added, with support at $r = a$ either retained or removed. Most of these problems can serve as test cases; many analytical solutions are tabulated [1.5, 7.3].

(a) Perform the analysis using shell of revolution elements, flattened into annular disks.

(b) Perform the analysis using general plate elements.

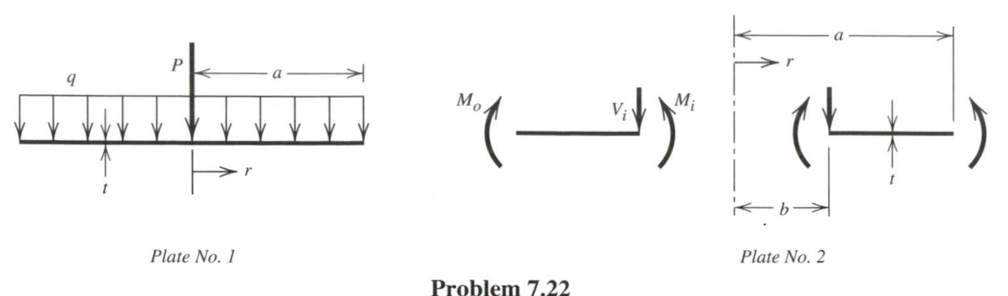

Plate No. 1 Plate No. 2

Problem 7.22

7.23 Circular plates (e.g., those in Problem 7.22) may be stiffened by the addition of n radially oriented stiffening beams, each of cross section b by h, either straddling the plate midsurface or attached to either lateral surface, and separated by equal angles in the circumferential direction.

(a) Analyze with "smeared" stiffeners, that is, with individual stiffeners replaced by an increase in plate thickness to a value that accounts for beam and plate stiffnesses together. (The revised t depends on both n and radial position.)

(b) Analyze with stiffeners maintained as discrete elements.

7.24 Remove the concentrated force loading from the plate discussed in Section 7.3. Instead, apply loading over a circular region of radius R as follows.

(a) A uniformly distributed lateral pressure p.

(b) A uniform line load q around the circle of radius R.

(c) The temperature gradient $\Delta T = (2z/t)T_o$ over the region within the circle, where T_o is the surface temperature relative to the initial stress-free temperature.

7.25 A thin shell of revolution in the shape of a truncated cone is simply supported around its base, as shown. A uniformly distributed line load q is applied to the top. *Note*: If the cone is rather flat, it is called a "Belleville spring" and has a nonlinear load versus deflection relation.

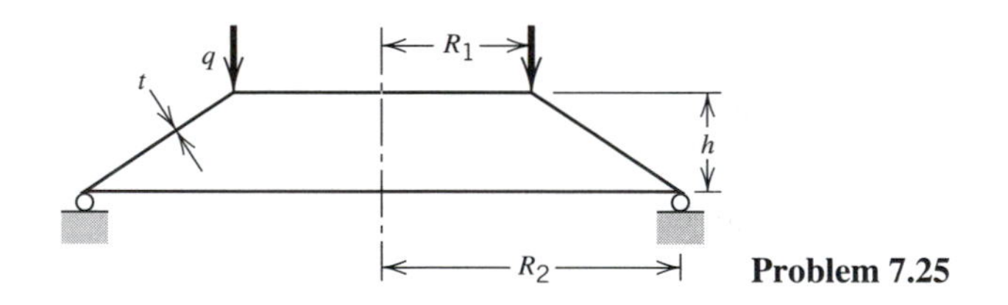

Problem 7.25

7.26 Possible modifications of Problem 7.25 include the following:

(a) Add a reinforcing ring (like that in Problem 7.29) to the top and/or the bottom.

(b) Uniformly heat the upper half or the lower half.

(c) Apply load q around only half of the top circumference.

(d) Uniformly heat only half of the circumference.

7.27 (a) Analyze the water tank depicted in Fig. 5.2-5.

(b) As an alternative load on the tank, let the temperature vary linearly through the wall thickness t.

7.28 (a) Analyze the Dewar flask described in Problem 7.10.

(b) As an alternative load on the Dewar flask, uniformly heat the inner shell only. Assume that temperature varies linearly around the toroidal knuckle that connects inner and outer shells.

7.29 Let a thin-walled shell of revolution consist of a cylindrical portion of radius R_c capped by a closure whose meridian is a circular arc of radius R_s, as shown. The juncture may be reinforced by a ring of rectangular cross section. Let the cylinder and the end cap intersect the ring (if present) at midwidth, a distance $b/2$ from each side. The base of the cylindrical shell (not shown) may be considered fixed. Internal pressure p is applied. Some special cases of possible interest are as follows:

(a) $a_1 = a_2 = 0$, $R_s = R_c$ (hemispherical end cap).
(b) $a_1 = \sqrt{3}\, R_c$, $a_2 = 0$, $R_s = 2R_c$ (end cap subtends a 60° angle).
(c) $a_1 = -\sqrt{3}\, R_c$, $a_2 = 0$, $R_s = 2R_c$ (like part (b), but with the end cap "dished in").

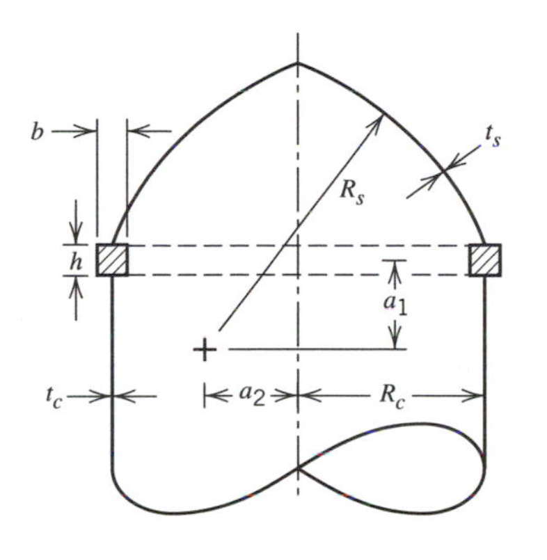

Problem 7.29

7.30 Repeat Problem 7.29, but load the shell by its own weight. Let the axis of the shell be vertical.

7.31 Repeat Problem 7.29, but load the shell by its own weight. Let the axis of the shell be horizontal.

7.32 Consider a uniform circular ring under diametral loads, as in Fig. 2.7-2. Model the ring with shell elements, using n uniform elements around a quadrant and one element to span the dimension normal to the plane of the ring. Questions such as the following may be studied:

(a) If n = 1, 2, 4, . . . , how quickly do displacements converge toward correct results?
(b) Does it matter whether the arcs of individual elements are straight or curved?
(c) Is there evidence that the bending moment has the possible behavior shown in Fig. 7.5-3?
(d) Does a reduction in t reduce accuracy? How small can t be in relation to r?
(e) If t becomes comparable to r, is transverse shear deformation taken into account, and how reliable are these results?

7.33 Solve the ring problem of Fig. 2.7-2 using shell of revolution elements and Fourier series to represent the loads.

7.34 Solve the problem depicted in Fig. 7.4-3. Use only enough support conditions to prevent rigid-body motion. Consider various values of L/R and R/t.

7.35 Consider a curved beam loaded by bending moment (as in Fig. 5.2-3). Some possible cross sections are shown in the sketch for the present problem. These cross sections are thin-walled, so that the effect of distortion of the cross section may be appreciable (as in Fig. 5.2-3).

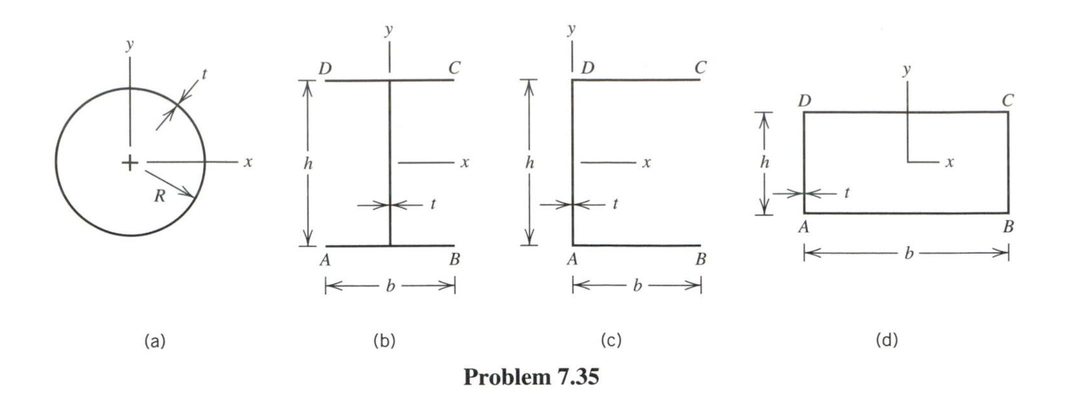

Problem 7.35

7.36 Analyze a straight thin-walled member, having cross section (b), (c), or (d) of Problem 7.35, under loads similar to those in Fig. 7.4-3. For example, apply z-direction forces P, $-P$, P, and $-P$ to points A, B, C, and D, respectively. Use supports that prevent rigid-body motion but apply no reactions to the member.

7.37 Analyze a straight thin-walled member, having cross section (b), (c), or (d) of Problem 7.35, under torsional load. Fix all d.o.f. at one end of a member of length L. At the other end apply x-direction forces P, P, $-P$, and $-P$ to points A, B, C, and D, respectively.

7.38 Consider cantilever beams that are slender, thin-walled, and tip-loaded by transverse force F. There are many possibilities for shape of cross section and placement of force F, three of which are shown in (a), (b), and (c) of the sketch.

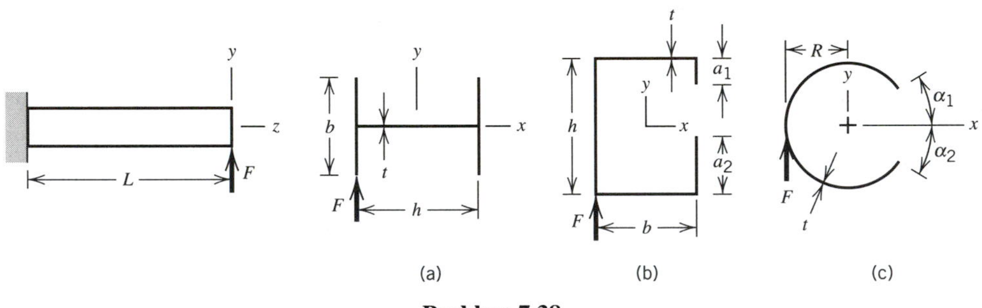

Problem 7.38

7.39 In Problem 7.38, determine the x coordinate of the shear center of cross sections (b) and (c).

7.40 Consider a cantilever beam of I section, as in Problem 7.38a. Apply a tip force, transverse to the beam axis and in the plane of the web. Study the effect on stresses of making the flange width b significantly larger than the depth h of the cross section.

7.41 Thin-walled pipes of circular cross section intersect at a right angle, as shown. The ellipse formed by the intersection may or may not be reinforced by a stiffening member on the outside of the shell. Stresses at and near the intersection are of interest. Obtain a separate solution for each loading. Six load cases are shown in the sketch. For convenience, loads may be applied to a circular disk on the end of the structure.

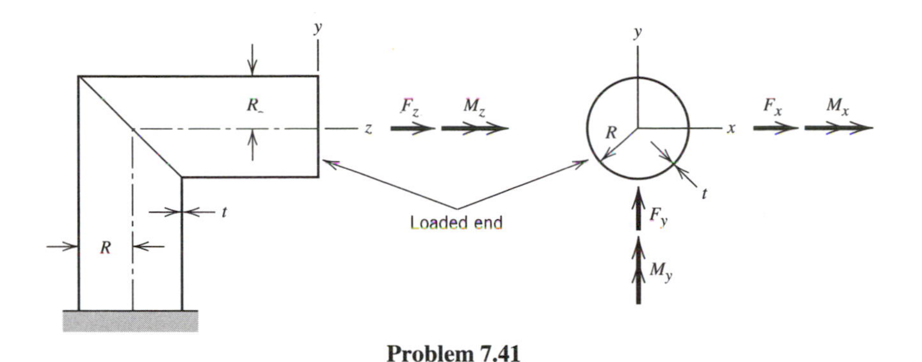

Problem 7.41

7.42 (a) Use shell of revolution elements to analyze a spherical shell under internal pressure. Make the elements conical and start with a coarse mesh (such as that in Fig. 7.5-3).

(b) Imagine that Fig. 7.5-3 is not a coarse-mesh model of a spherical shell, but instead depicts the actual shape of the vessel. Model each conical frustum by several shell of revolution elements, and apply internal pressure.

7.43 Apply the shell element (or elements) in the software you use to the test cases depicted in Fig. 7.5-2. Conduct convergence studies, using 1, 4, 16, and so on quadrilateral subdivisions per quadrant (or per octant) of the structure. (For triangular elements this means using 2, 8, 32, and so on elements per quadrant or per octant.)

7.44 If the software you use allows nonaxisymmetric loads on a shell of revolution, analyze cases (b) and (c) of Fig. 7.5-2 using this approach.

7.45 Apply torque about the axis to the slit tube depicted in Fig. 7.7-1. However, use support conditions sufficient only to prevent rigid-body motion.

7.46 Let the cylindrical shell of Fig. 7.5-2b be loaded by its own weight as it rests on a rigid horizontal surface parallel to the axis of revolution.

7.47 Apply a torque about the longitudinal axis to the twisted strip in Fig. 7.5-2d.

7.48 (a) Attach a reinforcing beam to each straight edge of the shell roof in Fig. 7.5-2a. Let each beam have length L, width b, and depth h.

(b) Let the shell roof be cantilevered from one end. Fix all d.o.f. on the supported curved edge. Leave all other d.o.f. free.

7.49 A square tube and a circular tube are connected by a conical transition section, as shown. Possible loads include internal pressure p, temperature gradient, bending, axial force, and torque about the longitudinal axis.

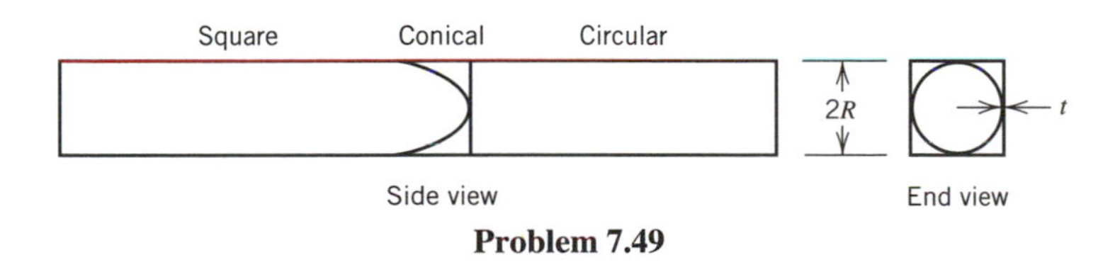

Problem 7.49

7.50 A segment of a circular cylindrical shell is fixed on two straight edges, as shown. Uniform line load q acts on one of the curved edges and is radially directed.

7.51 The sketch shows an end closure on a cylindrical pressure vessel of radius R_c. The closure consists of a toroidal "knuckle" of meridional radius R_t and a spherical cap of radius R_s. Investigate stresses produced by (a) internal pressure, and (b) temperature that varies linearly through the wall thickness.

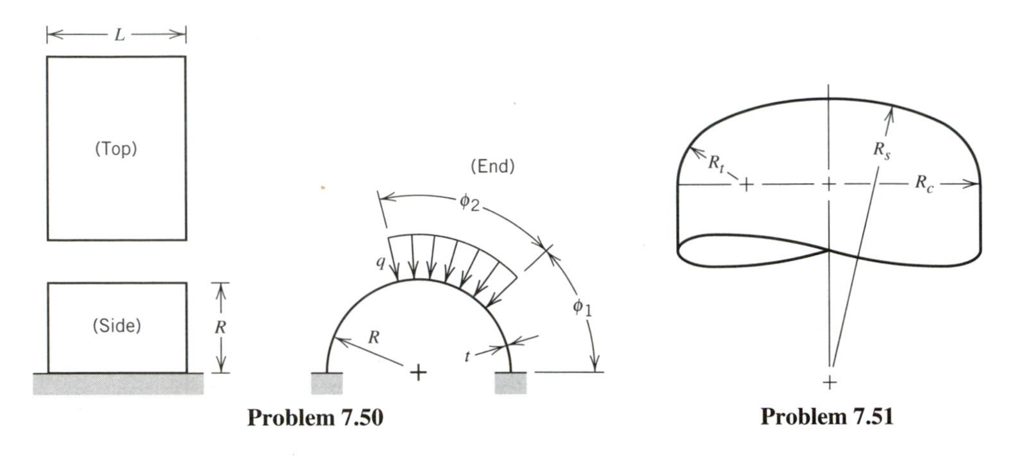

Problem 7.50 **Problem 7.51**

CHAPTER 8

Thermal Analysis

This chapter considers analysis tools for calculation of temperature distribution in a solid body. Concepts of heat transfer are summarized first, followed by a discussion of thermal loads and boundary conditions, nonlinear effects, thermal transients, and modeling considerations. Finally, an example FE analysis addresses the temperature field and the thermal stresses it produces.

8.1 INTRODUCTION. SOME BASIC EQUATIONS

Overview. This book is primarily concerned with stress analysis. Temperature and temperature gradients are an important cause of stress. Accordingly, for our purposes, "thermal analysis" means primarily the calculation of temperatures within a solid body. A by-product of temperature calculation is information about the magnitude and direction of heat flow in the body. This information may be useful in its own right.

Heat is transferred to or from a body by convection and radiation (Fig. 8.1-1). Heat flow across a boundary is analogous to surface load in stress analysis. In addition, there may be internal heat generation, produced by electric current, dielectric heating, or other sources. A distributed internal heat source is analogous to body force in stress analysis. At some points on the boundary or within, temperatures may be prescribed. Prescribed temperatures are analogous to prescribed displacements. Heat moves within the body by conduction. For a *steady-state* (time-independent) problem, the mathematics of all this leads to the global FE equation

$$\mathbf{K}_T\mathbf{T} = \mathbf{Q} \tag{8.1-1}$$

where matrix $\mathbf{K}_T$ depends on the conductivity of the material, $\mathbf{T}$ is a vector of node point temperatures of the solid body, and $\mathbf{Q}$ is a vector of thermal loads. If present, convection and radiation boundary conditions contribute terms to both $\mathbf{K}_T$ and $\mathbf{Q}$. We wish to solve for the unknown nodal temperatures in $\mathbf{T}$. There is obvious resemblance between Eq. 8.1-1 and global equation $\mathbf{KD} = \mathbf{R}$ of stress analysis. Indeed, the same element types, even the same FE mesh, can be used for both thermal analysis and stress analysis. Therefore, having determined nodal temperatures by FE, one can immediately use these temperatures for stress analysis without the trouble of preparing a new FE model.

Thermal conductivity and other properties may depend on temperature strongly enough that $\mathbf{K}_T$ in Eq. 8.1-1 must be regarded as a function of temperature rather than a matrix of constants. Thus the problem becomes nonlinear. The problem is *inherently* nonlinear if there is radiation heat transfer, because then heat flux across the boundary depends on differences of fourth powers of absolute temperatures rather than on simple

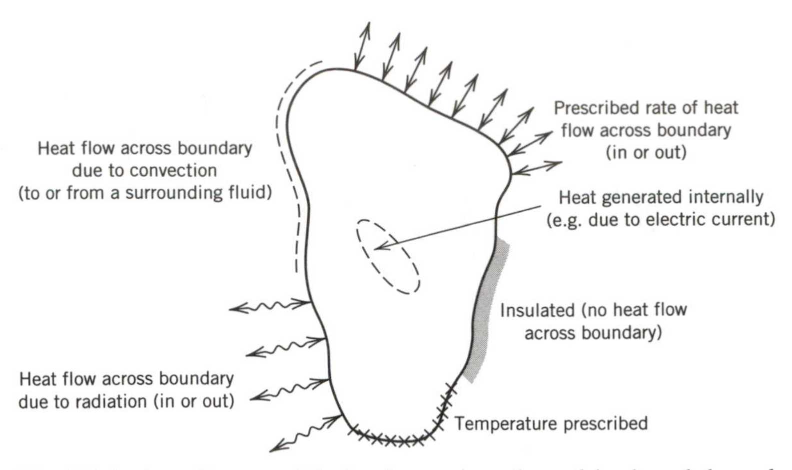

Fig. 8.1-1. An arbitrary solid, showing various thermal loads and thermal boundary conditions.

temperature differences. If steady-state conditions do not prevail, Eq. 8.1-1 is augmented by a "thermal mass" matrix and a vector of nodal rates of change of temperature. The resulting equation is analogous to the equation that describes structural dynamics.

Equations of Heat Flow. Consider an isotropic material and imagine that there is a temperature gradient in the x direction (Fig. 8.1-2a). According to the Fourier heat conduction equation,

$$f_x = -k\frac{\partial T}{\partial x} \tag{8.1-2}$$

where f_x is *heat flux per unit area* and k is the *thermal conductivity.* The negative sign means that heat flows in a direction opposite to the direction of temperature increase.

If the material is anisotropic and x is not a principal material direction, the direction of heat flow is in general not parallel to the temperature gradient (Fig. 8.1-2b). A more general form of Eq. 8.1-2 is

$$\begin{Bmatrix} f_x \\ f_y \\ f_z \end{Bmatrix} = -\boldsymbol{\kappa} \begin{Bmatrix} \partial T/\partial x \\ \partial T/\partial y \\ \partial T/\partial z \end{Bmatrix} \tag{8.1-3}$$

where x, y, and z are mutually perpendicular axes (not necessarily Cartesian) and $\boldsymbol{\kappa}$ is in general a full 3 by 3 matrix of thermal conductivities. If x, y, and z are principal axes of

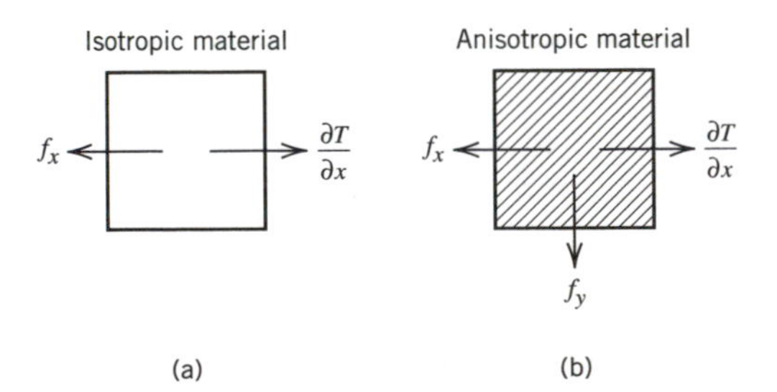

Fig. 8.1-2. Plane heat flow with $\partial T/\partial x > 0$ and $\partial T/\partial y = 0$. Fluxes f_x and f_y are shown in the negative sense (but in the correct direction if $\partial T/\partial x > 0$). Hatching in (b) suggests a principal material direction.

the material, then $\boldsymbol{\kappa}$ is a diagonal matrix. In the special case of isotropy, $\boldsymbol{\kappa}$ reduces to a diagonal matrix with $\kappa_{11} = \kappa_{22} = \kappa_{33} = k$.

By considering a differential element of volume and writing the energy balance equation (rate in) – (rate out) = (rate of increase within), we obtain [2.2, 8.1]

$$-\begin{bmatrix} \dfrac{\partial}{\partial x} & \dfrac{\partial}{\partial y} & \dfrac{\partial}{\partial z} \end{bmatrix} \begin{Bmatrix} f_x \\ f_y \\ f_z \end{Bmatrix} + q_v = c\rho \frac{\partial T}{\partial t} \tag{8.1-4}$$

where q_v is the rate of internal heat generation per unit volume, c is the specific heat, ρ is the mass density, and t is time. If the body is plane and there is convection and/or radiation heat transfer across its flat *lateral* surfaces, Eq. 8.1-4 must be augmented by flux terms (Eqs. 8.2-5 and 8.3-4). The problem becomes steady-state if $\partial T/\partial t = 0$. Heat fluxes f_x, f_y, and f_z can be removed from Eqs. 8.1-4 by substitution from Eq. 8.1-3. Thus, for the special case of material isotropy and steady-state conditions, Eq. 8.1-4 reduces to

$$\nabla \bullet (k\, \nabla T) = -q_v \tag{8.1-5}$$

where ∇ is the gradient operator; in two dimensions, for example, $\nabla = \mathbf{i}(\partial/\partial x) + \mathbf{j}(\partial/\partial y)$, where $\mathbf{i}$ and $\mathbf{j}$ are unit vectors in x and y directions, respectively. If, in addition, k is independent of the coordinates, Eq. 8.1-5 reduces to

$$k\, \nabla^2 T = -q_v \qquad \text{or} \qquad k\left(\frac{\partial^2 T}{\partial x^2} + \frac{\partial^2 T}{\partial y^2} + \frac{\partial^2 T}{\partial z^2}\right) = -q_v \tag{8.1-6}$$

A steady-state problem is solved by determining a function $T = T(x, y, z)$ that satisfies Eqs. 8.1-3 and 8.1-4 and also meets prescribed boundary conditions.

Nomenclature and Units. Quantities used in this chapter are as follows. The unit of heat is the same as the unit of energy, namely, the joule; 1 J = 1 N·m. The unit of power is the watt; 1 W = 1 J/s = 1 N·m/s.

A = cross-sectional area (m^2)
c = specific heat (J/kg·°C)
f = heat flux per unit area (W/m^2)
h = heat transfer coefficient; also called film coefficient (W/m^2·°C)
k = thermal conductivity (W/m·°C)
q_v = rate of internal heat generation per unit volume (W/m^3)
q = rate of heat flow; $q = Af$ or $q = Sf$ (W)
S = surface area (m^2)
T = temperature (°C, or °K when used with radiation)
T_f = temperature of adjacent fluid outside the boundary layer (°C)
$\dot{T}$ = $\partial T/\partial t$ (°C/s)
t = time (s)
ρ = mass density (kg/m^3)
σ = Stefan–Boltzmann constant ($\sigma = 5.670(10^{-8})$ W/m^2·°K^4)

In the foregoing units, actual fluids and solids display numerical values in the approximate ranges $10^2 < c < 10^4$, $10 < h < 10^5$, and $10^{-2} < k < 500$ [8.1]. In Eq. 8.1-1, terms in $\mathbf{K}_T$ have units W/°C, terms in $\mathbf{T}$ have units °C, and terms in $\mathbf{Q}$ have units W. Absolute temperature is measured in °K, where °K = °C + 273.

8.2 FINITE ELEMENTS IN THERMAL ANALYSIS

The development of finite elements and matrix equations in thermal analysis parallels what is done in stress analysis. An element conductivity matrix $\mathbf{k}_T$ can be generated by a direct method for very simple elements. Otherwise a formal procedure is required. The formal procedure makes use of shape functions, whose properties govern the capability of the element. In the present section we summarize the thermal FE method but postpone radiation heat transfer to Section 8.3.

Bar Element, Direct Method. An equation like Eq. 8.1-1, but pertaining to a single element, can be written directly for very simple elements. Consider a uniform bar whose lateral surface is insulated, Fig. 8.2-1. The rate of heat flow is $q = Af$. It is constant in this element because cross-sectional area A is constant and heat can flow only axially. Nodal heat flow rates q_1 and q_2 are considered positive when directed *into* the element. Hence $q_1 = -q_2$ whether the temperature gradient is positive or negative. Nodal temperatures are T_1 and T_2. Equation 8.1-2 yields q_1 and q_2, first for $T_1 \neq 0$ and then for $T_2 \neq 0$. Results are shown in Fig. 8.2-1. In matrix format, these results are

$$\begin{bmatrix} Ak/L & -Ak/L \\ -Ak/L & Ak/L \end{bmatrix} \begin{Bmatrix} T_1 \\ T_2 \end{Bmatrix} = \begin{Bmatrix} q_1 \\ q_2 \end{Bmatrix} \tag{8.2-1}$$

where the square matrix is $\mathbf{k}_T$, the *element conductivity matrix*. Clearly, Eq. 8.2-1 resembles the stiffness equation for a bar element, Eq. 2.2-2. In common software there is no thermal analogue of a beam element because temperature gradients are not used as nodal d.o.f.

An actual bar could be modeled by several of these elements if temperatures at several locations between its ends were required. A tapered bar could be analyzed conveniently in this way.

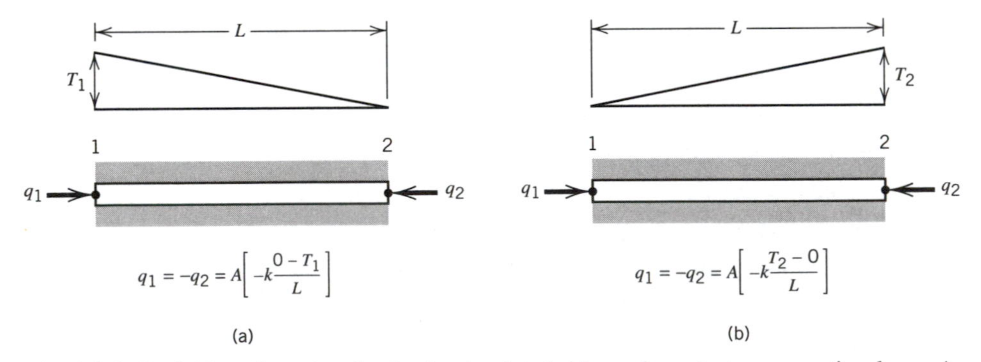

Fig. 8.2-1. Nodal heat flows in a bar having insulated sides and constant cross-sectional area A. (a) $T_1 > 0$, $T_2 = 0$. (b) $T_1 = 0$, $T_2 > 0$.

General Elements, Formal Procedure. The formal procedure parallels that used to obtain the stiffness matrix of a displacement-based finite element. One begins by interpolating temperature over an element from element nodal temperatures $\mathbf{T}_e$.

$$T = [N_1 \; N_2 \cdots N_n] \begin{Bmatrix} T_1 \\ T_2 \\ \vdots \\ T_n \end{Bmatrix} \quad \text{or} \quad T = \mathbf{N}\mathbf{T}_e \tag{8.2-2}$$

Individual shape functions in **N** are suited to the element type and can be exactly those used also to interpolate a displacement field: Eqs. 3.4-3 for a four-node rectangle, Eqs. 4.4-2 for an arbitrarily shaped four-node quadrilateral, and so on. The form of interpolation determines the complexity of the temperature field that an element can represent. In Cartesian coordinates, temperature gradients in a plane element are

$$\begin{Bmatrix} \partial T/\partial x \\ \partial T/\partial y \end{Bmatrix} = \begin{bmatrix} \partial N_1/\partial x & \partial N_2/\partial x & \cdots & \partial N_n/\partial x \\ \partial N_1/\partial y & \partial N_2/\partial y & \cdots & \partial N_n/\partial y \end{bmatrix} \begin{Bmatrix} T_1 \\ T_2 \\ \vdots \\ T_n \end{Bmatrix} \tag{8.2-3a}$$

$$\text{or} \qquad \mathbf{T}_\partial = \mathbf{B}\mathbf{T}_e \quad \text{where} \quad \mathbf{B} = \begin{Bmatrix} \partial/\partial x \\ \partial/\partial y \end{Bmatrix} \mathbf{N} \tag{8.2-3b}$$

For solids, a third row expressing $\partial T/\partial z$ is added. Analogous forms can be written in other coordinate systems, such as the cylindrical system used for solids of revolution. Regardless of the coordinate system, the expression for an element conductivity matrix can be shown to be

$$\mathbf{k}_T = \int \mathbf{B}^T \boldsymbol{\kappa} \mathbf{B} \, dV \tag{8.2-4}$$

where $\boldsymbol{\kappa}$ is the array of thermal conductivities from Eq. 8.1-3 and integration is over the element volume. For a bar element, **N** is given by Eq. 2.2-5, $\boldsymbol{\kappa}$ becomes the scalar k, and Eq. 8.2-4 yields the conductivity matrix in Eq. 8.2-1.

Remarks. Thermal finite elements are assembled in the same way as structural finite elements (e.g., Eq. 2.6-2; also Section 4.1). In contrast to stress analysis, thermal analysis is a *scalar* field problem because T has no direction associated with it. A thermal FE formulation has but one d.o.f. per node, namely, nodal temperature. A FE temperature field is continuous within elements and across interelement boundaries. Temperature gradients, like strains in stress analysis, are typically *not* interelement-continuous. The troublesome shear-locking phenomenon discussed in Section 3.6 does not appear in thermal analysis because the product $\mathbf{BT}_e$ is not required to yield a quantity analogous to shear strain.

Excepting when there is heat transfer by radiation, the rate of heat flow is proportional to temperature differences. Accordingly, nodal temperatures need not be absolute temperatures. Instead, they can be measured relative to any convenient reference level at which the body has a uniform temperature.

Upon assembly of elements, nodal rates of heat flow q_i from separate elements are combined at shared nodes and become Q_i, the net flow into node i of the structure. Thus

$Q_i = 0$, except at nodes where T_i is prescribed, nodes on a structure boundary across which heat is transferred, or at internal nodes where Q_i may arise from q_v. Usually Q_i is determined from convection or radiation boundary conditions, just as equivalent nodal forces are determined from applied surface pressure in stress analysis. A user-prescribed nodal Q_i is a point source or a sink and is analogous to a concentrated nodal force.

On an insulated boundary, nodal temperatures are unknown and the corresponding nodal flows Q_i are zero (as described in the foregoing paragraph). Prescribed *non*zero nodal temperatures are treated like prescribed nonzero nodal displacements in stress analysis. Thus if a nodal temperature T_j in **T** of Eq. 8.1-1 is prescribed, in every row i the known quantity $K_{ij}T_j$ is transferred to the right-hand side, that is, subtracted from Q_i, so that **T** continues to contain only *unknown* nodal temperatures. In this way equation j is effectively discarded, so that the number of "active" equations is reduced by one. Note that in Eq. 8.1-1 one can prescribe *either* T_j or Q_j but not both at once. Thus if T_j is prescribed and Q_j is needed, Q_j must be calculated by postprocessing.

The work of assembly, construction of appropriate Q_i terms, and imposition of nodal temperatures is carried out automatically by the software after the physical problem has been described by appropriate input data. It is likely that mesh layout will be governed by what is needed in subsequent stress analysis, as stress gradients are typically larger than thermal gradients and therefore demand more detail in the FE model.

Convection Boundary Conditions. The equation used to treat convection heat transfer was proposed by Newton:

$$f = h(T_f - T) \tag{8.2-5}$$

where f is flux normal to the surface and positive inward, T is the temperature of the surface of the solid, and T_f is the temperature of the surrounding fluid. Temperature in the fluid varies from T to T_f through the thickness of a boundary layer adjacent to the solid. Unfortunately, the heat transfer coefficient h depends on many factors, some of which are temperature dependent: velocity of the fluid stream, roughness and geometry of the surface, and density, viscosity, conductivity, and specific heat of the fluid. Accordingly, tabulated data of h for a specific fluid may state only a typical range of values.

The formal procedures that yield Eq. 8.2-4 [2.2] also show that Eq. 8.2-5 leads to both an element matrix and an element vector:

$$\text{matrix: } \int \mathbf{N}^T\mathbf{N}\, h\, dS \qquad \text{vector: } \int \mathbf{N}^T h\, T_f\, dS \tag{8.2-6}$$

where dS is an increment of the element surface subject to convection. The matrix combines with the element conductivity matrix and hence contributes to $\mathbf{K}_T$ in Eq. 8.1-1. The vector contributes to **Q**. In a plane problem with insulated lateral surfaces, integration in Eq. 8.2-6 spans only element edges that are also boundaries of the structure and subject to convection. In such a case there will be few contributions to the global equations from Eqs. 8.2-6. However, in general, convection may occur on lateral surfaces of plates as well as on plate edges, so that Eqs. 8.2-6 may contribute extensively to the global equations.

It happens that the second of Eqs. 8.2-6 has the same form as the equation that yields nodal forces from surface pressure in stress analysis. Therefore, for constant and linear flux distributions, nodal heat flows Q_i have the same relative magnitudes as do nodal forces in Figs. 3.9-1, 3.9-3, 3.9-5, and 6.2-3. In particular, note Fig. 6.2-3: in thermal analysis, its counterpart is a uniform flux directed into a rectangular element face, resulting in some nodal heat flows that are inward and some that are outward.

8.3 RADIATION AND OTHER NONLINEARITIES

Radiation Boundary Conditions. To introduce concepts and equations, consider two parallel planes, of infinite extent so that edge effects can be neglected. Imagine that the planes are ideal blackbodies: each is both a perfect absorber and a perfect radiator. Let one plane have temperature T and the other temperature T_r. The plane of temperature T receives heat flux σT_r^4 and radiates heat flux σT^4, where σ is the Stefan–Boltzmann constant and temperatures are *absolute*; that is, °K. Thus the *net* heat flux received by the surface of temperature T is

$$f = \sigma(T_r^4 - T^4) \tag{8.3-1}$$

Departing from the ideal, let the planes no longer be blackbodies, which means that they are now characterized by "emissivities" ε and ε_r. Emissivity is defined as the ratio of total emissive power to that of a blackbody at the same temperature, so that $0 < \varepsilon < 1$. The value of ε depends on the roughness, degree of oxidation (if metal), and temperature of a surface. As examples, $\varepsilon \approx 0.1$ for polished aluminum, and $\varepsilon \approx 0.9$ for black paint on metal and for paper at room temperature. For infinite parallel planes, Eq. 8.3-1 is replaced by [8.1]

$$f = \frac{\sigma}{(1/\varepsilon) + (1/\varepsilon_r) - 1}\left(T_r^4 - T^4\right) \tag{8.3-2}$$

The next complication is that practical surfaces may not be parallel, are often not flat, and are certainly not infinite. These geometric complications are accounted for by a shape factor (also called a view, angle, configuration, or interception factor). For two flat areas S and S_r, the shape factor depends on the distance between them, their magnitudes, and the angle between their normals. Putting all this together, for two surfaces of absolute temperatures T and T_r, the net average heat flux received by the surface of temperature T can be written in the form

$$f = F\sigma\left(T_r^4 - T^4\right) \tag{8.3-3}$$

where F is a factor that accounts for the geometries of the radiating surfaces and their emissivities. The calculation of F is sufficiently complicated that it may be done by a separate computer program. Flux f in Eq. 8.3-3 is an *average* value; clearly, the pointwise flux would not be constant around the perimeters of parallel pipes of temperatures T and T_r. For some surfaces an average value may be adequate, as, for example, when S is the area of one face of one finite element in a mesh that is not extremely coarse. Equation 8.3-3 can be written as many times as there are surfaces that exchange radiant heat with the face of temperature T. Finally, by factoring $T_r^4 - T^4$ in Eq. 8.3-3, we can write

$$f = h_r(T_r - T) \qquad \text{where} \qquad h_r = F\sigma(T_r^2 + T^2)(T_r + T) \tag{8.3-4}$$

Comparing this equation with Eqs. 8.2-5, we see that the flux expressions have the same form. Accordingly, a radiation boundary condition leads to matrix and vector expressions having the same form as for convection, Eq. 8.2-6, but with h and T_f replaced by h_r and T_r, respectively. Note that h_r is a *temperature-dependent* heat transfer coefficient. This makes the problem nonlinear and requires an iterative solution, as summarized in what follows.

The foregoing brief summary has omitted much, including any mention of radiation from flames and hot gases. Remember that for radiation heat transfer T and T_r must be *absolute* temperatures (measured in °K rather than °C). Some software, at the analyst's request, will add 273 to nodal temperatures expressed in °C. Thus, absolute temperatures are used for internal processing where needed but input and output temperatures will be in °C.

Solution of Nonlinear Problems. If temperatures at some locations in the body are appreciably higher or lower than the mean temperature, it may be necessary to regard conductivity k as temperature dependent. Thus $\mathbf{K}_T$ in Eq. 8.1-1 becomes a function of T. The problem is nonlinear because $\mathbf{T}$ is not directly proportional to thermal load $\mathbf{Q}$. If there is convection heat transfer with a temperature-dependent coefficient h, $\mathbf{K}_T$ *and* $\mathbf{Q}$ become functions of T. If there is radiation heat transfer, $\mathbf{K}_T$ and $\mathbf{Q}$ are *certain* to be functions of T. Equation 8.8-1 assumes the form

$$[\mathbf{K}_T(T)]\{\mathbf{T}\} = \{\mathbf{Q}(T)\} \tag{8.3-5}$$

in which we wish to solve for $\mathbf{T}$. Solution methods for nonlinear structural problems, discussed in Chapter 10, are also applicable to nonlinear thermal problems. Here we explain only the very straightforward method of direct substitution, which is as follows. Assume an initial temperature field T_0, which might be a uniform temperature throughout the body. Generate $\mathbf{K}_T$ and $\mathbf{Q}$ based on this temperature and solve Eq. 8.3-5 for $\mathbf{T}$. Use the newly computed temperatures to obtain new values of k, h, and h_r, generate a new $\mathbf{K}_T$ and a new $\mathbf{Q}$, and solve for a new $\mathbf{T}$. Thus we generate the sequence of solutions

$$[\mathbf{K}_T(T_0)]\{\mathbf{T}_1\} = \{\mathbf{Q}(T_0)\}, \quad [\mathbf{K}_T(T_1)]\{\mathbf{T}_2\} = \{\mathbf{Q}(T_1)\}, \ldots \tag{8.3-6}$$

Iteration is stopped when an iteration limit is reached or when a convergence test is satisfied. A possible convergence test involves the calculation

$$e_i = \frac{\|\mathbf{T}_i - \mathbf{T}_{i-1}\|}{\|\mathbf{T}_i\|} \tag{8.3-7}$$

where the "norm" symbol usually indicates Euclidean norm, that is, the square root of the sum of the squares. The ith iteration is deemed converged when e_i is less than a user-prescribed value or a default value coded in the software.

Equations 8.3-6 describe only one of many possible algorithms for solution of nonlinear problems. Algorithms may differ greatly in efficiency and propensity to become numerically unstable. Convergence is sometimes slow and may even fail if radiation effects are dominant because these effects contribute strong nonlinearities.

8.4 THERMAL TRANSIENTS

When steady-state conditions do not prevail, temperature change in a unit volume of material is resisted by "thermal mass" that depends on the mass density ρ of the material and its specific heat c. Equation 8.1-1 is augmented to become

$$\mathbf{K}_T\mathbf{T} + \mathbf{C}\dot{\mathbf{T}} = \mathbf{Q} \qquad \text{where} \qquad \mathbf{Q} = \mathbf{Q}(t) \tag{8.4-1}$$

in which $\dot{\mathbf{T}} = \partial\mathbf{T}/\partial t$. In general, thermal loads $\mathbf{Q}$ are time dependent. Matrix $\mathbf{C}$ may be called a "heat capacity" matrix. It is built by assembling element heat capacity matrices $\mathbf{c}$.

$$\mathbf{C} = \sum_{\text{(assemble)}} \mathbf{c} \quad \text{where} \quad \mathbf{c} = \int \mathbf{N}^T\mathbf{N}\,\rho c\,dV \tag{8.4-2}$$

Integration is over the element volume. Shape functions $\mathbf{N}$ are as described in connection with Eq. 8.2-2. Matrix $\mathbf{c}$ has the same form as mass matrix $\mathbf{m}$ of structural dynamics, Eq. 9.3-4. With $\mathbf{c}$, as with $\mathbf{m}$, the integral formulation can be replaced by an ad hoc "lumped" formulation. This matter is discussed in Sections 9.3 and 9.11.

For now we restrict attention to linear problems. We wish to calculate $\mathbf{T}$ as a function of time, when initial temperatures are prescribed, $\mathbf{Q}$ is a known function of time, and $\mathbf{K}_T$ and $\mathbf{C}$ are time independent. The calculation is usually done by direct integration in time. For this purpose, Eq. 8.4-1 is written in the form

$$\mathbf{K}_T\mathbf{T}_n + \mathbf{C}\dot{\mathbf{T}}_n = \mathbf{Q}_n \tag{8.4-3}$$

where n is the nth instant of time. For times $t = \Delta t$, $t = 2\Delta t$, and so on, where Δt is an increment of time, we seek the corresponding nodal temperatures $\mathbf{T}_1$, $\mathbf{T}_2$, and so on. Initial nodal temperatures $\mathbf{T}_0$ are presumed known and $\mathbf{Q}_n$ is presumed known for all n. Direct integration can be based on the formula [8.2]

$$\mathbf{T}_{n+1} = \mathbf{T}_n + \Delta t\{(1-\gamma)\dot{\mathbf{T}}_n + \gamma\dot{\mathbf{T}}_{n+1}\} \tag{8.4-4}$$

which resembles Eq. 9.9-6b of structural dynamics. The quantity γ is a number that can be chosen by the analyst. By manipulation [2.2], Eqs. 8.4-3 and 8.4-4 yield

$$\left[\gamma\mathbf{K}_T + \frac{1}{\Delta t}\mathbf{C}\right]\mathbf{T}_{n+1} = (1-\gamma)\mathbf{Q}_n + \gamma\mathbf{Q}_{n+1} - \left[(1-\gamma)\mathbf{K}_T - \frac{1}{\Delta t}\mathbf{C}\right]\mathbf{T}_n \tag{8.4-5}$$

If $\gamma = \frac{1}{2}$ the algorithm is called either the *Crank–Nicolson method* or the *trapezoidal rule*. For $\frac{1}{2} \le \gamma \le 1$ the algorithm is unconditionally stable in linear problems. This means that the numerical process will not "blow up" even if Δt is large. Results will not necessarily be *accurate* for large values of Δt because important changes may take place on a small time scale. For $\gamma = \frac{1}{2}$ the method is second-order accurate, which means that numerical error in $\mathbf{T}$ produced by the algorithm is approximately quartered when Δt is halved. This is a global estimate; at some locations there may be only first-order accuracy. If $\mathbf{Q}$ represents a thermal shock, the solution displays some spurious oscillation, which can be reduced by using a value of γ greater than $\frac{1}{2}$ so as to introduce "algorithmic damping." Spurious oscillation can be avoided entirely by a slightly more complicated form of the algorithm [8.3].

In a nonlinear problem, $\mathbf{K}_T$, $\mathbf{C}$, and $\mathbf{Q}$ may all depend on temperature, and the foregoing algorithm may become unstable when $\gamma = \frac{1}{2}$. A larger value of γ may preserve stability. However, accuracy suffers as γ increases. Unconditional stability in a nonlinear problem is guaranteed only if $\gamma = 1$. Iterations within each time step may be needed to keep a nonlinear analysis "on track."

8.5 MODELING CONSIDERATIONS

Element types, sizes, and shapes for a thermal FE model may be dictated less by thermal considerations than by an anticipated stress analysis based on the same mesh. The mesh demands of stress analysis are usually more severe. For example, a modest temperature gradient may create forces of constraint that produce large strain gradients, especially near holes, grooves, and other stress raisers. Accordingly, one should avoid three-node triangles and elements markedly elongated in a direction of little temperature change, even though such elements may work well for thermal analysis.

A temperature field that is linear in Cartesian coordinates produces deformation but no stress in an unrestrained body that is homogeneous and either isotropic or rectilinearly orthotropic. A temperature field that is linear in *cylindrical* coordinates *does* produce stress. This is the common situation in a pipe of standard dimensions with different temperatures inside and outside: temperature varies almost linearly with the radial coordinate.

In some software, convection and radiation conditions can be handled by use of "surface" elements that overlay the boundary surfaces of conducting elements. Surface elements serve to supply the necessary terms to $\mathbf{K}_T$ and $\mathbf{Q}$. They can also serve in the calculation of factor F in Eq. 8.3-4. Factor F is a function of position, so if it is taken as constant over each surface element, accuracy will be improved by mesh refinement.

As with any FE analysis, details must be treated with care if blunders are to be avoided. Dimensions, constants, and thermal loads must be stated in a consistent system of units. Orthotropic properties must not be confused as to direction, and local axes are needed if principal material directions are not parallel to global axes. Some boundary conditions on a solid or shell of revolution must be stated for the entire circumference or on a per radian basis, as the software requires. Absolute temperatures are needed when there is radiation heat transfer. Therefore, if a nonabsolute temperature scale is used, the software requires input of an "offset" that states the absolute temperature of the zero-degree mark in the scale used. Computed fluxes may be reported in a local coordinate system for some elements. Surface fluxes may be defined as positive inward or positive outward, depending on the software.

For time-independent analysis with neither convection nor radiation boundary conditions, at least one nodal temperature must be prescribed in order to avoid a singular matrix $\mathbf{K}_T$. Convection and radiation conditions contribute terms to $\mathbf{K}_T$ that make it nonsingular; then none of the nodal temperatures need be prescribed. In transient analysis a singular $\mathbf{K}_T$ is acceptable, but *all* nodal temperatures must be prescribed as initial conditions at time $t = 0$. Typically, these temperatures are not all zero (unlike structural mechanics, where initial displacements and velocities are often all zero).

Only half (or less) of the structure need be modeled by FE if there is symmetry of geometry, material properties, boundary conditions, and thermal loads with respect to one (or more) planes. Heat does not flow across a plane of symmetry, so nodes in a plane of symmetry of the structure become nodes on an insulated boundary of the FE model.

Boundary conditions are temperature dependent if heat transfer coefficients are temperature dependent. Similarly, in a transient problem, boundary conditions may be both temperature and time dependent. Sometimes boundary conditions are unclear [8.4]. Consider a rectangular plate, with two opposite edges at 0°C and the other two opposite edges at 500°C. Should corner temperatures be assigned as 0°C, 500°C, 250°C, or something else? This problem is ill-posed and probably results from overidealization of the physical situation. A similar difficulty arises if temperature is prescribed along one edge and flux is prescribed along an adjacent edge. At the corner where edges meet, which condition should be used? One can prescribe T or Q at the corner node but not both.

A critique of computed results should begin by comparing computed results with a *previously obtained* approximation. Computed temperatures and fluxes at boundaries should be checked to see that there are no disagreements with the boundary conditions intended. Temperature contours (isotherms) should be parallel to a boundary of constant

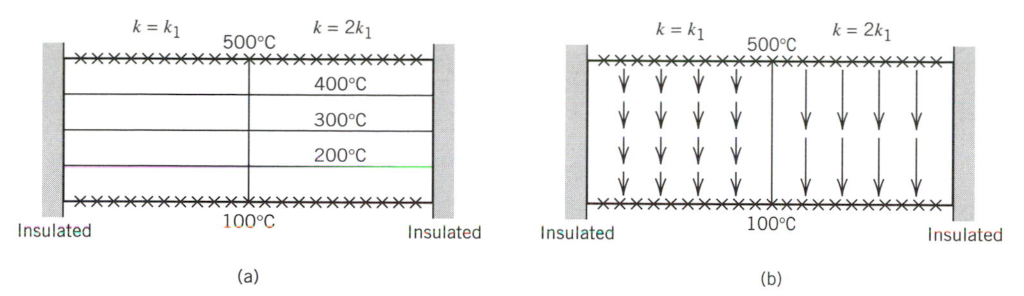

Fig. 8.5-1. (a) Temperature contours and (b) heat flux in a two-dimensional bimaterial block with two isothermal boundaries and two insulated boundaries. Each material is isotropic.

temperature and normal to a plane of symmetry. Heat flux (Eq. 8.1-2) can be plotted in vector fashion, so that one sees several arrows, each pointing in the local direction of heat flow, with each arrow length proportional to the local magnitude of flux. Heat flux should be parallel to an insulated boundary and parallel to a plane of symmetry. These remarks are illustrated by Fig. 8.5-1. Note that although the horizontal centerline is a line of geometric symmetry, the thermal problem is not symmetric with respect to this line because prescribed boundary temperatures are unequal. Significant disagreement between conditions expected and conditions obtained suggests an error in understanding, an error in modeling, or a need for mesh refinement. The reader may wish to review Chapter 5: although it is devoted to stress analysis, there are thermal counterparts for most items discussed.

Temperatures are interelement-continuous; fluxes are not. In this way temperatures resemble displacements and fluxes resemble stresses. Accordingly, concepts about discretization error in stress analysis can be transferred directly to thermal analysis. Flux contours should be plotted element by element, that is, *without nodal averaging*. Significant interelement discontinuities warn of a need for mesh refinement. The difference between the element by element flux field and the flux field based on nodal average fluxes can be regarded as an error measure and can be used to drive adaptive meshing as described in Section 5.16. In a good mesh, each element spans about the same number of flux contours.

Various physical phenomena are described by an equation having the form of Eq. 8.1-2 or 8.1-3. Accordingly, by appropriate definition of variables and physical constants, these phenomena can be analyzed by FE software intended for thermal analysis. The phenomena include deflection of an elastic membrane under lateral pressure, diffusion of moisture, electrostatic fields, fluid flow through a porous medium, incompressible and irrotational flow of an ideal fluid, and pure torsion of a prismatic shaft.

This brief chapter omits a great deal. Study of it alone cannot confer competence in thermal analysis. Like any other FE analysis, FE thermal analysis is likely to be unsuccessful if performed by someone who lacks a physical grasp of the problem area.

8.6 AN APPLICATION

Sections of a pipe are connected by a flanged joint (Fig. 8.6-1). Each flange has been slipped onto its section of the pipe and connected to it by two circumferential welds. Bolts draw the flanges together and compress a gasket between them. In the neighborhood of the joint, fluid in the pipe has temperature 0°C, and vapor condensing on the outside of the pipe has temperature 100°C. Heat transfer coefficients h are 5000 W/m$^2\cdot$°C

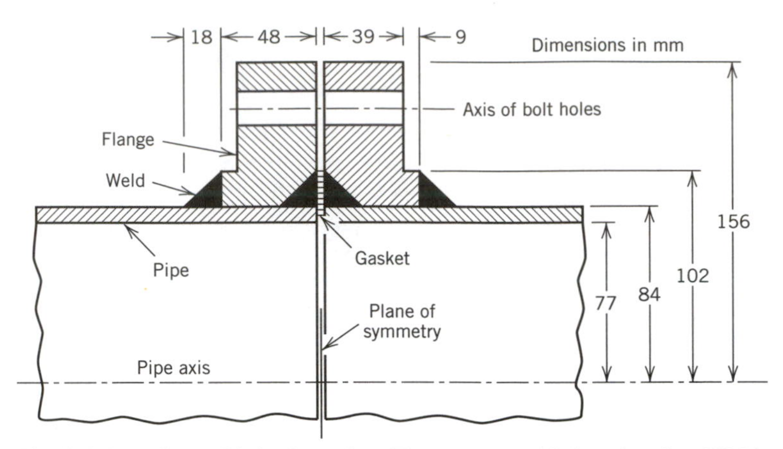

Fig. 8.6-1. A flanged joint in a pipe. Flanges are welded to the pipe. Welds are shown dark in the figure.

inside the pipe and 20,000 W/m$^2\cdot$°C outside. Material of the pipe and flange has thermal conductivity k = 20 W/m$\cdot$°C. The steady-state temperature field is required in the pipe and in the flange, for use in a subsequent stress analysis with E = 200 GPa, ν = 0.3, and $\alpha = 12(10^{-6})$/°C.

Preliminary Analysis. The inside of the pipe must be warmer than 0°C and the outside must be cooler than 100°C. In Fig. 8.6-2, locations where these limits are most closely approached should be along AB and along CD, respectively. Away from the relatively massive connection, where the wall is thin, we should see the warmest inside surface temperature and the coolest outside surface temperature. Nevertheless, heat flux should be large here because the wall is thin. The maximum possible value of flux through the pipe wall is easily approximated by regarding the wall as plane and using the limiting surface temperatures in Eq. 8.1-2.

$$f_{\text{lim}} \approx -k\frac{\Delta T}{\Delta r} = -20\frac{100-0}{0.084-0.077} = -286{,}000 \text{ W/m}^2 \tag{8.6-1}$$

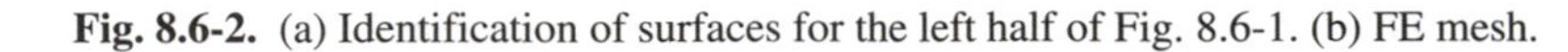

Fig. 8.6-2. (a) Identification of surfaces for the left half of Fig. 8.6-1. (b) FE mesh.

where the negative sign means that heat flows inward. In the flange, flux should be small because temperature gradients should be small, but flux should have a component directed toward *BC*.

Between the welds, along *IJ* in Fig. 8.6-2a, pipe and flange touch only at random and isolated points, if at all. Conductivity across this cylindrical interface is very low as compared with solid metal. We will assume that the conductivity is zero, and model *IJ* as an insulator. This is a pessimistic assumption for eventual stress analysis because it will increase temperature differences in the joint and therefore increase stresses. In thermal analysis we would expect to see large temperature gradients, and therefore large flux, near *I* and *J*.

FE Model and Analysis. The FE model is shown in Fig. 8.6-2b. The mesh shown was generated automatically by the software, based on data that locate boundary lines and state the desired element size near lettered points in Fig. 8.6-2a. Element sizes are smallest where greatest temperature gradients are expected. This is a rather coarse mesh, especially if it is also to be used for subsequent stress analysis. Elements are axisymmetric solids of revolution, seen in cross section in Fig. 8.6-2b. All are either six-node triangles or eight-node quadrilaterals. The apparent discontinuity along *IJ* is intentional. Sets of nodes on either side of *IJ* are left unconnected to model an interface that transfers no heat. Bolts and bolt holes are ignored because temperature gradients are expected to be low at bolt locations. The gap between adjacent flanges is also ignored because convection there is expected to be very low. Thus *BC* is taken as a symmetry plane of the structure, across which no heat flows. Temperatures are not prescribed at any node. Convection boundary conditions are prescribed along *AB* (inside) and *CDEFGH* (outside).

Critique of FE Results. Computed temperature contours are shown in Fig. 8.6-3. The lowest temperature is 3.59°C along the right half of inside boundary *AB*. The highest temperature is 99.99°C along the outer surface *CDEFG*. Temperature contours are interelement-continuous except along *IJ*, where discontinuity is expected. In the left portion of the model, temperature contours are parallel to the pipe axis, which indicates that this portion of the model is sufficiently long for thermal analysis. Abrupt changes in contour directions near *I* and *J* (as seen in Fig. 8.6-3b) suggest a need for mesh refinement in these areas. Also, contours are not quite normal to symmetry plane *BC*.

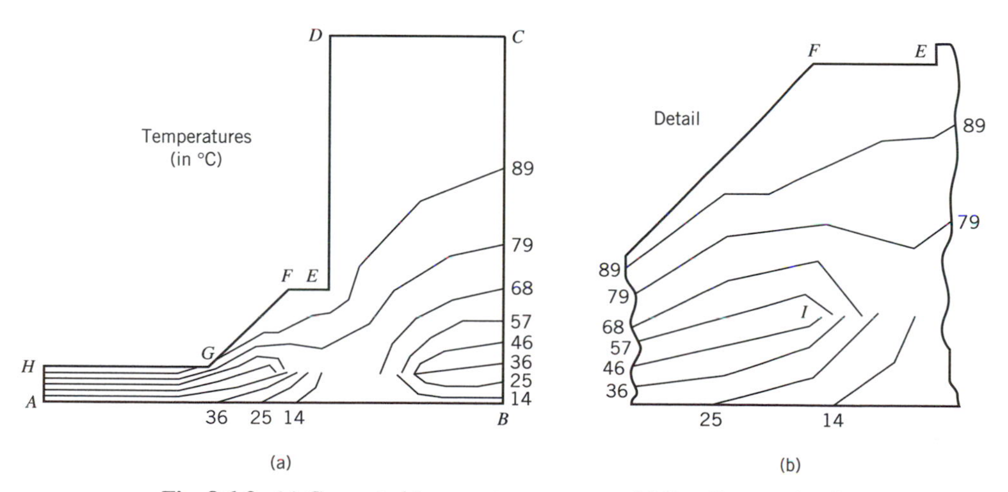

Fig. 8.6-3. (a) Computed temperature contours. (b) Detail near point *I*.

A vector plot of computed flux appears in Fig. 8.6-4a. Flux arrows are perpendicular to temperature contours because the material is isotropic. Arrows point in the direction of heat flow and arrow lengths are proportional to flux magnitude. Each arrow emanates from the center of an element. In the outer portion of the flange, some arrows are so short that they appear as dots. Radial flux near *AH* is 170,000 W/m^2, which, as expected, is less than the approximate limiting value in Eq. 8.6-1. Flux directions agree with expectations of preliminary analysis. Flux contours near *I*, Fig. 8.6-4b, show strong interelement discontinuities, again suggesting a need for local mesh refinement. Similar behavior appears near *J* and to a lesser extent near *G*. Nevertheless, the relative error (Section 5.16), applied to the heat flux field in the present application, is only $\eta = 0.03$.

The thermal results appear reliable except for a need for more detail near *G*, *I*, and *J*. Precisely *at* these locations there can never be enough detail because flux is theoretically infinite at sharp reentrant corners. We will ignore these difficulties and proceed, mainly in order to show how the same mesh behaves when used for stress analysis.

Subsequent Stress Analysis. The mesh shown in Fig. 8.6-2b is used again, now with computed nodal temperatures used to load the model. Load from bolt tensions could also be applied, as described in Section 5.7, but we will not do so here. Two significant questions about boundary conditions arise. First, should nodes along *BC* be fixed against axial motion or not? Allowing movement gives no credit to resistance offered by the gasket and the bolts, while fixity gives too much credit. Second, is it proper to let nodes along *IJ* move independently? If sides of the interface pull apart, the answer is yes. But if sides tend to overlap, the model should allow relative axial motion but not relative radial motion for element edges that abut one another across the interface.

We elect to run the stress analysis twice, first allowing axial motion along *BC* and second preventing it. Thus we expect to bracket the correct results. Before running the FE analysis we make the following displacement and stress predictions, which are applicable to either set of boundary conditions. The cooler inside surface tends to contract relative to the warmer material, thus pulling material toward the inside and opening a gap along *IJ*. As seen in Fig. 8.6-3a, material between *IJ* and the inner surface is significantly cooler than most of the remaining material. This implies that contraction of this region is largely

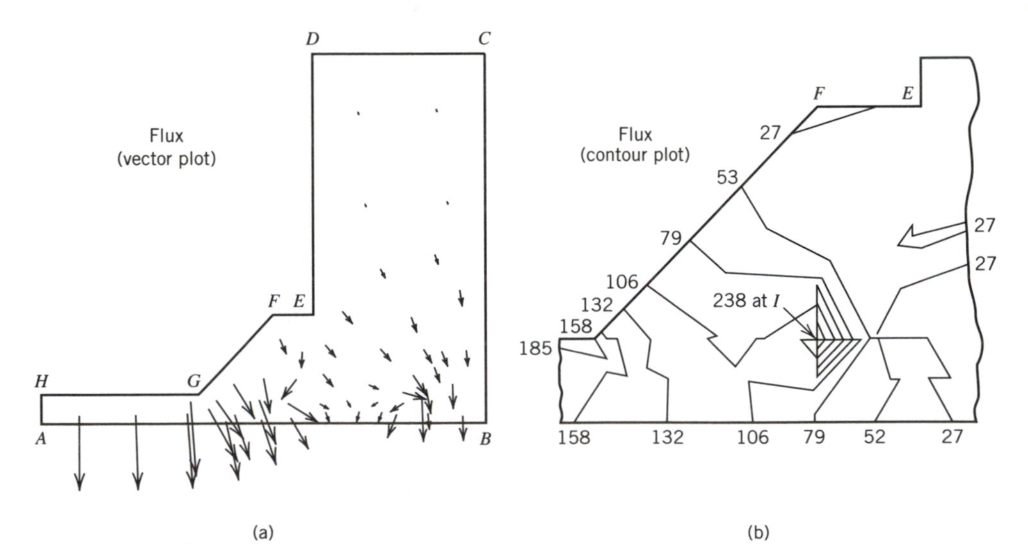

Fig. 8.6-4. (a) Vector plot of the computed flux field. (b) Detail of computed flux contours near point *I*. Flux units are thousands of watts per square meter (10^3 W/m^2).

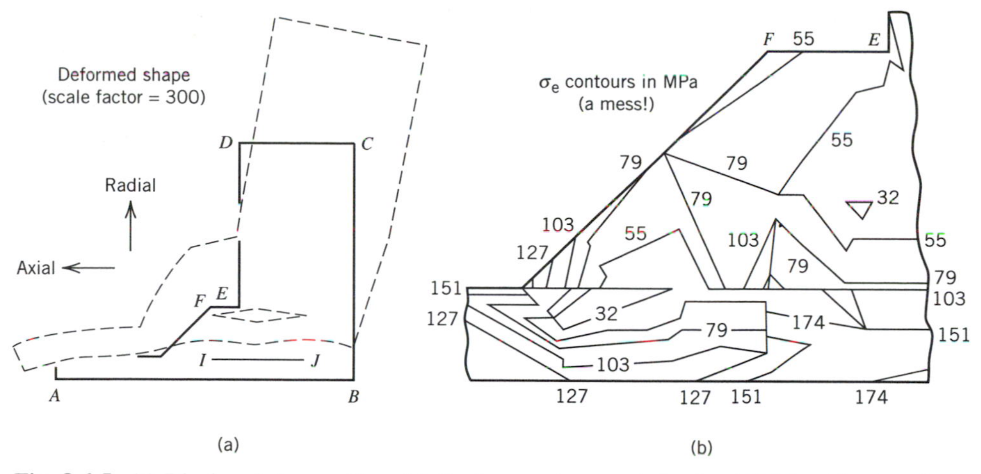

Fig. 8.6-5. (a) Displaced shape with axial displacement prevented only at B. Magnification factor 300. (b) Detail of computed contours of von Mises stress σ_e near point I. (The mesh is *much* too coarse.)

restrained by its comparatively massive surroundings. Material *fully* restrained, uniaxially stressed, and cooled 100°C relative to its supports would have the tensile stress $\alpha E \Delta T = 12(10^{-6})200(10^9)(100) = 240$ MPa. We expect that the maximum computed stress will be approximately this large. Similarly, temperatures at A and H are about 30°C and 90°C respectively, that is, each is 30°C different from the average temperature of about 60°C at the pipe midsurface. Therefore we expect to see circumferential and axial stresses of about $\alpha E\, \Delta T = 12(10^{-6})200(10^9)(30) = 72$ MPa, tensile inside and compressive outside.

In the first analysis, axial motion is prevented only at B. The deformed shape, exaggerated for plotting, is shown dashed in Fig. 8.6-5a, superposed on the undeformed shape. *The undeformed shape is assumed to prevail when all material is at the zero–degree reference*, which is 0°C in this example application. Point C moves less than 0.1 mm to the right, which is probably not enough to meet its neighboring point on the other flange and create a gap closure problem. In the first analysis and in the second, in which axial motion is prevented along BC, a gap opens along IJ, as expected (which incidentally confirms the assumption made in thermal analysis that there is no conduction across IJ). The left portion of the deformed FE model is not parallel to the pipe axis, which indicates that this portion is too short for accurate stress analysis. Circumferential and axial stresses at A and H are found to have approximately the values predicted. Other computed stresses appear in Table 8.6-1. Again the maximum stresses have approximately the values predicted. Unfortunately, stress contours have *gross* interelement discontinuities (Fig.

TABLE 8.6-1. Maximum and minimum stresses (in MPa) for different axial restraint along BC. Stresses are σ_r(radial), σ_θ(circumferential), σ_z(axial), and σ_e(von Mises). Locations are identified in Fig. 8.6-2.

	Only node B axially restrained				All nodes on BC axially restrained			
	σ_r	σ_θ	σ_z	σ_e	σ_r	σ_θ	σ_z	σ_e
Maximum stress	113	216	244	222	105	229	257	242
Location	*I,J*	*L-N*	*I,J*	*J*	*I,J*	*M-B*	*I,J,K*	*G,J*
Minimum stress	-68	-142	-199	8	-107	-225	-279	18
Location	*G*	*G*	*G*	*P*	*G*	*G*	*G*	*C*

8.6-5b). Similar results (not plotted) appear in the second analysis, when there is axial fixity of all nodes along BC. From the stress field, relative energy errors of the first and second stress analyses are both $\eta = 0.28$, far larger than the value $\eta = 0.03$ obtained in thermal analysis. We conclude that considerable mesh refinement is needed. Generalizing, we also conclude that a mesh adequate for thermal analysis may be quite inadequate for stress analysis.

ANALYTICAL PROBLEMS

8.1 Derive Eq. 8.1-4. Suggestions: If flux f_x enters the x-normal face of a differential volume $dV = dx\, dy\, dz$, flux $f_x + (\partial f_x/\partial x)dx$ exits the opposite face, and similarly for y-normal and z-normal faces.

8.2 Determine the 2 by 2 conductivity matrix $\mathbf{k}_T$ of a thin, flat, solid of revolution element, which is shown in cross section in the sketch. Lateral surfaces are insulated. The material is homogeneous and isotropic.

(a) Apply the direct method. Assume that the rate of heat flow has magnitude

$$q = 2\pi r_{\text{ave}} t f = \pi(a + b)t[k(T_2 - T_1)/(a - b)]$$

for all r in the range $b < r < a$.

(b) Apply the formal procedure, Eq. 8.2-4.

8.3 Lateral surfaces of the uniform bar shown are insulated. The bar is modeled by two elements, each of length L. Node 3 is maintained at temperature T_3. Constant heat flow q_1 is imposed at node 1. In terms of q_1, dimensions, and constants, what are T_1 and T_2 relative to T_3? Also, do T_1 and T_2 yield the expected value of q at node 3? Make use of the conductivity matrix in your solution.

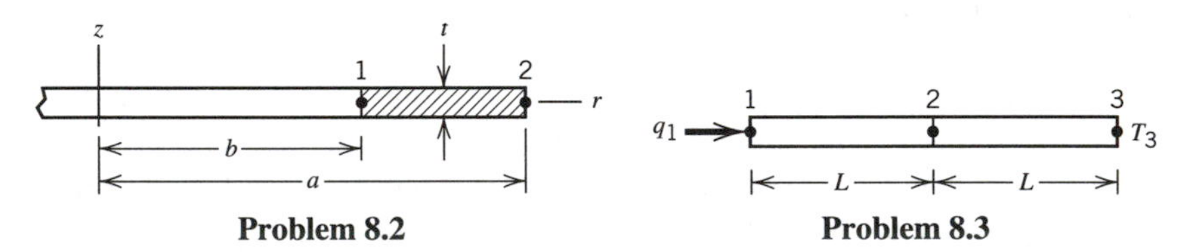

Problem 8.2 **Problem 8.3**

8.4 Repeat Problem 8.3, but use a stepped bar rather than a uniform bar. Let elements 1-2 and 2-3 have the respective cross-sectional areas A_o and $2A_o$.

8.5 Lateral surfaces of the uniform bar shown are insulated. The bar is modeled by three identical elements. Nodes 1 and 4 are maintained at the respective temperatures 0°C and 300°C. What are the temperatures at nodes 2 and 3? Make use of the conductivity matrix in your solution.

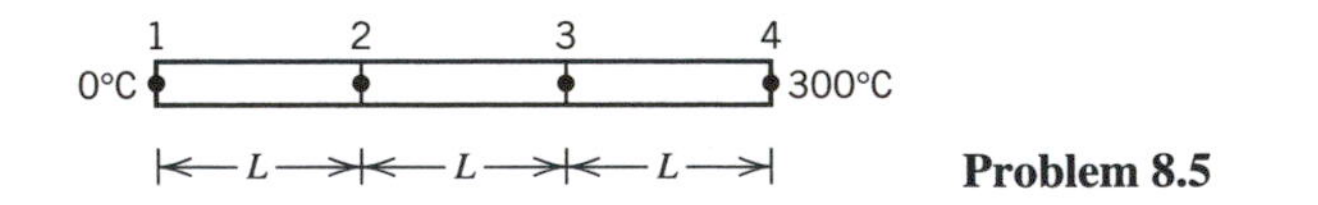

Problem 8.5

8.6 Repeat Problem 8.5, but use a stepped bar rather than a uniform bar. Let elements 1-2, 2-3, and 3-4 have the respective cross-sectional areas A_o, $2A_o$, and $3A_o$.

8.7 A uniform bar is modeled by two identical two-node elements, as shown. Node 1 is maintained at temperature $T_1 = 0$°C. The bar is surrounded by fluid of temperature

$T_f = 200°C$, which transfers heat to the bar across its cylindrical surface of area S. In the units listed at the end of Section 8.1, data are: $A = 300(10^{-6})$, $h = 600$, $k = 200$, and $S = 0.020$.

(a) In terms of h and S_e (the lateral surface area of one element), formulate the 2 by 2 element matrix and the 2 by 1 element vector whose general forms appear in Eqs. 8.2-6.

(b) Use the results of part (a) to form structure matrices, and use them to determine nodal temperatures T_2 and T_3.

(c) Repeat part (b), but alter the model so that the left element is half as long as the right element. The overall length of the model remains 0.300 m.

(d) Upon comparing results from parts (b) and (c), what conclusion do you draw?

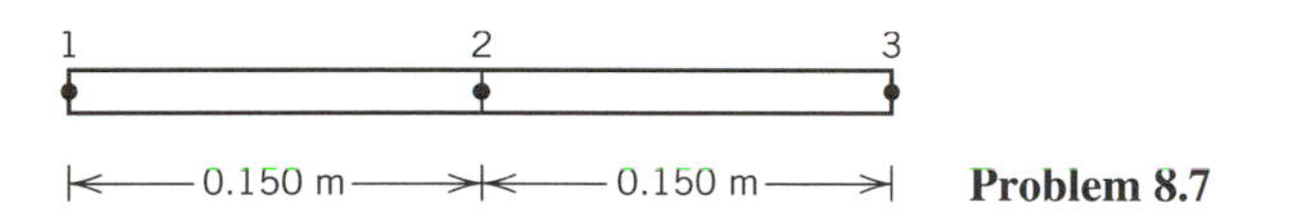

Problem 8.7

8.8 A pipe of outer radius b is covered by insulation of outer radius a, as shown in cross section. Convection heat transfer takes place at the outer surface of the insulation. It can be shown [8.1] that radial heat flow in a cylinder of length L is $q = 2\pi kL(T_b - T_a)/\ln(a/b)$. Assume that k and h are temperature independent and that T_b and T_f are independent of a and b. Show that q becomes maximum when $a = k/h$ (provided, of course, that $b < k/h$).

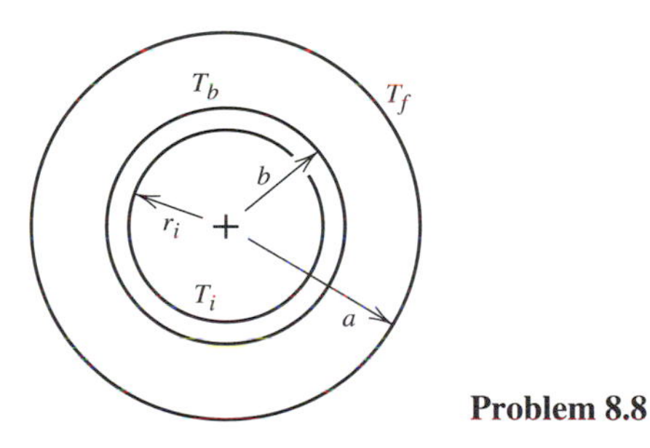

Problem 8.8

8.9 Over the range $0°C < T < 500°C$, the thermal conductivity of a certain metal in W/m · °C may be taken as $k = 73 - 0.06T$. Consider that lateral surfaces of the uniform bar shown are insulated and that ends are maintained at the respective temperatures 0°C and 500°C. Calculate the temperatures of nodes 2 and 3 and the heat flux per unit of cross-sectional area of the bar. Assume that conductivity k throughout an element is determined by the average of the element's two nodal temperatures. Carry out three iterations: assume that $T = T_0 = 0°C$ throughout the bar to start, and use Eq. 8.3-6 to solve successively for $\mathbf{T}_1$, $\mathbf{T}_2$, and $\mathbf{T}_3$.

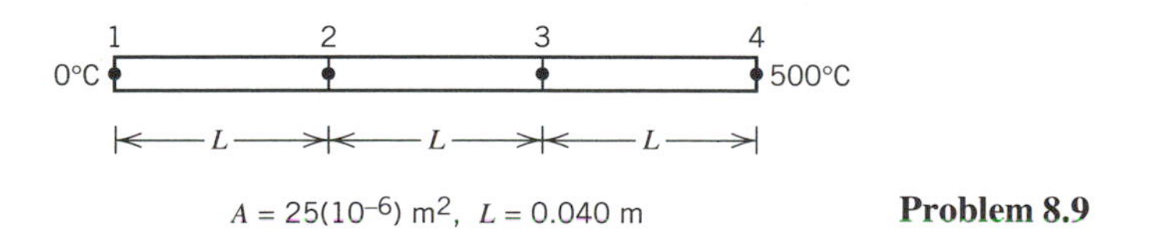

Problem 8.9

8.10 A 10-mm thick flat sheet of material ab is shown in cross section by the sketch. The sheet is parallel to a flat wall c that radiates. Assume that $k = 0.70$ W/m·°C and that

$\varepsilon_b = \varepsilon_c = 0.6$, all independent of temperature. If surface temperatures are maintained at $T_a = 20°C$ and $T_c = 600°C$, what is the temperature of surface b? Use the direct substitution method, even though there is but one unknown, namely, T_b. Assume $T_b = 100°C$ to start the process.

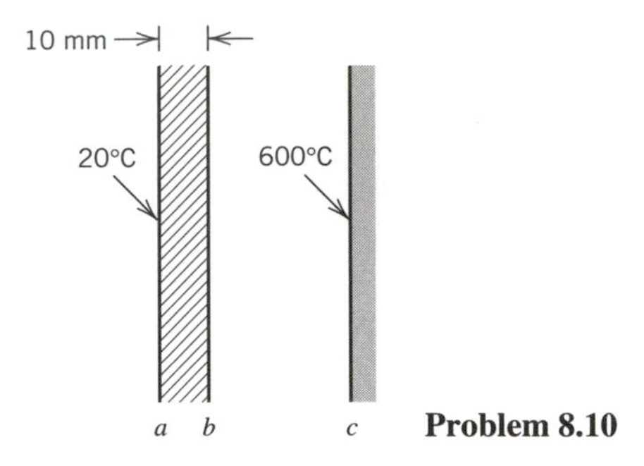

Problem 8.10

8.11 Imagine that a single-d.o.f. problem yields the differential equation $2T + 6\dot{T} = 8$. The initial condition is $T = 0$ at $t = 0$. Compute T as a function of time by the following methods.

(a) Obtain the exact solution.

(b) Use Eq. 8.4-5 with $\gamma = \frac{1}{2}$. Take eight steps with $\Delta t = 1.0$.

(c) Use Eq. 8.4-5 with $\gamma = \frac{1}{2}$. Take eight steps with $\Delta t = 10.0$.

COMPUTATIONAL PROBLEMS

In the following problems compute temperature and heat flux in the plane or solid bodies. Exploit symmetry if possible. Examine temperature contours, flux contours (unaveraged!), and vector plots of the flux field. If mesh refinement is undertaken, estimate the percentage error of FE results in the finest mesh. Unless otherwise stated, assume that plane models are homogeneous, isotropic, have unit thickness, and insulated lateral surfaces. Also assume that steady-state conditions prevail unless otherwise stated. Any of the steady-state problems can be converted to a transient problem by assuming that all nodal temperatures are initially uniform and change over time to a steady-state condition. Each thermal analysis can be followed by stress analysis if so desired.

Choose convenient numbers for quantities indicated by symbols. It is desirable to consult texts and handbooks for information about constants and coefficients and their temperature dependence. But for convenience the following approximate data are provided, in units used in the latter part of Section 8.1: $c = 460$ (steel), $c = 4200$ (water), $h = 10$ (air), $h = 1000$ (water), $k = 20$ (steel), $\rho = 7800$ (steel), $\rho = 1000$ (water). These values lie in wide ranges actually observed. The observed range of h is particularly large (two orders of magnitude for air; more for water if condensing steam is included).

A FE analysis should be preceded by an analytical approximation that is oversimplified if necessary. If these results and FE results have substantial disagreement we are warned of trouble somewhere.

8.12 Let the plane annular sector shown be homogeneous ($k_1 = k_2$); that is, ignore k_2, ϕ, β, c, and d in this problem.

(a) Let boundaries AB and CD be insulated. Impose temperature T_1 along AD and temperature T_2 along BC.

(b) Let boundaries AD and AB be insulated. Impose heat flow q along BC and temperature T_1 along CD except at corner C.

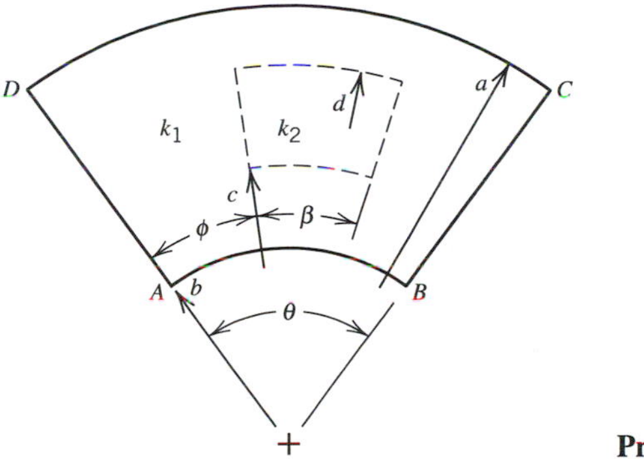

Problem 8.12

8.13 Repeat Problem 8.12, but let the sector be inhomogeneous ($k_1 \neq k_2$).

8.14 The plane annular sector shown is orthotropic. Principal material axes ns have constant orientation ϕ with respect to global Cartesian axes XY. Principal conductivities are k_n and $k_s = ck_n$, where c is a number greater than zero. Follow the instructions of parts (a) and (b) of Problem 8.12.

8.15 The plane annular sector shown has prescribed temperature T_1 along BC, CD, and DA, and prescribed temperature T_2 along AB. (This is the "undefined corner" problem noted in the text.) Evaluate the temperature field and, in particular, the temperature at point E, whose radial coordinate is $(2b + a)/3$. Use more than one mesh, starting with a coarse mesh. Choose T_A and T_B as follows.

(a) Prescribe $T_A = T_B = T_1$.
(b) Prescribe $T_A = T_B = T_2$.
(c) Prescribe $T_A = T_B = (T_1 + T_2)/2$.
(d) Let T_A and T_B be undefined. What temperature does the solution provide at A and B?

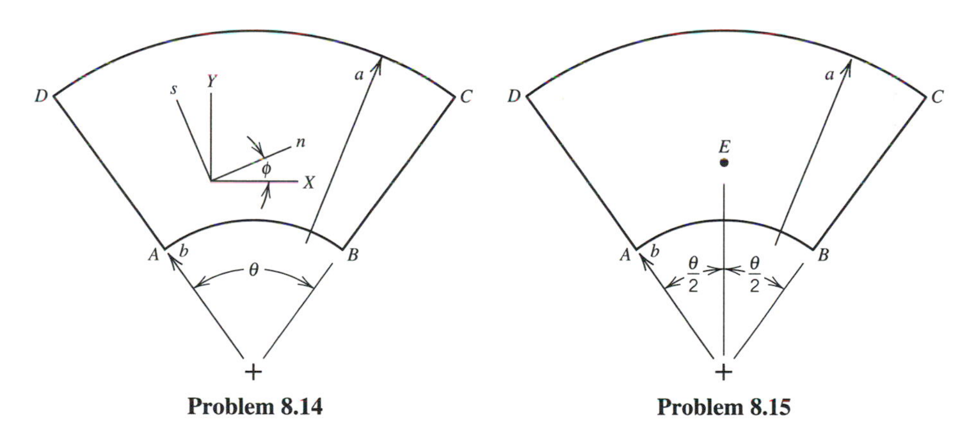

Problem 8.14 **Problem 8.15**

8.16 Repeat Problem 8.15, but prescribe T_1 along AB, BC, and CD, and T_2 along DA. Parts (a) through (d) now refer to corners A and D.

8.17 Consider a chimney of square cross section and uniform wall thickness. Gas inside has temperature T_1 and air outside has temperature T_2. Assume that conditions are the same from one cross section to the next.

8.18 A long pipe is covered by a layer of insulation (see the sketch for Problem 8.8). For the insulation one might use $k = 0.1$ W/m · °C.

(a) Impose temperatures T_i on the inside surface of the pipe and T_a on the outside surface of the insulation.

(b) Let the pipe carry water at temperature T_1 and let the insulation be surrounded by air at temperature T_f. Assume that conditions are axisymmetric.

8.19 A spherical vessel has a radially directed cylindrical outlet (see the sketch for Problem 6.24). Water at temperature T_1 fills the vessel and the outlet. The outside is surrounded by air at temperature T_f.

8.20 A glass cooking pot is filled with water at temperature T_1 and placed over a gas flame. Assume a reasonable shape and dimensions for the pot. Make a reasonable approximation for the value (or the variation) of T_f along sides of the pot. For glass, one might use $c = 800$ J/kg · °C, $k = 1$ W/m · °C, and $\rho = 2400$ kg/m^3.

8.21 A cylindrical shaft having a flat end is suddenly pushed against a flat surface and rotated with constant angular velocity, as shown. Let $P = 80$ kN, $R = 0.05$ m, and Ω correspond to 180 rpm. Assume that contact pressure is uniform, with a coefficient of friction of 0.2, and that the cylindrical surface of the shaft is surrounded by air at temperature T_f. Also assume that the flat surface is a perfect insulator.

(a) Compute the transient solution.

(b) Compute the steady-state solution.

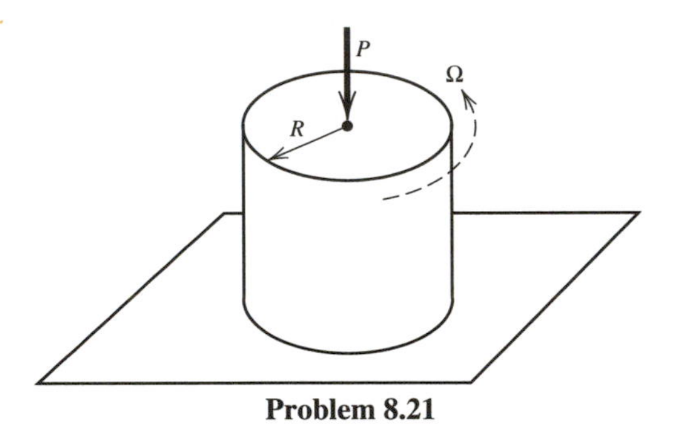

Problem 8.21

8.22 Repeat Problem 8.21, but now assume the flat surface is part of a very large 3D solid body, of the same material as the shaft and also exposed to air at temperature T_f on its surface.

8.23 A rectangular flat plate has dimensions a by b, as shown. Initially the plate is at temperature T_1. At time $t = 0$, a second closely spaced parallel plate, adjacent to portion $ABCD$ of the first plate but not shown, assumes and then maintains temperature T_2, and thus begins to exchange radiation with area $ABCD$ on one lateral surface of the first plate. On the remainder of this lateral surface and on the other lateral surface of the first plate, there is convection to air at temperature T_f. Assume that the emissivity of both plates is 0.3.

(a) Compute the transient solution.

(b) Compute the steady-state solution.

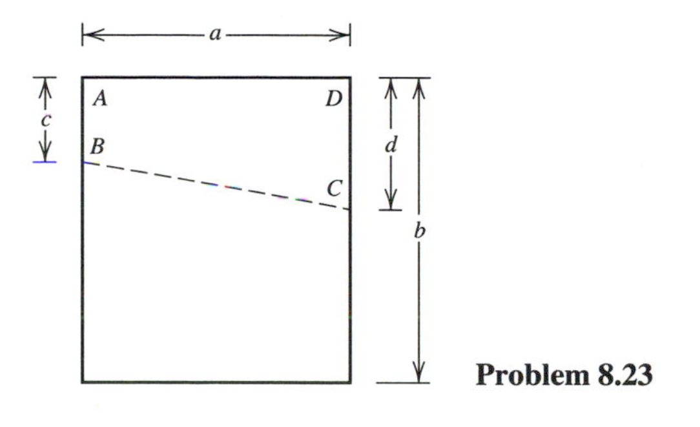

Problem 8.23

8.24 The sketch represents the cross section of a pipe. Initially the pipe and fluid flowing through it are at temperature T_1. At time $t = 0$, a closely spaced cylindrical surface spanning angle ϕ assumes and then maintains temperature T_2, and thus begins to exchange radiation with the adjacent outer surface of the pipe. Assume that conditions are the same from one cross section to the next.

(a) Compute the transient solution.

(b) Compute the steady-state solution.

8.25 The sketch represents a cross-sectional view of a large 3D solid body. Its flat surface AA, seen in edge view, is exposed to air at temperature T_1. Initially the solid body is at temperature T_2. At time $t = 0$, heat commences to be generated at a constant rate per unit volume in a cylindrical region of diameter D and height H.

(a) Compute the transient solution.

(b) Compute the steady-state solution.

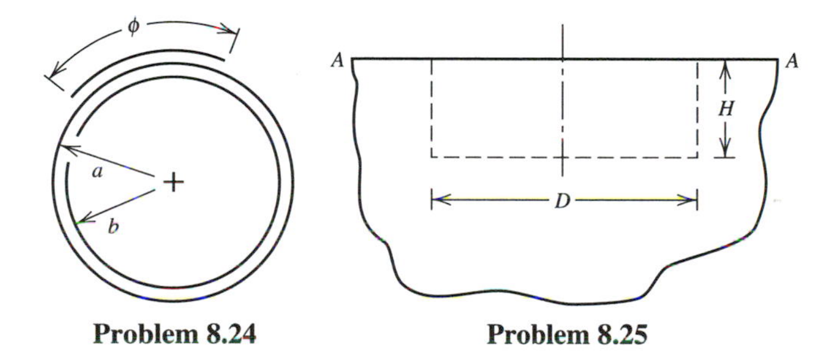

Problem 8.24 **Problem 8.25**

CHAPTER 9

Vibration and Dynamics

This chapter discusses how the FE method is used in problems of structural dynamics and vibrations. The more important aspects of theory and computational procedure are summarized. Application examples at the end of the chapter illustrate natural frequencies of vibration, harmonic response, the modal method for dynamic response, and response spectrum analysis.

9.1 INTRODUCTION

A structure moves as load is applied. If the load is cyclic, but with a frequency less than about one-third the structure's lowest natural frequency of vibration, the problem can probably be classed as static and analyzed by methods discussed in preceding chapters. If the load has a higher frequency, or varies randomly, or is applied suddenly, then a dynamic analysis is required. One may be interested in the largest acceleration in some part of a structure, the largest stresses, whether a structure will resonate with rotating machinery it supports, or other questions. Like static analysis, dynamic analysis uses a stiffness matrix, but it also uses a mass matrix and a damping matrix. Accordingly, FE modeling in dynamics includes most aspects of FE modeling in statics, and many additional concepts and options as well.

This chapter does not present a comprehensive treatment of structural dynamics, its underlying theory, and computational algorithms. For detailed development of these matters the reader is urged to consult specialty textbooks, of which there are many. Here we discuss in a nonrigorous way some types of dynamic analysis commonly performed by users of FE software. Specifically, we discuss natural frequencies of vibration, steady forced vibration, and transient response analysis. Included is a discussion of some tools needed to carry out these analyses, such as the use of vibration modes, reduction in the number of d.o.f., and integration in time. All tools discussed appear in one or more large-scale FE codes.

The reader who consults other books will find that there is little agreement on terminology. A given concept or procedure may be known by several names, some of which are noted in what follows.

9.2 BASIC EQUATIONS. VIBRATION

Single d.o.f. Figure 9.2-1a shows a single mass m and a single linear spring k. Motion is described by the single d.o.f. $u = u(t)$ and is governed by Newton's law $f = ma$.

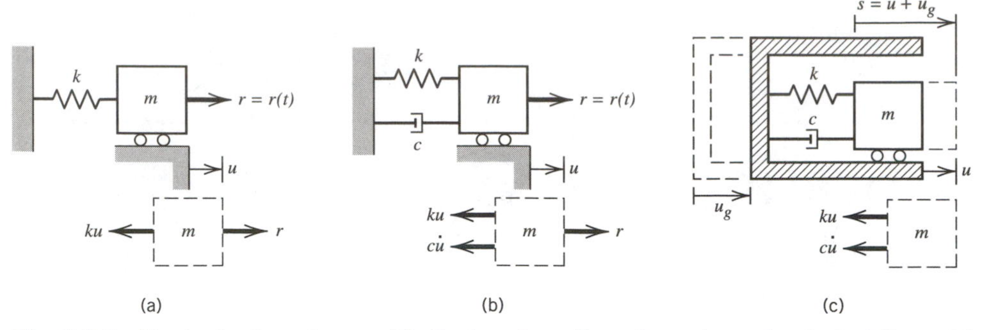

Fig. 9.2-1. Single d.o.f. systems, with displaced configurations shown by dashed lines. (a) Undamped system. (b) Damped system. (c) Support excitation $u_g = u_g(t)$.

Acceleration is $a = \ddot{u}$ (in common notation, a dot indicates differentiation with respect to time, so that $\dot{u} = du/dt$ and $\ddot{u} = d^2u/dt^2$). Accordingly, in Fig. 9.2-1a,

$$f = ma \qquad \text{hence} \qquad r - ku = m\ddot{u} \qquad \text{or} \qquad ku + m\ddot{u} = r \tag{9.2-1}$$

where $r = r(t)$ is an externally applied load or "forcing function" that varies with time in a known fashion.

In Fig. 9.2-1b, a viscous damper is added, using the standard "dashpot" symbol for it. A viscous damper resists velocity $\dot{u}$ with force $c\dot{u}$. Newton's law now yields

$$r - ku - c\dot{u} = m\ddot{u} \qquad \text{or} \qquad ku + c\dot{u} + m\ddot{u} = r \tag{9.2-2}$$

In Fig. 9.2-1c, the body that supports mass m is subjected to a prescribed motion $u_g = u_g(t)$. Displacement of the mass is u relative to the support and is $s = u + u_g$ relative to a fixed reference frame. Elastic and damping forces depend on the relative motion, but the inertia force depends on the absolute acceleration $\ddot{s} = d^2s/dt^2$. Accordingly, $f = ma$ yields

$$-ku - c\dot{u} = m(\ddot{u} + \ddot{u}_g) \qquad \text{or} \qquad ku + c\dot{u} + m\ddot{u} = -m\ddot{u}_g \tag{9.2-3}$$

The known forcing function is therefore $r = -m\ddot{u}_g$. This form of the equation of motion is convenient when measurements are made relative to the support, as, for example, when the support is the earth and u_g is the known motion of an earthquake.

Multiple d.o.f. If we deal with a FE structure, for which there is more than one d.o.f., u is replaced by a vector **D** of nodal d.o.f. and r is replaced by a load vector **R**, which may contain moments as well as forces and which is a known function of time. Also, k, c, and m are, respectively, replaced by a stiffness matrix **K**, a damping matrix **C**, and a mass matrix **M**. Thus the equation analogous to Eq. 9.2-2 *for a multi-d.o.f. structure* is

$$\text{multi-d.o.f.:} \quad \mathbf{KD} + \mathbf{C\dot{D}} + \mathbf{M\ddot{D}} = \mathbf{R} \qquad \text{where} \qquad \mathbf{R} = \mathbf{R}(t) \tag{9.2-4}$$

This is the governing equation of structural dynamics, written in its most common form. It states that externally applied forces **R** are resisted by the sum of three internal forces: stiffness forces **KD**, damping forces $\mathbf{C\dot{D}}$, and inertia forces $\mathbf{M\ddot{D}}$. All these forces are time dependent in a dynamics problem. If there are no nonlinearities, **K**, **C**, and **M** contain only constants; they are not functions of time or of displacement. In general, given **K**, **C**,

M, and **R**, one seeks to compute displacements **D**, velocities $\dot{\mathbf{D}}$, and accelerations $\ddot{\mathbf{D}}$ as functions of time. The goal may be more restricted, as, for example, when one seeks only the frequencies and modes of vibration. These computations, as well as mass and damping matrices, are considered in subsequent sections.

Single-d.o.f. Free Vibration. Vibration analysis is one of the most commonly performed dynamic analyses. Vibration of multi-d.o.f. systems is considered in Section 9.4. Here we consider a single d.o.f. The motion is called "free" vibration if the forcing function is omitted. Thus $r = 0$ in Eqs. 9.2-1 and 9.2-2.

If there is no damping, we use Eq. 9.2-1. After motion begins, as the result of initial conditions that do not concern us here, the mass vibrates with simple harmonic motion,

$$u = \bar{u} \sin \omega t \tag{9.2-5}$$

where $\bar{u}$ is the *amplitude* of motion and ω is the *natural frequency* of vibration. Another name for ω is *circular* frequency. Its units are radians per second. The *cyclic* frequency is $f = \omega/2\pi$, whose units are hertz (abbreviated Hz; the number of cycles per second). The *period* is $T = 1/f$. The motion is shown in Fig. 9.2-2a. Different initial conditions would shift the sine wave along the t axis. The value of ω is obtained by substitution of Eq. 9.2-5 into Eq. 9.2-1. Thus

$$k\bar{u} \sin \omega t - m\bar{u}\omega^2 \sin \omega t = 0 \qquad \text{from which} \qquad \omega = \sqrt{\frac{k}{m}} \tag{9.2-6}$$

Frequency ω is independent of whether amplitude $\bar{u}$ is large or small, provided that there are no nonlinearities. Nonlinearity could be introduced by yielding of the spring or by collision of the mass with a support.

If damping is present, free vibration is described by Eq. 9.2-2 with $r = 0$. It can be shown that if damping c is less than a "critical" value c_c the motion is oscillatory, but the amplitude of motion decays with time as shown in Fig. 9.2-2b. The damped natural frequency ω_d is less than the undamped natural frequency $\omega = \sqrt{k/m}$ and is given by

$$\omega_d = \omega\sqrt{1-\xi^2} \qquad \text{where} \qquad \xi = \frac{c}{c_c} \qquad \text{and} \qquad c_c = 2m\omega \tag{9.2-7}$$

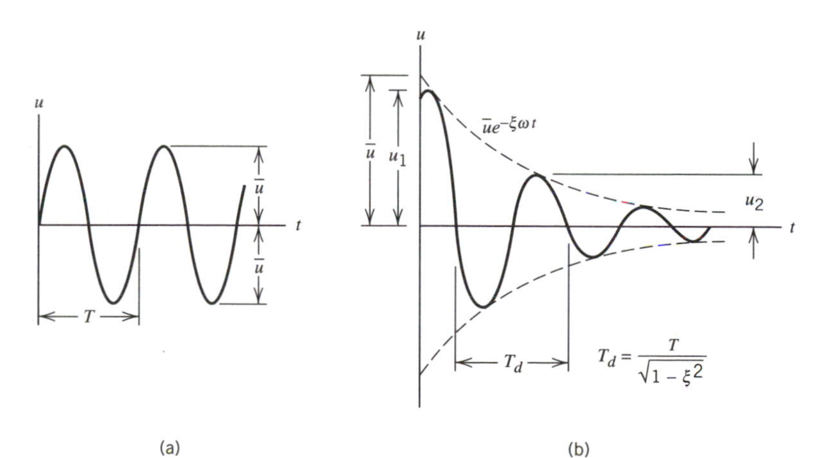

Fig. 9.2-2. (a) Undamped free vibration. (b) Damped free vibration ($c < c_c$).

where ξ, the fraction of critical damping, is called the *damping ratio.* Structural damping is usually small—typically ξ is less than 0.15—so that $\omega_d \approx \omega$. For small damping, the ratio of any two consecutive displacement peaks, u_2/u_1 in Fig. 9.2-2b, is related to ξ by the equation $\ln(u_2/u_1) \approx -2\pi\xi$. Thus, if $\xi = 0.1$, the amplitude of motion is reduced by about half in one cycle of vibration.

9.3 MASS MATRICES

The physical meaning of a mass matrix is analogous to the physical meaning of a stiffness matrix. The jth column of an element *stiffness* matrix is the vector of nodal loads that must be applied to the element in order to maintain the displacement field created by a unit value of the jth d.o.f. The jth column of an element *mass* matrix is the vector of nodal loads that must be applied to the element in order to maintain the *acceleration* field created by a unit value of the second time derivative of the jth d.o.f.

The simplest and historically earliest way of representing mass is by mass particles. The process is called "mass lumping" and results in a diagonal or "lumped" mass matrix. For example, consider lateral displacements of a two-node bar element of cross-sectional area A, length L, and mass density ρ. The element mass is therefore ρAL. As shown in Fig. 9.3-1a, mass lumping implies a discontinuous displacement field in which the two halves of the element translate separately. Accelerations $\ddot{v}_1$ and $\ddot{v}_2$ of the respective halves are associated with forces F_1 and F_2, each in accordance with Newton's law $f = ma$. Thus the lumped element mass matrix $\mathbf{m}$ is

$$\mathbf{m} = \begin{bmatrix} \rho AL/2 & 0 \\ 0 & \rho AL/2 \end{bmatrix} \qquad \text{so that} \qquad \mathbf{m}\begin{Bmatrix} \ddot{v}_1 \\ \ddot{v}_2 \end{Bmatrix} = \begin{Bmatrix} F_1 \\ F_2 \end{Bmatrix} \tag{9.3-1}$$

If general plane motion were allowed, so that element nodal d.o.f. became $\mathbf{d} = [u_1 \quad v_1 \quad u_2 \quad v_2]^T$, $\mathbf{m}$ would be a 4 by 4 *diagonal* matrix whose nonzero terms are $m_{ii} = \rho AL/2$, $i = 1, 2, 3, 4$.

It is more reasonable to assume a linear variation for the lateral displacement $v = v(x)$ of a two-node bar element. This results in a linear distribution of inertia force, Fig. 9.3-1b, whose intensity is $\rho A\ddot{v}_1$ and $\rho A\ddot{v}_2$ at the respective ends. Treating inertia force as a

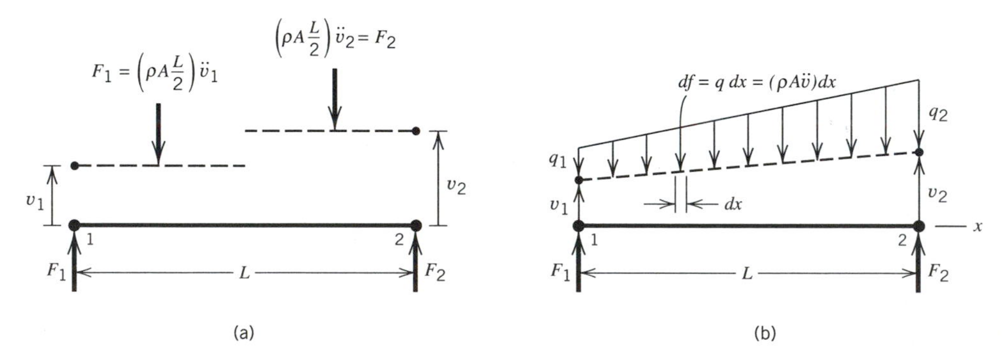

Fig. 9.3-1. Conceptual lateral displacements of a bar element and inertia forces for mass idealizations. (a) Provides diagonal mass matrix. (b) Provides nondiagonal mass matrix. $q_1 = \rho A\ddot{v}_1$, $q_2 = \rho A\ddot{v}_2$.

distributed load and applying equations of statics, we obtain

$$F_1 = \rho AL\left(\tfrac{1}{3}\ddot{v}_1 + \tfrac{1}{6}\ddot{v}_2\right) \qquad F_2 = \rho AL\left(\tfrac{1}{6}\ddot{v}_1 + \tfrac{1}{3}\ddot{v}_2\right) \tag{9.3-2}$$

Thus the element mass matrix **m** is

$$\mathbf{m} = \begin{bmatrix} \rho AL/3 & \rho AL/6 \\ \rho AL/6 & \rho AL/3 \end{bmatrix} \quad \text{so that} \quad \mathbf{m}\begin{Bmatrix} \ddot{v}_1 \\ \ddot{v}_2 \end{Bmatrix} = \begin{Bmatrix} F_1 \\ F_2 \end{Bmatrix} \tag{9.3-3}$$

For general plane motion, the m_{ij} in **m** of Eq. 9.3-3 would also apply to axial d.o.f. u_1 and u_2. The resulting **m** would be 4 by 4 and would contain eight nonzero coefficients.

For a plane *beam* element we must add rotational d.o.f. θ_{z1} and θ_{z2} to Fig. 9.3-1b; **m** then becomes a 4 by 4 matrix. Mass particles have no rotational inertia, so particle-lumping for a four-d.o.f. beam element produces a diagonal mass matrix whose only nonzero terms are $m_{11} = m_{33} = \rho AL/2$, associated with $\ddot{v}_1$ and $\ddot{v}_2$. If mass is also associated with rotational d.o.f., m_{22} and m_{44} become nonzero. Somewhat arbitrarily, we could say that $m_{22} = m_{44} = \rho AL^3/24$, which is the moment required to create unit angular acceleration of a bar of length $L/2$ pivoted at one end.

For elements in general the foregoing process is inadequate, in the same way that elementary direct methods fail to provide the stiffness matrix of an arbitrary element. By using inertia forces in virtual work arguments, it can be shown that a general formula for an element mass matrix is

$$\mathbf{m} = \int \mathbf{N}^T\mathbf{N}\rho \, dV \tag{9.3-4}$$

where ρ is mass density, V is the element volume, and **N** is the element shape function matrix. Equation 9.3-4 provides the "consistent" element mass matrix, so called because it uses the same shape functions as are used to produce the element stiffness matrix (Eq. 3.1-10). If applied to a two-node bar element with a linear lateral displacement field, Eq. 9.3-4 yields Eq. 9.3-3. If applied to a plane *beam* element (Fig. 2.3-1), whose lateral displacement field is cubic in x, Eq. 9.3-4 yields

$$\mathbf{m} = \frac{\rho AL}{420}\begin{bmatrix} 156 & 22L & 54 & -13L \\ 22L & 4L^2 & 13L & -3L^2 \\ 54 & 13L & 156 & -22L \\ -13L & -3L^2 & -22L & 4L^2 \end{bmatrix} \tag{9.3-5}$$

which operates on the accelerations $\ddot{\mathbf{d}} = [\ddot{v}_1 \quad \ddot{\theta}_{z1} \quad \ddot{v}_2 \quad \ddot{\theta}_{z2}]^T$.

All of the foregoing element mass matrices correctly represent resistance to translational acceleration. They differ in how resistance to angular acceleration is modeled. Whether a mass matrix is diagonal or full, and whether or not rotary inertia is associated with rotational d.o.f., proper convergence with mesh refinement is assured if element mass matrices provide the correct inertia force in response to all possible translational accelerations of the element. If elements are compatible and fully integrated and mass matrices are consistent, computed natural frequencies ω_i of the FE model are guaranteed to be upper bounds on natural frequencies of the mathematical model, that is, guaranteed to converge from above with mesh refinement. In other words, subject to the restrictions noted, discretization error raises computed natural frequencies.

The structure mass matrix **M** is formed by assembly of element mass matrices **m**, in the same way that the structure stiffness matrix is formed by assembly of element stiffness matrices. A consistent mass matrix **M** has the same pattern of zero and nonzero coefficients as the corresponding stiffness matrix **K**.

One must be careful with units. If lengths are in meters, $f = ma$ yields a force in newtons if m is in kilograms and a is in meters per second squared. If lengths are in millimeters, m (kilograms) must be divided by 1000 to obtain the force in newtons. Or, in terms of mass density, a ρ expressed as kg/m^3 must be divided by 10^{12} if length units are to be millimeters and force is still to be in newtons.

9.4 UNDAMPED FREE VIBRATION

We begin with Eq. 9.2-4. With no damping, $\mathbf{C} = \mathbf{0}$. Vibration is "free" if loads **R** are either zero or constant. Constant massless loads affect natural frequencies only if the loads create significant prestress, which modifies stiffness. Actual loads are usually associated with mass, which must be included in **M**. In vibration analysis we ask for the natural frequencies and modes of a structure without regard to which of them may be important in an application and without regard to how vibration is initiated. All d.o.f. move in phase with one another, and at the same frequency. Accordingly, all of the time-dependent displacements reach their maximum magnitudes at the same instant of time. Vibrational motion, calculated as described below, consists of displacements that vary sinusoidally with time relative to the mean configuration $\mathbf{D}_m$ created by constant loads $\mathbf{R}_c$. If $\mathbf{R}_c = \mathbf{0}$, the mean configuration is the unstressed configuration. Symbolically,

$$\mathbf{D} = \mathbf{D}_m + \overline{\mathbf{D}} \sin \omega t \qquad \text{where} \qquad \mathbf{D}_m = \mathbf{K}^{-1}\mathbf{R}_c \tag{9.4-1}$$

in which $\overline{\mathbf{D}}$ is the vector of nodal displacement amplitudes in vibration and ω is a natural frequency in radians per second. Hence $\ddot{\mathbf{D}} = -\omega^2\overline{\mathbf{D}} \sin \omega t$. Substituting all this information and $\mathbf{C} = \mathbf{0}$ into Eq. 9.2-4, we obtain

$$[\mathbf{K} - \omega^2\mathbf{M}]\overline{\mathbf{D}} = \mathbf{0} \tag{9.4-2}$$

as the governing equation of undamped free vibration. Mathematically, Eq. 9.4-2 is called an *eigenvalue problem.* It has the trivial solution $\overline{\mathbf{D}} = \mathbf{0}$. We are interested in nontrivial solutions, of which there are as many as the number of d.o.f. The ith nontrivial solution consists of a natural frequency of vibration and its associated mode $\overline{\mathbf{D}}_i$. A natural frequency may also be called a resonant frequency, and ω_i^2 is variously called an eigenvalue, latent root, or characteristic number. A mode may also be called an eigenvector, mode shape, normal mode, natural mode, characteristic mode, or principal mode. The smallest nonzero ω_i is called the *fundamental* frequency of vibration. Various algorithms are available for solving Eq. 9.4-2.

Some characteristics of a structural eigenvalue problem and the resulting frequencies and modes are as follows.

- Unlike a static problem, no support is necessary. A structure that is unsupported or partly supported has rigid-body modes; that is, displacement modes in which all strains are zero. For each rigid-body mode the natural frequency is zero. (Some solution algorithms fail if there are zero eigenvalues, unless an "eigenvalue shift" is invoked, either by the user or automatically by the software.)

- Equation 9.4-2 may be written in the form $\mathbf{K}\overline{\mathbf{D}}_i = -\omega_i^2 \mathbf{M}\overline{\mathbf{D}}_i$, which has the physical interpretation that a vibration mode is a configuration in which elastic loads are in balance with inertia loads.
- The relation among d.o.f. in $\overline{\mathbf{D}}_i$ defines the shape of mode i, but otherwise the magnitudes of the d.o.f. have no meaning. Thus, if c is an arbitrary number (including $c = -1$), $\overline{\mathbf{D}}_i$ and $c\overline{\mathbf{D}}_i$ are the same mode. For output, software may "normalize" each mode by scaling it so that its largest d.o.f. is unity. A more common normalization, at least for theory and internal processing, scales each mode $\overline{\mathbf{D}}_i$ so that

$$\overline{\mathbf{D}}_i^T \mathbf{M} \overline{\mathbf{D}}_i = 1 \tag{9.4-3}$$

  If stresses were computed directly from a displacement mode they would probably be unrealistically high because of scaling.
- Eigenvectors are orthogonal with respect to stiffness and mass matrices; that is, for $i \neq j$, $\overline{\mathbf{D}}_i^T \mathbf{K} \overline{\mathbf{D}}_j = 0$ and $\overline{\mathbf{D}}_i^T \mathbf{M} \overline{\mathbf{D}}_j = 0$.
- Usually, unless there is shock loading, only the modes of lowest frequency are important in the structural response. Accordingly, even if the model has a great many d.o.f., usually only the lowest frequencies and modes are extracted from Eq. 9.4-2.
- Lower modes exhibit fewer "waves" than higher modes. Accordingly, in lower modes there are more elements per wave and there is less discretization error.

Example. A uniform slender cantilever beam is modeled by a single element (Fig. 9.4-1a). Only flexural motion in the plane of the paper is to be considered. We will calculate frequencies and modes provided by two forms of the mass matrix. The calculation methods used in this example are suitable only for very small matrices and hand calculation.

Only the d.o.f. at node 2 are nonzero; therefore we use only terms from the lower right-hand corners of the element stiffness and mass matrices. Beginning with the consistent mass matrix, we use terms from Eqs. 2.3-2 and 9.3-5. Equation 9.4-2 becomes

$$\left(\frac{EI}{L^3} \begin{bmatrix} 12 & -6L \\ -6L & 4L^2 \end{bmatrix} - \omega^2 \frac{\rho A L}{420} \begin{bmatrix} 156 & -22L \\ -22L & 4L^2 \end{bmatrix} \right) \begin{Bmatrix} \bar{v}_2 \\ \bar{\theta}_{z2} \end{Bmatrix} = \begin{Bmatrix} 0 \\ 0 \end{Bmatrix} \tag{9.4-4}$$

For a nontrivial solution, the determinant of the complete matrix in Eq. 9.4-4 must vanish; that is,

$$\begin{vmatrix} 12 - 156a & -6L + 22La \\ -6L + 22La & 4L^2 - 4L^2 a \end{vmatrix} = 0 \qquad \text{where} \qquad a = \frac{\omega^2 \rho A L^4}{420 EI} \tag{9.4-5}$$

Solving the determinant for a and then solving for ω^2, we obtain the frequencies of mode

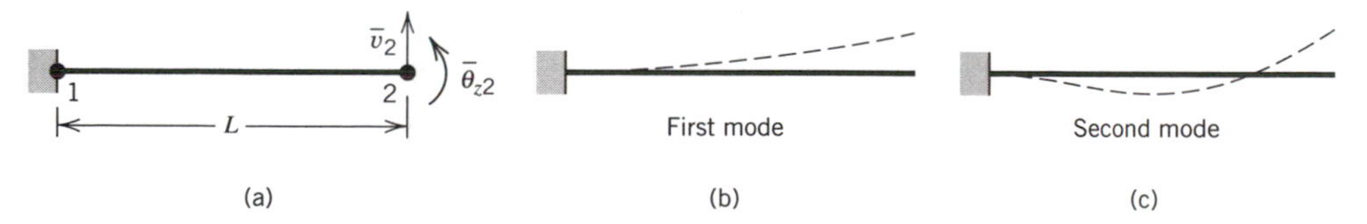

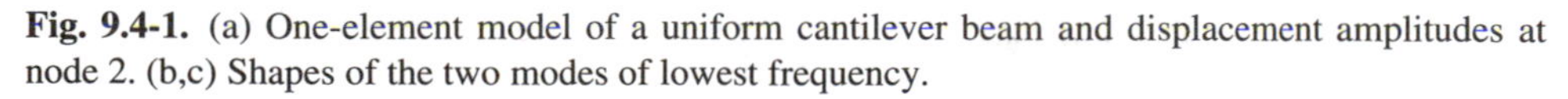

Fig. 9.4-1. (a) One-element model of a uniform cantilever beam and displacement amplitudes at node 2. (b,c) Shapes of the two modes of lowest frequency.

1 and mode 2:

$$\omega_1 = 3.533\left(\frac{EI}{\rho AL^4}\right)^{1/2} \qquad \omega_2 = 34.81\left(\frac{EI}{\rho AL^4}\right)^{1/2} \tag{9.4-6}$$

The shape of the ith mode can be obtained by substituting ω_i into either equation in Eq. 9.4-4, setting $\bar{v}_2 = 1$, and solving for $\bar{\theta}_{z2}$. In this way we obtain the mode shapes $\bar{v}_2 = 1$, $\bar{\theta}_{z2} = 1.38/L$ for the first mode and $\bar{v}_2 = 1$, $\bar{\theta}_{z2} = 7.62/L$ for the second mode. The displaced shape between nodes 1 and 2 is cubic, defined by the usual interpolation (Eq. 2.3-4). The two mode shapes are shown in Fig. 9.4-1b,c. Dashed lines show only one extreme position; the other is obtained by reversing the algebraic signs of all d.o.f. (Typical FE software plots only straight lines between nodes. Thus, in the present one-element example, software would plot each mode shape as a single straight line, even though the deformed beam centerline is actually a cubic curve.)

If we use instead lumped masses without rotary inertia, the only nonzero mass term in Eq. 9.4-4 becomes $\rho AL/2$, associated with $\bar{v}_2$. Computed frequencies are then

$$\omega_1 = 2.449\left(\frac{EI}{\rho AL^4}\right)^{1/2} \qquad \omega_2 \text{ not obtainable} \tag{9.4-7}$$

The only mode shape obtainable is associated with $\bar{v}_2 = 1$ and a value of $\bar{\theta}_{z2}$ consistent with a static transverse tip force sufficient to produce $\bar{v}_2 = 1$, namely, $\bar{\theta}_{z2} = 1.50/L$.

Vibration theory shows that the exact multipliers of $(EI/\rho AL^4)^{1/2}$ are 3.516 for mode 1 and 22.03 for mode 2. FE results are inexact because the assumed cubic deflection on which **k** and **m** are based cannot represent a mode shape exactly. As expected, FE results are more accurate for ω_1 than for ω_2, and the consistent mass matrix has overestimated the frequencies. In this example (but not in all problems), the consistent mass matrix outperforms a lumped mass matrix.

Only two modes and frequencies can be obtained in the foregoing example because it uses only two d.o.f. If length L were divided into (say) five elements, we would increase the number of d.o.f. to 10. Then 10 modes and frequencies would be obtainable. Also, we would obtain more accurate frequencies and mode shapes for the lower modes.

9.5 DAMPING

Damping dissipates energy and causes the amplitude of free vibration to decay with time (Fig. 9.2-2b). Viscous damping applies forces to the structure proportional to velocity but oppositely directed. Unfortunately, common sources of damping are not viscous and are neither easy to measure nor as easy to represent mathematically as viscous damping. These sources include "internal friction" in the material and Coulomb friction in connections, neither of which exerts forces proportional to velocity. Fortunately, damping in structural problems is usually small enough that it can be idealized as viscous, regardless of the actual damping mechanism, and represented by a form chosen for its mathematical convenience. "Small enough" usually means that damping forces $\mathbf{C}\dot{\mathbf{D}}$ are less than roughly 10% of forces $\mathbf{KD}$, $\mathbf{M}\ddot{\mathbf{D}}$, and $\mathbf{R}$ in Eq. 9.2-4. Typical values of damping ratio ξ range from about 0.02 for piping systems to about 0.07 for bolted structures and reinforced concrete. There are two common devices for including viscous damping in FE analysis: *proportional damping* and *modal damping*.

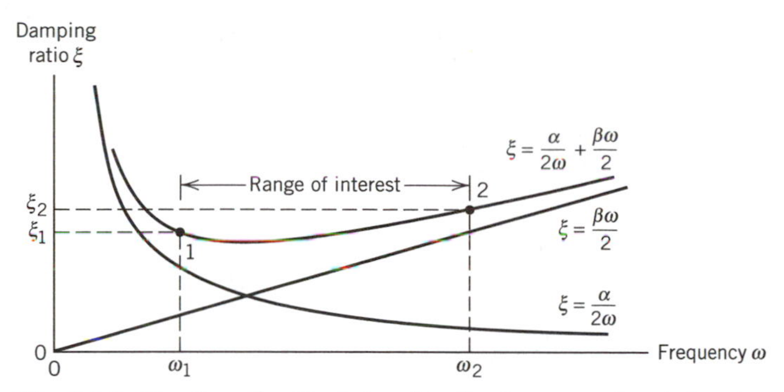

Fig. 9.5-1. Fraction of critical damping for the proportional damping scheme.

Proportional damping, also called Rayleigh damping, refers to the following arbitrary and nonphysical definition of the damping matrix:

$$\mathbf{C} = \alpha\mathbf{M} + \beta\mathbf{K} \tag{9.5-1}$$

The meaning of this definition, in terms of the damping ratio ξ, is shown in Fig. 9.5-1. Values of ω_1, ω_2, ξ_1, and ξ_2 are chosen by the analyst. By simultaneous solution of the two equations

$$\xi_1 = \frac{\alpha}{2\omega_1} + \frac{\beta\omega_1}{2} \qquad \xi_2 = \frac{\alpha}{2\omega_2} + \frac{\beta\omega_2}{2} \tag{9.5-2}$$

for α and β, matrix **C** of Eq. 9.5-1 is established. With this definition of **C**, one must accept the curve between points 1 and 2 as a satisfactory representation of damping over the frequency range of interest, $\omega_1 < \omega < \omega_2$. While this representation may be acceptable, it cannot be physically correct because it demands infinite damping at $\omega = 0$. The form $\mathbf{C} = \alpha\mathbf{M}$ damps lower modes more heavily; the form $\mathbf{C} = \beta\mathbf{K}$ damps higher modes more heavily.

Modal damping is another way of incorporating viscous damping. Like Eq. 9.5-1, it is an approximate representation, whose form yields equations that are comparatively easy to solve. It is discussed following Eq. 9.7-5.

Some software is capable of calculating damped natural frequencies. As shown by Eq. 9.2-7, these frequencies are only slightly smaller than undamped natural frequencies for common structures, for which $\xi < 0.15$.

9.6 REDUCTION

"Reduction" may also be called "condensation." It is a way of making a few d.o.f. represent *all* d.o.f. of the model. Thus the size of matrices is reduced and in some problems an analysis can be performed more cheaply and more quickly. The disadvantage of reduction is the introduction of some error and uncertainty into the analysis. Modeling guidelines that reduce error and uncertainty are presented in Section 9.11.

Theory. In reduction, some d.o.f., called *slaves*, have their motion dictated by other d.o.f., called *masters*. Slaves are discarded and masters are retained. The order of matrices in the reduced problem is equal to the number of masters. A commonly used method of reduction is called Guyan reduction [9.1]. Its principal assumption is that inertia forces on slaves are negligible in comparison with elastic forces transmitted to slaves by the motion of masters. For a mathematical statement of this assumption, we begin with the equation of free vibration, Eq. 9.4-2, partitioned according to masters $\overline{\mathbf{D}}_m$ and slaves $\overline{\mathbf{D}}_s$.

$$\left(\begin{bmatrix} \mathbf{K}_{mm} & \mathbf{K}_{ms} \\ \mathbf{K}_{ms}^T & \mathbf{K}_{ss} \end{bmatrix} - \omega^2 \begin{bmatrix} \mathbf{M}_{mm} & \mathbf{M}_{ms} \\ \mathbf{M}_{ms}^T & \mathbf{M}_{ss} \end{bmatrix}\right) \begin{Bmatrix} \overline{\mathbf{D}}_m \\ \overline{\mathbf{D}}_s \end{Bmatrix} = \begin{Bmatrix} \mathbf{0} \\ \mathbf{0} \end{Bmatrix} \tag{9.6-1}$$

The principal assumption is that the relation between $\overline{\mathbf{D}}_m$ and $\overline{\mathbf{D}}_s$ is dictated entirely by stiffness coefficients. Accordingly, we temporarily ignore all mass coefficients in the lower partition of Eq. 9.6-1 and obtain from it

$$\mathbf{D}_s = -\mathbf{K}_{ss}^{-1}\mathbf{K}_{ms}^T\mathbf{D}_m \tag{9.6-2}$$

Overbars have now been omitted from $\mathbf{D}_m$ and $\mathbf{D}_s$ because the master–slave transformation can be applied generally; its use need not be restricted to eigenvalue problems. The entire set of d.o.f. $\mathbf{D}$ is expressed in terms of masters by the equation

$$\mathbf{D} = \begin{Bmatrix} \mathbf{D}_m \\ \mathbf{D}_s \end{Bmatrix} = \mathbf{T}\mathbf{D}_m \qquad \text{where} \qquad \mathbf{T} = \begin{bmatrix} \mathbf{I} \\ -\mathbf{K}_{ss}^{-1}\mathbf{K}_{ms}^T \end{bmatrix} \tag{9.6-3}$$

where $\mathbf{I}$ is a unit matrix. Physically, the jth column of $\mathbf{T}$ represents the static displacement of the structure when the jth master has unit displacement and all other masters have zero displacement. This displacement state is sometimes called a *constraint mode.*

Substitution of Eq. 9.6-3 into Eq. 9.4-2, followed by premultiplication by $\mathbf{T}^T$, yields the reduced eigenproblem.

$$[\mathbf{K}_r - \omega^2\mathbf{M}_r]\overline{\mathbf{D}}_m = \mathbf{0} \qquad \text{where} \qquad \begin{cases} \mathbf{K}_r = \mathbf{T}^T\mathbf{K}\mathbf{T} \\ \mathbf{M}_r = \mathbf{T}^T\mathbf{M}\mathbf{T} \end{cases} \tag{9.6-4}$$

Similarly, the basic statement of matrix structural dynamics, Eq. 9.2-4, has the reduced form

$$\mathbf{K}_r\mathbf{D}_m + \mathbf{C}_r\dot{\mathbf{D}}_m + \mathbf{M}_r\ddot{\mathbf{D}}_m = \mathbf{R}_r \qquad \text{where} \qquad \begin{cases} \mathbf{C}_r = \mathbf{T}^T\mathbf{C}\mathbf{T} \\ \mathbf{R}_r = \mathbf{T}^T\mathbf{R} \end{cases} \tag{9.6-5}$$

After computing masters, as amplitudes by use of Eq. 9.6-4 or as functions of time by use of Eq. 9.6-5, the full set of d.o.f. may be needed for stress calculation. Slaves may be recovered from Eq. 9.6-2 or, for better accuracy in eigenvalue problems, from the lower partition of Eq. 9.6-1 with all mass coefficients retained.

Modeling guidelines for how masters and slaves should be chosen appear in Section 9.11. For now we simply state that d.o.f. having large mass to stiffness ratio are likely candidates for masters. Alternative methods of reduction are available; see the closing paragraph of Section 9.7.

Example. Consider a cantilever beam modeled by a single beam element (Fig. 9.4-1a). We will reduce the FE model to a single d.o.f., then calculate the fundamental frequency of vibration.

The full system is stated in Eq. 9.4-4. The d.o.f. $\bar{v}_2$ has the largest mass to stiffness ratio M_{ii}/K_{ii} and is therefore chosen as master. Equations 9.6-2 and 9.6-3 yield

$$\bar{\theta}_{z2} = \frac{3}{2L}\bar{v}_2 \qquad \text{and} \qquad \mathbf{T} = \begin{Bmatrix} 1 \\ 3/2L \end{Bmatrix} \tag{9.6-6}$$

The equation of the reduced system, Eq. 9.6-4, becomes

$$\left(\frac{3EI}{L^3} - \omega^2 \frac{33\rho AL}{140}\right)\bar{v}_2 = 0 \tag{9.6-7}$$

in which $3EI/L^3$ is recognized as the transverse stiffness of a tip-loaded cantilever beam when the tip is unrestrained. Equation 9.6-7 yields

$$\omega_1 = 3.568\left(\frac{EI}{\rho AL^4}\right)^{1/2} \tag{9.6-8}$$

This is the only frequency obtainable from the reduced system. It is larger than the frequency obtained from the full system, Eq. 9.4-6, because reduction has the effect of applying a constraint, which stiffens the system.

9.7 MODAL EQUATIONS

In Section 9.6, reduction is accomplished by expressing displacements of the full system in terms of static displacement modes associated with displacements of master d.o.f. In the present section, reduction is accomplished by expressing displacements of the full system in terms of a limited number of its vibration modes, specifically the modes of lowest frequency. The result is a set of *uncoupled* equations. The number of these equations is equal to the number of vibration modes retained in the transformation. Modal equations are useful in the analysis of harmonic response (Section 9.8) and dynamic response (Sections 9.9 and 9.10).

Theory. Some properties of vibration modes must be stated at the outset. First, *we require that each mode be scaled according to Eq. 9.4-3.* Now let Eq. 9.4-2 be written for mode i, then premultiplied by $\bar{\mathbf{D}}_i^T$, and finally solved for ω_i^2. Thus

$$\bar{\mathbf{D}}_i^T[\mathbf{K} - \omega_i^2\mathbf{M}]\bar{\mathbf{D}}_i = 0 \qquad \text{yields} \qquad \omega_i^2 = \bar{\mathbf{D}}_i^T\mathbf{K}\bar{\mathbf{D}}_i \tag{9.7-1}$$

It can be shown that vibration modes are orthogonal with respect to $\mathbf{K}$ and $\mathbf{M}$. That is, for $i \neq j$, $\bar{\mathbf{D}}_i^T\mathbf{K}\bar{\mathbf{D}}_j = 0$ and $\bar{\mathbf{D}}_i^T\mathbf{M}\bar{\mathbf{D}}_j = 0$. Columns of the *modal matrix* $\boldsymbol{\phi}$ are modes $\bar{\mathbf{D}}_i$ of the full system:

$$\underset{n\times n}{\boldsymbol{\phi}} = [\bar{\mathbf{D}}_1 \quad \bar{\mathbf{D}}_2 \quad \cdots \quad \bar{\mathbf{D}}_n] \tag{9.7-2}$$

where n is the number of d.o.f. in the full system. With this definition, Eq. 9.4-3, and the aforementioned orthogonality properties of the modes, we conclude that

$$\boldsymbol{\phi}^T\mathbf{K}\boldsymbol{\phi} = \boldsymbol{\omega}^2 \qquad \text{and} \qquad \boldsymbol{\phi}^T\mathbf{M}\boldsymbol{\phi} = \mathbf{I} \tag{9.7-3}$$

where $\boldsymbol{\omega}^2$ is a *diagonal* matrix of squared natural frequencies, $\boldsymbol{\omega}^2 = \lceil \omega_1^2 \quad \omega_2^2 \cdots \omega_n^2 \rfloor$, which is called the *spectral matrix*, and $\mathbf{I}$ is a unit matrix.

An arbitrary displacement vector $\mathbf{D}$ can be expressed as a linear combination of the vibration modes, that is, as $\mathbf{D} = \overline{\mathbf{D}}_1 z_1 + \overline{\mathbf{D}}_2 z_2 + \cdots + \overline{\mathbf{D}}_n z_n$, where z_i is the fraction of mode i that contributes to $\mathbf{D}$. For the time being we include all n modes in the transformation. Symbolically, the transformations for displacement, velocity, and acceleration are

$$\mathbf{D} = \boldsymbol{\phi}\mathbf{z} \qquad \text{hence} \qquad \dot{\mathbf{D}} = \boldsymbol{\phi}\dot{\mathbf{z}} \qquad \text{and} \qquad \ddot{\mathbf{D}} = \boldsymbol{\phi}\ddot{\mathbf{z}} \tag{9.7-4}$$

where $\mathbf{z}$ is the column vector $\mathbf{z} = [z_1 \quad z_2 \cdots z_n]^T$. The z_i may be called "principal coordinates" or "modal coordinates" to distinguish them from physical coordinates, which are the d.o.f. D_i themselves. Next, we substitute Eqs. 9.7-4 into Eq. 9.2-4, premultiply by $\boldsymbol{\phi}^T$, and take note of Eq. 9.7-3. Thus

$$\boldsymbol{\omega}^2\mathbf{z} + \mathbf{C}_\phi\dot{\mathbf{z}} + \ddot{\mathbf{z}} = \boldsymbol{\phi}^T\mathbf{R} \tag{9.7-5}$$

where, if the proportional damping defined by Eq. 9.5-1 is used, $\mathbf{C}_\phi$ is the *diagonal* matrix $\mathbf{C}_\phi = \alpha\mathbf{I} + \beta\boldsymbol{\omega}^2$. More often, modal equations make use of *modal* damping, which is a damping representation that has as much (or as little) justification as proportional damping, but works well for problems in which damping is small. Modal damping has a mathematically convenient form. To obtain it, we *arbitrarily* state that $\mathbf{C}_\phi$ is to be replaced by an alternative diagonal matrix whose ith diagonal coefficient is $C_{\phi ii} = 2\xi_i\omega_i$, where ξ_i is the damping ratio for mode i. Because $\boldsymbol{\omega}^2$ and the damping matrix are both diagonal, Eqs. 9.7-5 are *uncoupled.* With modal damping, the representative single-d.o.f. equation for any mode i is usually written in the form

$$\omega_i^2 z_i + 2\xi_i\omega_i\dot{z}_i + \ddot{z}_i = p_i \qquad \text{where} \qquad p_i = \boldsymbol{\phi}_i^T\mathbf{R} \tag{9.7-6}$$

where $\boldsymbol{\phi}_i$ is the ith column of $\boldsymbol{\phi}$; that is, $\boldsymbol{\phi}_i = \overline{\mathbf{D}}_i$, scaled according to Eq. 9.4-3. As justification for Eq. 9.7-6, note that Eq. 9.2-2 assumes the same form if it is divided by m and the substitutions $\omega = \sqrt{k/m}$ from Eq. 9.2-6 and $c/m = 2\xi\omega$ from Eq. 9.2-7 are made.

One may use the same ξ_i for all i if so desired. Frequencies ω_i and time-dependent modal loads p_i are in general different for each i. The z_i are computed as functions of time by solving Eq. 9.7-6 for each mode i. Hence Eqs. 9.7-4 yield displacements, velocities, and accelerations of the structural d.o.f. as functions of time.

Remarks. An important merit of modal equations is that, for many practical problems, we need not use all n modes of the system: *only the lowest few modes need be retained in the transformation.* In other words, if the lowest m modes are used, $\mathbf{D} = \boldsymbol{\phi}\mathbf{z}$ is replaced by

$$\mathbf{D} \approx \sum_{i=1}^{m} \overline{\mathbf{D}}_i z_i \qquad \text{where} \qquad \mathbf{D} \text{ is } n \times 1 \qquad \text{and} \qquad m < n \tag{9.7-7}$$

In effect, $\boldsymbol{\phi}$ becomes an n by m matrix rather than an n by n matrix, where $m << n$ is usually adequate to provide the required accuracy. Therefore an eigensolver need extract only a few frequencies and the corresponding modes, and Eqs. 9.7-6 are few in number. Conversely stated, modal equations are best suited to problems in which higher modes are unimportant; for example, they are better suited to earthquake loading than to shock loading. For shock loading, so many modes may be needed (especially to compute accelerations) that it may be better to compute response by a method other than modal equations.

Proportional damping and modal damping have both been mentioned in connection with Eq. 9.7-5. The two forms are both approximate but they are not equivalent. However, since structural damping is usually small, we expect that results will be similar and that either method will be adequate.

From Eqs. 9.6-3 and 9.7-4 we see that $\mathbf{T}$ and $\boldsymbol{\phi}$ play similar roles in transforming equations of the full system. Indeed, $\boldsymbol{\phi}$ can be regarded merely as a transformation matrix. Equation 9.6-3 leads to a fully coupled set of equations. Equation 9.7-4 leads to uncoupled equations but requires that an eigenproblem be solved. As an alternative to $\boldsymbol{\phi}$, a set of "Ritz vectors" can be generated without solving an eigenproblem. Several generalizations of these transformations are useful [2.2,9.2] but we will not pursue them here.

9.8 HARMONIC RESPONSE ANALYSIS

Harmonic response analysis seeks the *amplitude* of response when prescribed loads vary sinusoidally with time. Several loads may be simultaneously applied and they may or may not be in phase with one another. However, all loads must have the same frequency if methods discussed in the present section are to be applicable. Loads of different frequency require dynamic response analysis (Sections 9.9 and 9.10). Alternative names attached to harmonic response analysis include *harmonic loading, frequency response analysis,* and *forced vibration.*

When first applied, harmonic loading does not produce harmonic response. There is an initial transient that decays to zero over time because of damping. What then remains is steady-state motion of the same frequency as the loading. Harmonic response analysis addresses the steady-state response. An application is calculating the response of a building frame to machinery that runs at constant speed.

Single d.o.f. System. To introduce the subject, consider the spring–mass–damper system of Fig. 9.2-1b. Let it be loaded by the harmonic forcing function

$$r = F_0 \sin \Omega t \qquad (F_0 = \text{constant}) \tag{9.8-1}$$

which has circular frequency Ω. It can be shown [9.2,9.3] that the steady-state displacement response is

$$u = \bar{u} \sin(\Omega t - \alpha) \tag{9.8-2}$$

in which

$$\bar{u} = \frac{F_0/k}{\left[\left(1-\beta^2\right)^2 + \left(2\xi\beta\right)^2\right]^{1/2}} \qquad \alpha = \arctan\frac{2\xi\beta}{1-\beta^2} \qquad \beta = \frac{\Omega}{\omega} \tag{9.8-3}$$

where $\omega = \sqrt{k/m}$ and damping ratio ξ is defined by Eq. 9.2-7. We see that response u lags force r by the phase angle α. Steady-state displacement amplitude $\bar{u}$ is proportional to static displacement F_0/k, modified by a divisor that depends on ξ and β. These results are plotted in Fig. 9.8-1. Peaks of the curves occur slightly to the left of an ordinate at $\beta = 1.0$ because $\omega_d < \omega$, as shown by Eq. 9.2-7.

Static deflection F_0/k is greatly amplified if frequency Ω of the forcing function is near natural frequency ω of the system. If $\Omega = \omega$, a condition called *resonance*, the amplitude

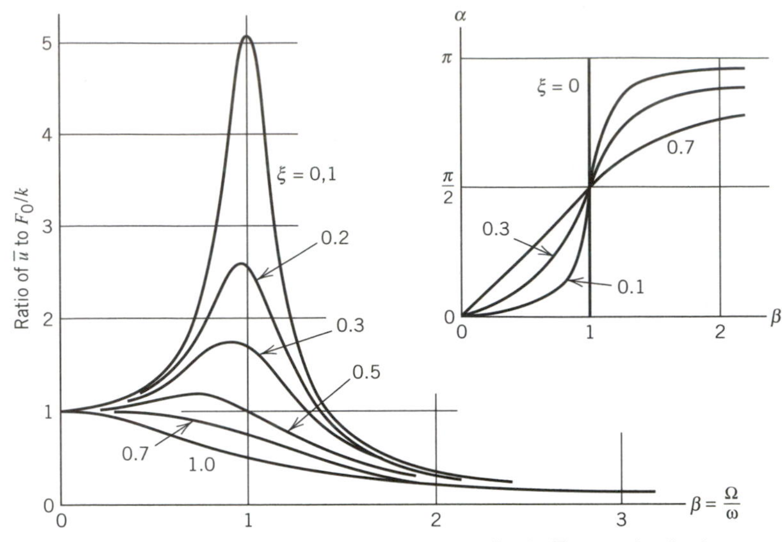

Fig. 9.8-1. Ratio of dynamic displacement amplitude $\bar{u}$ to static displacement F_0/k, and phase angle α, for a single-d.o.f. system (Eq. 9.8-3).

of displacement is limited only by damping and would be infinite if damping were zero. The displacement approaches zero as β becomes large: the force varies so rapidly that there is almost no time for the mass to respond.

Multi-d.o.f. structures behave qualitatively as described above for a single-d.o.f. system, but there are as many natural frequencies as d.o.f. There is a possibility of large-amplitude oscillations when the forcing function frequency coincides with any of the natural frequencies. Two methods of analysis for multi-d.o.f. structures are now described.

FE Structures: Modal Method. The first step is to solve the free vibration problem, Eq. 9.4-2, to obtain frequencies ω_i and modes $\overline{\mathbf{D}}_i$ of the system. Usually only the lower frequencies and modes are needed. Next, we must establish the z_i of Section 9.7 for each mode i retained in the analysis. For harmonic loading, each such equation has a solution of the form seen in Eqs. 9.8-2 and 9.8-3, namely,

$$z_i = \frac{p_i/\omega_i^2}{\left[\left(1-\beta_i^2\right)^2 + \left(2\xi_i\beta_i\right)^2\right]^{1/2}} \sin(\Omega t - \alpha_i) \tag{9.8-4}$$

in which

$$\alpha_i = \arctan\frac{2\xi_i\beta_i}{1-\beta_i^2} \qquad \beta_i = \frac{\Omega}{\omega_i} \qquad \Omega \text{ same for all } i \tag{9.8-5}$$

The next step is to convert back to physical coordinates $\mathbf{D}$ by applying Eq. 9.7-7 to the computed $\overline{\mathbf{D}}$ and z_i. Since phase angles α_i are different for different modes, the software must scan a cycle in order to determine the greatest magnitude (i.e., the amplitude) of displacement, velocity, or acceleration in the cycle. The amplitude may appear at a different time in the cycle for different d.o.f. Note, however, that α_i is a phase angle relative to p_i. Accordingly, if two or more forcing functions are applied and they are not in phase, an additional phase angle must enter into the calculations for each forcing function not in phase with the others. The analyst must consult the software documentation in order to state input data properly.

FE Structures: Direct Method. Some harmonic response calculations are simplified by the use of complex numbers. The following is a summary of results rather than an exposition of theory. The harmonic response equation can be written in the form

$$[\mathbf{K} + i\Omega\mathbf{C} - \Omega^2\mathbf{M}]\bar{\mathbf{D}} = \bar{\mathbf{R}} \tag{9.8-6}$$

where $i = \sqrt{-1}$ and $\bar{\mathbf{D}}$ and $\bar{\mathbf{R}}$ are complex amplitudes, respectively, of displacement and forcing function, both having circular frequency Ω. The coefficient of $\bar{\mathbf{D}}$ may be called a "dynamic stiffness matrix." If damping is zero and $\mathbf{M}$ is diagonal, the dynamic stiffness matrix can be regarded as $\mathbf{K}$ augmented by a spring of negative stiffness $\Omega^2 M_{ii}$ attached to each d.o.f. i. If damping is zero, Ω must not coincide with any natural frequency of the structure, lest the dynamic stiffness become zero. A disadvantage of Eq. 9.8-6 is that if Ω is *close* to a natural frequency, the dynamic stiffness matrix becomes ill conditioned. Another disadvantage is the expense of solving Eq. 9.8-6 several times if forcing functions of several different frequencies Ω must be investigated, as is common.

Both disadvantages are overcome if Eq. 9.8-6 is recast by use of the modal matrix $\boldsymbol{\phi}$ defined in Section 9.7 [9.4]. By substitution of Eq. 9.7-4 into Eq. 9.8-6 and premultiplication by $\boldsymbol{\phi}^T$, Eq. 9.8-6 becomes

$$[\boldsymbol{\omega}^2 + i\Omega\mathbf{C}_\phi - \Omega^2\mathbf{I}]\bar{\mathbf{z}} = \boldsymbol{\phi}^T\bar{\mathbf{R}} \tag{9.8-7}$$

where $\mathbf{C}_\phi$ is diagonal if proportional damping or modal damping is used. Solving for $\bar{\mathbf{z}}$ and using Eq. 9.7-4 to recover $\bar{\mathbf{D}}$, we obtain

$$\bar{\mathbf{D}} = \boldsymbol{\phi}[\boldsymbol{\omega}^2 + i\Omega\mathbf{C}_\phi - \Omega^2\mathbf{I}]^{-1}\boldsymbol{\phi}^T\bar{\mathbf{R}} \tag{9.8-8}$$

As is usual in modal analysis, only the several lowest modes are used in the transformation. The bracketed matrix in Eq. 9.8-8 is diagonal, so it is easy to avoid trouble if Ω is close to or equal to a natural frequency ω, and the inversion is trivial.

9.9 DYNAMIC RESPONSE ANALYSIS

Dynamic response analysis is the calculation of how a structure responds to arbitrary time-dependent loading. Alternative names include *transient response analysis* and *time-history analysis.* Available methods include *modal analysis* and *direct integration.* The term "direct" means that no transformation to a special form is required, in contrast to the principal mode form of modal analysis. For dynamic response analysis, Eq. 9.2-4 is written in the form

$$\mathbf{K}\mathbf{D}_n + \mathbf{C}\dot{\mathbf{D}}_n + \mathbf{M}\ddot{\mathbf{D}}_n = \mathbf{R}_n \tag{9.9-1}$$

where $\mathbf{R}_n$ is the known time-dependent forcing function at the nth instant. We seek $\mathbf{D}_n$, $\dot{\mathbf{D}}_n$, and $\ddot{\mathbf{D}}_n$ at particular instants of time. Thus $n = 0, 1, 2, \ldots$ corresponds to times $t = 0$, $t = \Delta t$, $t = 2\Delta t$, ... , where Δt is a time increment. After calculating response at instant n, time is advanced by Δt, and response is calculated at instant $n + 1$. Subsequently, a plot of displacement, velocity, or acceleration of any d.o.f. can be constructed by connecting computed points.

We will discuss two methods of direct integration and also discuss the mode superposition method. Each method has a different combination of advantages and disadvantages. Direct integration methods produce an equation of the form

$$\mathbf{A}\mathbf{D}_{n+1} = \mathbf{F}(t) \qquad \text{or} \qquad \mathbf{D}_{n+1} = \mathbf{A}^{-1}\mathbf{F}(t) \tag{9.9-2}$$

in which **A** is nonsingular, and independent of time in linear problems, while **F** depends on quantities at instant n and perhaps also at instant $n - 1$. The specific forms of **A** and **F** depend on the algorithm chosen. Of the many direct integration algorithms available we will summarize two: the *central difference method* and the *Newmark method.*

Direct Integration: Central Difference Method. The basis of the method is a set of finite difference formulas for first and second time derivatives, centered at instant n. For a single d.o.f. u, velocity $\dot{u}$ is approximated as $(u_{n+1} - u_{n-1})/2\,\Delta t$ and acceleration $\ddot{u}$ is approximated as $\Delta\dot{u}/\Delta t$, where $\Delta\dot{u}$ comes from velocities evaluated at instants $n - \frac{1}{2}$ and $n + \frac{1}{2}$, a time interval Δt apart. For a vector of d.o.f. rather than a single d.o.f., these formulas are

$$\dot{\mathbf{D}}_n \approx \frac{1}{2\,\Delta t}\{\mathbf{D}_{n+1} - \mathbf{D}_{n-1}\} \qquad \text{and} \qquad \ddot{\mathbf{D}}_n \approx \frac{1}{(\Delta t)^2}\{\mathbf{D}_{n+1} - 2\mathbf{D}_n + \mathbf{D}_{n-1}\} \tag{9.9-3}$$

From Eqs. 9.9-1, 9.9-2, and 9.9-3 we obtain

$$\mathbf{A} = \frac{1}{2\,\Delta t}\mathbf{C} + \frac{1}{(\Delta t)^2}\mathbf{M} \tag{9.9-4a}$$

$$\mathbf{F}(t) = \mathbf{R} - \left[\mathbf{K} - \frac{2}{(\Delta t)^2}\mathbf{M}\right]\mathbf{D}_n - \left[\frac{1}{(\Delta t)^2}\mathbf{M} - \frac{1}{2\,\Delta t}\mathbf{C}\right]\mathbf{D}_{n-1} \tag{9.9-4b}$$

Thus, for any n, $\mathbf{D}_{n+1}$ is calculated from known values of $\mathbf{D}_n$ and $\mathbf{D}_{n-1}$. Velocities and accelerations, if desired, are then available from Eqs. 9.9-3. Initial displacements $\mathbf{D}_0$ and velocities $\dot{\mathbf{D}}_0$ are known. To calculate $\mathbf{D}_1$ we need to know $\mathbf{D}_{-1}$. It may be determined from a Taylor series expansion about instant n, which for $n = 0$ and a backward Δt yields $\mathbf{D}_{-1} = \mathbf{D}_0 - \Delta t\,\dot{\mathbf{D}}_0 + (\Delta t)^2\ddot{\mathbf{D}}_0/2$, in which $\ddot{\mathbf{D}}_0$ is determined from Eq. 9.9-1 at instant $n = 0$.

The central difference method is *conditionally stable*: if Δt is too large, computed displacements become wildly inaccurate and grow without limit. To guarantee numerical stability, we must use

$$\Delta t < \Delta t_{cr} \qquad \text{where} \qquad \Delta t_{cr} = \frac{2}{\omega_{max}} \equiv \frac{T_{min}}{\pi} \tag{9.9-5}$$

where ω_{max} is the largest undamped natural frequency of the system. This frequency can be calculated from Eq. 9.4-2 or by use of more efficient alternatives [2.2].

Direct Integration: Newmark Method. The basis of the method is the following set of equations:

$$\mathbf{D}_{n+1} \approx \mathbf{D}_n + \Delta t\,\dot{\mathbf{D}}_n + \frac{(\Delta t)^2}{2}\left\{(1-2\beta)\ddot{\mathbf{D}}_n + 2\beta\ddot{\mathbf{D}}_{n+1}\right\} \tag{9.9-6a}$$

$$\dot{\mathbf{D}}_{n+1} \approx \dot{\mathbf{D}}_n + \Delta t\{(1-\gamma)\ddot{\mathbf{D}}_n + \gamma\ddot{\mathbf{D}}_{n+1}\} \tag{9.9-6b}$$

where β and γ are numbers that can be chosen by the analyst. Substitution of Eqs. 9.9-6 into 9.9-1 yields

$$\mathbf{A} = \mathbf{K} + \frac{\gamma}{\beta\Delta t}\mathbf{C} + \frac{1}{\beta(\Delta t)^2}\mathbf{M} \tag{9.9-7a}$$

$$\mathbf{F}(t) = f(\beta, f, \Delta t, \mathbf{R}, \mathbf{C}, \mathbf{M}, \mathbf{D}_n, \dot{\mathbf{D}}_n, \ddot{\mathbf{D}}_n) \tag{9.9-7b}$$

The expression for $\mathbf{F}(t)$ has not been written out because it is lengthy [2.2] and not essential to our discussion. Using Eqs. 9.9-2 and 9.9-7, $\mathbf{D}_{n+1}$ is calculated from known values of $\mathbf{D}_n$, $\dot{\mathbf{D}}_n$, and $\ddot{\mathbf{D}}_n$. Hence Eq. 9.9-6a yields $\ddot{\mathbf{D}}_{n+1}$, Eq. 9.9-6b yields $\dot{\mathbf{D}}_{n+1}$, and we are ready to calculate $\mathbf{D}_{n+2}$. To start the process we need to know $\ddot{\mathbf{D}}_0$. It may be determined from Eq. 9.9-1 at instant $n = 0$.

Certain choices of β and γ make the Newmark method *unconditionally stable*: the numerical process will not "blow up" no matter how large the value of Δt. This does *NOT* mean that results are guaranteed *accurate* if Δt is large! It can be shown [9.5] that $2\beta \geq \gamma \geq \frac{1}{2}$ guarantees stability. A slightly more stringent criterion for unconditional stability is [9.5]

$$0 \leq \xi < 1, \qquad \gamma \geq \tfrac{1}{2}, \qquad \beta \geq \tfrac{1}{4}\left(\gamma + \tfrac{1}{2}\right)^2 \tag{9.9-8}$$

which for $\gamma > \frac{1}{2}$ and $\beta = (\gamma + \frac{1}{2})^2/4$ provides algorithmic or "artificial" damping in the higher modes. A popular choice is $\gamma = \frac{1}{2}$ and $\beta = \frac{1}{4}$. This choice is called the "trapezoidal rule" or the "constant average acceleration method."

Mode Superposition Method. Equations of the modal method appear in Section 9.7. In review and summary, the dynamic response calculation procedure is this: solve a vibration problem to determine the lower frequencies and modes of the structure, evaluate the modal time-dependent loads p_i from the physical time-dependent loads $\mathbf{R}$, solve the uncoupled modal equations to determine $z_i = z_i(t)$ for each i, and finally recover the physical d.o.f. $\mathbf{D}$ by using Eq. 9.7-7. Most of the computational effort in the mode superposition method occurs in solving for frequencies and modes of the system.

The uncoupled equations, Eqs. 9.7-6, can be integrated exactly for certain loadings, such as a suddenly applied constant load, a ramp load, and a sinusoidal variation. More generally, one of the direct integration methods can be used. Thus, for each mode i, one might use the Newmark method to calculate z_i for time instants $n = 1, 2, 3,$ separated by time intervals Δt. For numerical integration, initial values of $\mathbf{z}$ and its time derivatives can be obtained from initial values of $\mathbf{D}$ and its time derivatives as follows: premultiply each of Eqs. 9.7-4 by $\boldsymbol{\phi}^T\mathbf{M}$ and take note of the second of Eqs. 9.7-3. Thus

$$\mathbf{z}_0 = \boldsymbol{\phi}^T\mathbf{M}\mathbf{D}_0 \qquad \dot{\mathbf{z}}_0 = \boldsymbol{\phi}^T\mathbf{M}\dot{\mathbf{D}}_0 \qquad \ddot{\mathbf{z}}_0 = \boldsymbol{\phi}^T\mathbf{M}\ddot{\mathbf{D}}_0 \tag{9.9-9}$$

Useful variants of the mode superposition method include the mode acceleration method and the use of Ritz vectors [2.2, 9.2].

Brief Comparison of the Methods. It is important to consider how many natural frequencies of the structure may be excited by the loading and how many of these modes may be needed to calculate the desired response. In *structural dynamics* problems, such as loading by an earthquake, only the lower frequencies and modes are significantly excited. Only the ten lowest modes may be important in some problems [3.1]. In *wave propagation* problems, such as shock or blast loading, higher modes and frequencies are also excited. The number of significant modes may be large; perhaps two-thirds the number of d.o.f., or more, if we wish to calculate accelerations as functions of time [3.1].

The central difference method is best suited to wave propagation problems: the small time step required for stability is not so great a disadvantage, because a small time step is needed anyway to capture the behavior of all the important modes, and the response probably needs to be calculated for only a short time span. The Newmark and modal methods are best suited to structural dynamics problems: contributions of higher modes are not represented, but these modes have almost no effect on the structural response. Additional remarks related to modeling appear in Section 9.11.

9.10 RESPONSE SPECTRUM ANALYSIS

Dynamic response analysis often seeks the *maximum* displacement, acceleration, or stress at a certain location in the structure, without regard to the time at which the maximum occurs. This is particularly true in early stages of design. To obtain the maximum one could calculate response as a function of time by one of the methods described in Section 9.9, then discard all results except the single largest value. As an alternative procedure, FE software often supports *response spectrum analysis*, by which a maximum can be calculated more cheaply, although usually not as accurately. When used for analysis of an impulse load, a response spectrum may be called a "shock response spectrum" or simply a "shock spectrum." In brief, response spectrum analysis first computes the maximum response of each separate vibration mode, then combines the modal maxima in a way that produces an estimate of the maximum response of the structure itself. In more detail, the calculation of maximum displacement response proceeds as follows.

Time-varying loads **R** on the structure are arbitrary, but must be known. A different **R** produces different results. A modal analysis is performed, and Eq. 9.7-6 is integrated for each mode retained in the analysis. The result is as many data sets as there are modes retained. The ith data set pertains to the mode of frequency ω_i and lists z_i as a function of time. Each data set is scanned to find the numerically largest z in the entire response history of mode i, $z_{i\max}$. For a multi-d.o.f. structure, let Δ_{ji} represent the physical value of d.o.f. j associated with $z_{i\max}$. From Eq. 9.7-4,

$$\Delta_{ji} = \phi_{ji} \, z_{i\max} \tag{9.10-1}$$

in which all three quantities are scalars. For example, $\Delta_{32} = \phi_{32} \, z_{2\max}$ is the physical value of the third d.o.f. produced by the maximum modal response in the second mode. It is *not* d.o.f. D_3 itself, because we have not formed $D_3 = \Sigma\phi_{3i} \, z_i$, as would be done in the mode superposition method. Indeed, such a summation with $z_{i\max}$ values would not make physical sense because in general the maximum z_i values appear at different instants of time. The method by which Δ_{ji} values from several modes i are combined to produce an actual maximum displacement D_j is in general only an approximation of the actual maximum. One method of combination is to form the sum of absolute magnitudes:

$$\left(D_j\right)_{\max} \leq \sum_i \left|\Delta_{ji}\right| \tag{9.10-2}$$

This method yields the largest magnitude D_j could possibly have, because it presumes that the maxima of all modes appear at the same instant of time. It is usually more accurate to use the SRSS method (square root of the sum of the squares):

$$\left(D_j\right)_{\max} \approx \sqrt{\sum_i \Delta_{ji}^2} \tag{9.10-3}$$

Additional combination methods have been proposed. They are usually more reliable but are more complicated.

We have described the calculation of estimated maximum displacement $(D_j)_{\max}$ for a single d.o.f. j, but of course any number of additional d.o.f. can also be treated, and at very little additional cost.

More About Response Spectra. In general, a response spectrum is a plot of the *maximum* value of some response (displacement, velocity, or acceleration) versus frequency ω (or versus period T), *for a single-d.o.f. system.* Here ω refers to the undamped natural frequency of the single-d.o.f. system, $\omega = \sqrt{k/m}$. Often the ordinate of a displacement response spectrum plot is in the form of an amplification factor, $S = S(\omega)$, which multiplies the static response. Thus, for displacement response,

$$S_i = \frac{u_{\max}}{u_{\text{st}}} \qquad \text{at frequency } \omega_i \tag{9.10-4}$$

Displacement $u_{\max}$ is a function of ω. It is computed by integrating Eq. 9.2-2 for many different values of $\omega = \sqrt{k/m}$, each time using the same prescribed forcing function $r = r(t)$, and selecting the $u_{\max}$ of each such solution. Static displacement u_{st} is independent of ω. It is computed from Eq. 9.2-2 with $\dot{u} = 0$ and $\ddot{u} = 0$ and using the maximum value of r over the time span of the forcing function. Thus $u_{\text{st}} = r_{\max}/k$. With the ordinate of the displacement response spectrum plot defined as an amplification factor, the magnitude of r in Eq. 9.2-2 does not matter; only its variation with time is of importance.

The foregoing calculation process may be repeated for other values of the damping ratio, thus providing response spectra (Fig. 9.10-1). The generation of response spectra may require hundreds of integrations of Eq. 9.2-2, but this equation contains only one d.o.f. Once response spectra have been calculated for a certain forcing function, they need never be recalculated, regardless of the number or variety of multi-d.o.f. structures to which the forcing function may be applied. Indeed, response spectra for a "standard" earthquake have already been calculated and are readily available.

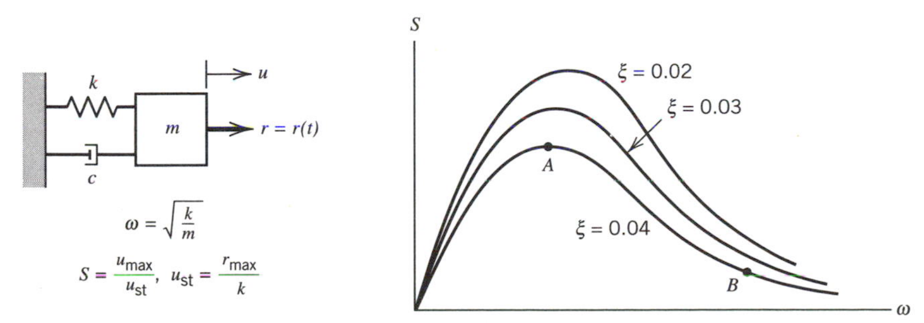

Fig. 9.10-1. Hypothetical displacement response spectra for a known (but here unspecified) forcing function $r = r(t)$.

Response spectra (of a single-d.o.f. system) are applicable to a multi-d.o.f. system loaded by the same forcing function, based on the proposition that each mode of the structure behaves like a single-d.o.f. system [9.6]. That is, with z_i for the mode and u for the single-d.o.f. system,

$$\frac{z_{i\max}}{z_{ist}} = \frac{u_{\max}}{u_{st}} \qquad \text{or} \qquad z_{i\max} = \frac{u_{\max}}{u_{st}} z_{ist} \tag{9.10-5}$$

Substituting from Eq. 9.10-4, and obtaining z_{ist} from Eq. 9.7-6 with $\dot{z} = 0$ and $\ddot{z} = 0$, we obtain

$$z_{i\max} = S_i \frac{p_{i\max}}{\omega_i^2} \tag{9.10-6}$$

where $S_i = S_i(\omega_i)$ is obtained by reading the response spectrum plot at the frequency of mode i of the multi-d.o.f. structure. We then use Eq. 9.10-1 to compute Δ_{ji}. These calculations are performed for all modes of importance. Finally, we use Eq. 9.10-2 or 9.10-3. An application of this method appears in Section 9.13.

Similar response spectrum calculations serve to estimate a maximum velocity $\dot{D}_{j\max}$ and a maximum acceleration $\ddot{D}_{j\max}$. Velocity and acceleration equations analogous to Eq. 9.10-6 are

$$\dot{z}_{i\max} = S_{vi} \frac{p_{i\max}}{\omega_i} \qquad \text{and} \qquad \ddot{z}_{i\max} = S_{ai} p_{i\max} \tag{9.10-7}$$

where S_{vi} and S_{ai} are multipliers that depend on ω_i. They are obtainable from the same data that yields S_i of Eq. 9.10-6. Then Eq. 9.10-1 yields

$$\dot{\Delta}_{ji} = \phi_{ji} \dot{z}_{i\max} \qquad \text{and} \qquad \ddot{\Delta}_{ji} = \phi_{ji} \ddot{z}_{i\max} \tag{9.10-8}$$

Finally, velocity and acceleration equations analogous to Eqs. 9.10-2 and 9.10-3 can be written.

9.11 REMARKS. MODELING CONSIDERATIONS

Dynamic analysis is more complicated than static stress analysis. Loadings have the additional dimension of time, there is greater variety in the possible goals of analysis, and there are a greater number of available procedures that may lead to each goal. Dynamic analysis is also more expensive in terms of the analyst's time and demands on computer resources. It is more difficult to anticipate structural response; hence it is more difficult to do a preliminary analysis, plan the numerical analysis, and judge the quality of computed results.

Nevertheless, general advice given in Chapter 5 remains applicable. Understand the physical problem and the concepts of analysis procedures. Study the software documentation. Start simply, with test cases, coarse-mesh models, and analyses of individual features of the problem. Expect to revise and improve models. Critically examine computed results. Keep records of the sources of data, assumptions made, and the status of the project. The reader may wish to review Chapter 5 for other detailed advice that may apply to dynamics as well as to statics.

Static and dynamic loads may create quite different responses in a given structure. In Fig. 9.11-1a, the right-hand part of the beam contributes nothing to the response due to a static load P. *Dynamically*, the right-hand part has great influence on the response. Also, for a given magnitude of P, dynamic displacements and stresses may exceed their static counterparts, and their distribution will probably not be obvious even when it is known how P varies with time. In Fig. 9.11-1b, static analysis may emphasize resistance to vertical loads. Yet the frame is more flexible horizontally than vertically, so earthquake loading may excite primarily horizontal motion. Similarly, a structure may have flat panels that carry static loads by membrane action, but under dynamic loading these panels may vibrate laterally with bending deformations.

One might first ask if dynamic analysis is necessary. A cyclic forcing function whose frequency is less than one-third the lowest natural frequency of the structure produces an undamped maximum response only about 10% greater than the static response to the amplitude of the load, so a static analysis may be adequate.

If a dynamic solution is elected, many questions must be posed and answered [5.4]. What is the goal of the analysis? How much accuracy is required? What simplifications are possible? Is damping important? If so, how should it be represented? Are there material or geometric nonlinearities? What frequencies are contained in the loads, and what frequencies are important in the structure? What computational procedures are appropriate? What are the capabilities of the software? How will the large volume of computed results be sorted, displayed, and checked?

Regardless of what else may also be done, an early computation in dynamic analysis is usually the extraction of natural frequencies and their associated modes. This computation is easily done, provides data used in other procedures, and provides physical insight into how the structure behaves.

Many of the following subsections bear on more than one aspect of dynamic analysis and therefore should not be studied in isolation. The subsections expand on the analytical procedures previously described and add modeling advice.

Mass Representation. The mass matrix for any kind of finite element can be written in various ways, such as lumped (diagonal), consistent, or some combination of the two. Which way is best? Some considerations that influence the choice are as follows.

If displacement fields are compatible, stiffness matrices are not softened by low-order integration rules, and mass matrices are consistent, then the FE model will yield undamped natural frequencies of vibration that are upper bounds on the exact frequencies of the mathematical model. Lumped mass matrices often (but not always) produce natural frequencies that are *lower* than exact. This observation suggests that accuracy might be

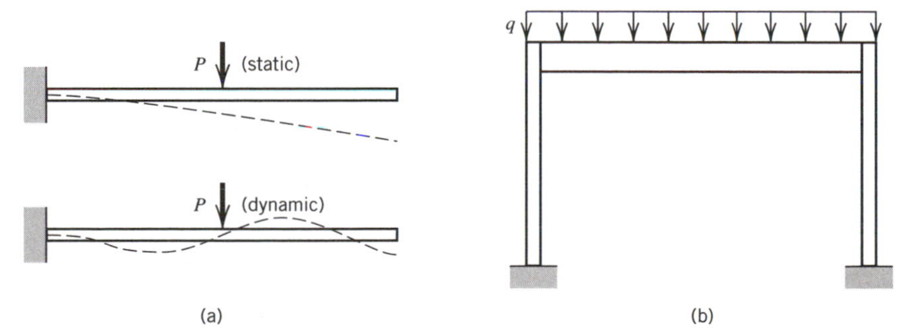

Fig. 9.11-1. (a) Cantilever beams under static and dynamic loads. (b) Building frame under static load q.

improved by forming both a consistent mass matrix $\mathbf{m}_c$ and a lumped mass matrix $\mathbf{m}_l$ for a given element, then combining them to produce the mass matrix $\mathbf{m}_{cl} = \beta\mathbf{m}_c + (1 - \beta)\mathbf{m}_l$, where $0 < \beta < 1$. While $\mathbf{m}_{cl}$ may yield more accurate natural frequencies than either $\mathbf{m}_c$ or $\mathbf{m}_l$, it offers neither the upper bound property of $\mathbf{m}_c$ nor the economy of storage (and perhaps also of execution time) offered by $\mathbf{m}_l$.

Some computational algorithms have trouble with a diagonal mass matrix that contains zeros on the diagonal, as will happen if there are rotational d.o.f. to which no rotary inertia has been assigned. One can preclude this possibility by always assigning rotary inertia to rotational d.o.f. The rule for choosing values of rotary inertia can be somewhat arbitrary because these inertias are relatively unimportant in the lower modes represented by the FE model.

These considerations are confusing to the less experienced, but may not be of great concern to the typical analyst because FE software may offer few choices. The choice may be between lumped masses and whatever other mass representation is automatically calculated by the software, whether it is a consistent mass matrix or not. At times it is quite proper to combine lumped masses with another mass representation as, for example, when a heavy machine rests on an elastic supporting structure. If elastic properties of the machine are unimportant, it is a "nonstructural" mass, and a FE model may use consistent mass matrices for the structure, augmented by mass lumps to account for the translational and rotary inertia of the nonstructural mass. A nonstructural mass that is offset from the axis of a structural element can be attached to an adjacent structural node i by a rigid massless link (Fig. 9.11-2). The transformation is described in Section 4.3. The resulting FE model has no node at the mass center of the machine and a mass matrix that is inherently nondiagonal. An ad hoc algorithm that diagonalizes it will place the mass at the structure node. This is a misrepresentation because it does not model the interaction between lateral translation and rotation [9.4].

Vibration Calculations. The eigenvalue problem, Eq. 9.4-2, can be solved by various algorithms. They differ in applicability and efficiency according to matrix sparsity, matrix topology, the number of frequencies to be calculated, and other factors. Commercial software usually offers more than one choice of algorithm. Commonly, one computes only the lowest m frequencies and modes, where m is chosen by the analyst. Often m includes only about 10% of all modes of the FE model, as no more are needed for many analyses. For details about eigensolvers, textbooks and software user information should be consulted.

If a structure is not fully supported or has mechanisms, $\mathbf{K}$ is singular and there will be a zero frequency for each possible rigid-body mode or mechanism. In such a case the user

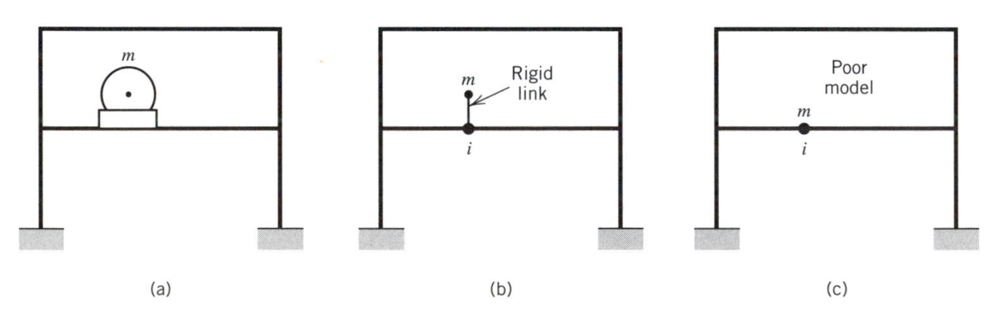

Fig. 9.11-2. (a) Machine of mass m attached to a building frame. (b) Acceptable model. (c) Poor model.

may have to ask the software to do an "eigenvalue shift" to enable computation. If **M** is singular because it is diagonal and has some zero coefficients M_{ii}, which is not recommended, there will be an infinite frequency for each zero mass coefficient.

Different modes may have the same frequency. A case in point is a straight cantilever beam of circular cross section lying along the x axis. For a displacement mode in (say) the xy plane there is a mode of identical shape in any other plane that contains the x axis. The modes have the same frequency, but because they are in different planes they are different modes. This may make the modes hard to interpret, and it is possible that the eigensolver will miss one of the repeated frequencies. Simple problems of this type may be used as test cases. Software may contain a "Sturm sequence" check that can discover if the eigensolver has missed any frequencies in the range for which the user seeks solutions, that is, in the range from zero frequency to a user-prescribed frequency. If so, the solution might be repeated, this time asking for a greater range of frequencies. The missed frequencies and modes may now appear [9.4].

Stabilization devices may be used to prevent mechanisms in certain elements (Section 4.6). These devices prevent zero frequencies but may introduce *low* frequencies instead, which contaminate vibration response in the frequency range that is usually of greatest interest.

Again, computation and critical inspection of frequencies and modes are desirable early steps in dynamic analysis. These steps can provide useful insight into structural behavior and can offer information needed in making decisions demanded by harmonic, dynamic, and response spectrum analyses.

Qualitative Prediction of Frequencies. An undamped natural frequency can be computed from the *Rayleigh quotient*, which can be obtained by premultiplying Eq. 9.4-2 by $\bar{\mathbf{D}}^T$ and solving for ω^2.

$$\omega^2 = \frac{\bar{\mathbf{D}}^T \mathbf{K} \bar{\mathbf{D}}}{\bar{\mathbf{D}}^T \mathbf{M} \bar{\mathbf{D}}} \tag{9.11-1}$$

where $\bar{\mathbf{D}}$ is the vibration mode. This is the same equation as Eq. 9.7-1 except for not identifying a mode by subscript i and not requiring the scaling described by Eq. 9.4-3. The numerator of Eq. 9.11-1 is twice the maximum strain energy $U_{\max}$ (see Eq. 3.1-9, which, however, is for a single element). The denominator (if multiplied by ω^2) is twice the maximum kinetic energy $V_{\max}$. It serves the following qualitative argument to regard the numerator and denominator of Eq. 9.11-1 as $U_{\max}$ and $V_{\max}$, respectively. In Fig. 9.11-3, $U_{\max}$ occurs in a dashed-line configuration, when the beam is instantaneously motionless, and $V_{\max}$ occurs in a solid-line configuration, when the beam is instantaneously undeformed.

The Rayleigh quotient can sometimes be used to rank the frequencies of possible vibration modes [9.4]. Since the magnitude of $\bar{\mathbf{D}}$ does not matter, it is convenient to imagine that the maximum displacement is the same in each mode. Thus, if we compare modes 1 and 2 in Fig. 9.11-3, we see that curvatures are greater for mode 2, hence the nu-

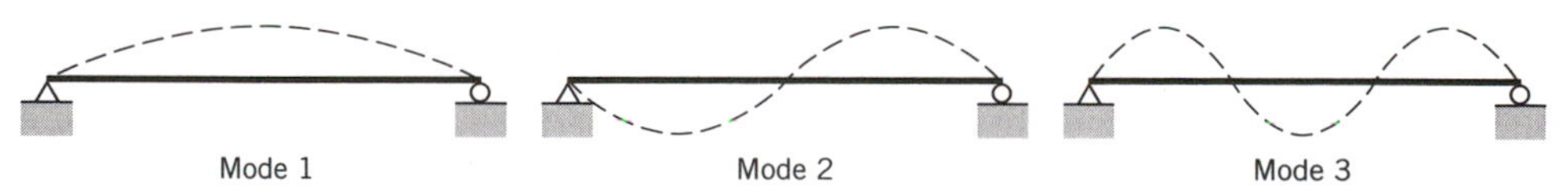

Fig. 9.11-3. The first three vibration modes of a simply supported beam. Displacements are exaggerated.

merator of Eq. 9.11-1 is greater for mode 2. Also, mode 2 has a point of zero displacement that mode 1 does not, so that the denominator of Eq. 9.11-1 is smaller in mode 2 than in mode 1. Hence Eq. 9.11-1 shows that $\omega_2 > \omega_1$. Similar comparisons involving mode 3 can also be made.

Quantitative Prediction of Frequencies. The frequency of a given mode, often the fundamental mode, can often be estimated from Eq. 9.2-6, $\omega = \sqrt{k/m}$. To use this equation we must conceptually reduce the structure to a system having a single d.o.f. and estimate k and m. An estimate of k can be obtained if the vibration mode is similar to the deflection produced by a static force. As an example, consider mode 1 in Fig. 9.11-3. Its shape resembles the deflected shape that would be produced by a transverse static load P at midspan, for which the midspan deflection is $v = PL^3/48EI$. Hence $k = P/v = 48EI/L^3$. Ends of the beam are at rest, so not all mass of the beam has displacement v. Let us guess that for a beam of mass m, mass $m/2$ participates in a vibration mode of amplitude v. Hence

$$\omega \approx \sqrt{\frac{48EI}{L^3}\frac{1}{m/2}} = 9.80\sqrt{\frac{EI}{mL^3}} \tag{9.11-2}$$

The exact multiplier for a slender simply supported beam is 9.87 rather than 9.80. The approximate frequency of mode 2 in Fig. 9.11-3 can be obtained similarly, by treating the left or right half in the same way the entire beam is treated in obtaining Eq. 9.11-2.

The fundamental frequency of a cantilever beam, Fig. 9.4-1b, can be estimated as follows. Apply a transverse tip load P, whose deflection is $v = PL^3/3EI$. Hence $k = P/v = 3EI/L^3$. Guess that $m/2$ is the effective mass. Hence $\omega_1 \approx \sqrt{(3EI/L^3)/(m/2)} = 2.45\sqrt{EI/mL^3}$. This happens to be the lumped mass approximation, Eq. 9.4-7, which is 30% low.

If stiffness k cannot be obtained by hand calculation, it can be computed from the FE model itself, by applying a static force P. The static force must be located and directed so as to produce a deflected shape that resembles the anticipated mode shape (if possible). If Δ is the computed deflection of the loaded point in the direction of P, the desired k is $k = P/\Delta$. There appears to be no such simple way to estimate the effective mass by use of standard FE software.

Reducing the Number of d.o.f. It may seem strange to create a large FE model and then reduce it rather than building a smaller model at the outset. However, a coarse-mesh model lacks the information content of a reduced model having an equal number of d.o.f. The reduced stiffness matrix $\mathbf{K}_r$ of Eqs. 9.6-4 and 9.6-5 is identical to the result of eliminating slave d.o.f. by steps of Gauss elimination in solving static equations $\mathbf{KD} = \mathbf{R}$ for unknowns $\mathbf{D}$. Thus, if the problem were static rather than dynamic, reduction would introduce no error.

Matrices produced by Guyan reduction are *full*, not sparse. Accordingly, if the anticipated computational economy is to be realized, the number of masters must be very much smaller than the total number of d.o.f. However, the smaller the number of masters, the greater the error introduced by reduction. Reduction has the effect of imposing displacement constraints on the full system, so that the frequency of a given mode is increased by reduction. An example appears in Fig. 9.11-4. We see that the lowest frequencies are least affected by reduction.

Any mode computed from the reduced system is a linear combination of the shapes described by columns of $\mathbf{T}$ in Eq. 9.6-3. Each column represents the displaced shape when a master d.o.f. is activated. Accordingly, if masters are too few in number or are badly cho-

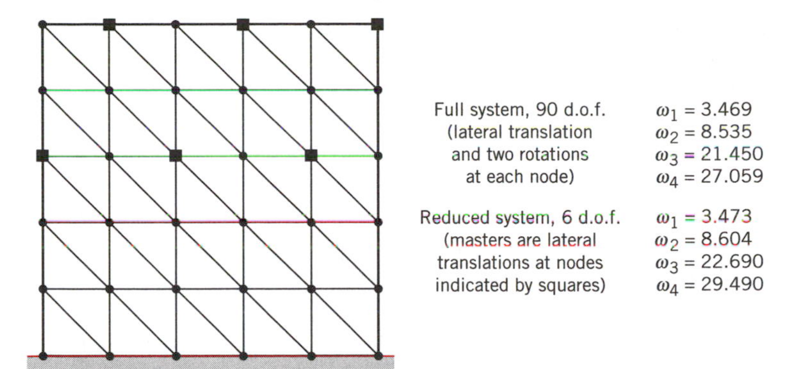

Fig. 9.11-4. A thin square cantilever plate that vibrates laterally, with the four lowest natural frequencies [9.7]. The reduced system is obtained by Guyan reduction.

sen, **T** is deficient in shape information, so that some modes computed from the reduced system will be inaccurate, and some modes of the full system may be skipped entirely. Deficiencies are usually greater in the higher modes. Accordingly, reduction should not be used when there is shock loading, for which higher frequency information in the model should be retained.

The d.o.f. chosen as masters for Guyan reduction should have large mass, large deflections in the modes of interest, or both. Rotational d.o.f. are rarely chosen as masters. Masters should be adequate in number and location to describe the modes of interest, especially the lower modes. Masters may be selected manually by the user, automatically by the software, or the two selection methods may be used in combination. One motivation for combined selection is to preserve as masters certain d.o.f. that might be classed as slaves by automatic selection. The choice of masters should be guided by the following rules [9.8]:

- Masters should have a large mass-to-stiffness ratio. A possible algorithm for automatic selection is to seek the d.o.f. having the *smallest* M_{ii}/K_{ii} ratio and eliminate it as first slave. Thus the system is reduced by one order. The next slave is the d.o.f. in the reduced system having smallest M_{ii}/K_{ii}. The process repeats until a prescribed number of masters remain.
- Masters should not be clustered in any one portion of the structure. If they are, some modes may be represented poorly and others not at all. Automatically chosen masters that form a tightly clustered set should be replaced or at least augmented by manually chosen masters.
- The d.o.f. at which time-varying forces or displacements are to be prescribed, or d.o.f. at which there are gap conditions in nonlinear problems, should be retained as masters.
- Masters should be d.o.f. in the direction of expected motion, for example, normal to the surface of a plate rather than tangent to its surface.

How many masters are enough? Opinions vary. Using one-tenth to one-half the total number of d.o.f. as masters has been recommended [3.1]. Another recommendation is that the number of masters should be at least twice the number of modes of interest [9.8]. The implication is that only the lower half of the computed modes and frequencies of the reduced system may be reliable. A way of deciding whether the number of masters is adequate is to redo the problem with (say) twice or half the number of masters and see if results for modes of interest are substantially the same.

Harmonic Response. The speed Ω of rotating machinery should pass quickly through a resonant frequency ω of the supporting structure, as sustained operation at $\Omega = \omega$ may produce destructive vibrations. Harmonic response is very sensitive to a forcing function *near* resonance. For this situation it is important that natural frequencies be calculated accurately. It is prudent to include damping so that harmonic response computations will not "blow up" should it happen that $\Omega = \omega$.

It is usual to calculate harmonic response for each of several frequencies Ω of forcing function. In each such analysis, calculations should include all modes up to and including the first mode whose frequency exceeds Ω. A plot of response amplitude versus Ω can be prepared for each location of interest on the structure. For such a series of analyses, the method of Eq. 9.8-8 would be much cheaper than the method of Eq. 9.8-6.

A forcing function having components of different frequencies must be resolved into component loading cases, each having a single frequency. Results produced by the separate components can be combined during postprocessing. Similarly, a forcing function that is periodic but not sinusoidal can be represented as the sum of its Fourier series components. An analysis can be performed for each component and the results combined.

Direct integration is an alternative way to calculate harmonic response. However, it is likely to be an expensive way because it automatically includes the transient response, which will usually not be damped out until the motion has been computed over several cycles.

Dynamic Response: Central Difference Method. The critical time step, Eq. 9.9-5, can be computed without the expense of solving an eigenproblem. Alternatives to eigensolution include estimating ω_{max} by a Gerschgorin bound and calculating Δt_{cr} as the time required for a sound wave to travel across the smallest element of the mesh [2.2]. For economy, one may use a Δt between 0.95 and 0.98 times Δt_{cr}. The very highest modes may not be computed accurately, but this usually does not matter. Nonlinearities, if present, may alter the critical time step.

The central difference method is most efficient if $\mathbf{A}$ of Eq. 9.9-2 is diagonal. This is easily achieved if $\mathbf{M}$ is lumped and without zero entries and $\mathbf{C}$ is either diagonal or zero. Additionally, the term $\mathbf{KD}_n$ in Eq. 9.9-4b can be efficiently computed by summing element contributions. Then the arrays that must be stored are all vectors, not matrices, and computational cost per time step in Eq. 9.9-2 is very low. Nevertheless, because Δt_{cr} is quite small, the method may be uneconomical for all but wave propagation problems.

A uniform mesh allows waves to propagate equally in all directions. Abrupt element size changes create numerical noise and artificial wave reflections. Low-order elements are better than higher-order elements at modeling shock wavefronts. A lumped $\mathbf{M}$ creates fewer spurious oscillations than a consistent $\mathbf{M}$ and also produces a larger Δt_{cr}. The trick of using a very stiff element as a support (Section 5.8) has the effect of raising ω_{max}, thus decreasing Δt_{cr}, and should therefore be avoided. Reduction (Section 9.6) should also be avoided because it discards higher-frequency information that is needed for wave propagation problems. (See also the "Cautionary remarks" at the end of Section 9.11.)

Dynamic Response: Newmark Method. In comparison with the central difference method, the Newmark method has the advantage that Δt can be large without collapse of the numerical process. This is a very great advantage for structural dynamics problems, for which the response may be needed over a considerable time period. A large Δt has the effect of discarding information associated with modes of higher frequency. Usually this does not matter in a structural dynamics problem. If Δt becomes *too* large, too little information remains, and even low-frequency contributions to the results become inaccurate.

In Eq. 9.9-2, **A** is not a diagonal matrix. For this reason the cost per time step is much greater than with the central difference method. Use of a consistent **M** is beneficial to accuracy and only slightly detrimental to economy. If there are nonlinearities **A** must be repeatedly revised, which is a considerable expense.

Dynamic Response: Modal Method. Like the Newmark method, the modal method is best suited to structural dynamics problems. For example, with earthquake loading, modes of frequency greater than 33 Hz are usually unimportant, and half the response may be associated with the first mode. Accordingly, few modes are needed in Eq. 9.7-7. In a wave propagation problem, the number of modes needed may be so large that the modal method would not be practical for such a problem. Further remarks on the number of modes needed appear in the next subsection.

The greatest computational expense is solving for the necessary modes and frequencies. This information is probably already available from a previous phase of the investigation. The uncoupled equations are cheaply integrated. Mode superposition is often the cheapest way of solving a structural dynamics problem. This is especially true if the effects of several loadings must be studied, because the same modes and frequencies are used for each different loading. In contrast, direct integration requires complete re-solution for each different loading.

Loading and Time Step. How many modes should be retained in modal analysis? What time step is appropriate for integration of the equations of motion, be they the modal equations (Eqs. 9.7-6) or the structural equations (Eqs. 9.9-7)? Opinions differ and reliable a priori rules are unavailable. Some considerations that may be helpful are as follows.

Consider first the applied loading. In Fig. 9.11-5, the spacing of points used to numerically represent the forcing function is Δt_s, so that T_Ω is the approximate period of the shortest wave represented by the loading. Structure modes whose cyclic frequencies are higher than $2/T_\Omega$ Hz are not likely to be significantly excited [5.4]. Thus the circular frequency Ω_u of the highest mode of loading that need be retained in the analysis can be taken as $\Omega_u = 4\pi/T_\Omega$.

In mode superposition analysis, the FE mesh should be able to represent modes up to frequency ω_u, where $\omega_u = 1.5\ \Omega_u$ (at least) to $\omega_u = 4\ \Omega_u$ (at the very most [3.1]). Another way to estimate ω_u is to examine response spectra produced by the loading [9.4].

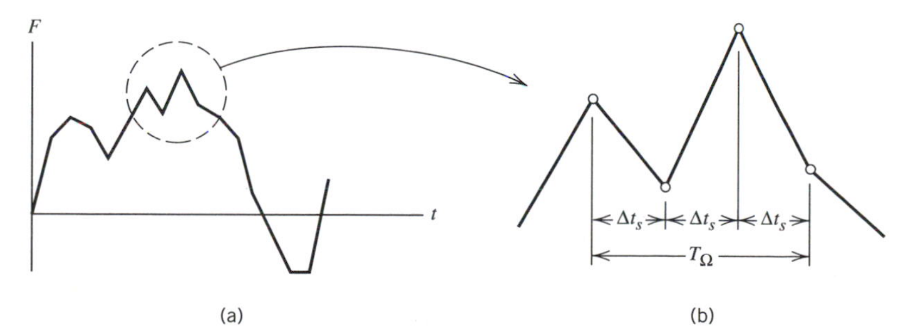

Fig. 9.11-5. (a) An arbitrary forcing function, represented by discrete data points. (b) Detail showing contribution of highest frequency.

Consider Fig. 9.10-1 with $\xi = 0.04$, for example. If $S_B = 0.1\ S_A$, one might say that frequencies larger than ω_B can be ignored, and therefore one may choose $\omega_u = \omega_B$ as the highest structural frequency that need be retained. In applying this procedure, one of course uses the response spectrum appropriate to the quantity of interest, be it displacement, velocity, or acceleration. However, the foregoing estimates of ω_u say nothing about the spatial nature of the loading. The number of modes required depends not only on the frequency content of the loads but also on how they are distributed over the structure. A concentrated load requires more modes than a uniform load distribution. A better estimate of ω_u that accounts for how the loading is distributed can be obtained from a full response spectrum analysis. For this the modes and frequencies must be available, so it does not provide an a priori estimate of ω_u. An example of the procedure appears in Section 9.13.

A mesh adequate to represent ω_u is one in which the actual physical mode shape associated with ω_u can be closely approximated by the FE mesh. This requires a few elements per wave of the mode, with "few" being a number that depends on the type of element. In general, the span and orientation of waves are not known in advance. Therefore it is likely that initial analyses will disclose that the mesh needs to be revised.

In a direct integration analysis of dynamic response by the Newmark method, time step Δt should be at least Δt_s (the load representation interval in Fig. 9.11-5 [5.4]), or from $T_u/10$ to $T_u/30$, where $T_u = 2\pi/\omega_u$. The time step in the central difference method is limited by the critical value, Eq. 9.9-5, and there is no merit in using a much smaller value. Similar remarks apply to direct integration of the uncoupled equations of the modal method, Eqs. 9.7-6. Here $\omega_u = \omega_i$ and therefore Δt can be different for each equation, with Δt being largest for the mode 1 equation. The practical range of Δt is enormous, from roughly 10^{-6} seconds for a wave propagation problem to roughly 10^5 years for a geologic problem involving viscous action and diffusion.

The appropriateness of values chosen for ω_u and Δt can be judged by repeating the analysis using other values. If results are little changed, the choices made are probably adequate.

Rotating Machinery. Rotation causes prestress that modifies stiffness. For example, centrifugal inertia force in a turbine blade produces a "stress-stiffening" effect that increases natural frequencies and whose influence may be comparable to that of the conventional stiffness matrix **K**. Stress stiffening is discussed in Section 10.3.

Another consequence of rotation is called "spin softening" [9.8]. Its nature can be explained with the aid of Fig. 9.11-6. A disk, which rotates in the plane of the paper at constant angular velocity Ω about point O, carries a rigid weightless bar ABC pivoted at B. Particle masses m are attached to the bar at A and C and to the disk via springs of stiffness k. Let ABC be rotated through a small angle θ relative to the disk, Fig. 9.11-6b. Together, inertia force and spring deformation produce net forces F_1 and F_2, which are essentially radial if $a << R$.

$$F_1 = m(R - a\theta)\Omega^2 + ka\theta \qquad F_2 = m(R + a\theta)\Omega^2 - ka\theta \tag{9.11-3}$$

Net torque about B equals the moment of inertia about B times angular acceleration $\ddot{\theta}$ of ABC.

$$F_2 a - F_1 a = 2ma^2\ddot{\theta} \tag{9.11-4}$$

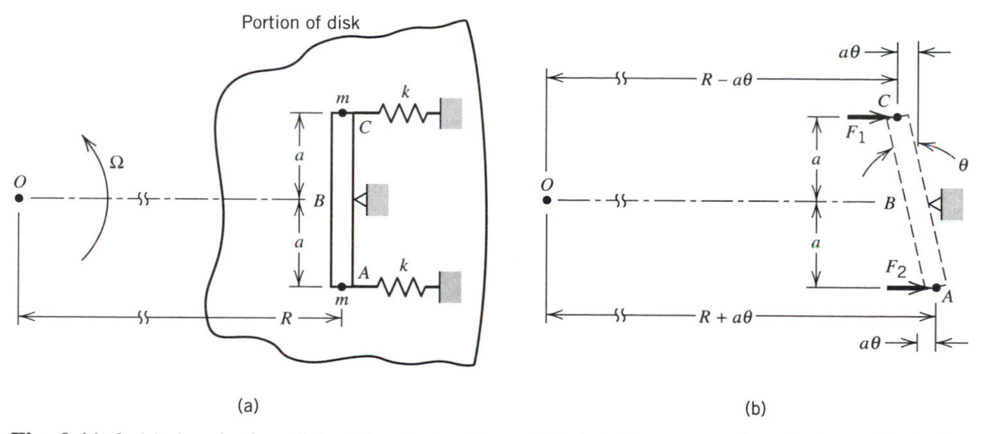

Fig. 9.11-6. (a) A spinning disk with attached bar *ABC* that illustrates spin softening. (b) Spring forces and inertia forces are represented together in F_1 and F_2. It is assumed that $a << R$ and $\theta << 1$.

If *ABC* oscillates about *B* with simple harmonic motion of amplitude $\bar{\theta}$ and circular frequency ω, then $\theta = \bar{\theta} \sin \omega t$ and $\ddot{\theta} = -\omega^2 \bar{\theta} \sin \omega t$. Combining this with Eqs. 9.11-3 and 9.11-4, we obtain

$$\omega = \left(\frac{k - m\Omega^2}{m} \right)^{1/2} \tag{9.11-5}$$

If Ω were zero, the frequency of vibration would be $\omega = \sqrt{k/m}$, as in Eq. 9.2-6. We see that the effect of Ω is to reduce ω. It is not hard to imagine that springs and mass particles in Fig. 9.11-6 have counterparts in an elastic body such as a fan blade, whose vibration frequency would be similarly affected by rotation.

Substructuring. Substructuring in dynamics has the same advantages and disadvantages as substructuring in statics (Section 4.11). It has the additional disadvantages of introducing an approximation, which static substructuring does not, and being more difficult to explain. Here, only the following very brief explanation is offered.

In dynamics, substructuring is known as *component mode synthesis* or simply *modal synthesis* [9.2]. Each component (substructure) is isolated and its natural modes and frequencies are calculated. Modes of the components do not combine to provide modes of the complete structure but, instead, lead to a transformation matrix analogous to **T** in Eq. 9.6-3. Condensed matrices analogous to those in Eqs. 9.6-4 and 9.6-5 are produced, so that the complete structure is represented by a smaller system of equations than would be present if all d.o.f. of the complete structure were retained.

Symmetry. In static analysis, symmetry of geometry, material properties, boundary conditions, and loads makes it possible to analyze part of a structure as representative of the whole (Section 4.12). Symmetry can also be exploited in vibration analysis, but it is easy to overlook some of the modes. For example, a FE model of the left half of a simply supported beam captures only antisymmetric vibration modes of the beam if the right end of the FE model is simply supported. Or, a shell of revolution might be modeled as axisymmetric in static analysis, but its fundamental frequency of vibration is almost certain to display waves around the circumference.

Stress Recovery. Stresses are computed from nodal d.o.f. in $\mathbf{D}$ by the same manipulations used in static stress analysis. However, if $\mathbf{D}$ is recovered from a condensed dynamic model there will be an accuracy loss, whether reduction was accomplished by Guyan reduction (Section 9.6) or by using the lower vibration modes (Section 9.7). The reason for the loss is that peak stresses are usually associated with large but isolated displacement gradients. Such displacement states require more than just the lowest displacement modes for their description, yet *only* the lowest modes are available because of reduction. Computed stresses will probably be underestimated. Note, however, that very large stresses might be computed by direct use of a vibration mode $\overline{\mathbf{D}}$, because of the way $\overline{\mathbf{D}}$ is scaled. Stresses computed in this way are not meaningful.

Test Cases. Test problems are useful in learning about structural dynamics, learning how to use software, and perhaps even validating the software. Problems for which answers are known can be found in textbooks, handbooks, and manuals that accompany software.

A cantilever beam of circular cross section is a test for the ability of an eigensolver to resolve different modes that have the same frequency. Properties of the beam can be perturbed slightly to provide a model with nearly equal frequencies. An unsupported structure tests the calculation of zero frequencies and rigid-body modes. It is instructive to solve a problem in different ways and compare results (e.g., vibration with and without reduction, or dynamic response by modal and by Newmark methods).

Pilot studies, described in Section 5.4, are particularly appropriate for dynamic response problems because appropriate computational choices may be difficult to make and structural behavior hard to foresee.

Modeling. Advice given in Chapter 5 regarding modeling and project planning in a static problem is largely applicable to a dynamics problem as well. In dynamics the mesh should be adequate to represent the highest mode of interest. What mesh is appropriate near a stress raiser? Stress concentrations create local disturbances but have little effect on the total strain energy in the structure. Hence, according to the argument made in connection with Eq. 9.11-1, stress concentrations have little effect on natural frequencies. Also, locations of high stress may differ from one mode to another. These considerations suggest that mesh refinement near probable stress raisers may sometimes be omitted, at least in initial analyses.

It can be shown that an approximate $\overline{\mathbf{D}}$ in Eq. 9.11-1 produces an ω that has much less error than $\overline{\mathbf{D}}$. Hence a finer mesh is needed to obtain an accurate mode shape than is needed to obtain an accurate frequency for that mode. (Note that stresses are calculated from the mode shape.) A consistent mass representation is better than a lumped mass representation in representing mode shapes. However, lumped masses tend to overestimate kinetic energy, and compatible and fully integrated elements overestimate strain energy, so that the overestimates tend to cancel in Eq. 9.11-1 and provide accurate frequencies as a result. If natural frequencies are computed twice, using consistent masses and then lumped masses, good agreement suggests that there is little discretization error in the computed frequencies.

For some structures the mass of the surrounding medium is significant. Fluid that surrounds a structure moves with it to some extent, adding nonstructural mass and also providing some damping. The added-mass effect can usually be ignored for a structure vibrating in air, but not for the aeroelastic problem of flutter in an airstream. Nor can it be ignored in ocean engineering, where a structure moves because of wave forces, but because it moves, the forces change. This is a *fluid–structure interaction* problem. The added-mass effect is also present for snow load on a roof and for a structure that rests on an elastic foundation. If the contact surface between a structure and the foundation that supports it can be idealized as a rigid circular disk, effective constants for mass, stiffness, and damping can be used [9.3]. A soil foundation has properties that are frequency and amplitude dependent, which can introduce significant nonlinearities [5.4].

In planning an analysis project the amount of output should be anticipated. In dynamic response analysis, 100 time steps produce as much output as 100 static analyses. It helps to anticipate what output may be required, and to know what sorting and plotting capabilities the software contains to help with the task of inspecting results.

Checking for Errors. Errors and blunders of static analysis, and more, are possible in dynamic analysis. As in statics one can check that nodes are suitably placed, that supports are of the proper type and properly located, and so on. Data blunders are more likely in dynamics because more data are needed. Mass must be specified, in consistent units, and ρ may mean mass density or weight density, depending on the software. Damping ratios must be provided, as fractions or percentages, as the software requires. Computed natural frequencies must not be interpreted as circular frequency if they are presented as cyclic frequency.

One can begin with a *static* analysis, using loads that seem likely to produce an approximation of the lowest vibration mode. Static analysis is comparatively cheap and easy and may disclose blunders or flaws in the model. One can also ask the software to compute total structure mass and the mass center location, and compare these results with expectations. As the first dynamic analysis, natural frequencies and modes of vibration can be computed. Zero frequencies indicate a lack of support (which may be intentional) or the presence of a mechanism. Absurdly high or low frequencies suggest an error in data. Modes should be plotted, animated, and viewed from different directions. Are they compatible with intended support conditions, and are they physically realistic?

In dynamics, it may be difficult for even an experienced analyst to predict the nature of the response. Modeling may therefore be difficult and it may even be hard to foresee what sort of response should be studied. This strongly suggests that analysis begin with pilot studies. One may also make use of a restart option in the software, so that results of a dynamic response analysis can be examined after a short time and computation resumed only if it seems to be working properly.

Error Estimation. Some static problems, such as a cantilever beam under tip load, can be modeled exactly by a single element. Such is not the case in dynamics. Polynomials, which form the shape functions of most elements in common use, cannot represent vibration mode shapes exactly. Thus there is always discretization error in stiffness representation. Mass representation also has discretization error, and the two errors may tend to either cancel or reinforce one another. Discretization error of stresses can be conveniently estimated from the difference between averaged and unaveraged stress fields (Section 5.16). Software does not as yet provide such convenient error measures for mass discretization error, natural frequencies [9.9], reduction, or time step Δt.

Comparison With Experiment. Experimentalists and analysts must work together to ensure that both groups address the same problem. The two groups must agree on geometry, elastic properties, damping, structural mass, nonstructural mass (magnitude, location, method of attachment), supports (hinged, elastic, or other), excitation (type, magnitude, location, direction), response to be measured (type, location, direction), and perhaps other considerations [9.4]. The structure is no doubt more complicated than its FE model, so that some vibration modes of the structure may be local and not obtainable from the FE model. Nodes of the FE model must be at the exact locations where measurements are to be taken. The FE model is quite probably linear, but the structure may not be. To find out, the structure could be excited at two or more levels of excitation at a given frequency and the measured response amplitude plotted to see if it is directly proportional to excitation level.

Cautionary Remarks. Remarks in a paper devoted to numerical modeling of impact phenomena [9.10] can also be applied, with slight modification, to numerical modeling in general. The paper advises that the primary requirement is a thorough understanding of the physics and mechanics of the problem, and that another requirement is a thorough knowledge of numerical modeling techniques. Otherwise the numerical problem may differ significantly from the physical problem. "In no way can today's computer programs for wave propagation and impact be treated as 'black boxes.' A minimum of 6 months to 2 years of experience is needed to be able to use such programs successfully. There is no shortcut, no royal road, to this process." The paper also warns against use of material properties inappropriate to the strain rate in the problem addressed, use of a mesh more suited to computational economy than to physics of the problem, and ascribing a physical cause to numerical instabilities that are not recognized as such.

9.12 AN APPLICATION: VIBRATION AND HARMONIC RESPONSE

The structure we consider is the right-angle frame *ABC* shown in Fig. 9.12-1a. The frame is uniform, pinned at *A*, and roller-supported at *C*. It is modeled by 50 beam elements, each of length 0.1 m. Selected nodes are numbered in Fig. 9.12-1a. Displacements are

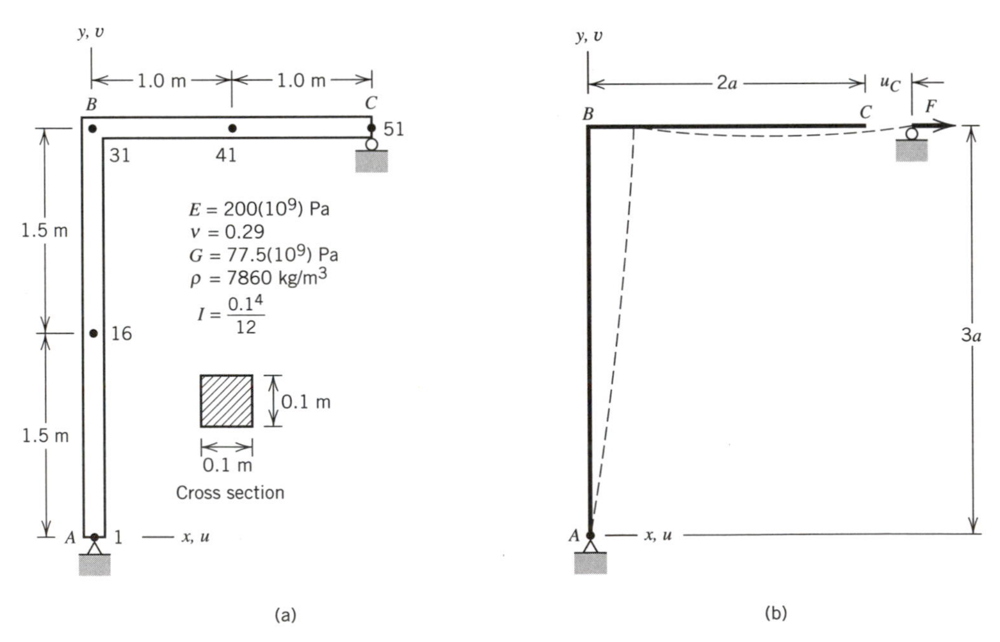

Fig. 9.12-1. (a) Plane frame structure and its properties. The numbers 1, 16, 31, 41, and 51 are node numbers used in the FE model. (b) Static loading used in preliminary analysis. Here $a = 1.0$ m.

confined to the plane of the paper. The simplicity of this structure allows us to concentrate on dynamics rather than geometric modeling and to display computed results easily. We will compute undamped natural frequencies of vibration, then the harmonic response for a range of forcing function frequencies with $\xi = 0.02$ as the damping ratio for all modes. Lumped masses with rotary inertia are used by the software, but the documentation offers no details about the formulation.

Preliminary Analysis. It is easy to estimate the lowest natural frequency of vibration by reducing the problem to a single degree of freedom, then calculating the frequency $\omega_1 = \sqrt{k/m}$. To obtain k we assume that the shape of the lowest vibration mode resembles the static deflected shape of the frame when a horizontal force F is applied at node C (Fig. 9.12-1b). Deflection u_C can be computed by applying static force F to the FE model. Or, with $a = 1.0$ m, simple methods of either energy or beam theory yield

$$u_C = \frac{15Fa^3}{EI} \qquad \text{hence} \qquad k = \frac{F}{u_C} = \frac{EI}{15a^3} = 111{,}100\ \text{N/m} \tag{9.12-1}$$

All the mass of BC has horizontal displacement u_C, but only some of the mass of AB. An accurate evaluation based on deflected shape can be made but it is not worth the trouble. Let us assume that *half* the mass of AB has displacement u_C. Thus, multiplying mass density by volume, we obtain the effective mass $m = 7860(2.0 + 1.5)(0.1)(0.1) = 275$ kg. Therefore

$$\omega_1 \approx \sqrt{\frac{k}{m}} = 20.1/\text{s} \qquad \text{and} \qquad f_1 = \frac{\omega_1}{2\pi} \approx 3.20\ \text{Hz} \tag{9.12-2}$$

Frequencies and shapes of higher modes become increasingly difficult to anticipate. As for harmonic response, we know that there will be peaks of response when the frequency of the forcing function coincides with a natural frequency of vibration. Magnitudes of the peaks can be computed by hand when frequency data are known (see Eq. 9.12-3).

Critique of FE Vibration Analysis. Computed results for the five lowest frequencies and modes are shown in Fig. 9.12-2. There is excellent agreement between the computed f_1 and the estimated f_1 in Eq. 9.12-2. Displacement amplitudes are exaggerated for plotting and have no significance. Recalling the argument that follows Eq. 9.11-1, mode shapes in Fig. 9.12-2 imply an ordering of frequencies that agrees with the ordering actually computed. Each wave of these modes is spanned by several elements, which implies

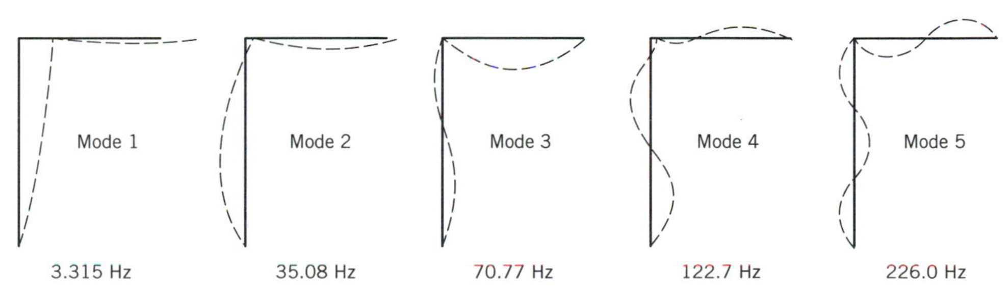

Fig. 9.12-2. Computed mode shapes and cyclic frequencies $f = \omega/2\pi$.

TABLE 9.12-1. Vibration of the frame in Fig. 9.12-1: selected terms of the mass-normalized eigenvector $\overline{\mathbf{D}}_i$ (i.e., the nodal amplitudes in mode i) for selected nodes, each multiplied by 1000.

D.o.f.	Node	Mode 1	Mode 2	Mode 3	Mode 4	Mode 5
$\bar{u}$	16	38.6	−81.5	23.8	−27.7	62.7
$\bar{\theta}_z$	41	1.0	7.0	−19.4	52.6	248.9
$\bar{u}$	51	62.3	33.1	−3.4	21.6	9.2

that the FE mesh is adequate. All in all, there is no apparent reason to disbelieve the computed frequencies. Additional computed frequencies in hertz (Hz) are $f_6 = 269.4$, $f_7 = 396.6$, $f_8 = 420.8$, $f_9 = 552.3$, and $f_{10} = 649.6$. The FE model has a total of 150 frequencies. Amplitudes of selected d.o.f. from the lowest modes appear in Table 9.12-1. Each mode (eigenvector) $\overline{\mathbf{D}}_i$ is scaled so that $\overline{\mathbf{D}}_i^T\mathbf{M}\overline{\mathbf{D}}_i = 1$. Note that algebraic signs in Table 9.12-1 agree with what is seen in Fig. 9.12-2.

Harmonic Response Analysis. Let node 51 be loaded by a horizontal force of 3000 sin Ωt, measured in newtons. Harmonic response is computed by the modal method for each of several frequencies Ω. The damping ratio is $\xi = 0.02$ for all modes. Figure 9.12-3a shows the forced vibration mode for $\Omega = 122$ Hz, which is almost the natural frequency of mode 4. As might be expected, the shape is very similar to that of mode 4, but not quite the same. Computed response data can also be used to plot the maximum response of any quantity versus frequency of the forcing function; that is, to plot a form of Fig. 9.8-1. Figure 9.12-3b is such a plot. It shows the amplitude of horizontal displacement at node 16 versus forcing function frequency in the arbitrarily chosen range 48 to 180 Hz. Much larger peaks may appear at lower resonant frequencies. The expected peaks are observed at $\Omega_3 = \omega_3$ and at $\Omega_4 = \omega_4$. Only five modes were retained in the analysis, so that results for $\Omega > \Omega_4$ are questionable. However, a reanalysis with only four modes retained produced almost identical results in the range plotted.

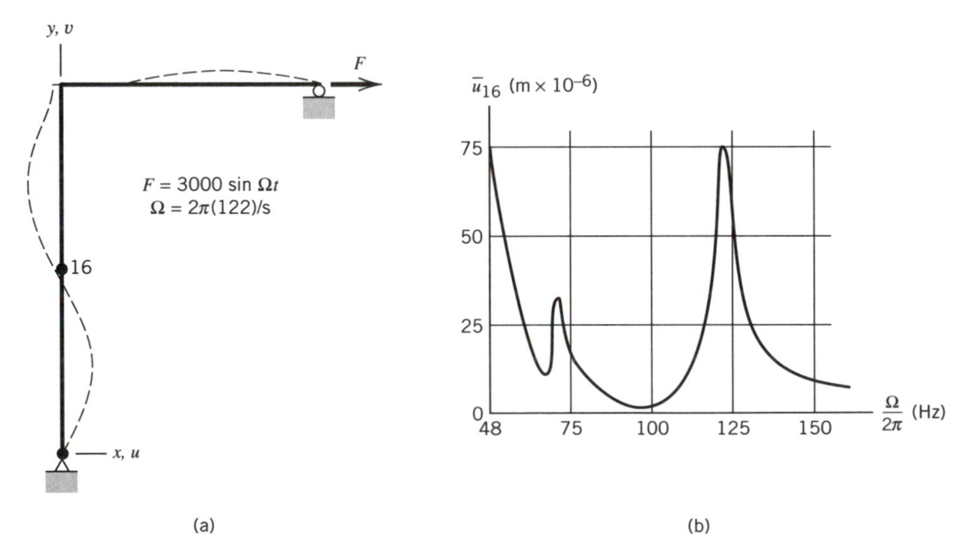

Fig. 9.12-3. Harmonic response. (a) Vibration shape for forcing function frequency $f = 122$ Hz. (b) Horizontal displacement amplitude at node 16 versus frequency of the forcing function.

Using modal data from Table 9.12-1, peak amplitudes in Fig. 9.12-3b can be checked. Consider the peak that corresponds to Ω_4. The only nonzero entry in the load amplitude vector **R** is 3000 N rightward at node 51. Therefore, from Eq. 9.7-6, the modal load in mode 4 is $p_4 = 0.0216(3000) = 64.8$. Also $\Omega_4 = \omega_4 = 2\pi(122.7) = 770.9/\text{s}$, $\xi_4 = 0.02$, and $\beta_4 = 1.0$. Equation 9.8-4 yields

$$z_{4\,\text{max}} = \frac{p_4/\omega_4^2}{2\xi_4} = \frac{64.8/770.9^2}{2(0.02)} = 0.00273 \qquad (9.12\text{-}3)$$

Hence $\bar{u}_{16} = 0.0277 z_{4\text{max}} = 75.5(10^{-6})\,\text{m}$, which agrees with the value plotted in Fig. 9.12-3b. *Static* deflection under the 3000-N load is over 300 times greater. However, the amplitude of horizontal reaction at node 1 was found to be about 4800 N when $\Omega_4 = \omega_4$; that is, 1.6 times the amplitude of the forcing function.

Note: Data in this section and the next come from three different computer programs because no one program provided all the results needed. All three programs provided modes and frequencies, but they did not quite agree. For this reason there are slight inconsistencies in some of the numerical results. Such difficulties are commonplace.

9.13 AN APPLICATION: DYNAMIC RESPONSE

The frame whose data appear in Fig. 9.12-1 is now subjected to an impulse loading. As shown in Fig. 9.13-1, a horizontal force of 100,000 N is applied at node 51 for 0.01 seconds. The frame is initially undeformed and at rest. We will calculate selected displacements, velocities, and accelerations as functions of time, and also seek maximum values of these quantities. Because load is suddenly applied it is a shock loading, *for which modal analysis is not well suited.* But we will use modal analysis anyway in order to show how it fares when applied to such a problem.

Preliminary Analysis. Let us continue the approximate analysis begun with Eq. 9.12-1, and now estimate the maximum horizontal acceleration, velocity, and displacement of node 51. Over the very short period of loading, displacements are so small that elastic forces have almost zero effect on the horizontal acceleration of portion BC. Therefore we use simple equations of particle dynamics. From Newton's law $f = ma$, acceleration at

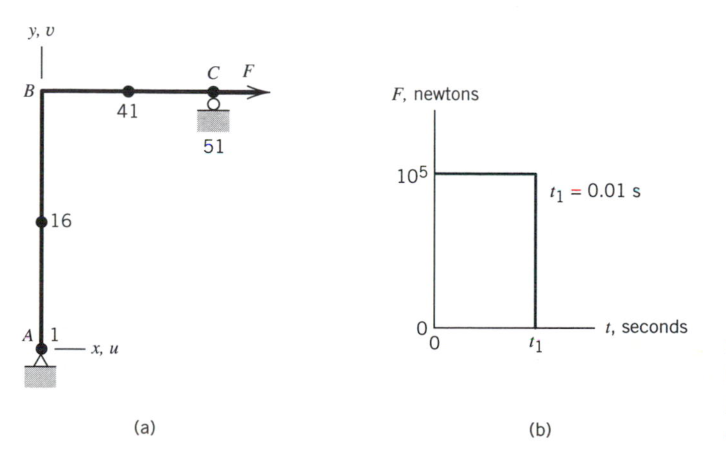

Fig. 9.13-1. (a) Frame loaded at point C by force F. (b) Prescribed variation of force F with time.

time $t = 0$ should be approximately

$$\ddot{u}_{51} \approx \frac{F}{m} = \frac{100{,}000}{275} = 364 \text{ m/s}^2 \tag{9.13-1}$$

The impulse Ft_1 is equal to the change of linear momentum, which we take to be $m\dot{u}_{51}$. Hence at $t = t_1$

$$\dot{u}_{51} \approx \frac{Ft_1}{m} = \frac{100{,}000(0.01)}{275} = 3.64 \text{ m/s} \tag{9.13-2}$$

To estimate maximum displacement, we assume that maximum kinetic energy is equal to maximum strain energy in mode 1, whose approximate stiffness k is available from Eq. 9.12-1. Hence

$$\tfrac{1}{2}m(\dot{u}_{51})^2 \approx \tfrac{1}{2}k(u_{51})^2 \tag{9.13-3a}$$

$$u_{51} \approx \left[\frac{275(3.64)^2}{111{,}100}\right]^{1/2} = 0.181 \text{ m} \tag{9.13-3b}$$

FE Model and Analysis. The FE model and its computed modes and frequencies are exactly those also used for the calculations of Section 9.12. To study the effect of retaining different numbers of modes in the analysis, dynamic response is analyzed several times by the mode superposition method, each time retaining one more mode in the analysis, up to a total of 15 modes used. The frequency of mode 15 is $f_{15} = 1268$ Hz. In each analysis the time step for integration of Eq. 9.7-6 is $\Delta t = 0.0001$ seconds, starting at time $t = 0$ and continuing to at least time $t = 0.40$ seconds.

Critique of FE Results. With 15 modes retained, computed maxima of horizontal acceleration, velocity, and displacement at node 51 are, respectively, 173 m/s^2, 4.27 m/s, and 0.181 m. The agreement of velocity and displacement with Eqs. 9.13-2 and 9.13-3b is good to excellent. The agreement of acceleration with Eq. 9.13-1 is not good and reflects the inappropriateness of the mode superposition method for acceleration calculation when there is shock loading.

We arbitrarily decide to present graphical results for displacement, velocity, and acceleration of the rotational d.o.f. at node 41. Figure 9.13-2a shows the variation of θ_{z41} with time when only two modes are retained in the analysis. We see that the effect of damping is to make mode 2 decay quickly, leaving mode 1 to decay over a larger time span. This is reasonable in view of the argument that follows Eq. 9.2-7: modes decay at a certain rate per cycle, and over a given time mode 2 executes more cycles that mode 1. The initial portion of the response is shown in Fig. 9.13-2b, but with 15 modes used in the analysis. The contribution of modes higher than the second is evident, but it is also clear that almost all of the *maximum* displacement response is represented by the first two modes. Similar results, now for velocity $\dot{\theta}_{z41}$ and acceleration $\ddot{\theta}_{z41}$ early in the response, appear in Fig. 9.13-3. Here we see that two modes yield only about one-third the angular velocity predicted by 15 modes. The maximum angular acceleration may not be represented accurately by even 15 modes; we can say only that two modes are utterly inadequate.

For the horizontal d.o.f. at node 16 and the rotational d.o.f. at node 41, Table 9.13-1

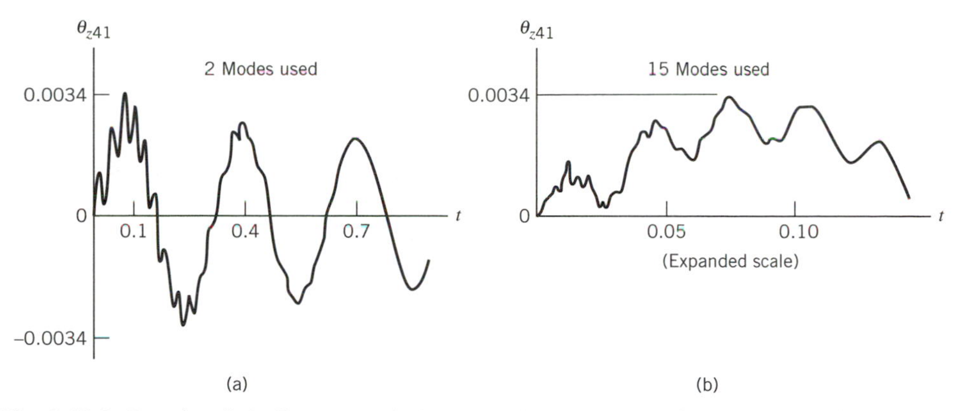

Fig. 9.13-2. Rotation θ_z (radians) at node 41 versus time t (seconds) for the loading of Fig. 9.13-1. (a) Results when two modes are included in the analysis. (b) Results when 15 modes are included in the analysis; initial portion of the plot.

lists maximum values that appear in the entire response history, using various numbers of modes in the analysis. Again it appears that two modes are adequate for displacement analysis. However, we cannot be sure that 15 modes are adequate for velocity analysis, and the data suggest that 15 modes are inadequate for acceleration analysis. *We emphasize that the mode superposition method is not faulty; the difficulty is that the modal method and shock loading are not well suited to one another.*

Response Spectrum Analysis. If we require only estimates of maximum magnitudes of response, response spectrum analysis can be used as an alternative to dynamic response analysis. As will be seen, response spectrum analysis can also be used to indicate the number of modes that will be needed if a dynamic response analysis by the mode superposition method is to be undertaken subsequent to computation of modes and frequencies.

Response spectrum analysis is supported by some software, but we will carry out the following calculations by hand for the sake of illustration. In contrast to the spectra in Fig. 9.10-1, the undamped spectrum for the loading shown in Fig. 9.13-1 *does not decay* with increasing ω. This in itself suggests that higher modes will be important. For undamped displacement response, with f the cyclic frequency of the single-d.o.f. system, factor S (defined in Fig. 9.10-1) is $S = 2\sin\pi t_1 f$ for $t_1 f < 0.5$ and $S = 2$ for $t_1 f > 0.5$ [9.3]. In Eq. 9.10-7, for velocity, $S_v = S$; for acceleration, $S_a = 1$ for all t_1. Consider the mode 1 contribution to maximum horizontal displacement at node 16 in Fig. 9.13-1. As explained above Eq. 9.12-3, but considering mode 1 and a force of 10^5 N, $p_1 = 0.0623(100{,}000) = 6230$. Also $S_1 = 2\sin\pi(0.01)(3.315) = 0.208$. Next, Eq. 9.10-6 becomes

$$z_{1\max} = 0.208\frac{6230}{(2\pi 3.315)^2} = 2.99 \tag{9.13-4}$$

Finally, with mode 1 data for u_{16} from Table 9.12-1, Eq. 9.10-1 gives $\Delta_{16,1} = 0.0386(2.99) = 0.115$, which is the first entry in Table 9.13-2. The remaining entries can be similarly calculated from data in Table 9.12-1 and equations in Section 9.10. The first row in Table 9.13-2 does not quite agree with the first row in Table 9.13-1 because our system is damped but the S factors used pertain to undamped motion, and because of software differences (see the *Note* that concludes Section 9.12).

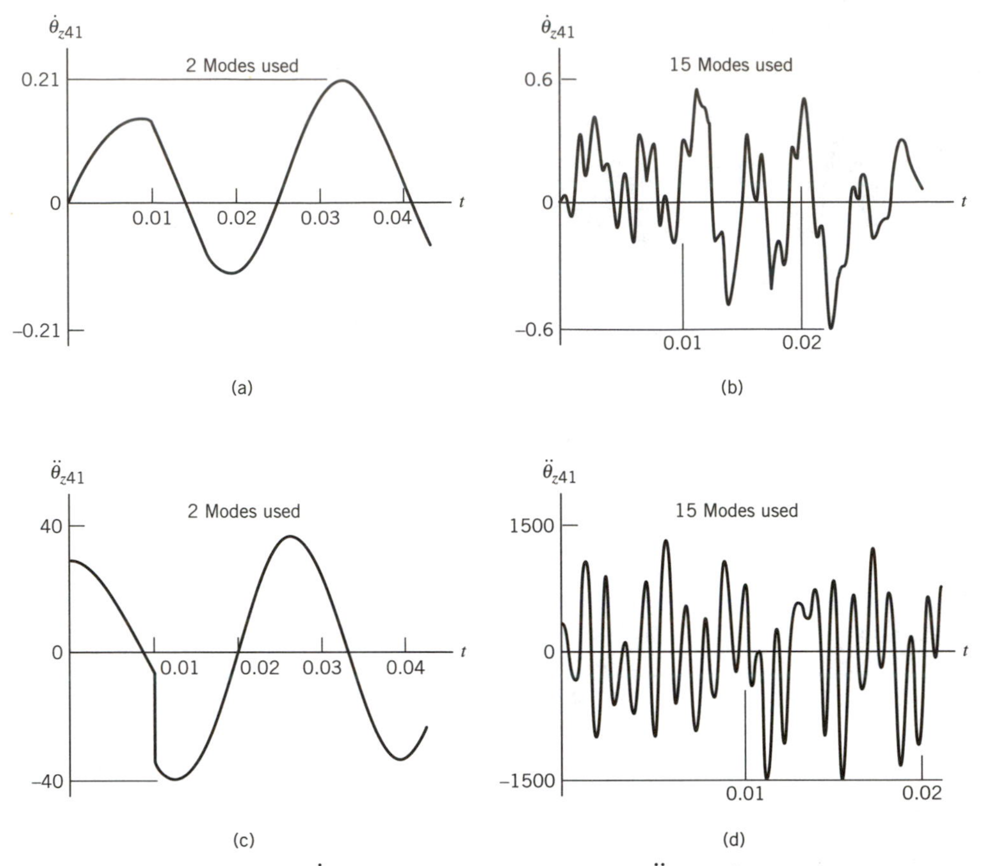

Fig. 9.13-3. Angular velocity $\dot{\theta}_z$ (rad/s) and angular acceleration $\ddot{\theta}_z$ (rad/s^2) at node 41 versus time t (s) for the loading of Fig. 9.13-1.

With data from Table 9.13-2, it is clear that Eq. 9.10-2 gives an upper bound on maximum displacement response reported in Table 9.13-1. The number of modes required to compute (say) u_{16} and its time derivatives, by either the response spectrum method or a dynamic response method, can be estimated from Table 9.13-2: two to five modes for displacements, two to perhaps ten modes for velocities, but perhaps almost all modes for

TABLE 9.13-1. Selected results of greatest magnitude for the problem of Fig. 9.13-1, computed by the modal method using different numbers of modes.

Modes used	$10^3 u_{16}$ (m)	$\dot{u}_{16}$ (m/s)	$\ddot{u}_{16}$ (m/s^2)	$10^3 \theta_{z41}$ (rad)	$\dot{\theta}_{z41}$ (rad/s)	$\ddot{\theta}_{z41}$ (rad/s^2)
1	112	2.38	242	2.80	0.060	6
2	119	4.33	−463	3.37	0.211	−40
3	119	4.31	−464	3.36	0.230	−44
4	119	4.31	529	3.44	0.349	−174
5	119	4.31	553	3.47	0.368	−441
10	119	4.37	830	3.47	0.539	−862
15	119	4.39	1341	3.47	−0.603	−1524

TABLE 9.13-2. Response spectrum analysis of the problem in Fig. 9.13-1: physical d.o.f. Δ_{ji} of Eq. 9.10-1 in each of several modes. Signs are omitted because they are not used in Eqs. 9.10-2 and 9.10-3.

Mode number	$10^3\Delta$ for u_{16} (m)	$\dot{\Delta}$ for $\dot{u}_{16}$ (m/s)	$\ddot{\Delta}$ for $\ddot{u}_{16}$ (m/s^2)	$10^3\Delta$ for θ_{z41} (rad)	$\dot{\Delta}$ for $\dot{\theta}_{z41}$ (rad/s)	$\ddot{\Delta}$ for $\ddot{\theta}_{z41}$ (rad/s^2)
1	115[a]	2.39[a]	241[a]	2.93[a]	0.061[a]	6[a]
2	10	2.17	269	0.84	0.186	23
3	0	0.04	8	0.07	0.030	7
4	0	0.15	60	0.38	0.293	113
5	0	0.08	57	0.22	0.318	226
10	0	0.06	129	0.03	0.128	261
15	0	0.09	375	0.01	0.089	352
	Estimated maximum from Eq. 9.10-3 using all modes up to 15					
1–15	116	3.25	719	3.09	0.780	1489

[a] Line 1 of this table does not quite agree with line 1 of Table 9.13-1; see the text.

accelerations. The latter conclusion is also evident from the second of Eqs. 9.10-7, which contains no ω_i divisor and for which $S_{ai} = 1$ for the shock loading used in the present example.

ANALYTICAL PROBLEMS

9.1 (a) In Fig. 9.2-1a, let r be a constant force F. What is the effect of F on the time-dependent displacement and the natural frequency of vibration?

(b) A weight whose mass is m is placed at the middle of a uniform beam of length L that is clamped at each end. The mass of the beam may be neglected. Estimate the natural frequency of vibration in terms of m, L, E, and I. Suggestion: First determine an effective k.

9.2 Use Eq. 9.3-4 to derive the following element mass matrices:

(a) Equation 9.3-3.

(b) The mass matrix of a two-d.o.f., linear-displacement-field element, as in Fig. 9.3-1b, but with a cross-sectional area that varies linearly from to A_1 to A_2.

9.3 If a differential element has mass dm and velocity v, its kinetic energy is $v^2 dm/2$. Show that the kinetic energy of a finite element is therefore $\dot{\mathbf{d}}^T\mathbf{m}\dot{\mathbf{d}}/2$, where $\mathbf{m}$ is the consistent element mass matrix and $\mathbf{d}$ is the vector of element nodal d.o.f.

9.4 Consider a uniform plane beam element of length L, whose nodal d.o.f. are a lateral translation and a rotation at each end (four d.o.f. altogether). If the element is to have the correct kinetic energies in the rigid-body motions of lateral translation and rotation about the mass center, what should be the terms in a diagonal element mass matrix? Does this matrix provide the correct kinetic energy of rotation about one end?

9.5 A uniform two-node bar element is allowed to displace three-dimensionally. The nodes have only translational d.o.f. What is the diagonal mass matrix of the element?

9.6 Consider the kinetic energies of a straight, two-node element in these three rigid-body motions in the plane of the paper: lateral translation, rotation about the mass center, and rotation about the left end. Do the following mass matrices provide the correct kinetic energy or not?

(a) Bar element, lumped mass matrix (Eq. 9.3-1).
(b) Bar element, consistent mass matrix (Eq. 9.3-3).
(c) Beam element, consistent mass matrix (Eq. 9.3-5).

9.7 Rigid and massless link AB in the sketch connects a particle mass m at A to node B, which is a node of a FE structure. In terms of the three d.o.f. shown at node B, what is the 3 by 3 mass matrix associated with mass m? Suggestion: Review Section 4.3. Physically and qualitatively, what sort of error would be produced by somehow diagonalizing this matrix?

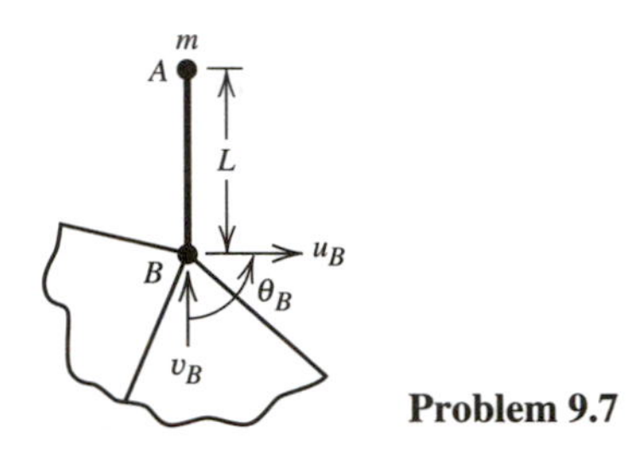

Problem 9.7

9.8 Using the diagonal mass matrix suggested below Eq. 9.3-3, in which all four m_{ii} are nonzero, solve for the natural frequencies and mode shapes of the cantilever beam in Fig. 9.4-1.

9.9 The sketch shows an unsupported uniform bar, modeled by one two-node element. Solve for the frequencies and modes of axial vibration using mass matrices as follows:

(a) Lumped mass matrix (Eq. 9.3-1).
(b) Consistent mass matrix (Eq. 9.3-3).
(c) A mass matrix that is the average of the lumped and consistent forms.

The exact fundamental frequency is $\omega_1 = (\pi/L)\sqrt{E/\rho}$.

9.10 The sketch shows a uniform bar, fixed at one end and modeled by two two-node elements of equal length. Solve for the frequencies and modes of axial vibration using mass matrices as follows:

(a) Lumped mass matrix (Eq. 9.3-1).
(b) Consistent mass matrix (Eq. 9.3-3).
(c) A mass matrix that is the average of the lumped and consistent forms.

The exact fundamental frequency is $\omega_1 = (\pi/2L)\sqrt{E/\rho}$.

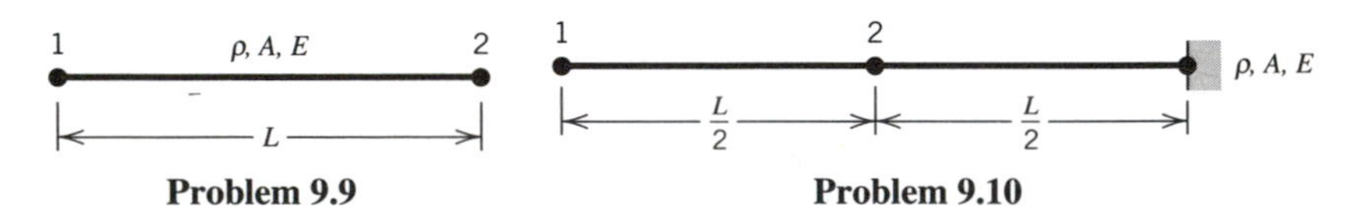

Problem 9.9 **Problem 9.10**

9.11 (a) For the example problem in Section 9.4, obtain the natural frequencies by using a mass matrix that is the average of the consistent mass matrix and the particle–lumped mass matrix (no rotary inertia in the lumped mass matrix).
(b) Obtain the associated mode shapes.

9.12 (a) Express the reduced stiffness matrix $\mathbf{K}_r = \mathbf{T}^T\mathbf{KT}$ in terms of $\mathbf{K}_{mm}$, $\mathbf{K}_{ms}$, and $\mathbf{K}_{ss}$ (see Eq. 9.6-1).
(b) Similarly, express the reduced mass matrix $\mathbf{M}_r = \mathbf{T}^T\mathbf{MT}$ in terms of stiffness and mass submatrices in Eq. 9.6-1.

9.13 Reconsider the example problem in Section 9.6. This time use $\bar{\theta}_{z2}$ as master and $\bar{v}_2$ as slave, and compute the fundamental frequency of vibration.

9.14 Apply reduction to Problem 9.11, and obtain the resulting fundamental frequency of vibration.

9.15 Apply reduction to Problem 9.10b. Obtain the fundamental frequency of vibration as follows:
(a) Use $\bar{u}_1$ as master, $\bar{u}_2$ as slave.
(b) Use $\bar{u}_2$ as master, $\bar{u}_1$ as slave.

9.16 For the example problem in Section 9.4, unscaled nodal displacements of mode 1 can be shown to be $\bar{v}_2 = 1.0$ and $\bar{\theta}_{z2} = 1.378/L$.
(a) Scale mode 1 so as to satisfy Eq. 9.4-3.
(b) Show that then $\bar{\mathbf{D}}_1^T\mathbf{K}\bar{\mathbf{D}}_1 = \omega_1^2$, as required by Eq. 9.7-3.

9.17 Determine the damping matrix **C** in Eq. 9.2-4 that is implied by the modal damping in Eq. 9.7-6. (The resulting **C** is full, and depends on $\boldsymbol{\phi}$ and a diagonal matrix whose ith diagonal term is $2\xi_i\omega_i$.)

9.18 The sketch shows three equal particle masses connected by three identical springs. Only axial motion is permitted. The vibration modes, scaled so that the largest d.o.f. in each is unity, are $\bar{\mathbf{D}}_1 = [1.000\ 0.802\ 0.445]^T$, $\bar{\mathbf{D}}_2 = [-0.802\ 0.445\ 1.000]^T$, and $\bar{\mathbf{D}}_3 = [-0.445\ 1.000\ -0.802]^T$.
(a) Sketch these modes, and show by multiplication that they are mass-matrix orthogonal. Are they also stiffness-matrix orthogonal?
(b) Let $\bar{\mathbf{D}}_1$ and $\bar{\mathbf{D}}_2$ be used in Eq. 9.7-7 (omit $\bar{\mathbf{D}}_3$). Also let $m = 1$ for each particle mass. What must be z_2 if $z_1 = 1$ and the displacement of node 1 is zero? Sketch the displaced shape of the structure thus obtained.
(c) For comparison, use u_1 and u_2 as masters in Eq. 9.6-3. For $u_1 = 0$ and $u_2 = 1$, sketch the displaced shape of the structure.

1 2 3
m k m k m k — x, u

Problem 9.18

9.19 A single-d.o.f. system without damping has natural frequency ω and is loaded by the force $r = F_0 \sin \Omega t$. For what ratio Ω/ω is the static displacement F_0/k amplified by no more than 10%?

9.20 (a) The finite difference expression for accelerations $\ddot{\mathbf{D}}$ can be written $\ddot{\mathbf{D}} = \Delta\dot{\mathbf{D}}/\Delta t$, where $\dot{\mathbf{D}} = \Delta\mathbf{D}/\Delta t$. Derive the second of Eqs. 9.9-3 by applying this information to two successive intervals.
(b) Show that Eqs. 9.2-4, 9.9-2, and 9.9-3 yield Eqs. 9.9-4.
(c) Why is $\gamma = \frac{1}{2}$ and $\beta = \frac{1}{4}$ in the Newmark method known as the "trapezoidal rule"? Suggestion: Sketch acceleration versus t and use Eq. 9.9-6b.

9.21 A particle of unit mass is supported by a spring of unit stiffness. There is no damping. At time $t = 0$, when the particle has zero displacement and is at rest, a constant unit force is applied. Use the central difference method, Eq. 9.9-4, to compute the displacement at successive time steps as follows:

(a) Use $\Delta t = 0.5$ and go to $t = 7.0$.
(b) Use $\Delta t = 1.0$ and go to $t = 7.0$.
(c) Use $\Delta t = 2.0$ and go to $t = 10.0$.
(d) Use $\Delta t = 3.0$ and go to $t = 15.0$.
(e) Obtain the exact solution and compare the results of part (a) with it.

9.22 For the example problem in Section 9.4, approximate ω_1 and ω_2 by use of Eq. 9.11-1. In each case, do so by guessing a reasonable but probably inexact displacement vector $\bar{\mathbf{D}}$.

9.23 Use Eq. 9.11-1 to obtain the natural frequencies of the spring–mass system whose vibration modes are stated in Problem 9.18.

9.24 Check the ordinate of the first peak in Fig. 9.12-3b (for which $f = 70.77$ Hz) by use of the method described in connection with Eq. 9.12-3.

9.25 Calculation of the first entry in Table 9.13-2 is explained in connection with Eq. 9.13-4. Proceeding in similar fashion, verify the following entries in Table 9.13-2:

(a) Δ terms for modes 2 through 5 in the u_{16} column.
(b) $\dot{\Delta}$ terms for modes 1 through 5 in the $\dot{u}_{16}$ column.
(c) $\ddot{\Delta}$ terms for modes 1 through 5 in the $\ddot{u}_{16}$ column.
(d) Δ terms for modes 1 through 5 in the θ_{z41} column.
(e) $\dot{\Delta}$ terms for modes 1 through 5 in the $\dot{\theta}_{z41}$ column.
(f) $\ddot{\Delta}$ terms for modes 1 through 5 in the $\ddot{\theta}_{z41}$ column.

COMPUTATIONAL PROBLEMS

When specific data are not stated in the following problems, choose convenient numbers and consistent units for material properties, dimensions, and loads. Unless directed otherwise, assume that thicknesses are uniform and the material is isotropic. Where possible without considerable effort and advanced knowledge, obtain preliminary analytical estimates of significant quantities to be computed.

9.26 Let displacements of the structure shown be confined to the plane of the paper. The structure may be regarded as a truss (bar elements and pinned member connections) or a frame (beam elements and welded member connections). For simplicity, assume that member cross sections are square, each h units on a side. One might, for example, use steel as the material, with $H = L = 6.0$ m and $h = 50$ mm. Using a single element to model each member, investigate the natural frequencies and modes under the following conditions:

(a) Truss structure, lumped mass formulation.
(b) Truss structure, consistent mass formulation.
(c) Frame structure, lumped mass formulation.
(d) Frame structure, consistent mass formulation.

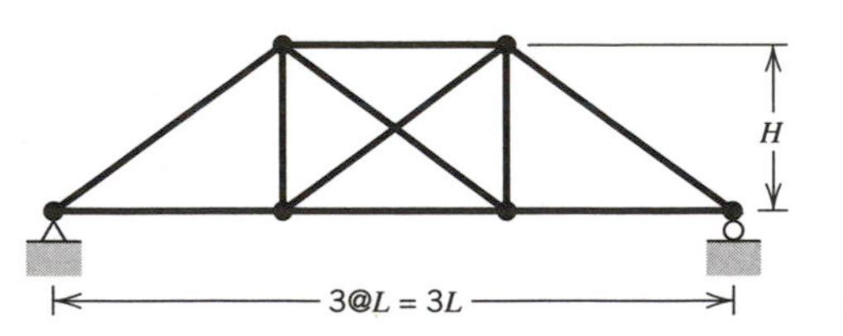

Problem 9.26

9.27 Repeat Problem 9.26, this time using two or more elements of equal length to model each member.

9.28 The structure shown consists of six identical slender rods of circular cross section. The rods are welded together with equal angles between them to form a plane structure. Investigate the first eight nonzero natural frequencies and their associated modes. Confine displacements to the plane of the paper, and consider that center C is:

(a) Unsupported.

(b) Allowed to rotate but not translate.

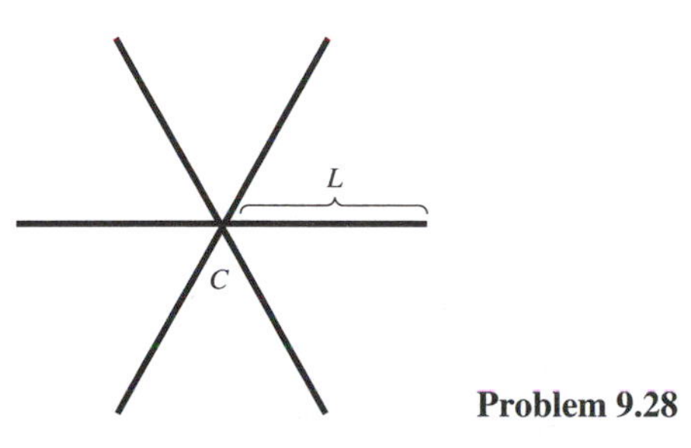

Problem 9.28

9.29 (a) Repeat Problem 9.28 with three-dimensional motion allowed.

(b) Repeat Problem 9.28 but with one of the rods doubled in mass.

9.30 Repeat Problem 9.28, but with different support conditions. Now provide simple support at the outer end of each rod and let center C be unsupported. Let displacements be:

(a) Confined to the plane of the paper.

(b) Allowed in any direction.

9.31 The uniform cantilever beam shown has total mass m_b and a solid circular cross section. A uniform *rigid* crossbar of total mass m_c is attached to the end of the beam, with its mass center offset a distance s from the end of the beam. Investigate the natural frequencies and modes for the following special cases:

(a) $m_c = 0$.

(b) $m_b = 0$, $s = 0$.

(c) $m_b = m_c$, $s = 0$.

(d) $m_b = m_c$, $h = 0.6L$, $s = 0.5h$.

9.32 The sketch shows a building frame of n stories, each h units high. Investigate the natural frequencies and modes.

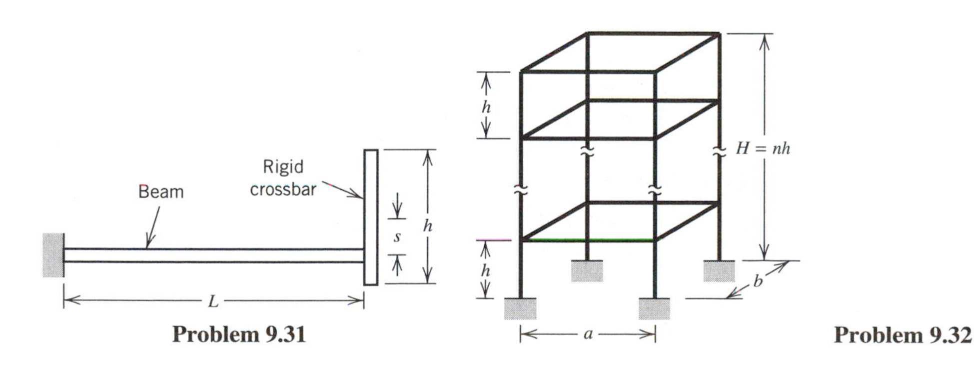

Problem 9.31 **Problem 9.32**

9.33 The spiral clock gong shown has centerline radius (in millimeters) $r = 26.65 + 0.684\theta$, where θ (in radians) is zero at A and 9π at B. Portion BC is straight and C is fixed. The material is steel and the cross section is a 1.69 mm by 2.80 mm rectangle with the smaller dimension in the plane of the paper. Let $E = 209$ GPa, $\nu = 0.3$, and $\rho = 7800$ kg/m^3. Investigate the first eight natural frequencies and modes (adapted from [9.11]).

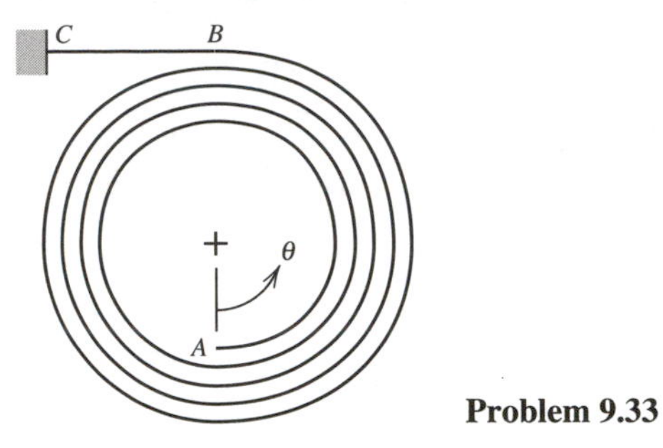

Problem 9.33

9.34 Investigate the natural frequencies and modes of a flat square membrane. Let displacements be confined to the plane of the membrane. As boundary conditions, consider that edges are:

(a) All free.
(b) All free but one.
(c) All fixed.

9.35 Investigate the natural frequencies and modes of lateral vibration of a flat square plate.

(a) Let the plate be unsupported.
(b) Let the plate be simply supported on all edges.
(c) Let the plate be clamped on one edge.
(d) Let the plate be clamped on all edges.

9.36 The sketch depicts two channel sections, welded together to create an angle ϕ between them. The structure is unsupported. Investigate the natural frequencies and modes.

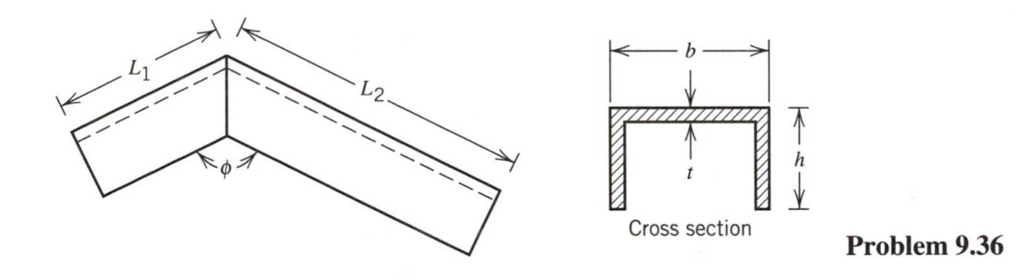

Problem 9.36

9.37 The sketch shows a portion of a cylindrical shell with fixed support along one straight edge. The shell becomes a cantilevered plate of length L if radius R is infinite. Investigate the natural frequencies and modes.

9.38 Repeat Problem 9.37, but free the plate/shell on the straight edge of length a, and instead provide fixed support along an adjacent edge.

9.39 Investigate the natural frequencies and modes of the conical shell shown.

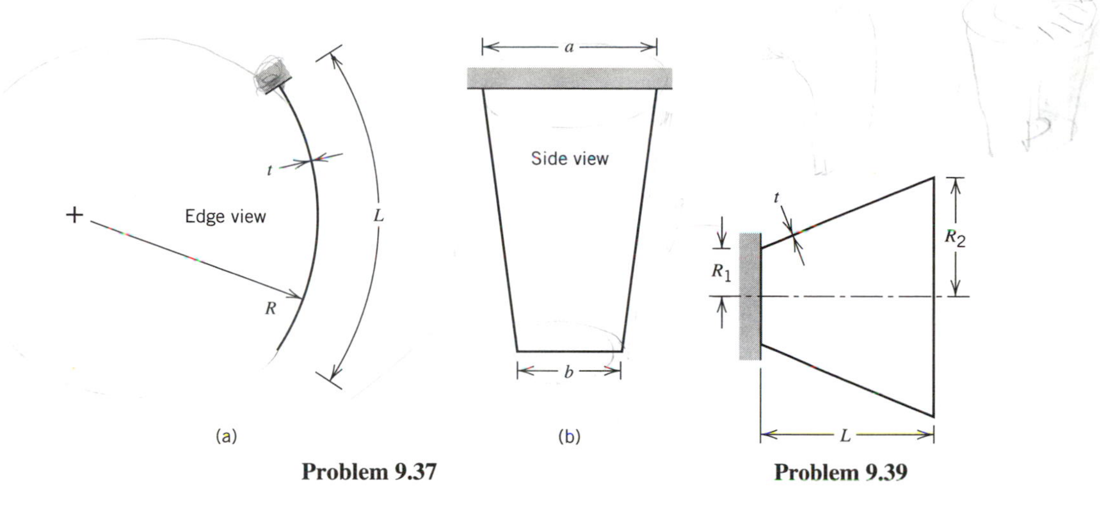

Problem 9.37 **Problem 9.39**

9.40 The following structures, described in preceding chapters, may be examined for their natural frequencies and modes. In cases where supports are not indicated, the structure may be left unsupported, provided with support just adequate to prevent rigid-body motion, or various other alternatives.

(a) Figure 5.2-3 (thin-walled curved beams).
(b) Figure 7.4-3 (cylindrical shell; as shown or slit open as in Fig. 7.7-1a).
(c) Figure 7.5-2a (shell roof).
(d) Figure 7.5-2c (hemispherical shell).
(e) Figure 7.5-2d (twisted strip).
(f) Problem 7.18 (trapezoidal plate, parallel edges free).
(g) Problem 7.41 (intersecting cylindrical shells).
(h) Problem 7.50 (cylindrical shell segment).

9.41 Investigate the fundamental frequency and vibration mode of a plate on an elastic foundation (e.g., the problem of Fig. 7.3-1).

(a) Ignore the mass of the foundation.
(b) Include the mass of the foundation. Assume, for example, that the elastic layer is ten times the thickness of the plate and one-quarter of its mass density. Approximate as necessary.
(c) Examine the effect of making other choices for the mass density of the foundation.

9.42 Let the left end of the beam in Problem 9.31 be connected to a rigid shaft by means of a ball and socket joint. The shaft has small diameter, rigid bearings and rotates at constant angular velocity about its axis, which is normal to the plane of the paper. Neglect the influence of gravity. Investigate the effects of stress stiffening and spin softening on the natural frequencies and modes of the beam. The crossbar may be rigid or flexible, as desired.

9.43 Reduction can be applied to most of the preceding problems in which natural frequencies are to be computed. One might examine the percentage change in computed frequencies as the number of masters is successively reduced. Also, a poor set of masters may be deliberately chosen and the results compared with those produced by a good choice and by the full set of d.o.f.

9.44 Harmonic response analysis can be applied to any of the preceding problems in which natural frequencies are to be computed. Great variety is afforded by options in the number of forcing functions, their directions and points of application, their frequency, and their phase angles.

9.45 The uniform straight bar shown has elastic modulus E, cross-sectional area A, mass density ρ, and is initially unloaded and at rest. At time $t = 0$ a tensile load of magnitude $P_0 = \sigma_0 A$ is suddenly applied and thereafter maintained. The plot shows the theoretical solution for the variation with time of axial displacement (dashed line) and axial stress (solid line) at the midpoint of the bar, with damping ignored. Time t_c is $t_c = (L/2)/\sqrt{\rho/E}$. Construct a FE model using two-node bar elements, and solve this problem numerically.

(a) Use the central difference method.

(b) Use the Newmark method.

(c) Use the mode superposition method.

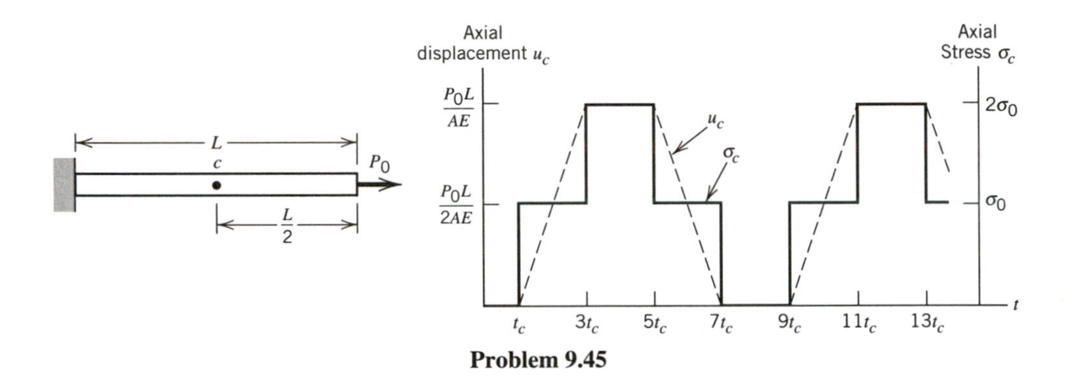

Problem 9.45

9.46 Repeat Problem 9.45 but with both ends of the bar free.

9.47 Repeat Problem 9.45 but with the cross-sectional area increased to $2A$ in the right-most 40% of the bar.

9.48 The uniform beam shown is initially at rest. The sketch includes a formula for the period of vibration T_n of mode n, in which $b_1 = 3.52$, $b_2 = 22.0$, and $b_n \approx (2n - 1)^2\pi^2/4$ for $n > 2$. Starting at time $t = 0$, load P increases linearly to P_0 and then drops suddenly to zero. Construct a FE model of two-node beam elements and compute the deflection versus time plot for the right-hand end of the beam. Consider:

(a) $t_0 = T_1/4$.

(b) $t_0 = T_2/4$.

(c) $t_0 = T_3/4$.

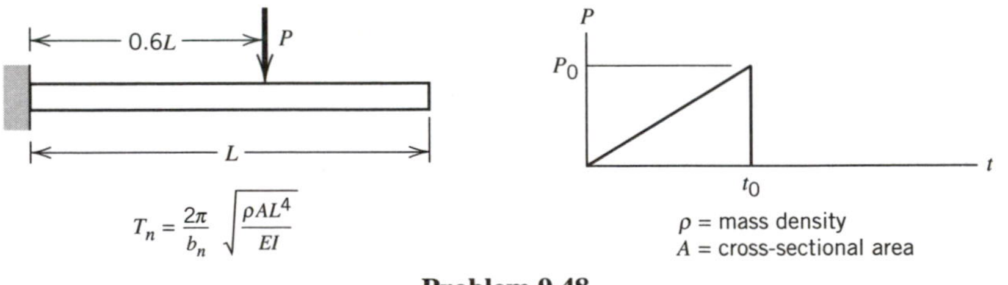

Problem 9.48

9.49 A uniform simply supported beam is loaded by two forcing functions at the locations shown. The functions are $F_1 = \sin 2\omega_1 t$ and $F_2 = \sin 6\omega_1 t$, where ω_1 is the fundamental vibration frequency of the beam. Compute, as a function of time, the center deflection, center acceleration, and support reactions.

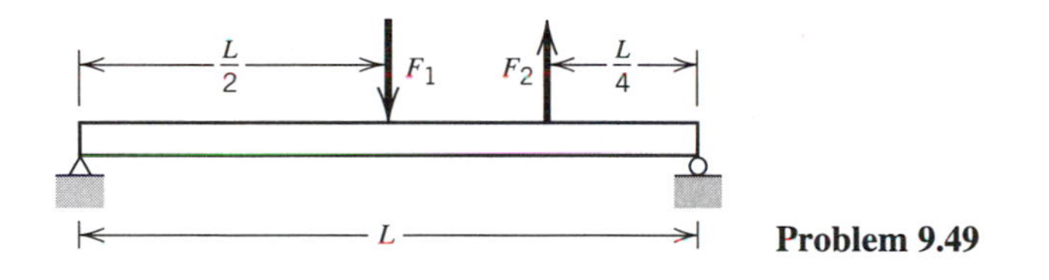

Problem 9.49

9.50 The cross-shaped structure shown supports heavy particles of slightly differing mass at A, B, C, and D. The structure consists of lightweight beams AE, BE, CE, DE, and FE. The beams are mutually orthogonal and are welded together at E. The beams are of equal length except for FE, which is comparatively short and is rigidly supported at F. Investigate the dynamic response of the structure after an initial velocity is imparted to the particle at A. Let this initial velocity be in (a) the y direction, and (b) the z direction.

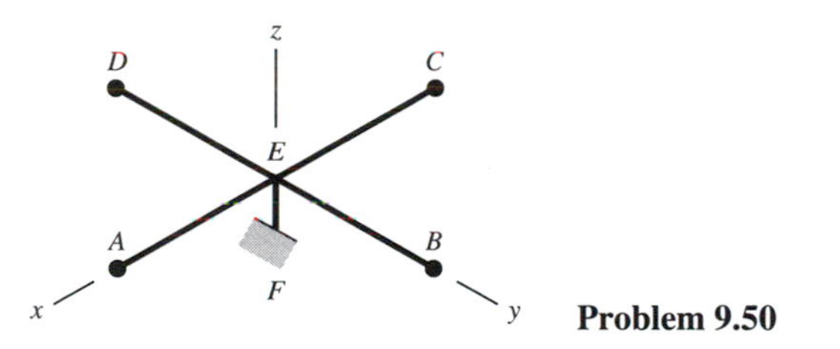

Problem 9.50

9.51 Dynamic response analysis can be applied to any of the preceding problems in which natural frequencies are to be computed. Great variety is afforded by options in the number of forcing functions, their directions and points of application, and their manner of time variation. Some possible load histories are shown in the sketch.

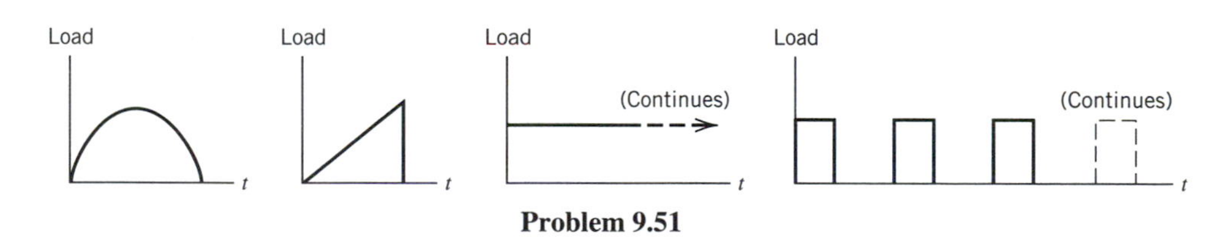

Problem 9.51

9.52 Response spectrum analysis can be applied to problems for which dynamic response analysis is requested, for example, Problems 9.48 and 9.49. Computed maxima can be compared if a problem has been solved by both methods.

CHAPTER 10

Nonlinearity in Stress Analysis

Concepts of nonlinearity are introduced by explaining simple solution algorithms. Stress stiffening is discussed, and its use in linear buckling analysis and nonlinear analysis is summarized. Geometric nonlinearity, material nonlinearity, and gap closure are explained as common sources of nonlinearity in stress analysis and are illustrated by simple example analyses.

10.1 INTRODUCTION

In *linear* analysis, response is directly proportional to load. Linearity may be a good representation of reality or may only be the inevitable result of assumptions made for analysis purposes. In linear analysis we assume that displacements and rotations are small, supports do not settle, stress is directly proportional to strain, and loads maintain their original directions as the structure deforms. Equilibrium equations $\mathbf{KD} = \mathbf{R}$ are written for the *original* support conditions, elastic stress–strain relations, load-free configuration, and load directions. Displacements $\mathbf{D} = \mathbf{K}^{-1}\mathbf{R}$ are obtained in a single step of equation-solving. We are fortunate that so many practical problems can be solved by so simple an approximation. However, any of the convenient assumptions that lead to a linear analysis may be at odds with reality. Adjacent parts may make or break contact. A contact area may change as load changes. Elastic material may become plastic, or the material may not have a linear stress–strain relation at any stress level. Part of the structure may lose stiffness because of buckling or failure of the material. Displacements and rotations may become large enough that equilibrium equations must be written for the deformed configuration rather than the original configuration. Large rotations cause pressure loads to change in direction, and also to change in magnitude if there is a change in the area to which they are applied. Thus, for various reasons, a problem may become nonlinear; for example, a plot of load versus displacement ceases to be a straight line.

Simple examples of nonlinear problems appear in Fig. 10.1-1. A slender beam, Fig. 10.1-1a, is loaded by a force P that acts normal to the beam axis at all times. This is an instance of a "follower force." The displacement shown is intended to represent actual displacement, not the scaling up of a linear small-displacement solution. In this case nonlinearity is *geometric*, meaning that nonlinearity arises because of significant changes in the geometry of the structure. *Material* nonlinearity causes the behavior in Fig. 10.1-1b: material of which the beam is made has a nonlinear stress–strain relation.

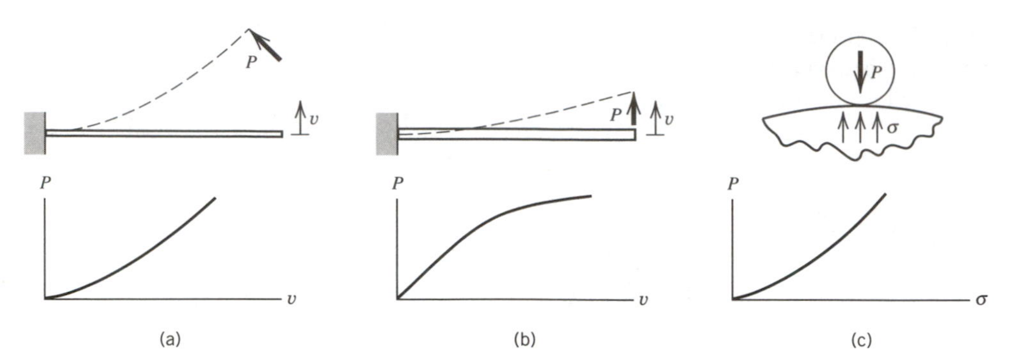

Fig. 10.1-1. (a) Slender elastic beam loaded by a follower force P. (b) Elastic–plastic beam loaded by a fixed-direction force P. (c) Contact stress in a roller bearing.

The construction in Fig. 10.1-1c exhibits "contact nonlinearity" because the area of contact between the two bodies grows as load P increases, whether the material yields or not. This is an instance of geometric nonlinearity in which displacements and strains are small.

Nonlinearity makes a problem more complicated because equations that describe the solution must incorporate conditions not fully known until the solution is known—the *actual* configuration, loading condition, state of stress, and support condition. The solution cannot be obtained in a single step of analysis. We must take *several* steps, update the tentative solution after each step, and repeat until a convergence test is satisfied. The usual linear analysis is only the first step in this sequence. Nonlinear analysis can treat a great variety of problems, but in a sense it is more restrictive than linear analysis because the principle of superposition *does not apply*; we cannot scale results in proportion to load or combine results from different load cases as in linear analysis. Accordingly, each different load case requires a separate analysis. Also, if a loading consists of component loads that are sequentially applied, results may not be independent of the order in which loads are applied.

In this chapter we discuss stress stiffening, large deflections, buckling, and how they are interrelated; time-independent material nonlinearity; and contact problems in which contact areas may change and gaps may open or close. In order to keep the discussion short and comparatively simple, we omit the subjects listed below, despite their practical importance. Our list of omitted subjects is not comprehensive but nevertheless suggests the broad range of nonlinear phenomena:

- Materials in which deformation depends on load rate as well as load level (creep, viscoelasticity, viscoplasticity).
- Problems in which strains are large, as in metal-forming processes (large strain analysis requires that terms be added to strain–displacement relations such as Eqs. 3.1-5).
- Linkages, mechanisms, and other problems that involve large rigid-body motion.
- Coupled problems, such as fluid–structure interaction, in which fluid forces cause a structure to move, but by moving it alters the fluid forces.
- Nonlinear dynamic problems, such as nonlinear vibrations and projectile impact.
- Nonlinearity in problems other than stress analysis.

10.2 SOLUTION ALGORITHMS. CONVERGENCE CRITERIA

How does software solve a nonlinear problem? It is important to understand the rudiments of solution algorithms because the analyst must make initial choices and know what to try next if a procedure fails. Many aspects of nonlinear solution methods can be discussed independently of the source of nonlinearity. So that simple 2D plots can be used for explanation, the following summary of solution methods considers a nonlinear equation having *a single d.o.f.* Methods discussed can also be applied to multi-d.o.f. nonlinear structures.

Consider a nonlinear spring loaded by a force P, Fig. 10.2-1a. In general, as stretch u increases, stiffness k of the spring may either increase (hardening structure) or decrease (softening structure). Multi-d.o.f. examples of these respective behaviors are a cable network that deflects laterally and a steel structure with spreading yield zones. Figure 10.2-1b depicts softening behavior. Let us say that the purpose of analysis is to determine the stretch of the spring for any value of load, that is, to construct a plot of P versus u. Stiffness k is a function of u and can be calculated for any value of u. However, we assume that the equation $ku = P$ cannot be solved explicitly for u as a function of P. This restriction is made so that the single-d.o.f. example resembles multi-d.o.f. equations $\mathbf{KD} = \mathbf{R}$ that are nonlinear because $\mathbf{K}$ is a function of $\mathbf{D}$. Numerical methods are unable to solve nonlinear equations explicitly for $\mathbf{D}$ as a function of $\mathbf{R}$. Instead, *a nonlinear problem is solved by taking a sequence of linear steps.*

Solution Algorithms. A simple way to solve nonlinear equations is called *direct substitution.* It is described for thermal analysis in Section 8.3. The method is slow unless nonlinearities are mild, and it is not often used for stress analysis. More often, various *incremental* methods are used. They use the *tangent stiffness*, which for a single-d.o.f. problem is the slope of the P versus u plot, $k_t = dP/du$ in Fig. 10.2-1. In a single-d.o.f. problem we could obtain the "correct" curve in Fig. 10.2-1b by calculating P for each of several values of u. The analogous procedure in a multi-d.o.f. problem would be to calculate load vectors $\mathbf{R}$ for each of several displacement vectors $\mathbf{D}$. However, this option is not avail-

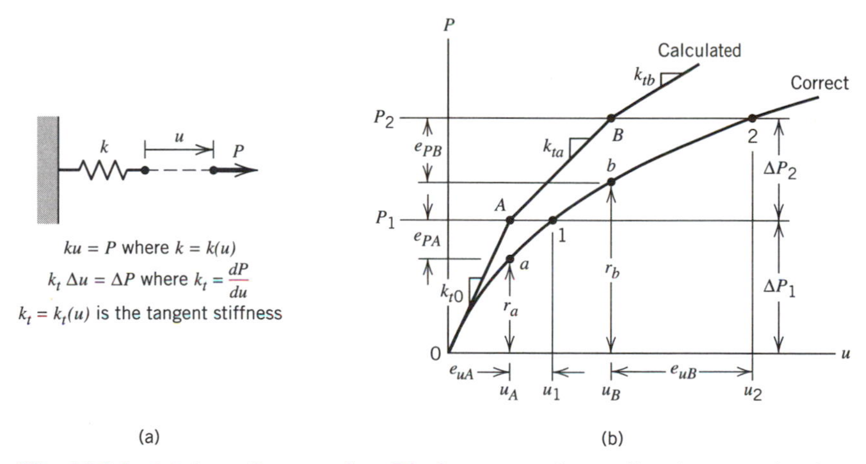

Fig. 10.2-1. (a) A nonlinear spring. Displacement u is *not* directly proportional to load P. (b) Load versus displacement relation of softening spring. Purely incremental solution, showing drift from the correct solution.

able because the proper relation among d.o.f. in **D** cannot be prescribed; it must be calculated. Only when the relation is known can we obtain from **D** the associated **R** and tangent stiffness matrix $\mathbf{K}_t$.

An *incremental solution* is the same as Euler's method of solving a first-order differential equation. Starting at $u = 0$, we obtain the initial tangent stiffness k_{t0}. Applying load increment ΔP_1 and solving a *linear* equation, we arrive at point A in Fig. 10.2-1b, for which the displacement is $u_A = u_a$. The tangent stiffness corresponding to this displacement is k_{ta}. Applying the next load increment, ΔP_2, we take another linear step at slope k_{ta} and arrive at point B. Symbolically, the process is

$$k_{t0}\,\Delta u_1 = P_1 - 0, \quad \text{solve for } \Delta u_1, \quad \text{then} \quad u_A = 0 + \Delta u_1 \tag{10.2-1a}$$

$$k_{ta}\,\Delta u_2 = P_2 - P_1, \quad \text{solve for } \Delta u_2, \quad \text{then} \quad u_B = u_A + \Delta u_2 \tag{10.2-1b}$$

and so on. The computed set of points A, B, and so on, can be connected by line segments to provide the calculated relation between P and u. The correct relation is unknown in practice, but we show it in Fig. 10.2-1b to illustrate that the calculated curve has progressive drift from the correct curve. Calculated displacements u_A, u_B, and so on are in error by the amounts e_{uA}, e_{uB}, and so on. A correction for this error is described next.

At point A in Fig. 10.2-1b, the applied force P_1 is greater than the resisting force of the spring, which is $r_a = ku_A$ when the stretch of the spring is u_A. The difference, $P_A - r_a = e_{PA}$, is a force imbalance that can be used to drive the displacement toward the correct value u_1 by doing an "equilibrium iteration" while applied force P_1 is held constant. In Fig. 10.2-2a, equilibrium iterations are performed by the Newton–Raphson method, which is explained in calculus books as a way to solve a nonlinear equation. In the first equilibrium iteration we use tangent stiffness k_{ta}, solve the equation $k_{ta}\,\Delta u = e_{PA}$ for Δu, then add it to u_A, thus arriving at point A'. At point A' we have a new force imbalance, specifically the small vertical distance between points A' and a', and a new tangent stiffness corresponding to displacement at A'. A step tangent to the curve from point a' now places us so close to the correct point 1 that we cannot see the difference on the plot. Each equilibrium iteration reduces the force imbalance. When it is considered small enough by some convergence test we are ready to increase the force to P_2, thus arriving at

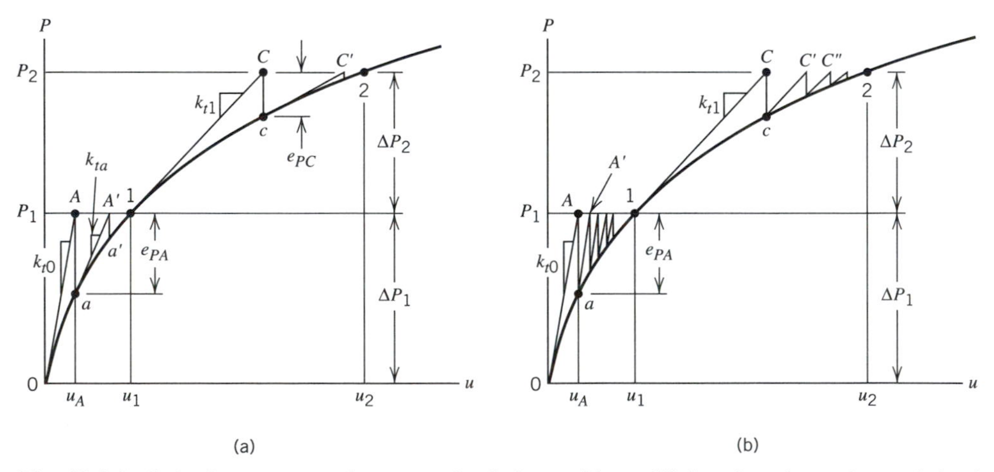

Fig. 10.2-2. Softening structure: incremental solutions with equilibrium iterations after each load step. (a) Newton–Raphson iterations. (b) Modified Newton–Raphson iterations.

point C, and commence equilibrium iterations with the load maintained at P_2. Thus we establish several points on the plot of P versus u and by connecting them we approximate the actual curve.

The foregoing method may be expensive in a multi-d.o.f. problem because tangent stiffness matrix $\mathbf{K}_t$ must be constructed and reduced for equation-solving in every iteration. An alternative, called *modified* Newton–Raphson iteration, is illustrated in Fig. 10.2-2b. Here iterations at each load level are all performed using the stiffness that prevails at the outset of the load step. Thus tangent stiffness matrix $\mathbf{K}_t$ needs to be constructed and reduced only once for all iterations at a given load level. However, as seen in Fig. 10.2-2b, the number of equilibrium iterations needed is considerably greater.

The same procedures may also be applied to a hardening structure, Fig. 10.2-3, but the convergence behaviors are different. The initial step from a tangent at 0 now yields too large a displacement u_A. Newton–Raphson equilibrium iterations yield negative correction increments Δu and converge to point 1. The initial step of modified Newton–Raphson iterations, from point a in Fig. 10.2-3b, either overcorrects so that subsequent convergence is slow, or misses the curve entirely so that convergence fails. Such troubles may be overcome or avoided by changing strategy (to Newton–Raphson iterations perhaps), reducing the magnitude of correction when hardening is detected, or dividing load increments ΔP_1, ΔP_2, and so on into subincrements. Software may be coded to monitor the solution and automatically take some of these actions when necessary, but the analyst must remain watchful.

The foregoing summary of solution procedures is far from exhaustive [10.1]. There exist "quasi-Newton" methods, which in effect update the inverse of the tangent stiffness matrix in each iteration rather than reconstructing the matrix itself. This can greatly reduce computation effort. In one dimension, a quasi-Newton step is a secant of the curve. In a rather different algorithm, called "dynamic relaxation," the structure is endowed with fictitious mass and damping, ideally such that the structure is critically damped. Dynamic response is computed. When the structure stops moving, the static solution has been obtained. Regardless of the solution method, one seeks a configuration $\mathbf{D}$ such that applied loads are in balance with resistance of the structure. Resistance depends on deformation and can be computed by summing element contributions, so that stiffness matrix $\mathbf{K}$ of the deformed structure need not be explicitly formed.

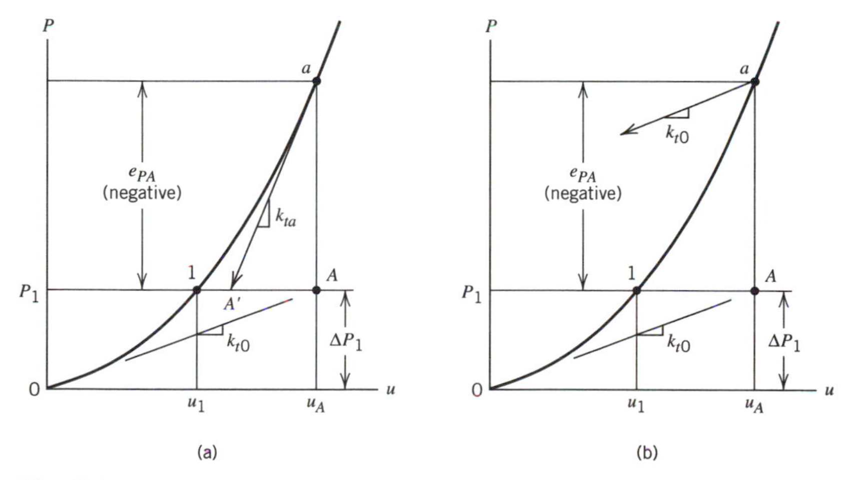

Fig. 10.2-3. Hardening structure: incremental solution with equilibrium iteration after the load step. (a) Newton–Raphson iteration. (b) Modified Newton–Raphson iteration.

In summary, nonlinear response is constructed by taking a number of linear steps. Each step represents a load level. The step from one load level to the next is broken into substeps if software is so instructed. Equilibrium is established by iteration at each step and each substep. Each step, substep, and equilibrium iteration requires solution of a set of linear equations. An iterative equation solver may be effective in a nonlinear problem because the solution at one substep is usually a very good approximation of the solution at the next substep.

Convergence Criteria. Equilibrium iterations at a given load level can be stopped when the solution is "close enough," as defined by the analyst. Assuming now that we deal with a multi-d.o.f. FE model, a possible definition of error is

$$err = \frac{\|\mathbf{e}_P\|}{\|\mathbf{R}\|} \tag{10.2-2}$$

where norm symbols usually indicate the Euclidean norm. Thus the numerator is the square root of the sum of the squares (SRSS) of current load imbalances for all d.o.f. of the model; the denominator is SRSS for current loads applied to all d.o.f. Other kinds of norm are sometimes used. Iteration may be terminated when *err* is reduced to (say) 0.001. Too small a tolerance wastes time in gaining unnecessary accuracy; too large a tolerance may not provide enough accuracy. In general, Eq. 10.2-2 contains both force and moment terms. Thus there is an awkward mix of units. For a given configuration, moment terms may dominate if length units are millimeters but not if length units are meters. This concern does not arise if the error measure algorithm incorporates length scaling. Alternatively, one may choose to monitor only force errors.

An error measure like Eq. 10.2-2 but based on displacements can be written. Thus the numerator involves displacement increments computed in the most recent iteration and the denominator involves current displacements. As with Eq. 10.2-2, there is a mixture of units (length units for displacements and radians for rotations), with similar awkwardness and similar remedy. A displacement error measure may cause premature termination of iteration merely because convergence is slow. On the other hand, it is possible that displacements have essentially converged while significant load imbalances remain. Probably one should always require that load error not exceed a tolerance limit. One may additionally require that displacement error not exceed a tolerance limit. Other error measures have also been proposed [10.2].

One may place a limit on the number of iterations allowed at each load level. Reaching the limit before achieving convergence is called a "convergence failure." When this happens, a software option may call for automatic restart from the previous load level but with a smaller load step. Otherwise the analyst must decide whether or not to proceed.

10.3 STRESS STIFFENING

The term "stress stiffening" refers to a coupling between membrane stress and lateral displacements associated with bending. The bending stiffness of a beam, arch, plate, or shell is increased by tensile membrane stress and is decreased by compressive membrane stress. Some of the stress may be produced by applied load and some may be residual stress from manufacturing or assembly processes. Sufficiently large compressive membrane stress reduces the bending stiffness to zero; that is, the structure buckles. Stress stiffening is usually negligible for comparatively massive bodies but may be important for thin-walled construction. A problem need not be nonlinear for stress stiffening to play a role: the effect is essential in linear buckling analysis and can strongly affect the vibration frequencies of a flexible rotating structure.

The rigid bar in Fig. 10.3-1a,b provides a simple analytical representation of stress stiffening. Lateral force F produces lateral displacement v_2. If $v_2 << L$, summation of moments about the left end yields the equilibrium equation

$$(k + k_\sigma)v_2 = F \qquad \text{where} \qquad k_\sigma = \frac{P}{L} \tag{10.3-1}$$

The effect of k_σ is to increase the net stiffness $(k + k_\sigma)$ when P is tensile and decrease it when P is compressive. Thus deflection v_2 produced by force F is decreased when $P > 0$ and increased when $P < 0$. If $k + k_\sigma = 0$, the net stiffness is *zero*. This happens when $P = -kL$, which is the buckling load according to linear theory in which P is independent of v_2 and $v_2 << L$.

For a multi-d.o.f. structure, the equation analogous to Eq. 10.3-1 is

$$[\mathbf{K} + \mathbf{K}_\sigma]\mathbf{D} = \mathbf{R} \tag{10.3-2}$$

where $\mathbf{K}$ is the conventional stiffness matrix, $\mathbf{D}$ is the vector of nodal d.o.f., $\mathbf{R}$ is the vector of applied loads, and $\mathbf{K}_\sigma$ is the *stress stiffness matrix*. Alternative names for $\mathbf{K}_\sigma$ include initial stress stiffness matrix, geometric stiffness matrix, differential stiffness matrix, and stability coefficient matrix. $\mathbf{K}_\sigma$ depends on membrane stresses and may either increase or decrease resistance of the structure to loads $\mathbf{R}$. Matrix $\mathbf{K}_\sigma$ is constructed by assembly of element matrices $\mathbf{k}_\sigma$, in the same way that $\mathbf{K}$ is constructed by assembly of element matrices $\mathbf{k}$. As an example of $\mathbf{k}_\sigma$, the stress stiffness matrix of a uniform simple plane beam element (Fig. 10.3-1c) is [2.2]

$$\mathbf{k}_\sigma = \frac{P}{30L}\begin{bmatrix} 36 & 3L & -36 & 3L \\ 3L & 4L^2 & -3L & -L^2 \\ -36 & -3L & 36 & -3L \\ 3L & -L^2 & -3L & 4L^2 \end{bmatrix} \tag{10.3-3}$$

which operates on d.o.f. $\mathbf{d} = [v_1 \quad \theta_{z1} \quad v_2 \quad \theta_{z2}]^T$. Force P is considered positive in tension. This form of $\mathbf{k}_\sigma$ is called "consistent" because it is based on the same shape functions as the conventional stiffness matrix $\mathbf{k}$, Eq. 2.3-2. Alternative forms of $\mathbf{k}_\sigma$ are available, even a form that is diagonal [10.3].

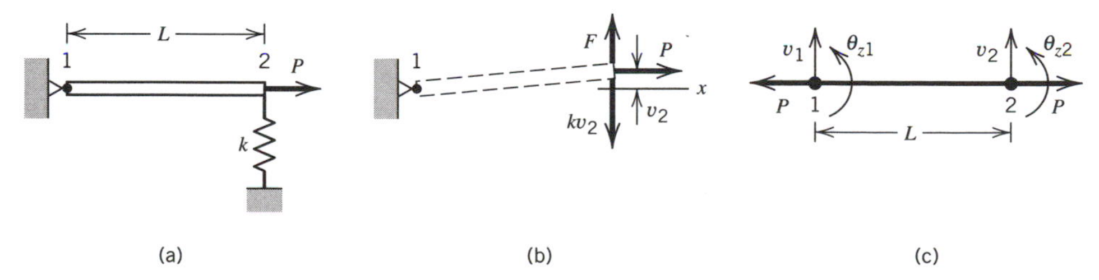

Fig. 10.3-1. (a) Rigid bar supported by a linear spring at one end and carrying axial force P. (b) The rigid bar displaced by lateral force F. (c) Simple plane beam element carrying axial force P.

Membrane stresses may be known at the outset and imposed like loads, or they may be initially unknown and develop as a result of deformation. In Fig. 10.3-2a, axial (membrane) force in the beam is zero before loading and remains almost zero after load q is applied if deflections are small. In Fig. 10.3-2b, a known axial force P is applied prior to application of load q and remains essentially constant if deflections are small. In Fig. 10.3-2c, immovable hinge supports make the beam stretch slightly even for small lateral deflection. Axial force is initially zero but grows as lateral deflection grows. *How* it grows must be calculated by nonlinear analysis, because it is a function of lateral deflection and in turn influences lateral deflection. Figure 10.3-2d is similar to Fig. 10.3-2c, but now the axial force is compressive and the possibility of buckling exists.

As another example, a turbine blade has tensile membrane stress proportional to the square of the angular velocity of the turbine. Tensile membrane stress increases the bending stiffness of the blade and is almost independent of bending deformation. Vibration frequencies of the blade are increased by stress stiffening, probably more than they are decreased by spin softening (Section 9.11).

Linear Buckling Analysis. The matrix equation for linear buckling analysis of a multi-d.o.f. structure is [2.2]

$$[\mathbf{K}+\lambda\mathbf{K}_\sigma]\{d\mathbf{D}\}=\mathbf{0} \tag{10.3-4}$$

where $\mathbf{K}_\sigma$ is calculated from an arbitrarily chosen level of membrane stress, and λ is the factor by which this level must be increased or decreased in order to produce buckling. At the critical (buckling) condition, there is a "bifurcation" in a load versus displacement plot: two infinitesimally close equilibrium states are possible—the unbuckled state and the buckled state—without any change in applied loads $\mathbf{R}$. Displacement increments $\{d\mathbf{D}\}$ are departures from the configuration $\mathbf{D}$ that exists just before buckling. The right-hand side of Eq. 10.3-4 is the corresponding change in applied loads and is therefore a null vector. Equation 10.3-4 is an *eigenproblem*, like Eq. 9.4-2 but with ω^2 replaced by λ. The computed value of λ may be positive or negative, depending on the state of membrane stress used to construct $\mathbf{K}_\sigma$.

As an example, imagine that a uniform one-element cantilever beam of length L is fixed at node 1, free at node 2, and loaded by a unit axial tensile force at node 2. With terms from Eqs. 2.3-2 and 10.3-3, Eq. 10.3-4 becomes

$$\left(\frac{EI}{L^3}\begin{bmatrix}12 & -6L\\ -6L & 4L^2\end{bmatrix}+\lambda\frac{1}{30L}\begin{bmatrix}36 & -3L\\ -3L & 4L^2\end{bmatrix}\right)\begin{Bmatrix}dv_2\\ d\theta_{z2}\end{Bmatrix}=\begin{Bmatrix}0\\0\end{Bmatrix} \tag{10.3-5}$$

This equation has only two d.o.f. and may therefore be solved by the same hand-calculation method used for Eq. 9.4-4. The smallest eigenvalue is $\lambda = -2.486\ EI/L^2$. Multiplying the eigenvalue by the reference load (unity in the present example), we obtain the critical axial load $P_{cr} = -2.486\ EI/L^2$, where the negative sign means that the critical load is re-

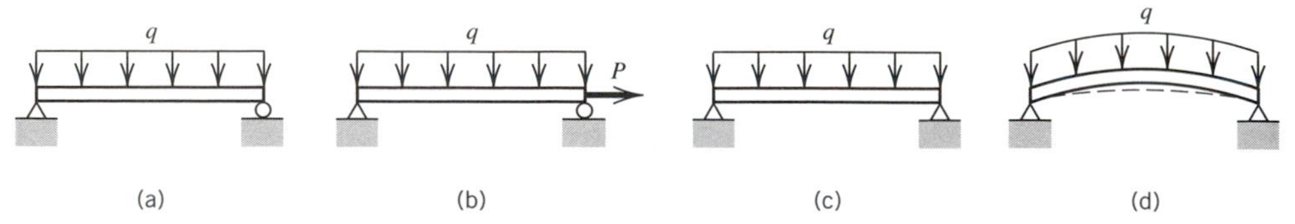

Fig. 10.3-2. Plane beams that differ in loading, support condition, or geometry.

versed from the direction used in writing Eq. 10.3-5. The exact value according to column theory is $P_{cr} = -2.467\ EI/L^2$.

Membrane stresses may be known at the outset, as in the foregoing example, or they may have to be computed. As an instance of the latter, imagine that a flat plate is subjected to a temperature field that is constant through the thickness and of known distribution in the plane of the plate. We wish to know how much the temperature field must be scaled (up or down) in order to reach the critical condition. Based on an arbitrary level of the temperature field, a plane stress analysis provides membrane stresses needed to construct $\mathbf{K}_\sigma$, after which Eq. 10.3-4 (which includes bending stiffness) can be solved for λ. The critical temperature field, according to linear buckling analysis, is obtained by multiplying the original temperature at all points by λ.

Linear buckling analysis uses $\mathbf{K}$ and $\mathbf{K}_\sigma$ based on the original, undeformed geometry of the structure. Membrane stresses do not change in distribution; the membrane stress field is merely scaled by λ to predict the critical condition. Linear analysis often overestimates the actual buckling load. It works well for straight columns and flat plates, which are assumed to remain free of bending until buckling occurs. In most thin-walled structures, membrane and bending stresses develop simultaneously and may interact before buckling occurs. Interaction may alter the distribution of membrane stresses and may cause them to vary nonlinearly with load. Additionally, $\mathbf{K}$ becomes a function of displacements if displacements are large or if there is yielding. Accordingly, most practical buckling problems are *nonlinear*, and buckling analysis should be based on the tangent stiffness that prevails at the instant of buckling. These considerations are automatically incorporated in the large-deflection analysis discussed next.

10.4 GEOMETRIC NONLINEARITY AND BUCKLING

Geometric nonlinearity (as opposed to material nonlinearity) arises when deformations are large enough to significantly alter the way load is applied or the way load is resisted by the structure. In a contact problem, such as steel rollers in contact (Fig. 10.1-1c), deformations are small, yet large enough to increase the contact area and hence make contact pressure on a roller a nonlinear function of applied load. A vaulter's pole experiences large deformations: load is resisted by little bending or a great deal, depending on the configuration. The general goal of analysis is to construct the nonlinear relation between applied load and the resulting deformation. The prominence given to buckling in the following remarks is mainly intended to illustrate the difference between linear and nonlinear buckling. In the remainder of this section we assume that structures are sufficiently thin that large displacements are possible without yielding of the material, although such need not be the case in practice. Also, we postpone nonlinearity of the contact type to Section 10.6.

We begin with a simple example of geometric nonlinearity. It illustrates aspects of nonlinear behavior that may appear in other structures as well. The beam of length L in Fig. 10.4-1a has flexural stiffness EI and axial stiffness AE/L. A frictionless roller at the right end is constrained to remain always in contact with a vertical wall. The linear spring has stiffness k. Displacements are confined to the plane of the paper. Bar and spring are unstressed when $F = 0$. For the case $c << L$, it can be shown that [10.1]

$$F = \left[\frac{AE}{2L^3}(2c - v)(c - v) + k \right] v \tag{10.4-1}$$

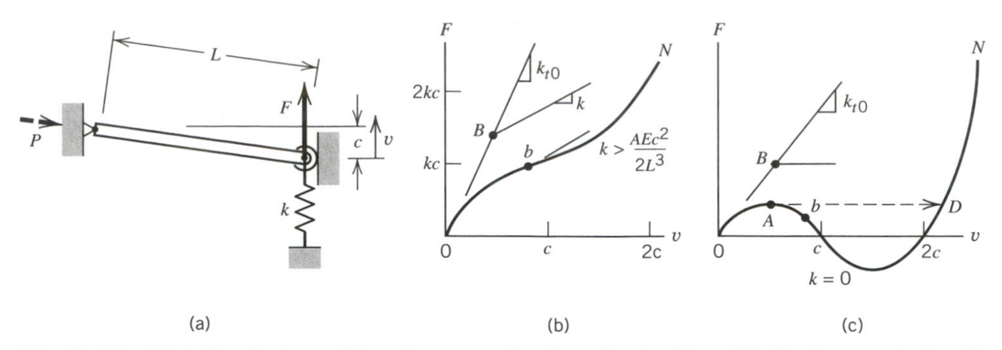

Fig. 10.4-1. (a) A slender beam, hinged at one end and confined without friction by a vertical wall at the other. (b,c) Possible load versus displacement behavior. In (b), k is such that $k_t > 0$ for all v. In (c), $k = 0$.

where v is the vertical displacement produced by applied force F. The bracketed expression is the stiffness k of the structure and is clearly displacement dependent. The tangent stiffness is

$$k_t = \frac{dF}{dv} = \frac{AE}{2L^3}\left(2c^2 - 6cv + 3v^2\right) + k \tag{10.4-2}$$

By summing vertical forces and using small angle approximations, we obtain axial compressive force P in the beam.

$$P = \frac{L}{c-v}(F - kv) = \frac{AE}{2L^2}(2c - v)v \tag{10.4-3}$$

Force P is zero when $v = 0$, grows to a maximum (in compression) at $v = c$, becomes zero again at $v = 2c$, then grows in tension for $v > 2c$.

In Fig. 10.4-1, a *linear* analysis is based on the undeformed configuration ($v = 0$) and yields a straight line whose slope is the initial tangent modulus, k_{t0}. The actual *non*linear response is curve 0N in Fig. 10.4-1b. Linear analysis indicates that P is directly proportional to F. When and if the beam buckles as a column, it carries the axial force $P = P_{cr} = \pi^2 EI/L^2$. Let B be the bifurcation point, at which buckling occurs according to linear theory. Actually, as v increases from $v = 0$, each additional increment dv produces less increase in P than the preceding increment. Accordingly, nonlinear theory shows that a v greater than that at B is required to produce P_{cr}. Therefore buckling may actually occur at a point such as b. Because F is smaller at b than at B, we conclude that linear buckling theory is unconservative in this problem. The postbuckling paths have positive slope because the linear spring is still active. A positive slope characterizes a structure that has postbuckling stability. In some structures the postbuckling path has negative slope, which means that the postbuckling configuration is unstable. (If equations of this example are to account for buckling of the beam, rotational d.o..f. must be added in order to describe bending of the beam. Thus the tangent stiffness k_t would become a tangent stiffness matrix $\mathbf{K}_t$.)

In Fig. 10.4-1c, the spring is absent ($k = 0$). As described above, linear theory may indicate bifurcation at B, while nonlinear theory indicates bifurcation at b. But buckling may not occur at either of these points: instead, displacement v may suddenly "snap through" from A to D. Point A is called a *limit point*, which is a point where the tangent stiffness is zero but there is no adjacent equilibrium configuration. A solution algorithm that cannot follow snap-through will display convergence failure at a limit point.

A thin cylindrical shell under axial compression displays an axial load versus axial deformation relation roughly similar to the F versus v relation in Fig. 10.4-1, but with k sufficiently large to prevent F from becoming negative. Because of imperfections of loading and geometry, perhaps too small to be seen by the unaided eye, a thin shell usually buckles at a load lower than predicted by linear theory: the shell often snaps through to a buckled configuration at a much lower load. Imperfections can be simulated in a FE model by deliberately making small alterations in geometry, loading, or support conditions. A nonlinear analysis can then be undertaken that provides a more realistic estimate of the actual collapse load than is provided by linear theory with ideal structure geometry.

Nonlinear solution methods summarized in Section 10.2 are able to check for bifurcation points and limit points as they track the load versus deformation response of a structure. The tangent stiffness matrix is repeatedly updated as the solution progresses. At a bifurcation point or a limit point, the stiffness becomes zero for the buckling displacement mode. This is signaled by the determinant of the tangent stiffness matrix becoming zero. At a load *greater* than a bifurcation or limit load the determinant is negative. Accordingly, software can be expected to alert the analyst whenever the determinant ceases to be positive. Also, if a solution approaches a limit point, computed displacement increments become very large or there may be convergence difficulties.

Additional structures that have geometric nonlinearity are sketched in Fig. 10.4-2. They illustrate aspects of geometric nonlinearity and considerations needed in order to obtain a numerical solution. Solutions are available for all of them, so they may be used as learning aids for program users or as test cases for software verification. Material nonlinearity is excluded from these examples.

- The rigid bar in Fig. 10.4-2a is supported at its lower end by a frictionless hinge and a rotational spring that exerts a moment proportional to rotation θ. Small-deflection theory yields the buckling load $P_{cr} = k_\theta / L$. When θ becomes large a load greater than P_{cr} can be supported, specifically $P = k_\theta \theta / (L \sin \theta)$.
- The elastic beam in Fig. 10.4-2b may carry tip loads F, P, and/or M. For $F = P = 0$, lateral tip displacement u is given by $u = (1 - \cos \alpha)(L/\alpha)$, where $\alpha = ML/EI$ and M can be arbitrarily large. Solutions for nonzero F, P, and M appear in [10.4], including cases of a follower force, for example, a force F that always acts normal to the beam rather than maintaining its orientation in space. The "elastica" problem is that of only P nonzero and with P always vertically directed [10.5]. It was published by Euler in 1744. In solving the elastica problem numerically, a small M or F is needed to initiate lateral deflection as P becomes greater than the column buckling load P_{cr}. When deflections of the elastica have become large the effect of stress stiffening is minor, although it does no harm to include it in calculations.
- Large deflections of plane frames are shown in Figs. 10.4-2c and 10.4-2d [10.2, 10.6, 10.7]. Buckling is not a possibility in the frame of Fig. 10.4-2c because membrane stresses are tensile. They are compressive in the frame of Fig. 10.4-2d, whose load versus deflection plot is of the type shown in Fig. 10.4-1c. There is a limit point and buckling of the snap-through type. Some solution algorithms have difficulty with behavior such as this.
- Figure 10.4-2e represents a flat rectangular membrane under distributed lateral load q. Solutions for uniform q appear in [10.8]. For example, the center deflection w_c of a square membrane of thickness t, side lengths $a = b$, Poisson ratio 0.3, and without initial membrane stress is $w_c = ct[(q/E)(b/t)^4]^{1/3}$, where $c = 0.2866$. If side lengths are

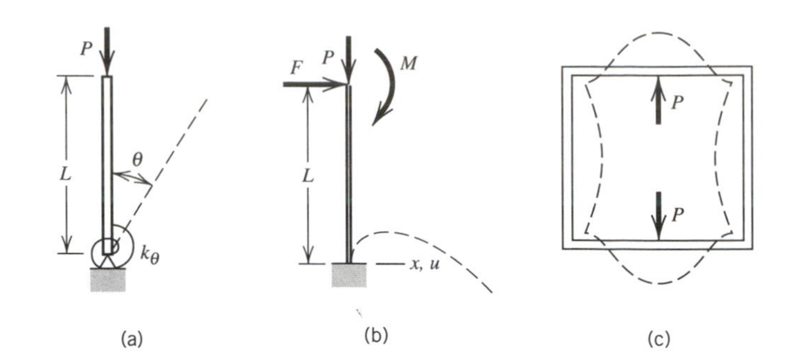

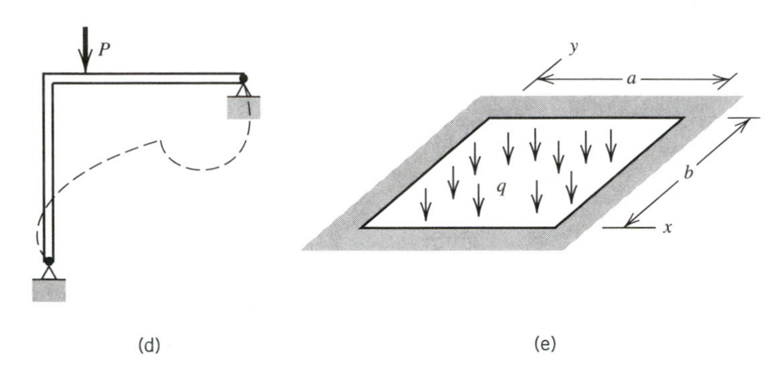

Fig. 10.4-2. Examples of structures in which geometric nonlinearity may be important. Dashed lines suggest actual deflected shapes, not scaled-up small-displacement solutions. (a) Rigid bar. (b) Elastic beam or column. (c,d) Plane frames. (e) Initially flat membrane or plate.

b and $a = 5b$ or greater, $c = 0.3458$. Despite the inclusion of membrane stresses and a nonlinear solution, these results are limited to "small" lateral deflection, that is, slopes $\partial w/\partial x$ and $\partial w/\partial y$ must be much smaller than unity. If deflections are truly large, the problem resembles blowing up a balloon, for which load q would be regarded as a pressure that acts as a follower force. Thus load on an element changes direction as the element rotates and becomes larger as the element grows in size. The initial flatness and lack of stress in the membrane means that it has no initial stiffness with which to resist q. To prevent a failure of the solution algorithm some lateral stiffness must be provided, if only for the initial step, in the form of an elastic foundation or a fictitious membrane stress. Similar considerations apply to an initially flat cable network.

- Figure 10.4-2e may also represent a flat plate under distributed lateral load q. Unlike a membrane, a plate has initial lateral stiffness because it resists bending. Nonlinear solutions, again limited to small lateral deflections, appear in [7.3].

FE analysis of large-deflection problems does not require the introduction of a new set of elements. The shape functions and element library used for linear small-deflection analysis can still be used in nonlinear analysis. Software with nonlinear capability is able to keep track of the deformations and rigid-body motions of elements, so that stiffnesses used in the solution process correspond to the deformed configuration rather than the original configuration.

10.5 MATERIAL NONLINEARITY

This section summarizes time-independent material nonlinearity, with emphasis on plastic action. Material behavior shown in Fig. 10.5-1 is characterized by a straight line, whose slope is elastic modulus E, and a curve, whose slope is tangent modulus E_t, where

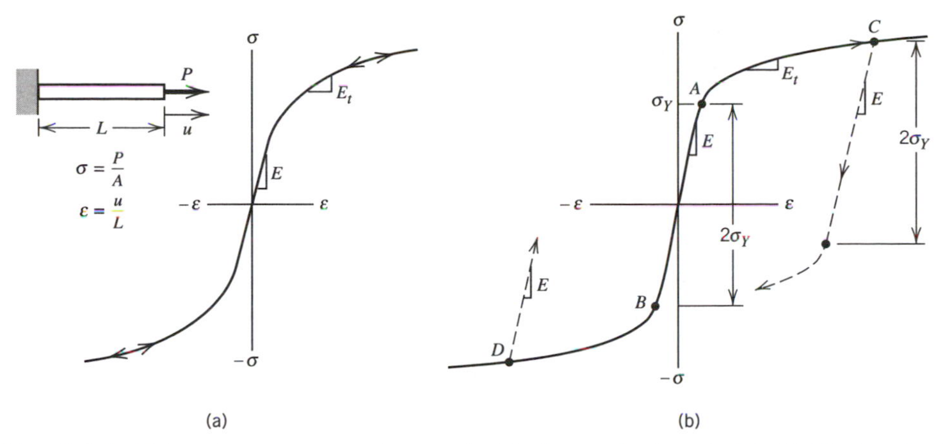

Fig. 10.5-1. Uniaxial stress–strain relations. (a) Nonlinear elastic material. (b) Elastic–plastic material.

$E_t = d\sigma/d\varepsilon$. Figure 10.5-1a shows nonlinear *elastic* behavior, which means that unloading from any stress level follows the same path as loading. In contrast, Fig. 10.5-1b displays *elastic–plastic* action: behavior is elastic only up to point A in tension and point B in compression. For higher stress levels there is plastic action, and unloading from points at which $|\sigma| > \sigma_Y$, such as C or D, follows a different path than loading, specifically a path of slope E, and results in permanent deformation when load has been removed. Stress σ_Y is the yield strength of the material. Usually σ_Y and the deformation-dependent tangent modulus E_t are determined from a tensile test. As input data to software, the curve can be described by data points that define a piecewise-linear stress–strain relation. The essential difference between nonlinear elastic and elastic–plastic materials is their behavior on unloading, so an input data switch can dictate which behavior the software is to use.

A structure in which there is yielding displays softening behavior (Figs. 10.2-1 and 10.2-2) rather than hardening behavior (Fig. 10.2-3). When part of a structure yields there is a transfer of load to other parts of the structure, so a plot of externally applied load versus displacement will continue to rise even if the material has a zero tangent modulus. However, if a collapse condition is approached, successive load increments will produce larger and larger displacement increments. If a final load state is reached by applying two or more different loads, the final state of stress and deformation may depend on the order in which loads are applied. Removal of all loads leaves the structure with a state of permanent deformation and residual stress.

The nature of an incremental solution is suggested by the one-dimensional example in Fig. 10.5-2, for which tangent stiffness k_t is seen to be stress dependent. One can concep-

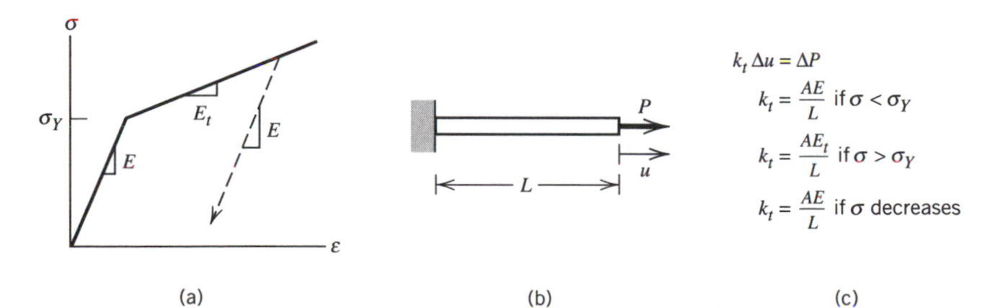

Fig. 10.5-2. (a) Uniaxial stress–strain relation, idealized as bilinear. (b) Bar in tension. (c) Incremental equation and stiffness for tensile load P.

tually extrapolate this example to a multi-d.o.f. FE formulation, in which matrix **E** of elastic constants is modified to incorporate a tangent modulus of the material and a way to detect unloading. Thus an elastic–plastic stiffness matrix emerges from Eq. 3.1-10.

For FE analysis one must have an understanding of the following three concepts from the theory of plasticity. First is the *yield criterion.* It relates the onset of yielding to the state of stress. For metals the von Mises criterion is most commonly used. Thus, when σ_e of Eq. 3.10-1 reaches σ_Y as determined from a uniaxial tensile test of the material, yielding is assumed to begin. The second concept is the *flow rule.* It relates stress increments, strain increments, and the state of stress in the plastic range. A flow rule known as the Prandtl–Reuss relation is commonly used for metals. The third concept is the *hardening rule.* It describes how the "yield surface" grows and moves as plastic strains accumulate. Various hardening rules are possible. The choice must be suited to the material. Metals are usually described well by "kinematic hardening," which is illustrated in Fig. 10.5-1b for a uniaxial state of stress: an elastic range $2\sigma_Y$ exists prior to yielding and is preserved after yielding. Further details of these concepts, and plasticity in general, may be found in [10.1, 10.2, 10.9] and many additional references.

A software user may find that the von Mises yield criterion and the Prandtl–Reuss flow rule are built into the plasticity algorithm and that any of various hardening rules may be chosen. Element stiffness matrices are formulated using the element geometries, nodal patterns, and shape functions that are also used in linear elastic analysis. An elastic–plastic algorithm must keep a record of the state of stress at each of many "sampling points" in the FE structure. These points may be element centroids or Gauss points of isoparametric elements. Load is applied in increments until the final load level is reached. The solution algorithm tracks the spread of yielding and reports the state of stress and deformation at each load level. An elastic–plastic solution may be deemed converged by the software when, in each element, the plastic strain increment in the most recent iteration is no more than a few percent of the elastic component of the total strain.

Plastic analysis may require a finer FE mesh than does linear elastic analysis. Consider Fig. 10.5-3. If the plane block is linearly elastic, a single four-node element can model exactly the relation between moment M and curvature $1/\rho$ of the block (see Section 3.6). As M becomes greater than M_Y, zones of yielding begin to move inward from top and bottom surfaces. Yielding will be detected only when it has spread as far as sampling points A and B in Fig. 10.5-3c. Clearly, an accurate representation of a plastic moment–curvature relation will require several sampling points. Therefore dimension h will proba-

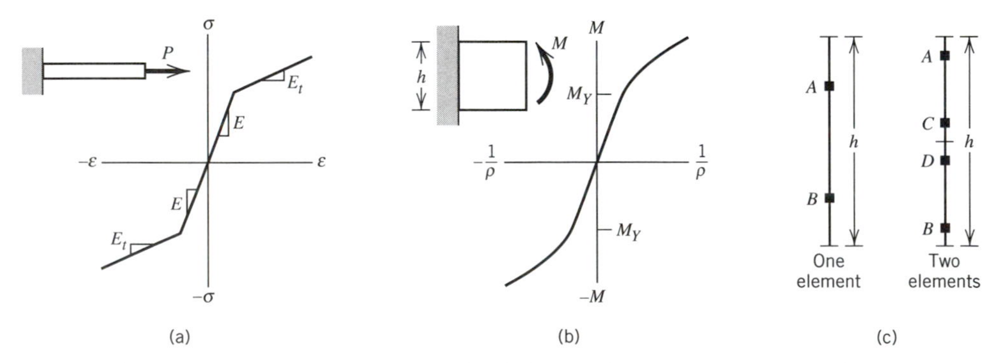

Fig. 10.5-3. (a) Idealized uniaxial stress–strain relation. (b) Moment–curvature relation for a block made of this material. (c) Locations of Gauss points of an order two rule if height h is spanned by one or two elements.

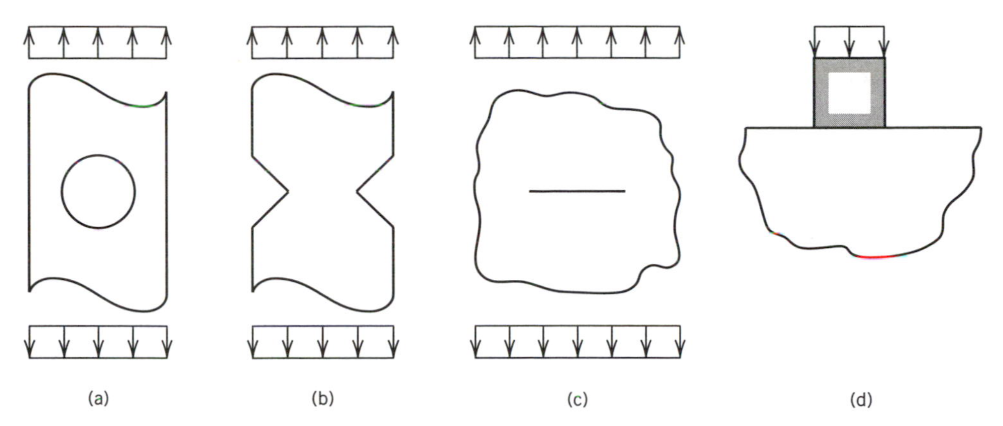

Fig. 10.5-4. Some plane problems for which elastic–plastic solutions are available. (a) Hole in a tensile strip. (b) Edge notches in a tensile strip. (c) Crack in an infinite plane under uniaxial stress. (d) Rigid, frictionless punch pressed against a semi-infinite plane.

bly be spanned by more than one element, and even then the representation will not be exact.

Solved problems that can be used as test cases and learning aids may be found in several books. Many solutions are for an elastic–perfectly plastic material ($E_t = 0$ in Fig. 10.5-3a), some are for a strain hardening material ($E_t > 0$), and a few are for a rigid–perfectly plastic material (infinite E, but $E_t = 0$). Some of these problems are as follows.

- Elementary textbooks discuss pure bending of beams and pure twisting of bars of circular cross section, including residual stresses on unloading. Fully plastic twisting of bars of noncircular cross section can be treated analytically by the sand-hill analogy, and some solutions are easily obtained [2.1, 10.9]. The material is elastic–perfectly plastic.
- A thick-walled cylinder under internal pressure is widely discussed: for example, in [2.1, 10.9, 10.10] for an elastic–perfectly plastic material and in [10.9] for a strain hardening material. The yield criterion used is Tresca, which states that yielding begins when SI of Eq. 3.10-2 becomes equal to σ_Y, where σ_Y is the yield strength in a tension test. For test-case purposes it is convenient to use plane stress conditions with a zero Poisson ratio in the elastic range. A thick-walled sphere is discussed in [10.9], with and without strain hardening.
- Plane bodies with cracks, holes, and notches (Fig. 10.5-4) have been analyzed with and without strain hardening. Plane stress conditions prevail for the hole and notch problems [10.11]. The solution of the crack problem allows either plane stress or plane strain conditions [10.9]. The solutions are numerical but are considered reliable.
- The rigid frictionless punch problem, Fig. 10.5-4d, uses the Tresca yield criterion, plane strain conditions, and a rigid–perfectly plastic material [10.9, 10.10].

10.6 PROBLEMS OF GAPS AND CONTACT

In some practical problems, two structures (or two parts of the same structure) may make contact when a gap closes, may separate after being in contact, or may slide on one another with friction. These problems are further explained with the aid of Fig. 10.6-1. In

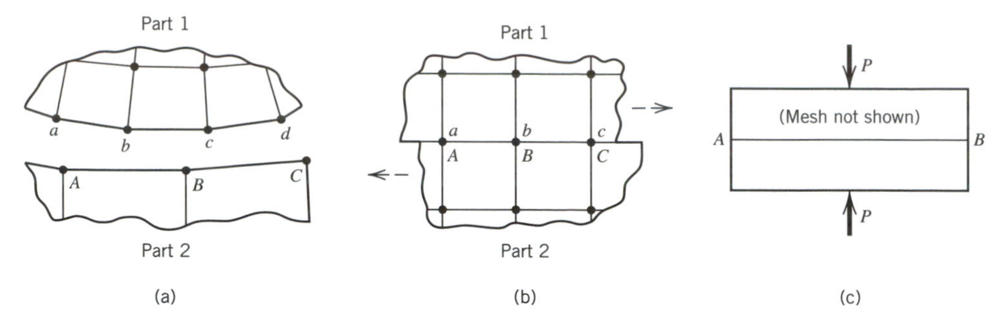

Fig. 10.6-1. Two parts of a FE model that may (a) come into contact, (b) slide on one another or come apart, and (c) lose contact near A and B.

Fig. 10.6-1a, parts 1 and 2 may make contact but because of relative motion between them it may not be known at the outset exactly where on parts 1 and 2 contact will occur. A solution algorithm must discover the contact location, then prevent the parts from interpenetrating. In Fig. 10.6-1b, it is known that parts 1 and 2 are in contact, but it is not known to what extent they will slide relative to one another or whether they will come apart. A solution algorithm must prevent sliding until friction is overcome, thereafter apply shear force proportional to the product of normal force and coefficient of friction, and allow no tensile contact force. In Fig. 10.6-1c, forces P press two elastic blocks together. To solve the problem a solution algorithm must compute the state of contact stress while allowing for possible separation near A and B.

Computationally, such problems are problems of constraints: a node may have one of its d.o.f. constrained against motion when it contacts a fixed support, or a node may be constrained to have the same motion as an adjacent node with which it comes in contact. Constraint conditions can be imposed exactly, by elimination or the Lagrange multiplier method, or approximately, by the penalty method (see Section 4.13). Current software is not unanimous in its choice of method, but the penalty method is a common choice: additional elements are introduced, whose stiffness is either zero or very small when a gap is open but large when a gap closes. A simple example will explain the computational process. The structure in Fig. 10.6-2a consists of two springs. The wall at B applies a support reaction only if gap g closes. The model used for computation is that of Fig. 10.6-2b. A third spring of comparatively large stiffness k_3 has been added. With u_A and u_B the displacement d.o.f. of the structure, the tangent stiffness matrix is

$$\mathbf{K}_t = \begin{bmatrix} k_1 + k_2 & -k_2 \\ -k_2 & k_B \end{bmatrix} \quad \text{where} \quad \begin{array}{lll} k_B = k_2 & \text{if} & u_B < g \\ k_B = k_2 + k_3 & \text{if} & u_B > g \end{array} \tag{10.6-1}$$

Matrix $\mathbf{K}_t$ is used to compute displacement increments: those associated with an incre-

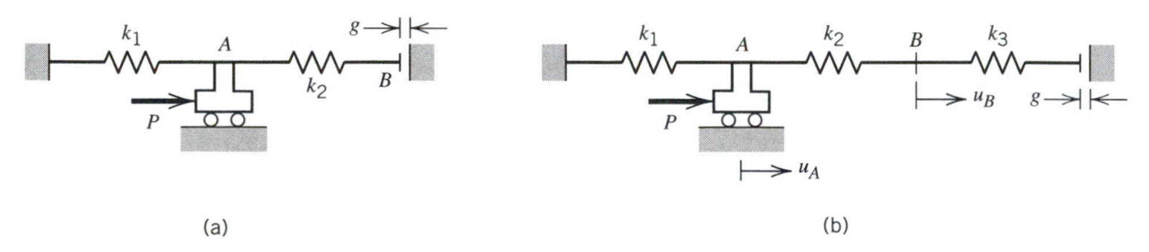

Fig. 10.6-2. (a) Simple structure with a gap g that may close. (b) Arrangement of the problem for numerical solution.

ment ΔP of externally applied load, and those associated with equilibrium iterations by the Newton–Raphson method while load P is maintained at its current level.

Applied load step: Equilibrium iterations at load P:

$$\mathbf{K}_t \begin{Bmatrix} \Delta u_A \\ \Delta u_B \end{Bmatrix} = \begin{Bmatrix} \Delta P \\ 0 \end{Bmatrix} \qquad \mathbf{K}_t \begin{Bmatrix} \Delta u_A \\ \Delta u_B \end{Bmatrix} = \begin{Bmatrix} P \\ 0 \end{Bmatrix} - \begin{Bmatrix} r_A \\ r_B \end{Bmatrix} \tag{10.6-2}$$

in which the resisting forces of the structure are given by

$$\begin{aligned} r_A &= (k_1 + k_2)u_A - k_2 u_B && \text{for all } u_B \\ r_B &= -k_2 u_A + k_2 u_B && \text{if } \quad u_B < g \\ r_B &= -k_2 u_A + k_2 u_B + k_3(u_B - g) && \text{if } \quad u_B > g \end{aligned} \tag{10.6-3}$$

The foregoing formulation can be obtained by defining the rightmost spring in Fig. 10.6-2b as nonlinear, as shown in Fig. 10.6-3a. The nonlinear spring is always attached to the structure but never has stiffness in tension and has no stiffness in compression until it shortens an amount g. For other applications, by defining an F versus e relation in piecewise-linear fashion, Fig. 10.6-3b, a software user can fashion a great variety of nonlinear springs. It is even possible to call for elastic unloading, as suggested by the dashed line in Fig. 10.6-3b.

The foregoing concepts have been elaborated into a variety of elements and algorithms. Some are as follows. For the situation in Fig. 10.6-1a, one can arrange for surfaces AB and BC to detect contact with any of the nodes a, b, c, or d, and then exert a resisting force normal to the contact surface. The resisting force is $f_n = k_n u_n$, where k_n is a large stiffness and u_n is the (small) amount of interpenetration. Contact friction, Fig. 10.6-1b, can be treated by a gap element that has large stiffnesses k_n for normal force and k_s for shear force. No forces are exerted when there is a gap between parts. When there is no gap, normal and shear forces are $f_n = k_n u_n$ and $f_s = k_s u_s$, except that f_s has the maximum magnitude $\mu |f_n|$, where μ is the coefficient of friction. Friction capability can also be included in the contact algorithm described for Fig. 10.6-1a.

Presently, spring stiffnesses used in gap and contact elements must be chosen by the analyst rather than by software defaults. If spring stiffnesses are too small the intended constraint will be poorly enforced. If they are too large, equations may become ill condi-

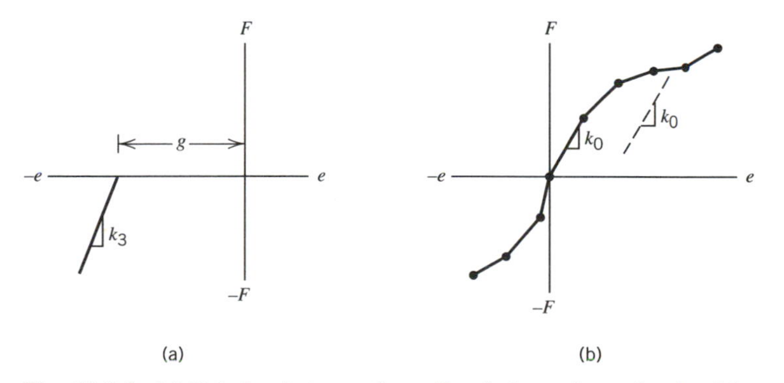

Fig. 10.6-3. (a) Relation between force F and elongation e for the rightmost spring in Fig. 10.6-2b. (b) An arbitrary, user-defined relation between F and e.

tioned and accuracy may be lost, as explained in Sections 4.13 and 5.10. Also, too large a stiffness may provoke a "bouncing" convergence failure, in which gap elements change status in each iteration (open, closed, open, closed,...). In many cases an appropriate spring stiffness is roughly 100 times the stiffness of an adjacent element, where the element stiffness may be approximated according to its type and how it is loaded by the force of constraint.

Classical solutions for contact stress may be used as test cases [1.5, 2.1, 6.1]. These solutions are for static elastic contact of two cylinders or two spheres of the same or different radii.

10.7 REMARKS. MODELING CONSIDERATIONS

Nonlinear analysis is more demanding than linear analysis, in terms of computer resources and the analyst's time and expertise. The goals of analysis may be more varied, and there are more computational paths that may lead to each goal. It is harder to foresee structural response. A good understanding of the response may develop only after performing several trial analyses. General advice given for linear problems remains applicable: try to understand the physical problem and the concepts that underlie analysis procedures; study software documentation; expect to use a sequence of models; critically examine computed results; and keep records of what is done in each analysis and what is learned from it. Before undertaking a nonlinear analysis one should be satisfied that it is really necessary. If it is, an initial linear analysis is usually appropriate, to better understand structural behavior and to test the FE model. These remarks should seem familiar. They resemble remarks in the first several paragraphs of Section 9.11. If "static" and "dynamic" in those paragraphs are replaced by "linear" and "nonlinear," the remarks are largely applicable to nonlinear problems.

Recall that the principle of superposition is not applicable to nonlinear problems. Double the load produces more, or less, than double the response. Results of separate load cases cannot be combined (e.g., in the manner of Fig. 4.12-3). The final state of stress and deformation may depend on the order in which loads are applied.

Strategy. Much more than in linear analysis, the nature of a problem may become clear only after solving it. At the outset the types and extent of nonlinearities may not be apparent. Even if they are, the appropriate elements, mesh layout, solution algorithms, and load steps may not be. Accordingly, an attempt to solve a nonlinear FE problem in "one go" is likely to fail, producing only confusion and frustration. As always, it is desirable to anticipate FE results by doing a simplified preliminary analysis. This may be particularly difficult when the problem is nonlinear, and some of it may be more qualitative than quantitative. A nonlinear analysis should make liberal use of test cases and pilot studies. Linear analysis should precede nonlinear analysis. For a given load, linear analysis can suggest the location and extent of yielding, or what gaps are likely to open or close. A linear buckling analysis may approximate the load and deformation state of the actual collapse. If different sources produce nonlinearity, it may be possible to add them one at a time [3.1], so as to better understand their effects and how to treat them. Initial models in a sequence may use a relatively coarse mesh, large load steps, and a liberal convergence tolerance. Subsequently, all of these can be refined. Usually it is necessary to achieve the final load in several steps, for computational reasons rather than physical reasons. Too large a load step may produce convergence failure. It may also produce an abrupt change in a load versus displacement plot that can be mistaken for actual physical behavior.

Once past early trials, the remainder of the analysis should be planned, as to what is to be done in each stage and why. Each load step can produce as much output as a linear static analysis, so one should anticipate what output to request and how it will be examined. Computed results should be examined after each load step. Status reports and warnings produced by the software should be taken seriously and understood. One may wish to examine results at each load level and go on to the next only if things appear to be going satisfactorily. Accordingly, sufficient data should be stored to allow a restart from the current load level. At a restart, the analyst may call for a change in load increment, convergence tolerance, or other aspects of the solution algorithm.

Modeling. Modeling considerations stated in Chapter 5 apply also to nonlinear problems. Preceding sections of the present chapter also contain some useful modeling information. Miscellaneous remarks are as follows. If there are follower forces (Fig. 10.1-1a), they must be identified as such so that they will be oriented with respect to the model as it deforms, rather than with respect to fixed global coordinates, as is more common. Probably software will always treat pressure loads as follower loads, but it is best to know for sure what software will do rather than make an assumption. Stress stiffening should be invoked if it is possible that its effect will be important. Stress stiffening can be produced by accumulated deformations under load, or by initial or residual stresses present before external load is applied. Residual stresses can be significant in standard steel beams. Similarly, initial stresses can be introduced by assembly before external load is applied. If a structure can be divided into linear and nonlinear parts, the linear parts can be represented by one or more substructures. Matrices that represent linear parts need not change from one iteration to the next. As compared with a full model, fewer d.o.f. are needed, and most of the d.o.f. retained can be in the nonlinear part, whose matrices *must* change from one iteration to the next.

Symmetry and antisymmetry conditions must be used with caution or avoided altogether. As an example, the structure in Fig. 10.7-1a has initial symmetry of geometry, supports, elastic properties, and loads. Any of these symmetries may disappear as load increases, depending on what details are prescribed for the model and its subsequent loading. Despite initial symmetry, the linear bifurcation buckling load will be unsymmetric, Fig. 10.7-1c. An unsymmetric shape may describe the actual large-deformation behavior of this structure under (initially) symmetric load, although in computation a small asymmetry of loading or geometry may be needed to make the numerical solution depart from a symmetric mode. Thereafter, a follower-force load will have asymmetry associated with the asymmetric displaced shape. Antisymmetry conditions should not be imposed when deformations are greater than infinitesimal. A case in point is the arch of Fig. 10.7-1b. The deformed shape is that of Fig. 10.7-1c. When deformations are greater than infinitesimal, midpoint A has nonzero vertical displacement, which violates a small-displacement antisymmetry condition (see Section 4.12).

Large strains may sufficiently deform a mesh that an initially adequate mesh becomes

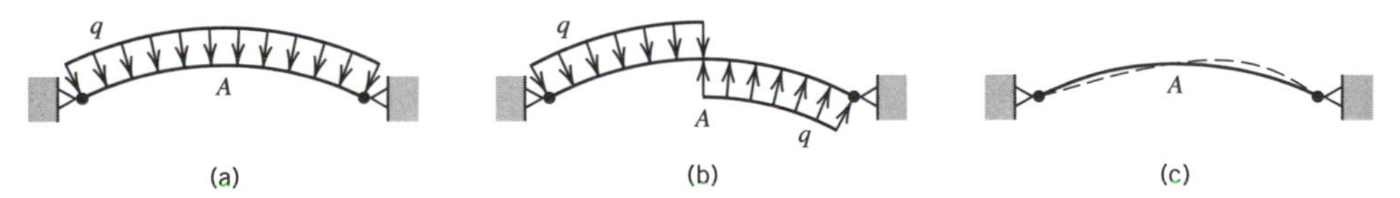

Fig. 10.7-1. Shallow arch whose initial geometry is symmetric about center point A. (a) Symmetric load. (b) Antisymmetric load. (c) A buckled shape for symmetric load or the deflected shape for antisymmetric load.

too coarse or element shapes become too distorted (as in Fig. 5.3-2). In the latter situation one can start with element shapes that are perhaps poor but deform into shapes that are good when the load of interest is reached. For large-strain analysis, software may require that material properties be defined in terms of true stress and true strain. If rotations and/or strains are indeed large, output stresses and strains may be defined differently from what one is accustomed to in conventional linear analysis. Software documentation and theoretical discussions must be consulted to discover what output is presented and what it means. Matters can get complicated and the analyst should not forego expert advice.

In elastic–plastic analysis the initial load step can be large if it takes the structure to the initiation of yield but not beyond. Detecting the onset of yield and tracking its spread demand an adequate distribution of sampling points and accordingly an adequately refined mesh. The plastic strain increment should not exceed about 5% in any substep of loading [9.8]. In examining results it is usually informative to plot the deformed structure and also plot the spread of plastic action as load increases.

In contact stress problems (e.g., Fig. 10.1-1c), a very refined mesh in the region of contact is usually needed in order to determine the extent of contact and especially to determine contact stresses. When using nonlinear springs, one should recall that spring stiffnesses should be related to element edge lengths and node patterns (see Fig. 5.8-2). Accordingly, as a rule nonlinear springs should not be attached to elements that have side nodes. In a structure having large displacements, the orientation of a spring or a bar that links nodes is defined by positions of the nodes to which it is connected (Fig. 10.7-2). Thus if it is intended that the link exert a force normal to the surfaces in contact, such may no longer be the case after significant deformation. The difficulty can be avoided by using a type of contact element that resists penetration of a surface by a node regardless of the orientation of the surface. (Note that a *linear* analysis is based on the initial vertical orientation of the link in Fig. 10.7-2, even when the result of this analysis shows that the link has rotated.)

Elements can "die" or be "born" [9.8]. An element can be made to "die," that is, have its stiffness set to almost zero, if it is overstressed. The element still has mass unless mass is also set to zero, as might be done if part of the structure melts and flows away. Conversely, one might call for elements to be "born" when a structure grows, due to solidification, spray deposition, filament winding, and so on.

Nonlinear dynamic analysis is performed by direct integration, because most software does not incorporate devices needed to enable modal methods to deal with nonlinearity.

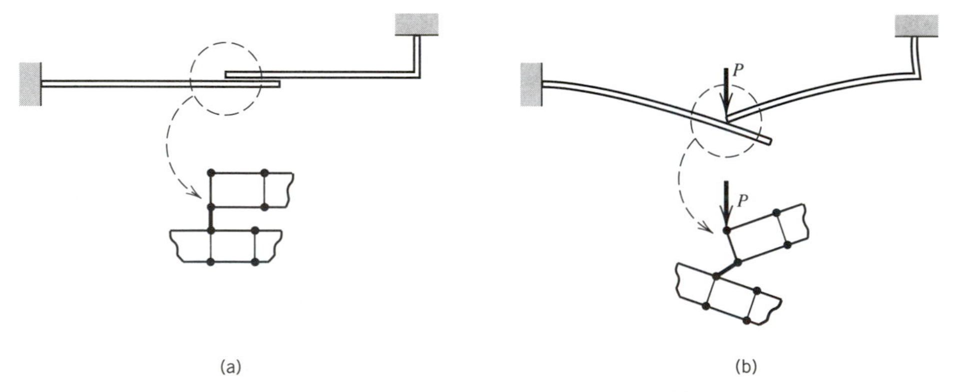

Fig. 10.7-2. A spring or a link connects nodes across a small gap. (a) Before loading: link normal to model surfaces. (b) After large deformation: link not normal to model surfaces.

In the dynamic analysis of a contact problem by the central difference method, the critical time step is substantially reduced if large spring stiffnesses are present, because they produce high natural frequencies. Regardless of the method of direct integration, perhaps 30 time steps per period of the highest frequency may be needed in order to prevent the algorithm from dissipating energy in a collision [9.8].

Convergence. The analyst sets the number of equilibrium iterations allowed at a given load level. The number may be roughly 20, or perhaps over 80 for a contact problem. A convergence failure occurs if the iteration limit is reached before the convergence tolerance is met. An unconverged solution violates equilibrium conditions. Convergence failure may indicate numerical troubles, such as the "bouncing" noted near the end of Section 10.6. Or convergence failure may indicate that the structure has collapsed because it has reached its load-carrying capacity, perhaps due to buckling, the absence or exhaustion of strain hardening capacity in plasticity, or a gap that opens permanently so that part of the structure "floats away." In the case of buckling, the structure may or may not have postbuckling strength, and the solution algorithm may or may not be able to "jump," as from A to D in Fig. 10.4-1c. As a limit point or a collapse condition is approached, computed displacement increments become large. *At* a limit point or a collapse condition, the tangent stiffness matrix $\mathbf{K}_t$ ceases to be positive definite, and an error message such as "singular stiffness matrix" or "the stiffness matrix has a zero or negative determinant" may be expected from the software. After a convergence failure, analysis can be restarted from the previous converged solution using a smaller load increment. Some software will do so automatically.

A converged solution may not be a physically correct solution. For a given load there may be two or more equilibrium configurations, and a computed equilibrium configuration may not be physically realistic because it is unstable. For example, let $F = M = 0$ in Fig. 10.4-2b. If P exceeds P_{cr} the vertical position is physically unstable but may be computationally stable unless lateral displacement is prompted by a geometric imperfection or a small lateral force. An equilibrium configuration for $P > P_{cr}$, shown dashed in Fig. 10.4-2b, may lie on either side of the vertical.

10.8 APPLICATIONS

The following examples are chosen mainly for their simplicity. They emphasize aspects of physical behavior more than the multitude of computational and modeling choices that may be invoked. These problems, and problems cited previously as possible test cases, may be used as computational exercises in which the effects of changes in structure geometry, mesh layout, algorithm, load levels, substeps, and convergence tolerance are explored.

Geometric Nonlinearity in a Frame. Bars of frame $ABCD$ in Fig. 10.8-1a all have a square cross section 8 mm on a side. The material is assumed to remain linearly elastic at all times, with $E = 200$ GPa. Beam elements are used for the FE model: 5 along AB, 2 along BC, and 12 along CD. Node A is pinned, node D may displace horizontally, and displacements are confined to the xy plane. The behavior produced by a horizontal load P at node C is to be investigated.

Because the frame is slender, buckling appears possible. An approximate preliminary analysis might regard CD as an inverted column of length $L = 480$ mm, fixed at C and free at D. In the original configuration, summation of moments about A shows that the

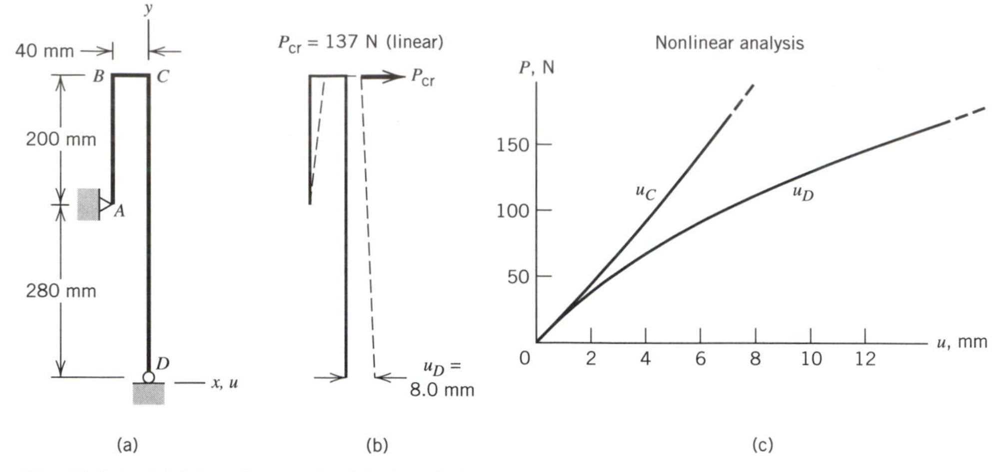

Fig. 10.8-1. (a) Plane frame. (b) Displaced shape when $P = P_{cr}$ by linear analysis. (c) Load versus displacement relations computed by nonlinear analysis.

vertical force at D is $5P$. Thus, using a column buckling formula, we obtain $5P_{cr} = \pi^2 EI/4L^2$ and $P_{cr} = 146$ N. In FE computation, linear static analysis yields axial forces that prestress members of the frame; then the state of prestress is used in a linear eigenvalue analysis, which yields $P_{cr} = 137$ N. The eigenmode that corresponds to $P_{cr} = 137$ N is shown in Fig. 10.8-1b. Results of a *non*linear analysis, in which the geometry is updated as load increases, are shown in Fig. 10.8-1c. We see that the frame does not buckle at all. Instead, displacements simply continue to increase as load increases.

To illustrate the effect of a geometric imperfection on a slender structure such as this, the initial (unloaded) geometry is slightly altered (Fig. 10.8-2a). The lowermost element along CD is now slightly inclined, so that node D is placed 2 mm to the left of its previous location. Linear static and buckling analyses now yield $P_{cr} = 114$ N and a buckling mode in which node D is displaced leftward (Fig. 10.8-2b). A nonlinear analysis, Fig. 10.8-2c, shows that node D moves to the right at first, but then reverses direction as load

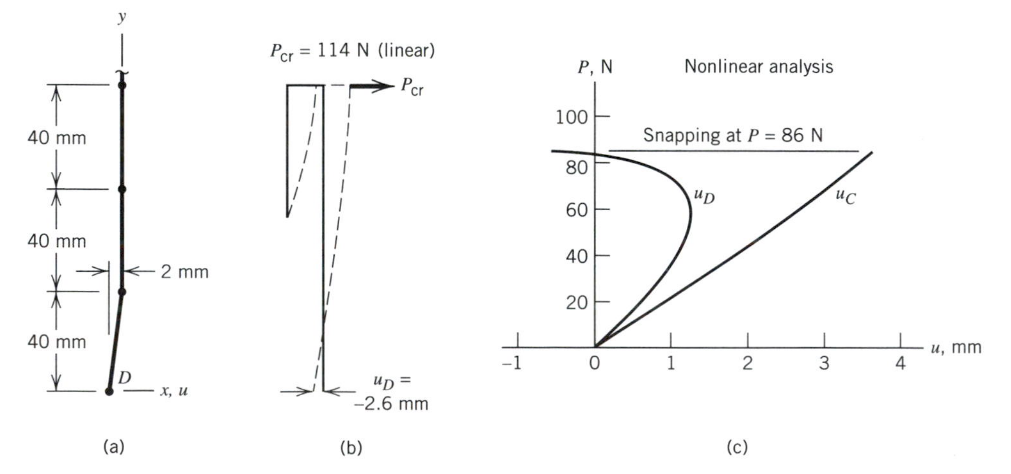

Fig. 10.8-2. (a) Detail of altered initial geometry for the frame of Fig. 10.8-1a. (b) Displaced shape when $P = P_{cr}$ by linear analysis. (c) Load versus displacement relations computed by nonlinear analysis.

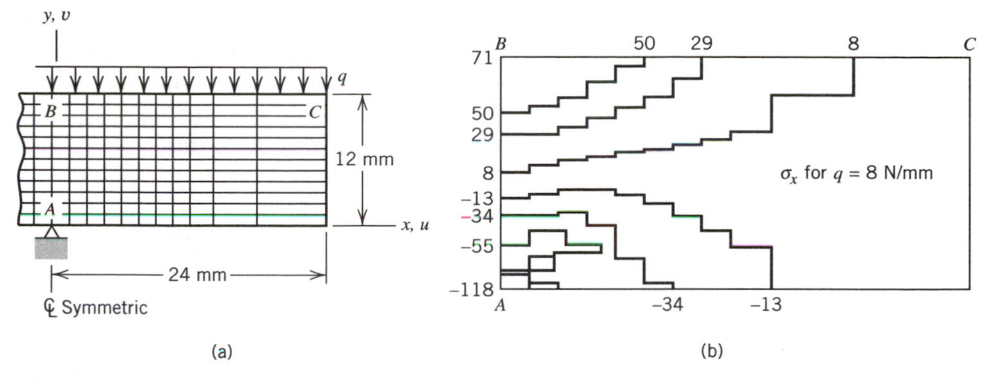

Fig. 10.8-3. (a) Geometry, loading, and mesh of four-node elements for a plane stress problem. (b) Unaveraged contours of axially directed stress σ_x (in MPa) at full load, $q = 8$ N/mm.

P increases. Collapse occurs at P = 86 N, when node D suddenly jumps leftward. With the software used, there is a convergence failure of the iterative solution process at the limit point, where $P = 86$ N.

Plastic Action In a Beam. A plane beam of unit thickness is uniformly loaded, as shown in Fig. 10.8-3a. A bilinear stress–strain relation is used, as in Fig. 10.5-2a, with $\sigma_Y = 70$ MPa, $E = 50$ GPa, and $E_t = 7$ GPa. Displacements of nodes are confined to the xy plane, with x-direction motion prevented along AB and all motion prevented at A. We seek stresses and displacements as load q is increased from zero to 8 N/mm and then removed.

A preliminary analysis based on the flexure formula $\sigma_x = Mc/I$ indicates that $\sigma_x = \sigma_Y$ when $q = 5.83$ N/mm. Clearly, the full load $q = 8$ N/mm will produce yielding, and therefore residual stresses will remain after unloading. It is also obvious there will be severe concentrations of stress and strain near A, so the present analysis must be regarded as an initial approximation.

Element-by-element contours of axial stress σ_x under full load are shown in Fig. 10.8-3b. As expected, near A stresses are high and contours are confused. At B, $\sigma_x = 71$ MPa and $\sigma_y = -8$ MPa; hence Eq. 3.10-1 yields $\sigma_e = 75$ MPa. Since $\sigma_e > \sigma_Y$ we see that there has been yielding near B, and of course near A as well. Residual σ_x stresses, which remain after unloading, are shown in Fig. 10.8-4a. The stress pattern along AB—tension,

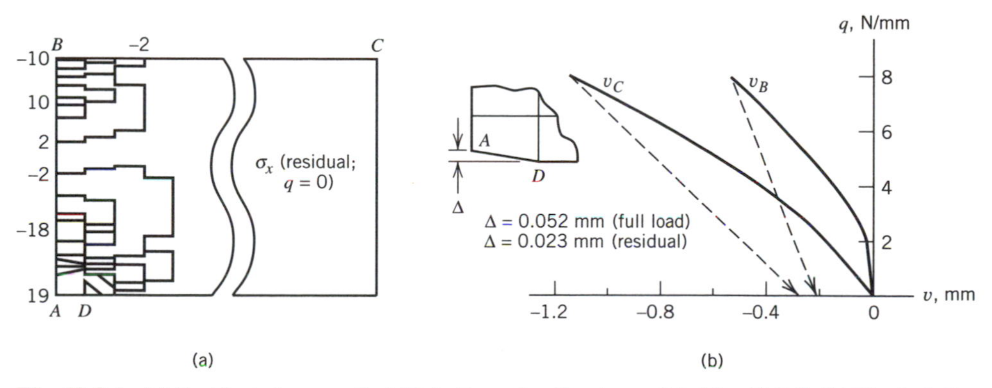

Fig. 10.8-4. (a) Residual stress σ_x (in MPa) after unloading to $q = 0$ in Fig. 10.8-3. (b) Vertical displacements of points B and C for loading (solid lines) and unloading (dashed lines). Inset shows deformation near A.

compression, tension, compression—agrees with residual stress calculations that may be found in books about elementary mechanics of materials. Contours of residual stress are confused because the mesh is coarse and the residual stress pattern is more complicated than the stress pattern under full load. Computed displacements, Fig. 10.8-4b, vary linearly with load up to about $q = 2$ N/mm, then vary nonlinearly. Unloading is elastic, but because of previous yielding there are permanent deformations.

Beam With Contact Nonlinearity. The beam in Fig. 10.8-5a has a square cross section 4 mm on a side. The material is assumed to remain linearly elastic at all times, with $E =$ 200 GPa. The beam is modeled by 15 beam elements, each 20 mm long. Supports at C and D can apply only upward force to the beam. Initially, there is contact at D and a 0.5-mm gap between the beam and the support at C. The behavior produced by a load P that increases from zero to 20 N is to be investigated.

The entire problem can be solved by elementary beam theory. However, let us apply beam theory to only the final condition, at load $P = 20$ N. At this load, as subsequent FE results will show, there is support at A and C and lift-off at D. The midspan displacement of a simply supported beam 200 mm long is $PL^3/48EI = 0.78$ mm. There is an additional midspan displacement of 0.25 mm due to the 0.50 mm displacement at C, for a total of 1.03 mm. Additional results can be computed by beam theory, and all are found to be in agreement with results subsequently computed by FE.

In planning the FE model we consider the possible gap and contact conditions at C and D. It is apparent that the beam will be simply supported at A and D until contact is made at C; then be statically indeterminate while there is contact at A, C, and D; then be statically determinate again with supports at A and C after contact is lost at D. This behavior is accommodated in the FE model by inserting gap elements between the beam and rigid supports at C and D. The gap elements are both compression-only springs of stiffness 5000 N/mm. The initial gap at C is 0.5 mm, so that the spring exerts no force on the beam until the spring has shortened 0.5 mm. The initial gap at D is zero.

The dashed line in Fig. 10.8-5a shows the displaced shape under the full load $P = 20$ N. Load versus deflection relationships for points B, C, and D are shown in Fig. 10.8-5b. Apparently the stiffness seen by load P changes very little when contact is lost at D. As expected, point C moves 0.5 mm downward. Numerical output shows that actually $v_C =$ -0.502 mm when $P = 20$ N. The extra 0.002 mm is the deflection of the gap element at C after contact is made.

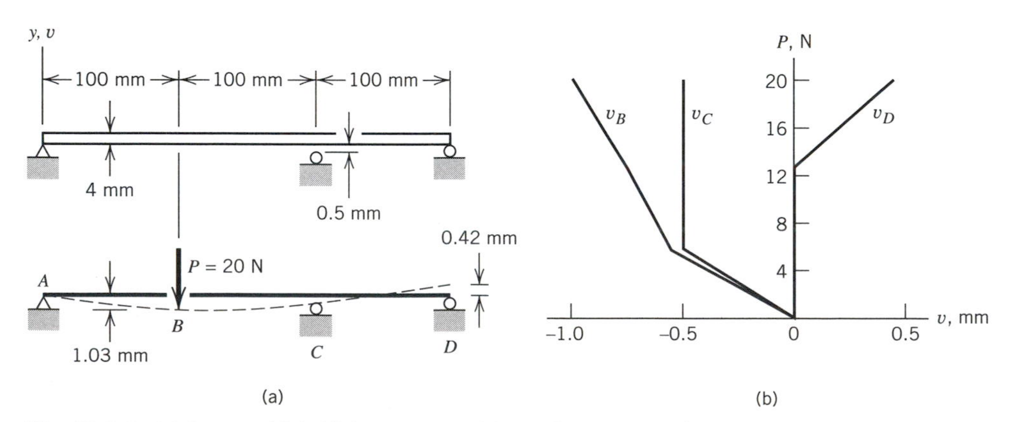

Fig. 10.8-5. (a) Beam with initial gap at C and loss of contact possible at D. (b) Vertical displacements at B, C, and D versus load P.

ANALYTICAL PROBLEMS

10.1 Frame *ABC* is loaded by moment M_B at *B*, as shown. Assume that there is no yielding. Make the following sketches with sufficient care that they look different (if you think they *are* different).

(a) Sketch the deflected shape you would expect to see after directing the software to scale up a linear solution by a large amount, so that corner *B* appears to rotate about 60°.

(b) Sketch the deflected shape you would expect to see if M_B is large enough to produce an *actual* rotation of about 60° at corner *B*.

(c) Repeat parts (a) and (b), now assuming that a frictionless vertical wall prevents horizontal displacement at *B*.

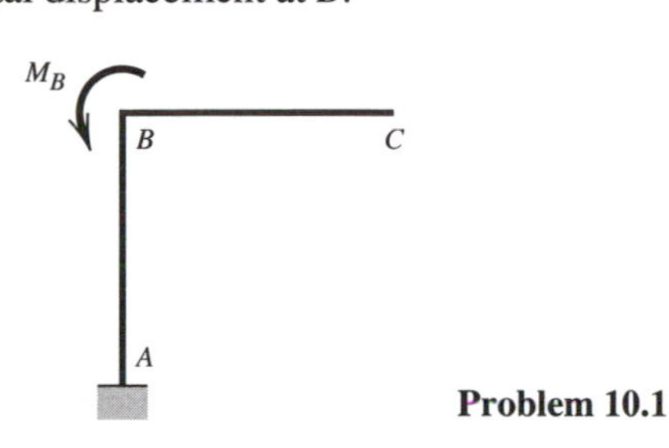

Problem 10.1

10.2 Let the nonlinear spring in Fig. 10.2-1a have stiffness $k = 5/(2 + u)$. Assume that physical constants and loads are in consistent units.

(a) Sketch the *P* versus *u* curve for positive *P*.

(b) Show that the tangent stiffness is $k_t = 10/(2 + u)^2$.

(c) Use the purely incremental method, as in Fig. 10.2-1b, to calculate the displacements that correspond to $P = 2.0$ and to $P = 4.5$.

(d) What is the correct displacement at each of these two loads?

(e) Calculate the displacements that correspond to $P = 2.0$ and to $P = 4.5$ by carrying out two Newton–Raphson equilibrium iterations after reaching each of the two load levels (follow the calculation scheme in Fig. 10.2-2a).

(f) Repeat part (e), but use modified Newton–Raphson iterations (follow the calculation scheme in Fig. 10.2-2b).

10.3 Repeat Problem 10.2, but let the nonlinear spring have stiffness $k = 2 + u^2$, for which $k_t = 2 + 3u^2$. However, in part (f), carry out five iterations at load $P = 2$ only, and show what is happening on the sketch of part (a).

10.4 Use Eq. 10.3-3 to calculate linear buckling loads for uniform columns under the conditions described below. Compare results with buckling loads obtained from elementary column theory.

(a) Both ends are simply supported. Use one element.

(b) Both ends are fixed. Use two elements of equal length.

(c) One end is fixed and the other is simply supported. Use one element.

10.5 Do the following for the device shown in Fig. 10.4-1a:

(a) Show that Eq. 10.4-2 follows from Eq. 10.4-1.

(b) Show that support reaction *P* is as stated in Eq. 10.4-3.

(c) Show that the tangent stiffness is always positive if *k* is as stated in Fig. 10.4-1b.

(d) Show that limit point A in Fig. 10.4-1c has the displacement coordinate $v = (1 - \sqrt{3}/3)c$.

(e) Let F be increased in a single step from zero to a value slightly greater than the ordinate at point A in Fig. 10.4-1c. On this plot, without calculations, sketch the behavior of a few Newton–Raphson equilibrium iterations (as is done for a different curve in Fig. 10.2-2a).

(f) Repeat part (e), but consider the modified Newton–Raphson method (as in Fig. 10.2-2b).

10.6 When load F is absent, displacement v in the sketch is zero and the springs are collinear and unstressed. In this problem, assume that displacements are small ($v << L$).

(a) Determine F as a function of v, k, and L.

(b) Obtain an expression for the tangent stiffness.

(c) Qualitatively sketch the relation between F (ordinate) and v (abscissa).

(d) Let $F = 1$ N, $L = 10$ mm, and $k = 800$ N/mm. Then $v = 0.5$ mm. Now increase F by $\Delta F = 7$ N and apply three Newton–Raphson equilibrium iterations to approximate the resulting v.

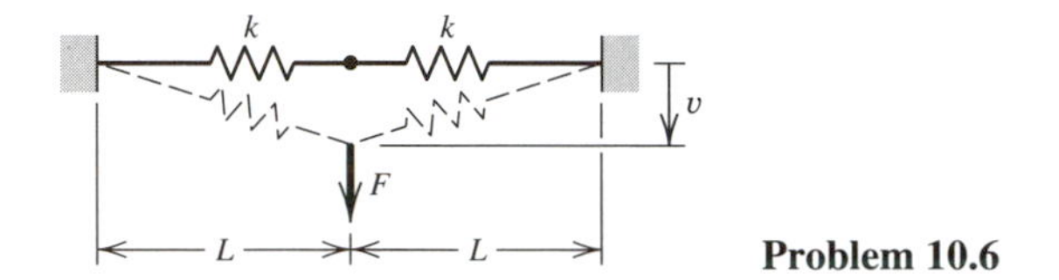

Problem 10.6

10.7 A formula for tip deflection of a beam caused by tip moment M in Fig. 10.4-2b is stated in the text. Derive this formula, and show that for small M it reduces to the familiar linear formula $u = ML^2/2EI$.

10.8 (a) A cable of weight q per unit length is stretched between two supports at the same elevation and a distance L apart. The center sag relative to the supports is w_c, where $w_c << L$. Assume that the deflected shape is parabolic, $w = (4w_c/L^2)(Lx - x^2)$, where $0 < x < L$. As can be shown, the arc length L_a of the cable is then $L_a = L + 8w_c^2/3L$. Use this information to derive Eq. 7.1-5.

(b) Compare the center deflection given by Eq. 7.1-5 with the center deflection stated in the text of the present chapter for a rectangular membrane, Fig. 10.4-2e, when the aspect ratio a/b of the membrane is large.

(c) Let tension in the cable be T, which can be expressed in terms of q, L, and w_c. The stretching stiffness of the cable is partly due to elasticity (i.e., AE/L) and partly due to its sag (i.e., dT/dL) with L_a constant. Note that dT and dL_a can each be expressed in terms of both dL and dw_c. Derive the net (nonlinear) stretching stiffness, in terms of A, E, L, w_c, and q.

10.9 Apply the method of Problem 10.8a to a uniformly loaded, initially flat circular membrane. Thus derive an expression for its center deflection, analogous to Eq. 7.1-5. Let $\nu = 0.3$.

10.10 Let the block of material in Fig. 10.5-3b have a rectangular cross section. Draw the curve in Fig. 10.5-3b to scale for the case $E_t = 0$.

10.11 The rigid block shown is constrained to translate horizontally. It is connected to a rigid wall by three identical uniform bars that carry only axial load. Each bar has cross-sectional area A, elastic modulus E, yield point σ_Y, and does not strain harden.

(a) Plot to scale the relation between load P and its horizontal displacement, as P increases from zero to $P = 3A\sigma_Y$.

(b) Determine the state of residual stress in each bar if load P now drops to zero.
(c) If load P is then reapplied, but in the opposite sense, for what value of P will yielding begin again?
(d) What is the initial range of P (tension to compression) for which there is no yielding? What is this range after unloading from the maximum load in part (a)?

10.12 In Fig. 10.6-2, let $k_1 = k_2 = 1$ N/mm, $k_3 = 100$ N/mm, and $g = 1$ mm. Apply load $P = 3$ N in a single step, then carry out as many Newton–Raphson equilibrium iterations as seem necessary to solve for the displacement at A.

10.13 The rigid block shown is constrained to translate horizontally.
(a) For this structure, write equations analogous to Eqs. 10.6-1 and 10.6-3.
(b) Let $k_0 = 8$ N/mm, $k_1 = 10$ N/mm, $k_2 = 12$ N/mm, $g_1 = 0.2$ mm, and $g_2 = 0.4$ mm. Apply a load $P = 4$ N in a single step, then carry out equilibrium iterations to obtain the corresponding horizontal displacement u of the block.

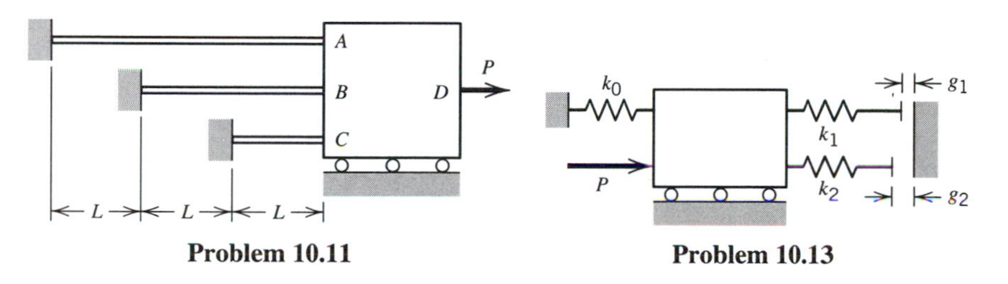

Problem 10.11 **Problem 10.13**

COMPUTATIONAL PROBLEMS

In the following problems, choose convenient numbers and consistent units for physical properties, dimensions, and loads. Exploit symmetry if possible. Unless directed otherwise, assume that thicknesses are uniform and the material is isotropic. Assume that pressure acts as a follower force but that other loads are fixed in direction. Assume that there is no yielding unless elastic–plastic behavior is mentioned. When additional assumptions are required, clearly state what they are.

Although not noted in what follows, problems previously cited in this chapter as possible test cases and the preceding analytical problems may also be posed as computational problems. Similarly, many problems cited in preceding chapters and problem sets may be reexamined with attention to geometric and/or material nonlinearity.

For comparison purposes, a FE analysis should be preceded by an alternative analysis, probably based on statics and mechanics of materials, and oversimplified if necessary. In a nonlinear problem such analysis may be feasible for only one or two stages of the response.

10.14 Determine the relation between load q and the center deflection of the beam in Fig. 10.3-2c (see also Eq. 7.1-6).

10.15 In Fig. 10.7-1a, determine the value of P for buckling or collapse. Include a linear buckling analysis as part of the study. Consider that the structure is (a) a shallow circular arch or (b) a shallow spherical dome. The problem may be repeated with other support conditions.

10.16 The conical shell shown in the sketch for Problem 7.25 is called a "Belleville spring" if the cone is rather flat. Determine the relation between load q and its vertical deflection. If the cone is so nearly flat that "snapping" is possible, how well does the actual collapse load agree with the critical load predicted by linear buckling analysis?

10.17 The cross section of a steel carpenter's tape is a shallow arc of small thickness. Imagine that an arbitrary length of the tape is subjected to a bending moment M whose vector representation is parallel to the span of the arc. What value of M causes buckling or collapse?

10.18 (a) Consider a beam loaded by force P, as shown. The beam rests on an elastic foundation that can push against the beam but cannot pull on it. Determine the extent and the intensity of the contact pressure between the beam and the foundation.

(b) Repeat part (a) but consider a plate, as shown, that rests on an elastic foundation.

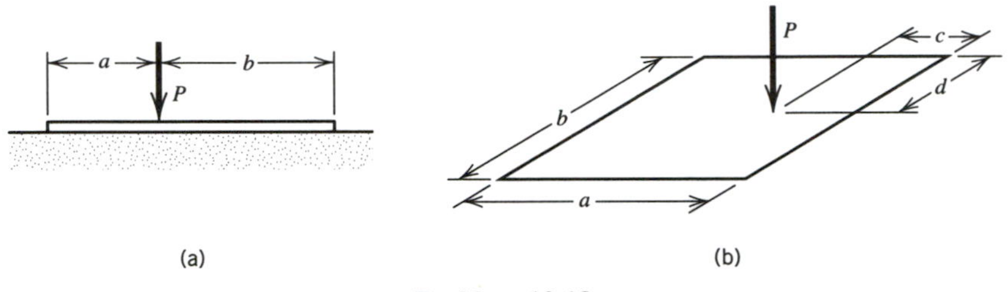

Problem 10.18

10.19 (a) Buckling loads for rectangular flat plates and circular cylindrical shells under various loadings have been obtained by classical linear analysis [10.5]. These problems may be used as test cases.

(b) Any of the cases of part (a) may be perturbed by imperfections and addressed by nonlinear analysis methods. Imperfections may take the form of small lateral loads or small lateral misplacements of one or more nodal coordinates.

10.20 A thin, flat, circular plate rests on a soft elastic foundation. A central force P presses the plate against the foundation. Will the plate ever buckle? If so, is the calculated linear buckling load a good estimate of the actual buckling load?

10.21 The sketch shows a large circular hole in a flat plate under tension. The "ligament" dimension h may become very small in comparison with width H. What is the stress concentration factor? Does it depend on the magnitude of the load?

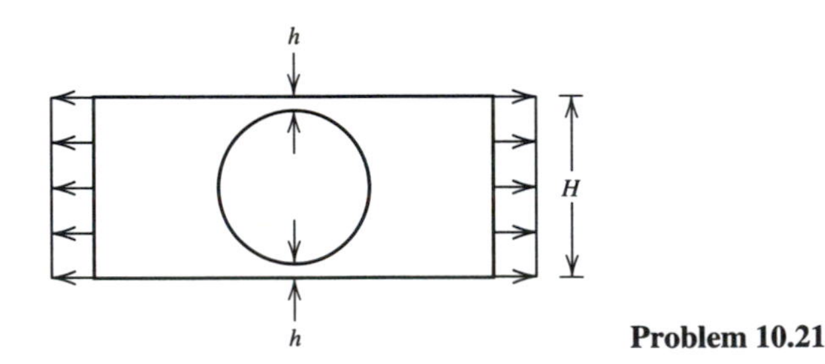

Problem 10.21

10.22 Repeat Problem 10.21, now regarding the sketch as representing a spherical cavity in an otherwise solid cylinder of circular cross section.

10.23 Reverse the direction of the load in Problem 10.21, and investigate the magnitude of load required to produce buckling or collapse. Consider the following support conditions:

(a) Supports adequate to keep the horizontal centerline of the left side collinear with the horizontal centerline of the right side.

(b) Supports on the left side just adequate to prevent rigid-body motion, with no supports on the right side.

10.24 Apply the instructions of Problem 10.23 to Problem 10.22.

10.25 (a) Examine stresses and deformations for the geometry of Problem 10.21 if axial load is absent but uniform pressure is applied to the boundary of the hole.

(b) Repeat part (a) but apply pressure to the outside lateral boundaries of the strip.

10.26 The sketch shows a thin, flat piece of metal fixed to a rigid support at $x = 0$. Initially, the structure midsurface and force P lie in the xy plane. Displacements are not confined to the xy plane. Investigate large-deflection behavior and possible buckling.

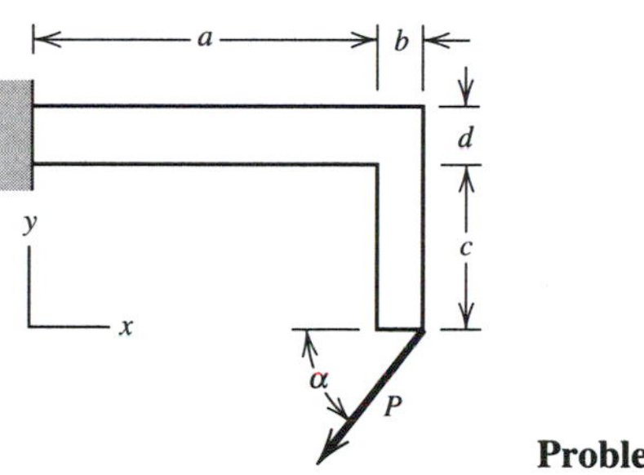

Problem 10.26

10.27 The sketch shows a flat piece of metal in a vertical plane. Its thickness t is small and the vertical slots are very narrow. Lower ends of the outer legs are fixed and a vertical distributed load q is applied to the bottom of the center leg. Allow out-of-plane deformations, and investigate the buckling and collapse behavior. Load q may act either downward, as shown, or upward.

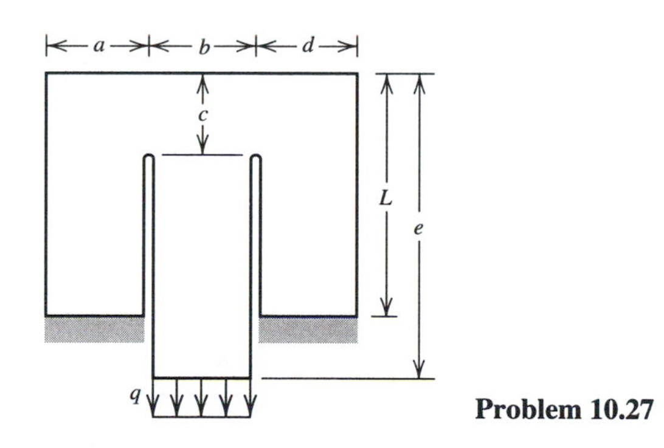

Problem 10.27

10.28 (a) The sketch represents the cross section of a horizontal cylindrical tank, partly full of water, and having supports that run lengthwise. Assume that the tank is long, so that end effects can be neglected in the analysis of a cross section near the middle of the tank. Analyze for the deflected shape, stresses, and possible buckling. Use $E = 200$ GPa and $R/t \approx 600$.

(b) The analysis may be repeated with supports that apply only vertical forces to the tank.

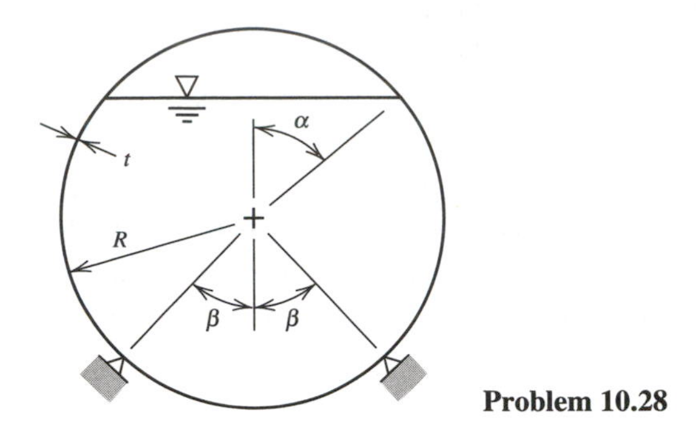

Problem 10.28

10.29 Determine the force versus deflection relation of an archery bow.

10.30 A chain is built of many identical links and hangs under its own weight. Investigate the natural frequencies and modes of vibration.

(a) Let the chain hang from a fixed support at one end.

(b) Let the chain hang from fixed supports at both ends. Choices are possible for the elevation of one end with respect to the other and the vertical sag relative to the higher support.

10.31 A thin, flat circular disk spins with angular velocity Ω about a central axis normal to the plane of the disk. Investigate the effect of Ω on:

(a) the fundamental frequency of vibration.

(b) the temperature T_0 that causes buckling, if temperature T varies linearly with the radial coordinate, from zero at the center to T_0 at the outside edge.

10.32 Straight elastic bars are welded together to form the T-shaped bar shown. It rotates at angular velocity Ω about an axis normal to the plane of the paper at A. Investigate the effects of angular velocity, stress stiffening, and spin softening on the natural frequencies of vibration. Clearly specify the support conditions you decide to use at A.

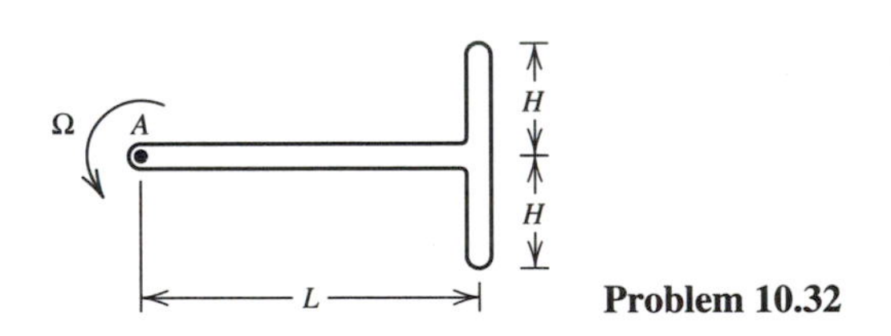

Problem 10.32

10.33 (a) Consider a long straight bar of narrow rectangular cross section, loaded by torque about the longitudinal axis of the bar. If the angle of twist may become large, how is the torque related to the angle of twist per unit length? Also, is buckling a possibility?

(b) The problem may be repeated using one of the cross sections depicted for Problem 7.38.

10.34 Each of the plane bodies shown may be loaded by force F and/or moment M. Use a stress–strain relation like that in Fig. 10.5-2a. Apply a load sufficient to produce yielding. Determine the extent of yielding and the residual stresses upon unloading.

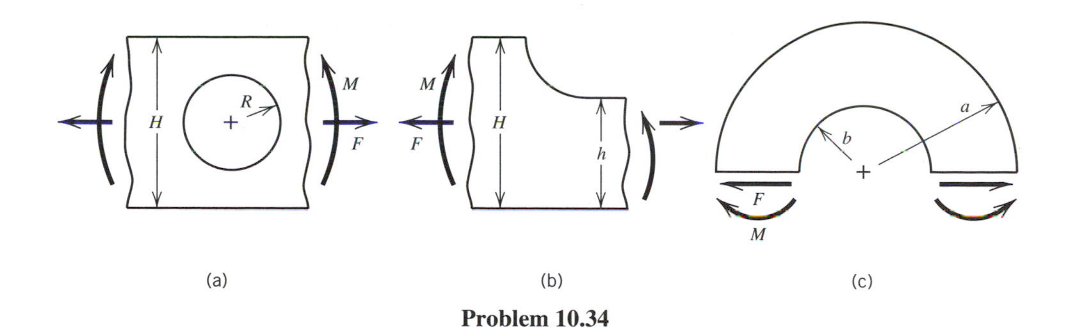

Problem 10.34

10.35 In Fig. 10.4-2b, let $M = 0$ and $F = 0.001P$. Choose dimensions such that P_{cr}/A is slightly greater in magnitude than σ_Y. Here $P_{cr} = \pi^2 EI/4L^2$, from elastic column buckling theory. Choose a specific stress–strain relation for which $E_t > 0$, and solve for the plastic buckling load [2.1, 10.5].

10.36 Consider plane beams or plane arches made of an elastic–perfectly plastic material (i.e., $E > 0$, $E_t = 0$). Let cross sections be rectangular. For each of the cases shown, calculate and plot load versus displacement relations. Also determine the maximum value of P and/or q that the structure can carry.

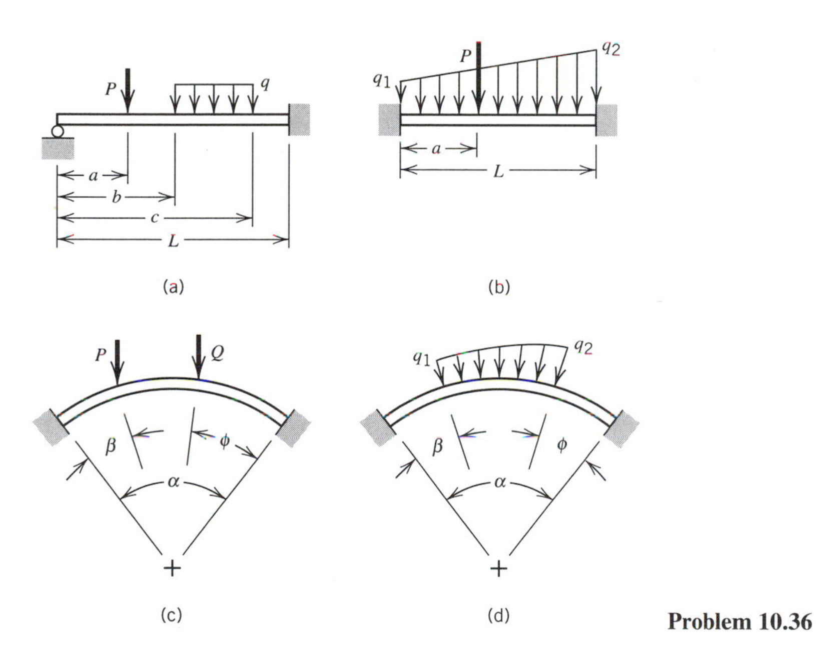

Problem 10.36

10.37 Consider tip-loaded cantilever beams having the cross sections shown for Problem 7.35. Let the material have a bilinear stress–strain relation, as in Fig. 10.5-2a. Compute the maximum x- or y-direction tip load that can be sustained if $E_t = 0$.

10.38 Reconsider the Belleville spring problem posed as Problem 10.16. Now let the spring rest directly on a flat horizontal surface whose coefficient of friction is μ.

10.39 The upper beam shown in the sketch has an initial curvature, so that a small gap g exists between the two beams when no load is applied. The lower beam is simply supported and initially straight. Examine the relation between load P and its displacement. Additional exercises result if other support conditions are prescribed at A and B.

10.40 Two plane, elastic blocks of material are pressed together, as shown. Examine the intensity and extent of the contact stress between the blocks. Assume that the coefficient of friction between blocks is zero.

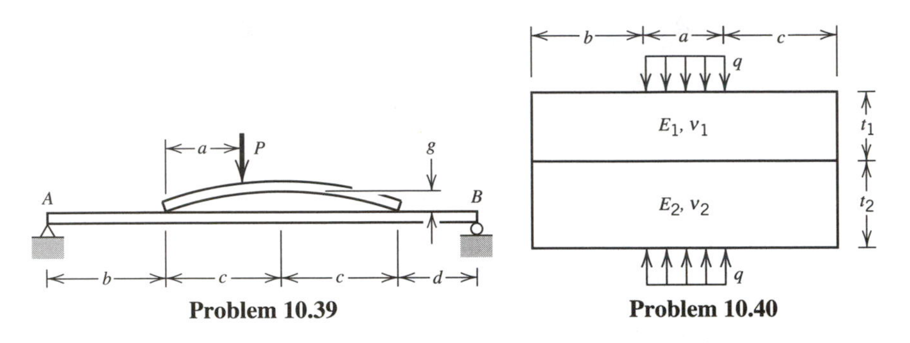

Problem 10.39 **Problem 10.40**

10.41 The sketch represents a thin-walled tube around a solid circular cylinder. The coefficient of friction is μ.

(a) Imagine that the inside diameter of the tube is slightly smaller than the diameter of the cylinder when both are unstressed and at the same temperature. The tube is heated, slipped over the cylinder, and allowed to cool. If heat transfer between tube and cylinder is neglected, what is the final state of stress in the tube?

(b) A force q is uniformly distributed around the end of the tube and is gradually increased from zero. What is the relation between q and axial displacement at each end of the tube? The starting condition is the result of part (a) or, as an option, a shrink fit without tangential interface traction.

(c) Repeat part (b) with the direction of load q reversed.

10.42 The sketch represents a simplified drum brake problem. A curved elastic bar AB spans an angle α, is pivoted at A, and is pressed against a rigid brake drum by force F. The drum rotates with angular velocity Ω. Investigate how the torque required to rotate the drum is related to the coefficient of friction and other variables of the problem. The problem may be repeated with the direction of rotation reversed.

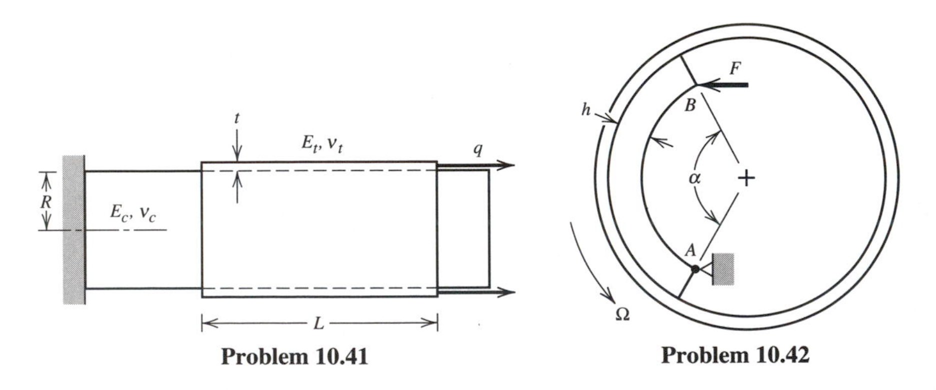

Problem 10.41 **Problem 10.42**

10.43 Before loading, the slender elastic cantilever beam shown is horizontal and just in contact with the rigid wall at C. The wall is inclined to the vertical at a small angle

α. The coefficient of friction for the contact at C is μ. Plot the relation between load P and the vertical deflection components of B and C. Also determine the maximum load that can be sustained.

10.44 A slender, elastic, circular ring is to be pulled through a smaller opening, as shown. The coefficient or friction between the ring and the opening is μ. Determine the relation between force P and its displacement. The problem may be repeated with P applied at A rather than at B.

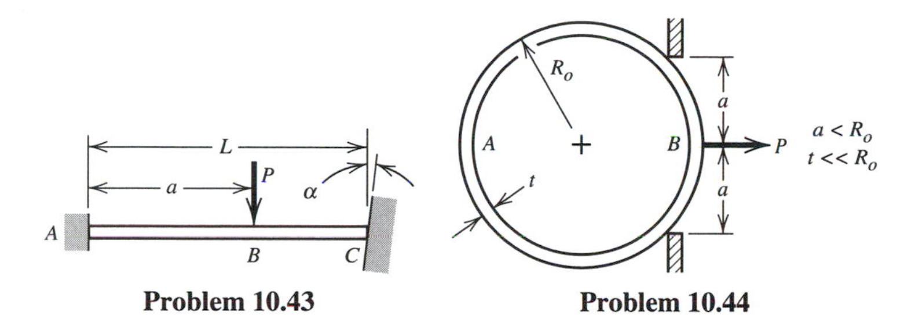

Problem 10.43 **Problem 10.44**

10.45 The sketch represents a centrally loaded, simply supported beam, built by nailing together two boards for which $E = 10$ GPa. Nails are uniformly spaced, 80 mm apart, and are each 3 mm in diameter. Assume that the shear force F in a nail is related to the relative slip Δ between boards at the nail location by the equation $F = (710 - 11\Delta)\Delta^{0.3}$, where F is in newtons and Δ is in millimeters. Investigate nail forces F, beam deflection, and beam stresses as load P increases. (This problem is a simplification of work reported in [10.12])

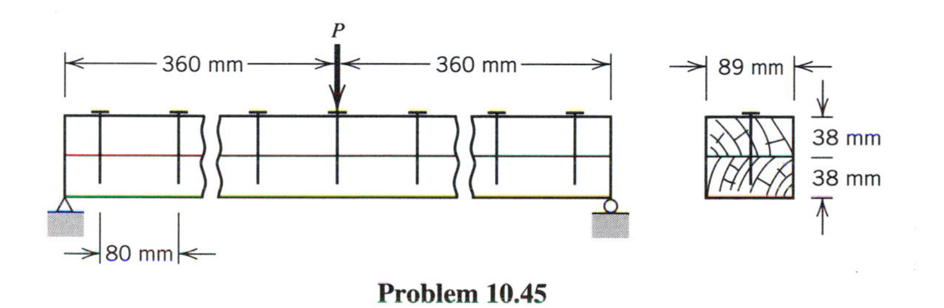

Problem 10.45

10.46 A long, flat strip of metal is placed on a smooth, rigid, horizontal surface. What compressive force, directed parallel to the long edges and uniformly distributed across the short edges, will produce buckling? (The problem is easily demonstrated by placing a sheet of paper on a desk and pushing the shorter edges towards one another.)

REFERENCES

Chapter 1

1.1 R. Courant, "Variational Methods for the Solution of Problems of Equilibrium and Vibrations," *Bulletin of the American Mathematical Society*, Vol. 49, 1943, pp. 1–23.

1.2 M. J. Turner, R. W. Clough, H. C. Martin, and L. J. Topp, "Stiffness and Deflection Analysis of Complex Structures," *Journal of the Aeronautical Sciences*, Vol. 23, No. 9, 1956, pp. 805–823.

1.3 R. W. Clough, "The Finite Element Method After Twenty-Five Years: A Personal View," *Computers & Structures*, Vol. 12, No. 4, 1980, pp. 361–370.

1.4 J. Robinson, *Early FEM Pioneers*, Robinson & Associates, Dorset, UK, 1985.

1.5 W. C. Young, *Roark's Formulas for Stress and Strain*, 6th ed., McGraw-Hill, New York, 1989.

1.6 P. S. Symonds and T. X. Yu, "Counterintuitive Behavior in a Problem of Elastic–Plastic Beam Dynamics," *ASME Journal of Applied Mechanics*, Vol. 52, No. 3, 1985, pp. 517–522.

1.7 G. E. Smith, "The Dangers of CAD," *Mechanical Engineering*, Vol. 108, No. 2, February 1986, pp. 58–64.

Chapter 2

2.1 R. D. Cook and W. C. Young, *Advanced Mechanics of Materials*, Macmillan, New York, 1985.

2.2 R. D. Cook, D. S. Malkus, and M. E. Plesha, *Concepts and Applications of Finite Element Analysis*, 3rd ed., John Wiley & Sons, New York, 1989.

2.3 C. J. Burgoyne and R. Dilmaghanian, "Bicycle Wheel as Prestressed Structure," *ASCE Journal of Engineering Mechanics*, Vol. 119, No. 3, 1993, pp. 439–455.

Chapter 3

3.1 K. J. Bathe, *Finite Element Procedures in Engineering Analysis*, Prentice-Hall, Englewood Cliffs, NJ, 1982.

3.2 R. H. MacNeal, *Finite Elements: Their Design and Performance*, Marcel Dekker, New York, 1994.

3.3 R. L. Taylor, P. J. Beresford, and E. L. Wilson, "A Non-Conforming Element for Stress Analysis," *International Journal for Numerical Methods in Engineering*, Vol. 10, No. 6, 1976, pp. 1211–1219.

3.4 R. D. Cook, "Beam Cantilevered from Elastic Support: Finite Element Modeling," *Communications in Applied Numerical Methods*, Vol. 7, No. 8, 1991, pp. 621–623.

3.5 E. L. Wilson and A. Ibrahimbegovic, "Use of Incompatible Displacement Modes for the Calculation of Element Stiffness or Stress," *Finite Elements in Analysis and Design*, Vol. 7, No. 3, 1990, pp. 229–241.

3.6 J. Pittr and H. Hartl, "Improved Stress Evaluation Under Thermal Load for Simple Finite Elements," *International Journal for Numerical Methods in Engineering*, Vol. 15, No. 10, 1980, pp. 1507–1515.

3.7 R. H. MacNeal and R. L. Harder, "A Proposed Standard Set of Problems to Test Finite Element Accuracy," *Finite Elements in Analysis and Design*, Vol. 1, No. 1, 1985, pp. 3–20.

3.8 R. E. Peterson, *Stress Concentration Factors*, John Wiley & Sons, New York, 1974, p. 174.

3.9 R. H. MacNeal, "On the Limits of Finite Element Perfectibility," *International Journal for Numerical Methods in Engineering*, Vol. 35, No. 8, 1992, pp. 1589–1601.

3.10 W. J. O'Donnell, "The Additional Deflection of a Cantilever Due to the Elasticity of the Support," *ASME Journal of Applied Mechanics*, Vol. 27, No. 3, 1960, pp. 461–464.

Chapter 4

4.1 H. Kardestuncer, ed., *Finite Element Handbook*, McGraw-Hill, New York, 1987.

4.2 R. E. Miller, "Reduction of the Error in Eccentric Beam Modeling," *International Journal for Numerical Methods in Engineering,* Vol. 15, No. 4, 1980, pp. 575–582.

4.3 A. H. Stroud and D. Secrest, *Gaussian Quadrature Formulas*, Prentice-Hall, Englewood Cliffs, NJ, 1966.

4.4 J. Barlow, "Optimal Stress Locations in Finite Element Models," *International Journal for Numerical Methods in Engineering*, Vol. 10, No. 2, 1976, pp. 243–251 (discussion: Vol. 11, No. 3, 1977, p. 604).

4.5 J. M. M. C. Marques and D. R. J. Owen, "Infinite Elements in Quasi-Static Materially Nonlinear Problems," *Computers & Structures*, Vol. 18, No. 4, 1984, pp. 739–751.

4.6 P. Bettess, *Infinite Elements*, Penshaw Press, Sunderland, UK, 1992.

4.7 P. D. Mangalgiri, B. Dattaguru, and T. S. Ramamurthy, "Specification of Skew Conditions in Finite Element Formulation," *International Journal for Numerical Methods in Engineering*, Vol. 12, No. 6, 1978, pp. 1037–1041.

4.8 R. S. Barsoum, "On the Use of Isoparametric Finite Elements in Linear Fracture Mechanics," *International Journal for Numerical Methods in Engineering*, Vol. 10, No. 1, 1976, pp. 25–37.

4.9 N. A. B. Yahia and M. S. Shephard, "On the Effect of Quarter-Point Element Size on Fracture Criteria," *International Journal for Numerical Methods in Engineering*, Vol. 21, No. 10, 1985, pp. 1911–1924.

Chapter 5

5.1 H. Bleich, "Die Spannungsverteilung in den Gurtungen gekrümmter Stabe mit T- und I- förmigem Querschnitt," *Der Stahlbau* (appendix to *Die Bautechnik*), Vol. 6, No. 1, 1933, pp. 3–6.

5.2 W. G. Dodge and S. E. Moore, "Stress Indices and Flexibility Factors for Moment Loadings on Elbows and Curved Pipes," *Welding Research Council Bulletin 179*, December 1972.

5.3 K. J. Bathe and C. A. Almeida, "A Simple and Effective Pipe Elbow Element—Linear Analysis," *ASME Journal of Applied Mechanics*, Vol. 47, No. 1, 1980, pp. 93–100.

5.4 C. Meyer, ed., *Finite Element Idealization*, American Society of Civil Engineers, New York, 1987.

5.5 R. D. Henshell, D. Walters, and G. B. Warburton, "A New Family of Curvilinear Plate Bending Elements for Vibration and Stability," *Journal of Sound and Vibration*, Vol. 20, No. 3, 1972, pp. 381–387 (discussion and authors' closure: Vol. 23, No. 4, 1972, pp. 507–513).

5.6 L. Z. Emkin, "Computers in Structural Engineering Practice: The Issue of Quality," *Computers & Structures*, Vol. 30, No. 3, 1988, pp. 439–446.

5.7 *The Standard NAFEMS Benchmarks* (Revision), National Agency for Finite Element Methods and Standards, Glasgow, UK, 1989.

5.8 T. Slot and W. J. O'Donnell, "Effective Elastic Constants for Thick Perforated Plates with Square and Triangular Penetration Patterns," *ASME Journal of Engineering for Industry*, Vol. 93, No. 4, 1971, pp. 935–942.

5.9 Anonymous, *A Finite Element Primer*, National Agency for Finite Element Methods and Standards, Glasgow, UK, 1986.

5.10 J. M. Stallings and D. Y. Huang, "Modeling Pretensions in Bolted Connections," *Computers & Structures*, Vol. 45, No. 4, 1992, pp. 801–803.

5.11 B. M. Irons, "Roundoff Criteria in Direct Stiffness Solutions," *AIAA Journal*, Vol. 6, No. 7, 1968, pp. 1308–1312.

5.12 I. Taig, "Finite Element Analysis in Industry—Expertise or Proficiency?," in *Accuracy, Reliability, and Training in FEM Technology* (Proceedings of Fourth World Congress and Exhibition on Finite Element Methods), J. Robinson, ed., Robinson and Associates, Wimborne, UK, 1984, pp. 56–70.

5.13 B. Szabo, "Estimation and Control of Error Based on *p* Convergence," in *Accuracy Estimates and Adaptive Refinements in Finite Element Computations*, I. Babuska et al., eds., John Wiley & Sons, Chichester, UK, 1986, pp. 61–78.

5.14 O. C. Zienkiewicz and J. Z. Zhu, "A Simple Error Estimator and Adaptive Procedure for Practical Engineering Analysis," *International Journal for Numerical Methods in Engineering*, Vol. 24, No. 2, 1987, pp. 337–357.

5.15 O. C. Zienkiewicz and J. Z. Zhu, "The Superconvergent Patch Recovery and a posteriori Error Estimates. Part 1: The Recovery Technique," *International Journal for Numerical Methods in Engineering*, Vol. 33, No. 7, 1992, pp. 1331–1364.

5.16 J. Z. Zhu and O. C. Zienkiewicz, "Adaptive Techniques in the Finite Element Method," *Communications in Applied Numerical Methods*, Vol. 4, No. 2, 1988, pp. 197–204.

5.17 A. R. Rizzo, "Quality Engineering with FEA and DOE," *Mechanical Engineering*, Vol. 116, No. 5, 1994, pp. 76–78.

Chapter 6

6.1 S. P. Timoshenko and J. N. Goodier, *Theory of Elasticity*, 3rd ed., McGraw-Hill, New York, 1970, p. 429.

Chapter 7

7.1 J. L. Batoz, "An Explicit Formulation for an Efficient Triangular Plate Bending Element," *International Journal for Numerical Methods in Engineering*, Vol. 18, No. 7, 1982, pp. 1077–1089.

7.2 M. P. Rossow, "Efficient C^0 Finite-Element Solutions of Simply-Supported Plates of Polygonal Shape," *ASME Journal of Applied Mechanics*, Vol. 44, No. 2, 1977, pp. 347–349.

7.3 S. P. Timoshenko and S. Woinowsky-Krieger, *Theory of Plates and Shells*, 2nd ed., McGraw-Hill, New York, 1959.

7.4 B. P. Naganarayana and G. Prathap, "Force and Moment Corrections for the Warped Four-Node Quadrilateral Plane Shell Element," *Computers & Structures*, Vol. 33, No. 4, 1989, pp. 1107–1115.

7.5 N. Carpenter, H. Stolarski, and T. Belytschko, "Improvements in 3-Node Triangular Shell Elements," *International Journal for Numerical Methods in Engineering*, Vol. 23, No. 9, 1986, pp. 1643–1667.

7.6 T. Belytschko, B. K. Wong, and H. Stolarski, "Assumed Strain Stabilization Procedure for the 9-Node Lagrange Shell Element," *International Journal for Numerical Methods in Engineering*, Vol. 28, No. 2, 1989, pp. 385–414.

Chapter 8

8.1 A. Chapman, *Heat Transfer*, 4th ed., Macmillan, New York, 1984.

8.2 T. J. R. Hughes, *The Finite Element Method*, Prentice-Hall, Englewood Cliffs, NJ, 1987.

8.3 R. E. Cornwell and D. S. Malkus, "Improved Numerical Dissipation for Time Integration Algorithms in Conduction Heat Transfer," *Computer Methods in Applied Mechanics and Engineering*, Vol. 97, No. 2, 1992, pp. 149–156.

8.4 K. E. Barrett, D. M. Butterfield, J. H. Tabor, and S. Ellis, "Quadratic Elements for Heat Transfer—a Cautionary Tale," *International Journal of Mechanical Engineering Education*, Vol. 18, No. 1, 1990, pp. 59–74.

Chapter 9

9.1 R. J. Guyan, "Reduction of Stiffness and Mass Matrices," *AIAA Journal*, Vol. 3, No. 2, 1965, p. 380.

9.2 R. R. Craig, Jr., *Structural Dynamics*, John Wiley & Sons, New York, 1981.

9.3 R. W. Clough and J. Penzien, *Dynamics of Structures*, McGraw-Hill, New York, 1975.

9.4 D. Hitchings, ed., *A Finite Element Dynamics Primer*, National Agency for Finite Element Methods and Standards, Glasgow, UK, 1986.

9.5 T. Belytschko and T. J. R. Hughes, eds., *Computational Methods for Transient Analysis*, North-Holland, Amsterdam, 1983.

9.6 M. L. James et al., *Vibration of Mechanical and Structural Systems*, Harper & Row, New York, 1989.

9.7 R. G. Anderson, B. M. Irons, and O. C. Zienkiewicz, "Vibration and Stability of Plates Using Finite Elements," *International Journal of Solids & Structures*, Vol. 4, No. 10, 1968, pp. 1031–1055.

9.8 Anonymous., *ANSYS User's Manual for Revision 5.0: Volume 1*, Swanson Analysis Systems Inc., Houston, PA, 1992.

9.9 J. Avrashi and R. D. Cook, "New Error Estimation for C^0 Eigenproblems in Finite Element Analysis," *Engineering Computations*, Vol. 10, No. 3, 1993, pp. 243–256.

9.10 J. A. Zukas, "Some Common Problems in the Numerical Modeling of Impact Phenomena," *Computing Systems in Engineering*, Vol. 4, No. 1, 1993, pp. 43–58.

9.11 R. Perrin, T. Charnley, and G. M. Swallowe, "Modes of the Spiral Clock Gong," *Journal of Sound and Vibration*, Vol. 162, No. 1, 1993, pp. 1–12.

Chapter 10

10.1 M. A. Crisfield, *Non-linear Finite Element Analysis of Solids and Structures, Vol. 1: Essentials*, John Wiley & Sons, Chichester, UK, 1991.

10.2 E. Hinton, ed., *NAFEMS Introduction to Nonlinear Finite Element Analysis*, National Agency for Finite Element Methods and Standards, Glasgow, UK, 1992.

10.3 S. J. Shah and W. D. Pilkey, "Lumped Parameter Approach to Stability Analysis," *ASCE Journal of Engineering Mechanics*, Vol. 119, No. 10, 1993, pp. 2109–2129.

10.4 J. H. Lau, "Large Deflections of Beams with Combined Loads," *ASCE Journal of the Engineering Mechanics Division*, Vol. 108, No. EM1, 1982, pp. 180–185.

10.5 S. P. Timoshenko and J. M. Gere, *Theory of Elastic Stability*, 2nd ed., McGraw-Hill, New York, 1961.

10.6 K. Mattiasson, "Numerical Results for Large Deflection Beam and Frame Problems Analyzed by Means of Elliptic Integrals," *International Journal for Numerical Methods in Engineering*, Vol. 17, No. 1, 1981, pp. 145–153.

10.7 S.-L. Lee, F. S. Manuel, and E. C. Rossow, "Large Deflections and Stability of Elastic Frames," *ASCE Journal of the Engineering Mechanics Division*, Vol. 94, No. EM2, 1968, pp. 521–547.

10.8 P. Seide, "Large Deflections of Rectangular Membranes under Uniform Pressure," *International Journal of Non-Linear Mechanics*, Vol. 12, No. 6, 1977, pp. 397–406.

10.9 A. Mendelson, *Plasticity: Theory and Application*, Macmillan, New York, 1968.

10.10 J. P. Den Hartog, *Advanced Mechanics of Materials*, McGraw-Hill, New York, 1952.

10.11 O. C. Zienkiewicz, *The Finite Element Method*, McGraw-Hill, London, 1977.

10.12 D. R. Bohnhoff, "Modeling Horizontally Laminated Beams," *ASCE Journal of Structural Engineering*, Vol. 118, No. 5, 1992, pp. 1393–1406.

INDEX

T